Animal Breeding and Livestock Management

Animal Breeding and Livestock Management

Edited by Carlos Hassey

SYRAWOOD
PUBLISHING HOUSE

New York

Published by Syrawood Publishing House,
750 Third Avenue, 9th Floor,
New York, NY 10017, USA
www.syrawoodpublishinghouse.com

Animal Breeding and Livestock Management
Edited by Carlos Hassey

International Standard Book Number: 978-1-68286-438-8 (Hardback)

Cataloging-in-publication Data

Animal breeding and livestock management / edited by Carlos Hassey.
    p. cm.
Includes bibliographical references and index.
ISBN 978-1-68286-438-8
1. Animal breeding. 2. Animal breeds. 3. Livestock. I. Hassey, Carlos.
SF105 .A55 2017
636.082--dc23

Printed in the United States of America.

TABLE OF CONTENTS

PREFACE

Livestock management is understood as rearing and raising of domesticated animals in an agricultural setting so that the domesticated animals produce commodities like food, fiber, etc. Animal breeding falls under the animal science and addresses the genetic value of livestock. This book is compiled in such a manner, that it will provide in-depth knowledge about the theory and practice of animal breeding and livestock management. It aims to elaborately put together the various sub-fields of livestock management to make the readers aware about the importance of livestock management and its different aspects to get a holistic understanding of the larger discipline of animal breeding. This book includes contributions of experts and scientists which will provide innovative insights into this field.

The world is advancing at a fast pace like never before. Therefore, the need is to keep up with the latest developments. This book was an idea that came to fruition when the specialists in the area realized the need to coordinate together and document essential themes in the subject. That's when I was requested to be the editor. Editing this book has been an honour as it brings together diverse authors researching on different streams of the field. The book collates essential materials contributed by veterans in the area which can be utilized by students and researchers alike.

Each chapter is a sole-standing publication that reflects each author's interpretation. Thus, the book displays a multi-facetted picture of our current understanding of applications and diverse aspects of the field. I would like to thank the contributors of this book and my family for their endless support.

Editor

CNS-Targeted Production of IL-17A Induces Glial Activation, Microvascular Pathology and Enhances the Neuroinflammatory Response to Systemic Endotoxemia

Julian Zimmermann[1], Marius Krauthausen[1], Markus J. Hofer[2], Michael T. Heneka[1,3], Iain L. Campbell[4], Marcus Müller[1,4]*

1 Department of Neurology, Universitätsklinikum Bonn, Bonn, Germany, **2** Department of Neuropathology, University Clinic of Marburg and Giessen, Marburg, Germany, **3** Clinical Neuroscience Unit, University of Bonn, Bonn, Germany, **4** School of Molecular Bioscience, University of Sydney, Sydney, Australia

Abstract

Interleukin-17A (IL-17A) is a key cytokine modulating the course of inflammatory diseases. Whereas effector functions of IL-17A like induction of antimicrobial peptides and leukocyte infiltration could clearly be demonstrated for peripheral organs, CNS specific effects are not well defined and appear controversial. To further clarify the functional significance of IL-17A in the CNS, we generated a transgenic mouse line with astrocyte-restricted expression of the IL-17A gene. GFAP/IL-17A transgenic mice develop normally and do not show any signs of neurological dysfunction. However, histological characterization revealed astrocytosis and activation of microglia. Demyelination, neurodegeneration or prominent tissue damage was not observed but a vascular pathology mimicking microangiopathic features was evident. Histological and flow cytometric analysis demonstrated the absence of parenchymal infiltration of immune cells into the CNS of GFAP/IL-17A transgenic mice. In GFAP/IL-17A mice, LPS-induced endotoxemia led to a more pronounced microglial activation with expansion of a distinct $CD45^{high}/CD11b^+$ population and increased induction of proinflammatory cytokines compared with controls. Our data argues against a direct role of IL-17A in mediating tissue damage during neuroinflammation. More likely IL-17A acts as a modulating factor in the network of induced cytokines. This novel mouse model will be a very useful tool to further characterize the role of IL-17A in neuroinflammatory disease models.

Editor: Ralf Andreas Linker, Friedrich-Alexander University Erlangen, Germany

Funding: This work was supported by a start up grant from the University of Sydney to ILC. MM was a post-doctoral fellow from the Deutsche Forschungsgemeinschaft (DFG, Mu17-07/3-1) and was also supported by the fund "Innovative Medical Research" of the University of Münster Medical School, Germany. JZ was funded by the fund "Bonfor" from the University of Bonn Medical School, Germany and the DFG (KFO177, University of Bonn). The funders had no role in study design, data collection and analysis, decision to publish, or preparation of the manuscript.

Competing Interests: The authors have declared that no competing interests exist.

* E-mail: marcus_m.mueller@ukb.uni-bonn.de

Introduction

Recently, a number of studies point toward a central role for the interleukin-17 (IL-17) cytokine family in various CNS diseases [1]. The IL-17 cytokine family consists of six members named IL-17 (IL-17A), IL-17B, IL-17C, IL-17D, IL-17E (IL-25) and IL-17F [2]. The most prominent members are IL-17A and IL-17F which form functional homo- or hetero-dimers with largely overlapping proinflammatory effects bridging the adaptive and innate immune response [3-5]. Effector functions of IL-17A are considered pivotal in the host response against extracellular and intracellular pathogens [6-8] and are associated with the pathogenesis of many autoimmune inflammatory diseases [9-14]. There is a convincing body of evidence that IL-17A plays an important role in inflammatory brain disorders including multiple sclerosis [15], infectious CNS diseases [16] and stroke [17,18] as well as in the pathophysiology of vascular inflammation and arteriosclerosis [19,20]. In these pathological conditions, the source of IL-17A can vary from infiltrating hematogenous immune cells like Th17 polarized CD4[+] T-cells [21,22], CD8[+] T-cells, gammadelta T-cells [23], NK-cells [24], and granulocytes [25,26] to CNS resident cells. In particular astrocytes have been demonstrated to secrete

IL-17 in pathological conditions like multiple sclerosis and ischemic brain injury [15,17,].

Th17 polarized T-cells came into focus of research after the pivotal role of IL-23 in the induction of EAE was described almost a decade ago [27] (reviewed in [28]). This finding resolved contradicting results that challenged the concept that organ specific autoimmunity was a Th1 driven condition: mice genetically-deficient in IFN-γ and IFN-γ receptor, as well as mice with impaired Th1 differentiation were not protected from EAE but developed more severe disease [29,30]. IL-23 induces the proliferation of a IL-17 secreting independent T-cell subset subsequently named Th17 cells [10,31,32]. To induce Th17 lineage commitment, stimulation of naïve T-cells with a combination of TGF-β and IL-6 [33–35] or with a combination of IL-21 and TGF-β [36] is required.

The receptor for IL-17A and IL-17F consists of a heterodimeric complex of IL-17RA and IL-17RC and is expressed in the CNS on astrocytes, microglia and endothelial cells [37,38]. Its stimulation induces NFkappaB and MAP kinase activation via TRAF6 and the adaptor protein Act-1 signaling [39,40] thus leading to the expression of many proinflammatory cytokines, chemokines and antimicrobial peptides. Particularly IL-17A is involved in the expansion and recruitment of neutrophils through the induction of G-CSF and the ELR+ members of the CXC family of chemokines CXCL1 and CXCL2 [41–43]. However, though effector functions of IL-17A are well characterized outside the brain, the direct CNS effector functions remain vague. *In vitro* data suggests an activation of microglia and synergistic effects of IL-6 stimulation on astrocytes through IL-17A signaling [44,45]. Furthermore, IL-17A is thought to disrupt the blood brain barrier by release of reactive oxygen species [39,46]. *In vivo*, there are few and partly controversial data regarding the impact of IL-17A on CNS autoimmune diseases. Whereas in EAE, genetic deletion or neutralization of this cytokine resulted in an attenuated disease course in some studies [47–50], it was shown recently in other studies that mice lacking IL-17A and IL-17F were still susceptible to EAE [51]. Disruption of IL-17A signaling pathways by genetic knockout of the IL-17 receptor subunit IL-17RC [52] or astrocyte targeted deletion of Act1 is highly capable of ameliorating EAE disease course [53].

Furthermore, knowledge about the impact of IL-17A on CNS infections is limited and comparably contradictory [16,54,55].

As outlined above, it is clear that IL-17A is a multifunctional cytokine with direct effects on CNS resident and infiltrating cells. However, at present most data addressing CNS effector functions of IL-17A are from *in vitro* experiments and we lack appropriate models for dissecting the functional properties *in vivo*. To examine the functional impact of IL-17 on the CNS *in vivo*, we have generated a transgenic mouse with astrocyte-specific production of IL-17A under the transcriptional control of a well-described glial fibrillary acidic protein (GFAP) genomic expression vector [56-58].

Materials and Methods

Animals

The method used for the generation of transgenic mice with astrocyte-targeted gene expression under the transcriptional control of a GFAP promoter was described in detail previously [57]. In brief, the coding sequence for the murine *Il17a* gene (bases 58-534; GenBank accession no. NM_010552) was amplified by RT-PCR from RNA isolated from the spleen of mice suffering from EAE at peak clinical disease. The oligonucleotide primers aagtgcacccagcaccag (5') and cgcgggtctctgtttagg (3') were used for the PCR. After digestion with the appropriate restriction enzymes and subsequent sequence verification, the amplified *Il17a* cDNA fragment was cloned into a GFAP expression vector containing a human growth hormone polyadenylation signal sequence downstream of the insert as described previously [58–59]. The resulting fusion gene construct was microinjected into the germline of (C57Bl/6×C3H/HeN) F1 mice. Genotyping of the animals was accomplished by PCR analysis of genomic tail DNA using primers targeted at the human growth hormone sequence and the *Il17a* sequence included in the transgene construct. Hemizygote transgenic founder mice were backcrossed to the C57BL/6 background for at least 8 generations before experiments were performed. Transgene negative mice served as wild-type littermate controls. All mice were maintained under pathogen-free conditions in the closed breeding colony of the University Hospital of Muenster, Germany. All animal experiments were approved by the Animal Care Commission of Nordrhein-Westfalen.

Tissue processing for histology

The tissue for analysis was obtained from GF/IL17 hemizygote animals and littermate controls at different ages as indicated in the results and figures. Brains were removed for routine histological and immunohistochemical examination. Immediately after euthanasia, brains were removed and brain halves (cut along the sagittal midline) were fixed overnight in PBS-buffered 4% paraformaldehyde at 4 °C, washed in PBS and subsequently embedded in paraffin. Sections (5 μm) were prepared from paraffin-embedded tissue. For immunohistochemistry on cryosections, tissue was embedded with Tissue Tek (Sakura Finetek, Staufen, Germany) after cryoprotection in 30 % (w/v) sucrose. For fluorescence microscopy 8 μm sections, for confocal laser scanning microscopy 50 μm sections were prepared. To ensure the same staining conditions sections from transgenic GF/IL17 animals were always mounted together with sections from control mice on a single glass slide.

Primary astrocytes cell culture

Primary mixed glial cell cultures were prepared from single 1 d old hemizygote GF/IL17 and wildtype littermate mice as previously described with modifications [60]. Briefly, mixed glial cultures were prepared from newborn mice and cultured in DMEM supplemented with 10% FCS and 100 U/ml penicillin/ streptomycin. After 10–14 d of primary cultivation floating microglial cells were discarded after intensive shaking. Remaining astrocytes were extensively washed with culture medium and kept in culture for 12 hours. Supernatants were harvested and snap frozen in liquid nitrogen, whereas astrocytes were dislodged by mild trypsinization and pelleted for RNA extraction using RNAeasy Columns (Qiagen, Hilden, Germany) according to the Manufacturer's instructions.

Routine histology and immunohistochemistry

Paraffin-embedded sections were stained with H&E and Luxol fast blue for routine histological analysis and evaluation of myelination. For the detection of calcium deposits Alizarin red S staining (Sigma) was performed [61]. For immunohistochemistry of paraffin embedded tissue, sections were rehydrated in graded ethanol series after deparaffination in xylene. Slides were then incubated over night at 4°C with primary antibodies (primary antibodies and corresponding protocols for immunohistochemistry are summarized in Table 1). After washing in PBS, a corresponding biotinylated secondary antibody (Axxora, Lörrach, Germany; 1/200) and HRP-coupled streptavidin (Axxora; 1/200) was used. Nova Red (Vector Labs) was applied as the immunoperoxidase substrate according to the manufacturer's instructions. Sections were counterstained with haematoxylin (Sigma-Aldrich). Fluorescent immunohistochemistry was applied to frozen sections. Primary antibodies were incubated over night at 4° C. After washing in PBS, an A594 or A488 fluorescence-conjugated secondary antibody (Invitrogen, Darmstadt, Germany; 1/200, 60 minutes) was used to visualize the primary antibody. Sections were counterstained with DAPI (Sigma-Aldrich, Munich, Germany). Conventional and immunofluorescence-stained sections were examined under a Nikon eclipse 800 bright-field and fluorescence microscope (Nikon, Düsseldorf, Germany). Bright field images and monochrome fluorescent images were acquired using a Spot flex camera and SPOT advanced 4.5 software (Diagnostic instruments, Sterling, MI).

Analysis of vascular density by confocal laser microscopy

For confocal analysis of vascular density free-floating sagittal sections (50 μm) were permeabilized by 1% Triton X-100 for 2 hours at room temperature. Consecutive antibody incubation steps were performed in 1% Triton X-100 overnight at 4°C. Microvasculature was stained using rabbit anti-laminin antibody

Table 1. Antibodies/Lectin used for histology.

Antibody/Lectin (source)	Specificity	Dilution Paraffin	Cryostat
Polyclonal rabbit anti-Iba1 reactive with human, mouse and rat Iba1 (Wako Chemicals, Neuss, Germany)	Microglia/ macrophages	–	1/500
Monoclonal rat anti- mouse CD68 (Serotec, Düsseldorf, Germany)	Microglia/ macrophages	–	1/200
Polyclonal rabbit anti-Laminin reactive with human and mouse Laminin (Sigma-Aldrich, Munich, Germany)	Basal lamina	–	1/50
Polyclonal rabbit anti -human GFAP (Dako, Hamburg, Germany)	Glial fibrillary acidic protein	1/200	–
Biotin-conjugated tomato lectin, *L. esculentum* (Axxora, Lörrach, Germany)	Microglia/ macrophages, endothelial cells	1/50	–
Monoclonal mouse anti-mouse NeuN (Chemicon, Schwalbach, Germany)	Neurons	1/200	–
Monoclonal mouse anti-mouse PLP (Serotec, Düsseldorf, Germany)	Proteo-Lipid-protein	1/500	–

with a subsequent incubation with an A594-conjugated secondary antibody. Sections from mutant and wild-type animals were processed in parallel. 20 µm z-stacks were captured from corresponding regions (corpus callosum, hippocampus, white matter) using a Zeiss LSM510 laser scanning microscope. Capillaries were identified by a lumen diameter < 10 µm and microvascular network was analyzed using ImageJ software.

Electron microscopy and analysis of basement membrane thickness

Brains were dissected after fixation in 3% glutaraldehyde. Corpus callosum dissections were later progressively dehydrated in a graded series of ethanol (30–100%) and embedded in Epon. Ultrathin sections (50–60 nm) were cut using an Ultracut microtome and placed on copper grids for analysis. Grids were stained and contrasted with uranyl acetate and lead citrate, followed by examination with Zeiss TEM900 electron microscope as described in detail before [62]. Basement membrane thickness was analyzed using ImageJ. Thickness was measured from saggital sectioned capillaries at three different positions where duplications of basement membrane were absent and mean values were calculated.

Cytokine and Chemokine mRNA determination by qRT-PCR

RNA of whole brain hemispheres, dissected forebrains and cerebelli or spinal cord was isolated using Trizol reagent (Invitrogen) according to the manufacturer's instructions. Total RNA (2 µg) was reverse-transcribed into cDNA using SuperScript^{TM} III Reverse Transcriptase (Invitrogen). Real-time quantitative PCR assays were performed using Taqman reagents (Applied Biosystens, Darmstadt, Germany). Samples were analyzed simultaneously for *Gapdh* mRNA as the internal control. The mRNA levels for each target were normalized to mRNA levels of *Gapdh* and expressed relative to that of nontransgenic littermate controls. Each sample was assayed in duplicates. Predesigned Taqman assays (Applied Biosystems) were used to amplify the following targets: *Gapdh, Il17a, Tnf, Il1b, Il6, Cxcl1, Cxcl2, Ccl2, Cxcl10, Csf2* and *Mmp9*. cDNA samples from neuroinflammatory disease models (EAE, murine West Nile virus encephalitis (WNV), and experimental cerebral malaria (ECM)) were used to compare the CNS production of *Il17a* in the GFAP/IL17A transgenic

mouse model with non-transgenic disease models. EAE and ECM were induced as described in detail before [63].

CNS leukocyte isolation and flow cytometry

CNS microglia and parenchymal infiltrating leukocytes were isolated from whole brain homogenates as described previously [64] with modifications. In brief, mice were perfused transcardially with ice cold PBS until flow through was completely clear to remove intravascular leukocytes. After dissection brains were grinded in Hank's Balanced Salt Solution (HBSS, Gibco, Eggenstein) using a tissue homogenizer (glass Potter, Braun, Melsungen) followed by a needle (0.6×25) and a syringe (5 ml) before passing through a 70 µm cell strainer (BD biosciences, Heidelberg). After pelleting, homogenates were resuspended in 75 % isotonic Percoll (GE-healthcare, Uppsala, Sweden) at 4°C. A discontinuous Percoll density gradient was layered as follows: 75 %, 25 % and 0% isotonic Percoll. The gradient was centrifuged for 25 min, 800 g at 4°C. Microglia, leukocytes, and astrocytes were collected from the 25 % / 75 % interface.

For surface marker staining the collected cells were directly washed in PBS, and blocked with CD16/CD32 (Fc block; eBioscience, Frankfurt/Main, Germany) antibody. Isolated leukocytes were incubated with fluorochrome-conjugated antibodies (eBioscience) to detect CD3e (PerCP-Cy5.5), CD11b (APC), CD11c (PE-Cy7), CD45 (FITC), CD45 (eFluor 450), Ly6G (PerCP-Cy5.5), B220 (APC-eFluor 780) and NK1.1 (PE-Cy7).

For intracellular cytokine staining cells were seeded in a 12 well plate at a density of 1×10^6 cells / ml in DMEM (Gibco, Eggenstein) containing glutamine and 10 % FCS. Cells were incubated for 4 hours in the presence of LPS (100 ng/ml, Sigma) and Brefeldin A to block cytokine secretion according to manufacturer's instructions (GolgiPlug, BD Biosciences). After blocking of Fc-receptors cells were surface stained with anti-CD45 (APC-Cy5.5) and anti-GLAST (APC, Miltenyi-Biotech, Bergisch Gladbach, Germany). Subsequently cells were washed, permeabilized using the Cytofix/Cytoperm kit (BD Biosciences) according to instructions, blocked again with Fc-block, and stained intracellulary with anti-TNFα (FITC, BD Biosciences).

After washing, bound Ab was detected using a BD FACSCanto II (BD Biosciences), and the acquired data were analyzed using the flow cytometry software, FlowJo (TreeStar, San Carlos, CA).

Protein Lysates and Western Blot

Tissue was homogenized using a Precellys 24 tissue homogenizer (Bertin Technologies, Saint-Quentin-en-Yvelines Cedex, France) in lysis buffer described elsewhere [65]. Samples were centrifuged at 12000 rpm / 13200 g for 15 min and supernatants were taken. The protein concentrations were determined using the BCA Protein Assay Kit (Pierce, Rockford, IL). Protein lysates (50 µg) were separated by 10 % SDS-PAGE gel using NuPage MES SDS running buffer (Invitrogen) at 150 V. PageRuler Prestained Protein Ladder (Fermentas, St. Leon-Rot, Germany) was used as standard. Proteins were transferred to 0.2 µm nitrocellulose membranes (Whatman, Dassel, Germany). Membranes were blocked for 30 min in TBST containing 5% skim milk. Immunoblotting were performed using monoclonal anti-GFAP (Chemicon, Schwalbach, Germany) and antibody CP06 (Oncogene Science, Cambridge, MA) detecting α-tubulin followed by incubation with the appropriate horseradish-peroxidase conjugated secondary antibodies (Jackson ImmunoResearch, Newmarket, UK). Immunoreactivity was detected by chemiluminescence reaction (Millipore, Schwalbach, Germany) and luminescence intensities were analyzed using Chemidoc XRS imaging system (BioRad, Munich, Germany). With the Quantity One (BioRad) program bands density were determined for each lane and the intensity ratio for the detected GFAP were calculated to α-tubulin.

Determination of serum and cell culture IL-17A protein level

Serum IL-17A was determined by ELISA (eBioscience) after cardiac puncture and centrifugation from GF/IL17 mice and littermate controls according to the manufacturer's protocoll. For the in vitro determination of IL-17A-secretion, primary astrocyte cultures were incubated in fresh DMEM containing 10 % FCS for 12 hours and supernatant was harvested. A standard curve was generated according to the Manufacturer's protocol.

Induction of endotoxemia

Endotoxemia was induced by a double intraperitoneal injection of 100 µg LPS (Escherichia coli 026:B6; Sigma, Munich, Germany) 24 hours and 4 hours before killing in 5 GF/IL17 transgenic mice and 5 WT littermate controls. Mice were perfused with ice-cold PBS and subsequently brains were dissected in the sagittal midline immediately. One half was immediately snap-frozen in liquid nitrogen and stored at -80°C until RNA isolation. The other half was homogenised for FACS-analysis as described above.

Analysis of blood brain barrier (BBB) integrity with Evans blue dye (EBD)

BBB integrity was determined with EBD as described previously [66,67] with modifications. 3 hours before scarification 3 mice per group received intraperitoneal injections of EBD (2 % w/v in isotonic NaCl, 4 mg / kg body weight, AppliChem, Darmstadt, Germany). To remove intravascular EBD before removal of CNS tissue mice were perfused with 60 ml of ice-cold PBS supplemented with 2 mM EDTA (Sigma). The effective clearance of dye by the perfusion step was confirmed by the appearance of a colorless perfusate. Brain was dissected into forebrain and cerebellum, spinal cord and portions of liver were removed, washed briefly with double distilled H2O and weighed. Tissue from LPS treated mice served as positive control. Tissue was homogenized in a three-fold volume of 50% trichloroacetic acid (w/v, AppliChem) solution and pelleted. Supernatants were diluted with ethanol (1:3), and fluorescence was quantified using a microplate fluorescence reader (Tecan infinite 200 M, Crailsheim, Germany), (excitation: 620 nm, emission: 680 nm). Sample value calculations were based on Evans blue dye standards mixed with the same solvent.

Statistical analysis

For statistical analysis, GraphPad Prism was used. Real time PCR data, ELISA, Western blot data, Evans blue dye extravasation data, or flow cytometry data were analyzed where appropriate by a two-tailed Student's t test with $p < 0.05$ considered to be statistically significant.

Results

Generation of GFAP-IL-17 transgenic mice and analysis of transgene expression: chronic CNS IL-17A stimulation neither induces major tissue damage, neurodegeneration nor demyelination

To characterize the effects of IL-17A in the CNS we used a well-established approach, targeting the expression of cytokine transgenes into astrocytes [68]. We detected 18 transgene positive founder mice and generated two independent mouse lines (GF/IL17–15 and GF/IL17–45) with stable genomic integration of the GFAP-IL-17A construct. A comparison of peripheral organs from GF/IL17 and wild-type mice revealed no significant differences of Il17a mRNA between the two groups (liver, spleen, kidney, gut, lung, heart, hamstring muscle, and sciatic nerve). IL-17A protein was not detectable in serum of both transgenic mice and littermate controls, respectively (data not shown). Hemizygous transgenic animals up to the age of 18 months developed normally without showing clinical signs of neurological disease or obvious behavioral abnormalities. In addition, none of the founder mice showed a clinical or histopathological phenotype (data not shown).

Analysis of relative transgene expression by quantitative PCR compared to WT whole brain lysates exhibited a comparable distribution of transgene encoded Il17a mRNA in forebrain (110.1 ± 22.6), cerebellum (153.9 ± 7.6), and spinal cord (137.9 ± 44.2) for both transgenic mouse lines (Fig. 1A). We further examined Il17a mRNA induction in viral and autoimmune inflammatory disease models to relate their Il17a mRNA induction with the level in our transgenic model. Compared with peak EAE, WNV encephalitis and ECM in C57Bl/6 WT mice CNS Il17a expression was significantly higher in the GF/IL17A mice (peak EAE: 10.6 ± 1.8, WNV 7.8 ± 4.6, ECM: 8.8 ± 5.1, for all p < 0.0001 compared to GF/IL17 mice). To confirm astrocyte IL-17A transcription and translation into protein with subsequent secretion we generated primary astrocyte cultures from GF/IL17–15 hemizygote mice and littermate controls. The presence of Il17a mRNA was detected using real time PCR (Fig. 1B) along with the confirmation of IL-17A protein secretion into the supernatant of GF/IL17 primary astrocyte cultures using ELISA (Fig. 1C). The further studies described below were conducted using the GF/IL17–15 line.

Routine histologic analysis of H&E-stained paraffin embedded sections was applied to detect histopathological alterations or cellular infiltrations in the CNS of GF/IL17 mice (Fig. 1E, 1K, 1Q) at different ages (1 – 12 month) compared with age- and sex-matched wild-type littermate controls (Fig. 1D, 1J, 1P). Chronic IL-17A production did not induce demyelination in GF/IL17 mice shown by luxol fast blue staining (Fig. 1F, 1G, 1L, 1M) or anti-PLP immunohistochemistry (Fig. 1N, 1O, 1R, 1S). Numbers and distribution of neurons were normal in GF/IL17 mice compared with wildtype controls using anti-NeuN immunohistochemistry (Fig. 1H, 1I, 1T, 1U). Taken together these findings indicate that the chronic astrocyte production of IL-17A in GF/

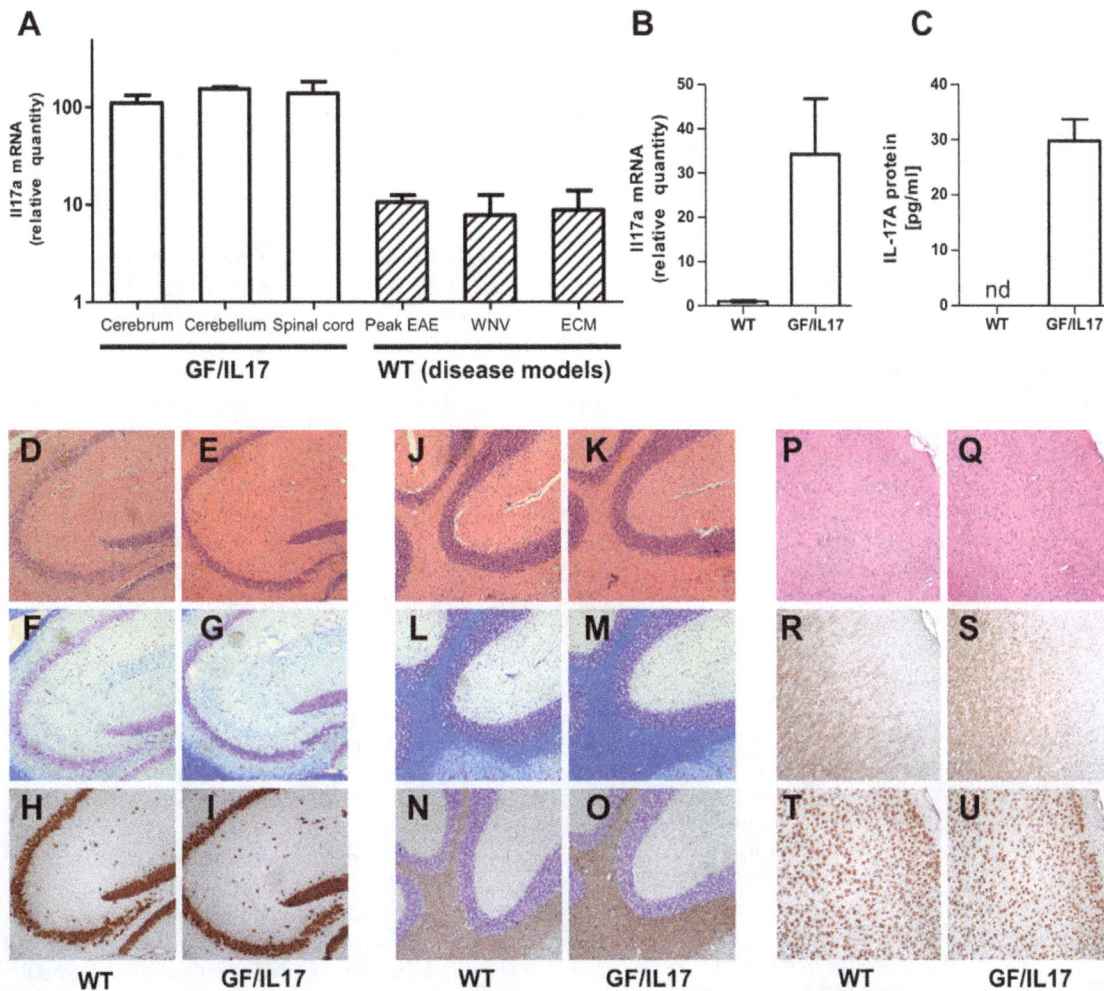

Figure 1. GF/IL17 mice express IL-17A mRNA and protein without major histological defects or leukocyte infiltration. (A) Relative expression of *Il17a* mRNA in GF/IL17 mice is equally distributed between forebrain, hindbrain and spinal cord. In comparison to disease models in WT mice (peak EAE, West Nile Virus encephalitis, experimental cerebral Malaria) CNS expression levels of *Il17a* in otherwise untreated GF/IL17 mice exceed levels of all tested disease models irrespective of the cellular source. **(B)** *Il17a* mRNA expression from GF/IL17 primary astrocytes and WT controls was quantified using real time PCR. **(C)** Astrocyte secretion of IL-17A protein was confirmed using ELISA. Supernatants of GF/IL17 and WT control primary astrocytes were analyzed after 12 hours of culture. In supernatant of WT control astrocytes IL-17A protein was not detectable. **(D–U)** Routine histological characterization of mice excluded major tissue damage or leukocyte infiltration. Representative areas of hippocampus **(D–I)** , cerebellum **(J–O)**, or cortex **(P–U)** are shown in WT controls **(D, F, H, J, L, N, P, R, T)** and GF/IL17 transgenic mice **(E, G, I, K, M, O, Q, S, U)** by HE **(D, E, J, K, P, Q)**, LFB **(F, G, L, M)**, anti murine NeuN mAb **(H, I, T, U)**, or anti murine PLP mAb **(N, O, R, S)** staining (age 9 month).

IL17 transgenic mice is neither associated with a clinical phenotype nor with overt histological alterations.

CNS specific IL-17A production neither alters the expression levels of inflammation-related genes nor promotes leukocyte infiltration into the CNS

We next determined whether the expression of a variety of IL-17A inflammation-related genes might be altered in the brains of GF/IL17 mice. With the exception of the transgene-encoded IL-17A, cerebral expression of the proinflammatory cytokines *Tnf*, *Il1b*, *Il6*, *Csf2*, the chemokines *Cxcl1*, *Cxcl2*, *Ccl2*, and the matrix degrading enzyme *Mmp9* were not induced in GF/IL17 mice compared with matched wild-type littermates (Table 2).

To further exclude leukocytic infiltrates in the brain of GF/IL17 mice, flow cytometric analysis was performed (Table 3). No differences were found in the cellular ratios of $CD45^{dim}/CD11b^+$ microglia, $CD45^+/Ly6G^+$ granulocytes, $CD45^+/CD3^+$ T-cells,

$CD45^+/B220^+$ B-cells, $CD45^+/NK1.1^+$ NK-cells, or $CD45^+/CD11c^+$ dendritic cells in GF/IL17 mice compared with WT mice.

Chronic CNS IL-17A production induces astrocytosis and microglial activation

To further examine the effect of chronic IL-17A production on the glial cell population, we characterized the phenotype of astrocytes and microglia by immunohistochemistry. We found evidence for substantial astrocyte activation in GF/IL17 transgenic mice, respectively. In all brain areas from GF/IL17 mice, strong GFAP-immunoreactivity was observed and astrocytes had a swollen cell body and hypertrophic processes characteristic for so called "reactive astrocytes" compared with astrocytes from wild-type control animals (Fig. 2A). The induction of astrocytosis by chronic IL-17A stimulation was confirmed by immunoblotting against GFAP. Corresponding to the reactive astrocytosis observed

Table 2. Regulation of inflammation related genes in GF/IL17 mice relative to littermate controls (arbitrary units).

Gene	Il17a	Tnfa	Il1b	Il6	Csf2	Cxcl1	Cxcl2	Ccl2	Mmp9
Rel. expression	111 ± 11***	1.0 ± 0.07	1.2 ± 0.11	0.9 ± 0.05	0.9 ± 0.05	1.0 ± 0.04	1.1 ± 0.06	0.9 ± 0.05	1.0 ± 0.03

by immunohistochemistry, western-blots of brain homogenates from GF/IL17 mice showed increased GFAP protein levels compared with wild-type control animals (Fig. 2B). Densitometry revealed a significant increase of GFAP in GF/IL17 mice compared with aged-matched non-Tg littermates (WT: 1.00 ± 0.06 arbitrary units versus GF/IL17: 1.48 ± 0.07 arbitrary units, p ≤ 0.01) (Fig. 2C).

Tomato lectin staining was used to analyse the microglial morphology and state of activation. Typical ramified microglial cells were found in wild-type mice, whereas GF/IL17 mice displayed more intensively labeled microglia with a hypertrophic cell body but still displaying mostly ramified processes (Fig. 2D). To further substantiate the microglial changes observed in GF/IL17 mice, we colocalized microglia with the lysosomal activation marker CD68 and found an increased CD68 staining in microglia from GF/IL17 mice compared with controls (Fig. 2E, arrows). FACS surface marker analysis of freshly isolated microglia confirmed the pronounced microglial activation state in GF/IL17 mice revealing an upregulated surface expression of CD11b compared with littermate controls (Fig. 2F). Statistical analysis of the mean fluorescence intensity of isolated microglia showed only minor effects of IL-17A production on CD45 surface marker expression (MFI - WT: 940.3 ± 59.9 versus GF/IL17: 987.7 ± 67.8, p: n.s.) whereas CD11b surface marker expression was significantly increased by IL-17A stimulation (MFI – WT: 3075 ± 72.1 versus GF/IL17: 3440 ± 89.6, p ≤ 0.05) (Fig. 2G). In summary chronic IL-17 production induces a substantial astrocytosis and microglial activation.

Chronic CNS IL-17A production induces a vascular pathology with calcification, capillary rarefaction, and thickening of basement membrane but without disruption of the blood brain barrier

In addition to the detection of microglial cells, the staining of microglia with tomato-lectin allows the examination of capillaries. Tomato-lectin staining revealed a stronger staining and a different staining pattern with strikingly more prominent vessels in GF/IL17 brain tissue compared to controls (Fig. 2D). Previous studies demonstrated that IL-17A stimulation induces the disruption of blood brain barrier integrity in cerebral microvascular endothelium [39,46]. We therefore further analyzed the cerebral microvasculature and found numerous small vascular calcifications in aged GF/IL17 transgenic mice located mostly in the thalamic region. No such calcifications were observed in wild-type

littermate controls (Fig. 3A) or younger animals of both genotypes suggesting that these bodies appear in an age-dependent manner. Anti-GFAP immunohistochemistry revealed a peri- and intravascular localization of these concrements, which were accompanied by perilesional astrogliosis (Fig. 3B). Lectin histochemistry verified the association of these calcifications to blood vessels (Fig. 3C). The calcium-specific Alizarin Red S staining confirmed that the observed concrements were calcifications (Fig. 3D).

Ultrastructural studies of capillaries within the corpus callosum revealed a marked thickening of the basement membrane in GF/IL17 mice compared with wild-type littermate controls, respectively. Moreover, basement membrane in GF/IL17 mice exhibited profuse duplications marking the perivascular space filled with pericyte processes (Fig. 3E and 3F). Statistical analysis of basement membrane thickness confirmed the significant enlargement of this layer in transgenic mice (WT: 80.7 nm ± 6.7 versus GF/IL17: 158.3 nm ± 21.3, p ≤ 0,05) (Fig. 3G).

To further examine the impact of chronic IL-17A stimulation on the microvascular network, we examined anti-laminin immunofluorescence staining specific for the basement membrane surrounding blood vessels in 50 μm thick brain sections. Predominately in the white matter of forebrain and cerebellum capillary density was reduced in GF/IL17 mice in comparison with controls (Fig. 3H and 3I). To assess the permeability of the blood brain barrier in GF/IL17 transgenic mice, we injected Evans blue dye (EBD) and determined the release of albumin bound EBD into the brain and spinal cord. None of the brains from either GF/IL17 mice or wild-type controls were macroscopically stained blue whereas tissue distribution of EBD was confirmed by a blue stained liver. Quantifying extravasated EBD from brain lysates further excluded a significant impairment of blood brain barrier function in GF/IL17 mice (Fig. 3J). Again equal tissue distribution of EBD was confirmed by quantifying extravasated EBD from liver lysates of corresponding mice. Additionally we performed an anti-mouse immunoglobulin immunohistochemistry excluding the extravasation of mouse immunoglobulins into the CNS in both GF/IL17 mice and wild-type controls (data not shown).

Taken together these findings indicate that chronic IL-17A production induces a prominent vascular pathology in the CNS with vascular calcification, capillary rarefaction and thickening of the basement membrane. In addition, we could not find evidence for a loss of integrity of the blood-brain barrier in GF/IL17 mice.

Table 3. FACS quantification of infiltrating leukocytes in GF/IL 17 mice and controls relative to all CD45 positive cells.

	Granulocytes	T-cells	B-cells	Nk-cells	Dendritic cells
	(CD45⁺/Ly6G⁺)	(CD45⁺/CD3⁺)	(CD45⁺/B220⁺)	(CD45⁺/Nk1.1⁺)	(CD45⁺/CD11c⁺)
WT	0.3 % ± 0.05	0.5 % ± 0.05	2 % ± 0.7	0.04 % ± 0.01	0.4 % ± 0.07
GF/IL17	0.1 % ± 0.02	0.5 % ± 0.2	1.6 % ± 0.2	0.1 % ± 0.03	0.3 % ± 0.03

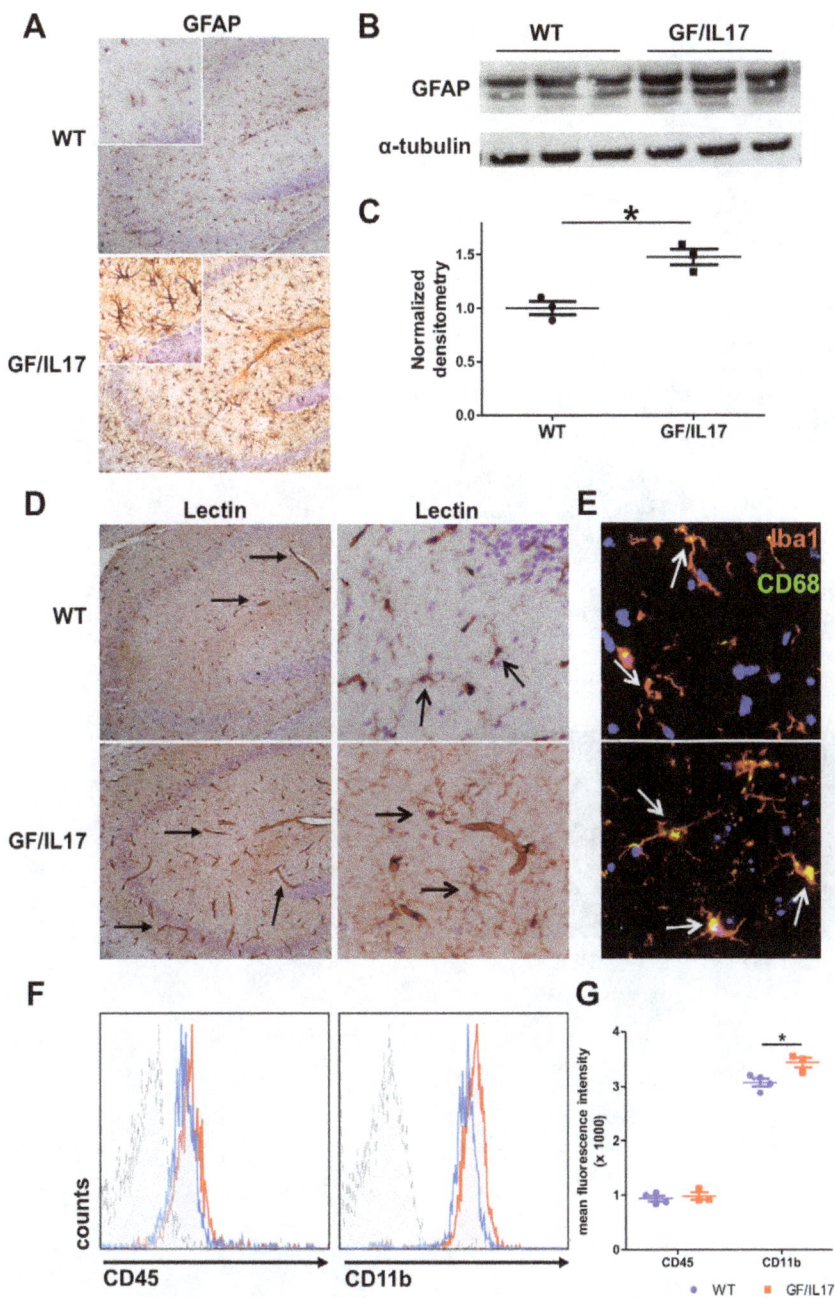

Figure 2. Transgenic CNS expression of IL-17A induces glial activation. **(A)** IHC for GFAP in the hippocampus of WT and GF/IL17 transgenic mice at 9 month. GFAP-staining revealed a strong astrocytic activation by morphological criteria in GF/IL17 mice. **(B)** Astrocytosis was confirmed by anti-GFAP immunoblotting. Whole brain lysates were analyzed by immunoblotting for the presence of GFAP. Anti-tubulin immunoblotting served as internal loading control on the same membrane. **(C)** Densitometric quantifications (arbitrary densitometry units) from immunoblots of B after normalization by tubulin densitometry units obtained from the same immunoblot. (*$p < 0.05$). **(D)** Tomato-lectin-staining in the hippocampus revealed an activated microglial morphology in GF/IL17A transgenic animals characterized by rounded cell bodies and microglial clustering (open arrows). In addition Lectin staining displayed prominent microvasculature in GF/IL17 mice compared with WT controls (closed arrows; see also Figure 3 for vascular pathology). **(E)** IHC of frozen brain sections for Iba1 (red), CD68 (green) and Dapi (blue). GF/IL17 mice showed a strong immunoreactivity for the activation marker CD68 in Iba1 stained microglia (white arrows indicating colocalisation of the lysosomal markes CD68 and Iba1, age: 9 month). (F) Representative flow cytometric analysis of surface marker expression from freshly isolated microglia in GF/IL17 mice (red) and WT littermate controls (blue). Dashed histogram: isotype control. Histograms were gated on microglial population according to forward/side scatter profile. GF/IL17 mice displayed similar surface expression levels for CD45 compared with WT whereas CD11b expression was upregulated in GF/IL17 mice compared with WT. (G) Statistical analysis of mean fluorescence intensity of freshly isolated microglia in GF/IL17 mice (red) and WT littermate controls (blue). Comparable expression levels of CD45 in GF/IL17 and WT mice whereas CD11b expression levels were significantly upregulated in GF/IL17 mice compared with WT controls (*$p < 0.05$).

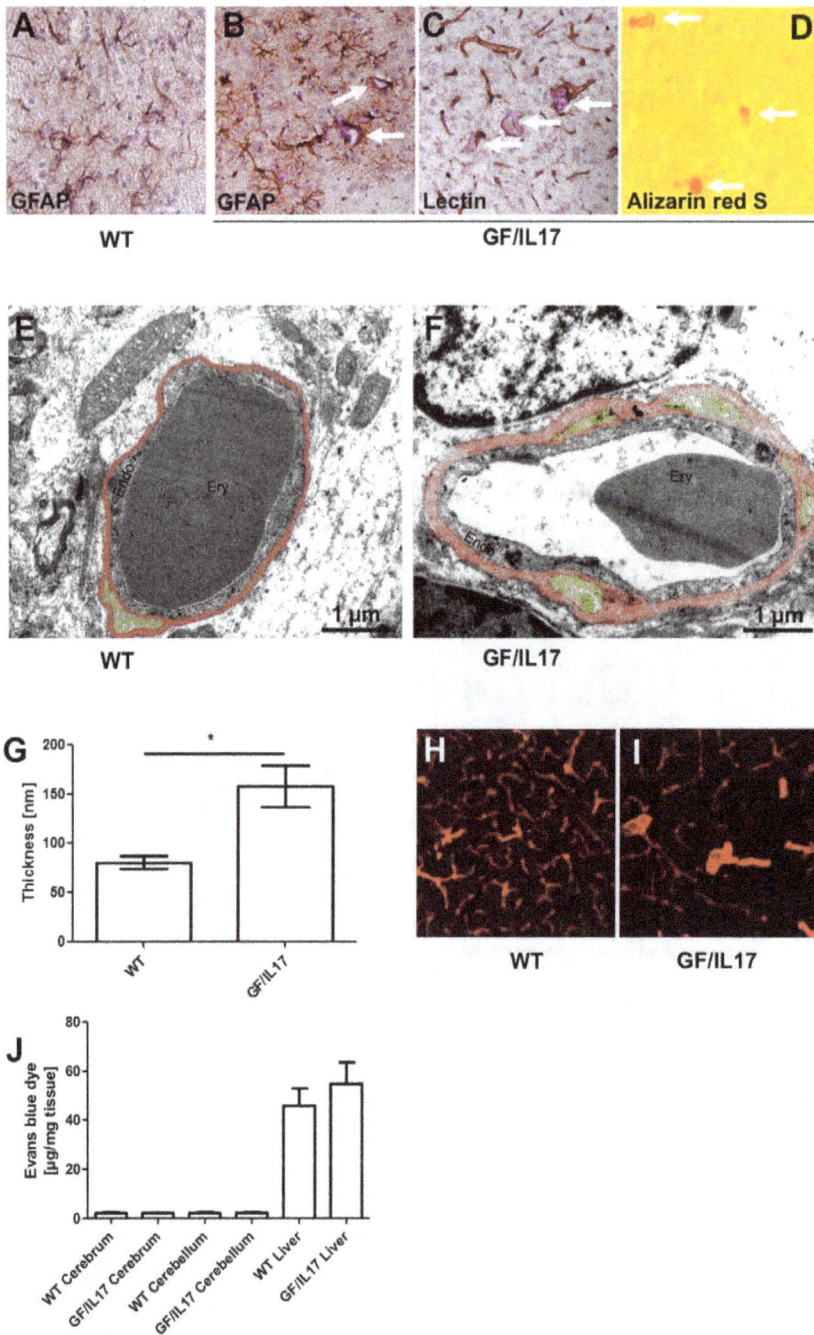

Figure 3. Astrocytic expression of IL-17A induces a vascular pathology with capillary calcifications, microvascular rarefaction and thickening of endothelial layer and basement membrane of vessel walls without disturbing blood-brain barrier integrity. GFAP immunohistochemistry (**A–B**), lectin (**C**) and Alizarin red S staining (**D**) of the thalamus in WT mice (**A**) and GF/IL17 transgenic animals (**C–D**) at the age of 09 month. Microvessels surrounded by GFAP positive astrocytic endfeet and laminin stained endothelia are filled with hematoxillin positive material (white arrows). Around calcified microvessels astrocytes display an activated morphology. Deposits are labelled orange/red in Alizarin red S staining (**D**) confirming vascular calcifications. Transmission electron microscopy of capillaries in the corpus callosum of (**E**) wild-type mice and (**F**) GF/IL17 transgenic mice at the age between 10 and 12 month exhibit morphological criteria of microangiopathy: the endothelial cell layer (*Endo*) appears enlarged compared with WT. Basement membrane (*red-colorored*) surrounding the vascular endothelium appears heavily thickened in TG animals. GF/IL17 mice display numerous duplications of the basement membrane spanning the perivascular space harbouring pericytes (*green colored*). Inside capillaries erythrocytes (*Ery*) are detectable. Scale bar represents 1 μm. (**G**) Measurement of basement membrane thickness revealed a significant diameter increase in GF/IL17 mice (p < 0,05) (**H**) Confocal microscopy of 50 μm sections labelled with anti-laminin displayed a dense microvascular network in the white matter of WT animals. (**I**) Rarefaction of microvasculature in corresponding areas in GF/IL17 mice. Furthermore arterioles appear thickened. (Age: 10–12 month of both transgenic and wild-type mice) (**J**) To examine blood-brain barrier integrity Evans blue dye (EBD) extravasation into tissue was quantified. Levels of tissue EBD in brains and spinal cord are equal in WT and GF/IL17 mice. Liver tissue served as positive controls.

Enhanced upregulation of inflammation related genes and increased numbers of a distinctive population of CD45^{high}/CD11b$^+$ activated microglia / monocytes in GF/IL17 mice in response to systemic LPS administration

The lack of spontaneous tissue destruction and immune cell recruitment into the brain of GF/IL17 mice led us to question if the central production of IL-17 would enhance the local effects of a systemic proinflammatory stimulus. To address this question, we subjected GF/IL17 mice and wild-type littermates to a peripheral immune challenge by intraperitoneal injections of LPS, a model inducing a neuroinflammatory response [69–71]. As evidence for modulatory effects of chronic CNS IL-17A production during immune responses, brain homogenates of GF/IL17 mice exhibited a significantly enhanced upregulation of *Tnf* (WT: 9.01 ± 0.74 versus GF/IL17: 17.21 ± 2.02, p ≤ 0.01) and *Il1b* (WT: 5.01 ± 1.41 versus GF/IL17: 10.85 ± 2.03, p ≤ 0.05) mRNA levels after peripheral immune challenge with LPS compared with mock treated wild-type littermates (Fig. 4A). Furthermore mRNA levels of the chemokine *Ccl2* were more pronounced upregulated in GF/IL17 mice compared with LPS treated WT controls, respectively. However, this difference failed to reach statistical significance. Transgenic IL-17A production had no influence on *Il6* mRNA levels in LPS-induced upregulation of inflammatory cytokines in the CNS.

Additional phenotypic analysis of the immune cells in the brain of LPS challenged mice revealed an increase in the numbers of a distinctive CD45^{high}/CD11b$^+$ population, indicative for activated microglia or accumulating monocytes/macrophages, in GF/IL17 mice compared with wild-type littermates (Fig. 4B). Numbers of CD45$^+$/Ly6G$^+$ granulocytes, CD45$^+$/CD3$^+$ T-cells, CD45$^+$/B220$^+$ B-cells, CD45$^+$/NK1.1$^+$ NK-cells, or CD45$^+$/CD11c$^+$ dendritic cells showed no difference, respectively. Statistical analysis of the ratio of the distinctive CD45^{high}/CD11b$^+$ immune cell population to CD45^{dim}/CD11b$^+$ resting microglia indicated a significant increase of activated microglia / infiltrating monocytes in GF/IL17 transgenic mice compared with non-transgenic mice (WT: 6.42 % ± 1.36 % versus GF/IL17: 10.33 % ± 0.99 %, p ≤ 0.05) (Fig. 4C).

The characterization of microglial activation by morphological criteria after LPS-induced endotoxemia revealed more activated microglia in GF/IL17 mice compared with wildtype controls (Fig. 4D), which was mostly pronounced in the periventricular region.

To elucidate the cellular source of the increased TNF-α secretion in GF/IL17 mice after LPS-induced inflammation, we performed intracellular cytokine staining and flow cytofluorometric analysis. TNF-α could be detected in CD45 positive microglia/leukocytes but not in GLAST positive astrocytes (Fig. 4E). The level of intracellular TNF-α after LPS treatment in CD45$^+$ microglia/leukocytes was increased in GF/IL17 mice compared with wild-type controls consistent with the *Tnf* RNA levels *in vivo*.

Taken together these results point to an augmented neuroinflammatory response in GF/IL17 mice following LPS challenge with increased secretion of proinflammatory cytokines, in particular an increased TNF-α production by microglia, and accumulation of a distinctive CD45^{high}/CD11b$^+$ leukocyte population in the inflamed CNS.

Discussion

The cytokine IL-17A has been implicated as an important effector cytokine in various CNS autoimmune and infectious diseases as well as in neurodegenerative processes. To further clarify the functional significance of IL-17A in the CNS, we generated a transgenic mouse line with astrocyte-targeted expression of the murine *Il17a* gene. This approach has been widely and successfully used previously to investigate the function of numerous cytokines and chemokines in the CNS (reviewed in [68]). In several of those transgenic models the local cytokine production by astrocytes is sufficient to initiate and maintain glial activation, leukocyte accumulation and severe CNS tissue injury and functional impairment [57,58,72,73]. However, some of these transgenic models displayed only a very mild or even no phenotype [59,74]. Thus, the widely varying and unique phenotypes associated with the transgene-encoded production of different cytokines in the CNS highlights not only the specificity of this approach but also the highly selective actions evoked by these factors.

We found that, despite clear evidence of the production of IL-17A in the CNS of GF/IL17 mice, these animals did not develop spontaneous leukocyte infiltration or major tissue destruction. The amount of IL-17A in the CNS of GF/IL17 mice is comparable to IL-17A CNS levels in other neuroinflammatory diseases like EAE, viral or protozoal encephalitis though the cellular sources of this cytokine might differ between our transgenic mouse model and the tested disease models. Consistent with the mild histological phenotype, the transgenic animals did not show any behavioral or physical abnormalities. In addition, no significant alterations in inflammation related gene expression were detected in the brain tissue from transgenic GF/IL17 mice compared with wild-type control littermates. Our findings concerning the impact of chronic IL-17A production on the CNS are in sharp contrast to studies in transgenic mice with overexpression of IL-17 in other organs such as skin or lung [32]: here, IL-17A induced a severe pathology with tissue destruction, which was also observed in a transgenic model with ubiquitous IL17-A overexpression [51]. The differences from our findings likely reflect organ specific factors like the blood brain barrier impeding leukocyte infiltration into the CNS or a reduced response of CNS-resident cells to the IL-17 stimulus. In particular, expression levels and distribution of the IL-17 receptors IL-17RA and IL-17RC might play a role. For IL-17RA highest expression levels have been observed in spleen or kidney whereas brain expression levels were weaker than in the lung [75]. Furthermore lung and skin expression levels of IL-17RC exceed CNS expression levels [76]. Interestingly, the highest IL-17RC expression levels have been shown for vascular endothelium in every examined organ. Our finding, that IL-17A mediates only modest CNS effects whereas severe detrimental effects for this cytokine have clearly been demonstrated for several peripheral organs has also been described for other cytokines as well: e.g. CCL21, which mediates chemotaxis of lymphocytes, induces thyroiditis if expressed under the promotor for thyroglobulin [77] whereas the expression of bioactive CCL21 under the GFAP promoter neither induced lymphocytic infiltration nor glial activation [78]. Therefore our findings underline the unique immunological milieu of the brain concerning cytokine actions.

Although chronic IL-17A production in the CNS does not induce major tissue damage, we found pronounced glial activation in otherwise healthy GF/IL17 mice. The lack of CNS tissue damage is in line with the findings of Haak et al., who demonstrated that transgenic production of IL-17A by T-cells during EAE failed to augment neuroinflammation and tissue damage [51]. In contrast to our findings, Haak et al. did not report any differences in glial activation. However, the transgenic production of IL-17A by T-cells as applied by Haak et al. leads only to increased IL-17 levels at the site of inflammation, where glial cells are already highly activated. Therefore our model of widespread IL17 production in the brain is much more suitable to

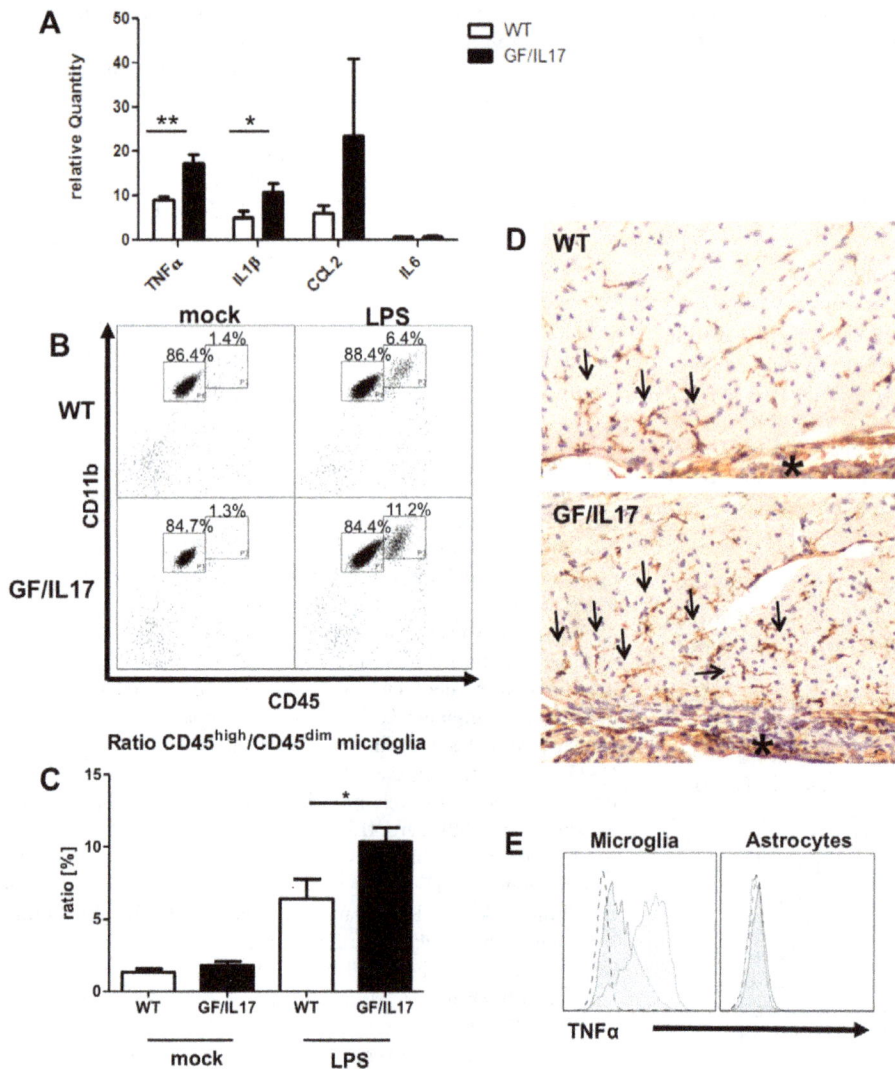

Figure 4. Transgenic IL-17A acts synergistically with other inflammatory stimuli and potentiates LPS induced microgial activation. GF/IL17 transgenic animals and littermate controls between 2 and 3 month were injected twice with 50 µg LPS i.p. in 24 hours or treated with mock injections of PBS. (**A**) Quantitative rt-PCR revealed a strong upregulation of the expression of inflammatory cytokines *Tnf*, *Il1b*, and *Ccl2* by LPS treatment whereas *Il6* was not induced following endotoxemia. This effect was markedly pronounced in GF/IL17 transgenic animals (*Tnf*: $p < 0,01$; *Il1b*: $p < 0,05$). Furthermore *Ccl2* expression was strikingly upregulated in some of the LPS treated GF/IL17 mice compared with LPS treated wild-type controls but due to the high interindividual variance not considered as significant. (**B**) Representative flow cytometry profiles from mock- or LPS-treated GF/IL17 and WT mice. LPS treatment induced a population of CD45^{high}/CD11b^{+} activated microglia in both WT and GF/IL17 mice. Chronic IL-17A stimulation strikingly augmented this effect, respectively. The numbers above the indicated gate show the mean percentages of FSC/SSC gated populations. (**C**) For the statistical analysis of infiltrating cell numbers a ratio between CD45^{high}/CD11b^{+} activated microglia and CD45^{dim}/CD11b^{+} resting microglia was calculated for each individual mouse.GF/IL17 transgenic animals exhibited a significantly elevated ratio of CD45^{high}/CD11b^{+} activated to CD45^{dim}/CD11b^{+} microglia after LPS treatment, respectively ($p < 0,05$). (**D**) Lectin immunohistochemistry revealed a pronounced accumulation of activated microglia (arrows) in the periventricular regions (asterisk: choroid plexus) after LPS treatment in GF/IL17 mice. (**E**) Intracellular staining of TNF-α after LPS treatment. Only CD45 positive microglia and monocytes/macrophages are stained by anti TNF-α antibody (left histogram) whereas GLAST positive astrocytes are negative for TNF-α (right histogram). In GF/IL17 transgenic mice (light gray) compared with wildtype mice (dark gray) CD45 positive microglia and monocytes/macrophages exhibit a stronger intracellular TNF-α staining after LPS treatment.

detect differences in glial cell populations. Astrocytes have been shown to constitutively express the IL-17 receptor, allowing a direct effect of IL-17A on astrocytes [38]. A critical effect of IL-17A on astrocytes was demonstrated by a recent study, identifying astrocytes as a major target cell population of IL-17A stimulation during EAE. This effect was mediated via Act1, a key component in IL-17 signalling [53]. In addition to astrocytosis, GF/IL17 mice displayed microglial cell activation. Microglia have also been shown to constitutively express the IL-17 receptor and microglial cells *in vitro* are known to directly respond to IL-17A stimulation by upregulating proinflammatory cytokines and chemokines [27,38]. However, we could not find any modulation of these inflammation-related genes in the CNS of the GF/IL17 mice possibly arguing for a counter regulation of IL-17A signalling after chronic stimulation or the necessity for a second stimulus, mediating synergistic effects in combination with IL-17A. Synergistic effects of IL-17 are well described for other cytokine systems like the IL-6 signalling cascade [45] or the IFN-γ dependent iNOS induction in astrocytes [79].

Unexpectedly, GF/IL17 mice developed a vascular phenotype with capillary rarefaction and thickening of the basement membrane, resembling pathological hallmarks of microangiopathy [80,81]. However, ischemic white matter lesions, which are typical for microangiopathy [80], were not observed in our transgenic model. IL-17A plays an important role in chronic and acute vascular inflammation (reviewed in [20]). A possible link between CNS vascular pathology and IL-17A has recently been published demonstrating an association between bleedings from arteriovenous malformations and a polymorphism in the IL-17A gene in human [82]. There are numerous reports associating IL-17A to the development of arteriosclerotic lesions outside the CNS. In human carotid arteriosclerotic plaques IL-17A is present and upregulated in ruptured plaques of symptomatic patients [83]. In apoE-deficient mice, a well-characterized model for arteriosclerosis, inhibition of IL-17A signalling reduced atherosclerotic lesion formation, prevented plaque rupture, and diminished levels of circulating proinflammatory cytokines [84-86]. Furthermore aged GF/IL17 mice displayed vascular calcifications predominately in thalamic regions. This pathological finding is strikingly similar to the thalamic calcifications in vitamin D receptor knockout mice [87], which just recently have been shown to exhibit elevated serum IL-17A levels [88,89]. Interestingly, a transgenic mouse model with astrocyte targeted expression of TGF-β develops a similar phenotype to GF/IL17 mice: in low expressing TGF-β1 mice there is an age dependent development of perivascular astrocytosis [90,91] and thickening of the endothelial basement membrane [92]. These findings are suggestive of a common mechanism of IL-17 and TGF-β signalling in mediating vascular pathology, especially as it is known that TGF-β is the most prominent upstream regulator of Th17 differentiation [34-37]. Surprisingly the BBB in GF/IL17 mice remained intact. Our finding seemingly contradicts a previous report showing that IL-17A induces reactive oxygen species (ROS) production in brain endothelial cultures, thereby downregulating the tight junction molecule occludin [46]. This discrepancy might be explained by a predominately abluminal secretion of IL-17A in brain vessels in GF/IL17A mice.

Systemic endotoxemia induces cerebral inflammation with upregulation of proinflammatory cytokines and the accumulation of a distinct CD45^{high}/CD11b^{+} population. Morphologically these cells seem to be activated microglia but the clear discrimination of this CD45^{high}/CD11b^{+} population to CD45^{dim}/CD11b^{+} microglia could possibly argue for a recruitment of hematogenous monocytes/macrophages into the CNS. Some studies suggest recruitment of neutrophils and monocytes/macrophages into the CNS in a dose and LPS-strain dependent manner [70,93,94]. In GF/IL17 mice as well as in littermate controls we could observe upregulation of proinflammatory TNF-α, IL-1β, and MCP-1 mRNA levels in brain lysates. This induction was significantly enhanced for TNF-α and IL-1β in GF/IL17 transgenic mice arguing for a synergistic effect of IL-17 signalling in neuroinflammatory responses. Additive effects of IL-17A and in example TNF-α have already been shown for numerous cell types [95-98].

In summary we have developed a novel transgenic model for the chronic, astrocyte targeted secretion of IL-17A. The findings provide evidence for 1) a minor role of IL-17 to directly induce CNS inflammation; 2) a direct CNS effect of IL-17A by activating glial cells and modulating CNS inflammatory responses and 3) a direct induction of a vascular pathology mimicking pathological hallmarks of microangiopathy. In conclusion, we have generated a novel transgenic model for the CNS directed production of IL-17A which provides a valuable tool for ongoing investigations into the CNS pathobiology of this cytokine.

Acknowledgments

We thank Karin Wacker, Carina Folger and Karen Tolksdorf for their expert technical assistance. Daniel Getts is gratefully acknowledged for providing WNV-cDNA. We further thank Jens Reimann for his support in routine histological procedures and Martin Fuhrmann for his support conducting the laser scanning microscopy studies.

Author Contributions

Conceived and designed the experiments: JZ MK MJH ILC MM. Performed the experiments: JZ MK MM. Analyzed the data: JZ MK MM. Contributed reagents/materials/analysis tools: MTH ILC MM. Wrote the paper: JZ MM.

References

1. Zepp J, Wu L, Li X (2011) IL-17 receptor signaling and T helper 17-mediated autoimmune demyelinating disease. Trends Immunol 32:232–239.
2. Rouvier E, Luciani M, Mattei M, Denizot F, Golstein P (1993) CTLA-8, cloned from an activated T cell, bearing AU-rich messenger RNA instability sequences, and homologous to a Herpesvirus saimiri gene. J Immunol 150:5445–5456.
3. Weaver CT, Hatton RD, Mangan PR, Harrington LE (2007) IL-17 family cytokines and the expanding diversity of effector T cell lineages. Annu Rev Immunol 25:821–852.
4. Cua DJ, Tato CM (2010) Innate IL-17-producing cells: the sentinels of the immune system. Nat Rev Immunol 10:479–489.
5. Puel A, Cypowyj S, Bustamante J, Wright JF, Liu L, et al. (2011) Chronic Mucocutaneous Candidiasis in Humans with Inborn Errors of Interleukin-17 Immunity. Science 332:65–68.
6. Kolls JK, McCray PB, Chan YR (2008) Cytokine-mediated regulation of antimicrobial proteins. Nat Rev Immunol 8:829–835.
7. Curtis MM, Way SS (2009) Interleukin-17 in host defence against bacterial, mycobacterial and fungal pathogens. Immunology 126:177–185.
8. Peck A, Mellins ED 2010) Precarious balance: Th17 cells in host defense. Infect Immun 78:32–38.
9. Bettelli E, Korn T, Oukka M, Kuchroo VK (2008) Induction and effector functions of T(H)17 cells. Nature 453:1051–1057.
10. Langrish CL, Chen Y, Blumenschein WM, Mattson J, Basham B, et al. (2005) IL-23 drives a pathogenic T cell population that induces autoimmune inflammation. J Exp Med 201:233–240.
11. Fujino S, Andoh A, Bamba S, Ogawa A, Hata K, et al. (2003) Increased expression of interleukin 17 in inflammatory bowel disease. Gut 52:65–70.
12. Lock C, Hermans G, Pedotti R, Brendolan A, Schadt E, et al. (2002) Gene-microarray analysis of multiple sclerosis lesions yields new targets validated in autoimmune encephalomyelitis. Nat Med 8:500–508.
13. Teunissen MB, Koomen CW, de Waal Malefyt R, Wierenga EA, Bos JD (1998) Interleukin-17 and interferon-gamma synergize in the enhancement of proinflammatory cytokine production by human keratinocytes. J Invest Dermatol 111:645–649.
14. Ziolkowska M, Koc A, Luszczykiewicz G, Ksiezopolska-Pietrzak K, Klimczak E, et al. (2000) High levels of IL-17 in rheumatoid arthritis patients: IL-15 triggers in vitro IL-17 production via cyclosporin A-sensitive mechanism. J Immunol 164:2832–2838.
15. Tzartos JS, Friese MA, Craner MJ, Palace J, Newcombe J, et al. (2008) Interleukin-17 Production in Central Nervous System-Infiltrating T Cells and Glial Cells Is Associated with Active Disease in Multiple Sclerosis. Am J Pathol 172:146–155.
16. Guiton R, Vasseur V, Charron S, Arias MT, Van Langendonck N, et al. (2010) Interleukin 17 Receptor Signaling Is Deleterious during Toxoplasma gondii Infection in Susceptible BL6 Mice. J Infect Dis 202:427–435.
17. Li G-Z, Zhong D, Yang L-M, Sun B, Zhong Z-H, et al. (2005) Expression of Interleukin-17 in Ischemic Brain Tissue. Scand J Immunol 62:481–486.
18. Shichita T, Sugiyama Y, Ooboshi H, Sugimori H, Nakagawa R, et al. (2009) Pivotal role of cerebral interleukin-17-producing [gamma][delta]T cells in the delayed phase of ischemic brain injury. Nat Med 15:946–950.
19. Smith E, Prasad K-MR, Butcher M, Dobrian A, Kolls JK, et al. (2010) Blockade of Interleukin-17A Results in Reduced Atherosclerosis in Apolipoprotein E-Deficient Mice. Circulation 121:1746–1755.
20. von Vietinghoff S, Ley K (2010) Interleukin 17 in vascular inflammation. Cytokine & Growth Factor Rev 21:463–469.

21. Brucklacher-Waldert V, Stuerner K, Kolster M, Wolthausen J, Tolosa E (2009) Phenotypical and functional characterization of T helper 17 cells in multiple sclerosis. Brain 132:3329.

22. Hofstetter H, Gold R, Hartung H-P (2009) Th17 Cells in MS and Experimental Autoimmune Encephalomyelitis. Int MS J 16:12–18.

23. Sutton CE, Lalor SJ, Sweeney CM, Brereton CF, Lavelle EC, et al. (2009) Interleukin-1 and IL-23 Induce Innate IL-17 Production from gammadelta T Cells, Amplifying Th17 Responses and Autoimmunity. Immunity 31:331–341.

24. Rachitskaya AV, Hansen AM, Horai R, Li Z, Villasmil R, et al. (2008) Cutting Edge: NKT Cells Constitutively Express IL-23 Receptor and RORγt and Rapidly Produce IL-17 upon Receptor Ligation in an IL-6-Independent Fashion. J Immunol 180:5167–5171.

25. Li L, Huang L, Vergis AL, Ye H, Bajwa A, et al. (2010) IL-17 produced by neutrophils regulates IFN-γ–mediated neutrophil migration in mouse kidney ischemia-reperfusion injury. J Clin Invest 120:331–342.

26. Hoshino A, Nagao T, Nagi-Miura N, Ohno N, Yasuhara M, et al. (2008) MPO-ANCA induces IL-17 production by activated neutrophils in vitro via its Fc region- and complement-dependent manner. J Autoimmun 31:79–89.

27. Cua DJ, Sherlock J, Chen Y, Murphy CA, Joyce B, et al. (2003) Interleukin-23 rather than interleukin-12 is the critical cytokine for autoimmune inflammation of the brain. Nature 421:744–748.

28. Korn T, Bettelli E, Oukka M, Kuchroo VK (2009) IL-17 and Th17 Cells. Annu Rev Immunol 27:485–517.

29. Tran EH, Prince EN, Owens T (2000) IFN-γ Shapes Immune Invasion of the Central Nervous System Via Regulation of Chemokines. J Immunol 164:2759–2768.

30. Zhang G-X, Gran B, Yu S, Li J, Siglienti I, et al. (2003) Induction of Experimental Autoimmune Encephalomyelitis in IL-12 Receptor-β2-Deficient Mice: IL-12 Responsiveness Is Not Required in the Pathogenesis of Inflammatory Demyelination in the Central Nervous System. J Immunol 170:2153–2160.

31. Park H, Li Z, Yang XO, Chang SH, Nurieva R, et al. (2005) A distinct lineage of CD4 T cells regulates tissue inflammation by producing interleukin 17. Nat Immunol 6:1133–1141.

32. Harrington LE, Hatton RD, Mangan PR, Turner H, Murphy TL, et al. (2005) Interleukin 17-producing CD4+ effector T cells develop via a lineage distinct from the T helper type 1 and 2 lineages. Nat Immunol 6:1123–1132.

33. Veldhoen M, Hocking RJ, Flavell RA, Stockinger B (2006) Signals mediated by transforming growth factor-[beta] initiate autoimmune encephalomyelitis, but chronic inflammation is needed to sustain disease. Nat Immunol 7:1151–1156.

34. Bettelli E, Carrier Y, Gao W, Korn T, Strom TB, et al. (2006) Reciprocal developmental pathways for the generation of pathogenic effector TH17 and regulatory T cells. Nature 441:235–238.

35. Mangan PR, Harrington LE, O'Quinn DB, Helms WS, Bullard DC, et al. (2006) Transforming growth factor-[beta] induces development of the TH17 lineage. Nature 441:231–234.

36. Korn T, Bettelli E, Gao W, Awasthi A, Jager A, et al. (2007) IL-21 initiates an alternative pathway to induce proinflammatory TH17 cells. Nature 448:484–487.

37. Sarma JD, Ciric B, Marek R, Sadhukhan S, Caruso ML, et al. (2009) Functional interleukin-17 receptor A is expressed in central nervous system glia and upregulated in experimental autoimmune encephalomyelitis. J Neuroinflammation 6:14.

38. Kebir H, Kreymborg K, Ifergan I, Dodelet-Devillers A, Cayrol R, et al. (2007) Human TH17 lymphocytes promote blood-brain barrier disruption and central nervous system inflammation. Nat Med 13:1173–1175.

39. Chang SH, Park H, Dong C (2006) Act1 adaptor protein is an immediate and essential signaling component of interleukin-17 receptor. J Biol Chem 281:35603–35607.

40. Qian Y, Liu C, Hartupee J, Altuntas CZ, Gulen MF, et al. (2007) The adaptor Act1 is required for interleukin 17-dependent signaling associated with autoimmune and inflammatory disease. Nat Immunol 8:247–256.

41. Ouyang W, Kolls JK, Zheng Y (2008) The biological functions of T helper 17 cell effector cytokines in inflammation. Immunity 28:454–467.

42. Carlson T, Kroenke M, Rao P, Lane TE, Segal B (2008) The Th17-ELR+ CXC chemokine pathway is essential for the development of central nervous system autoimmune disease. J Exp Med 205:811–823.

43. Fossiez F, Djossou O, Chomarat P, Flores-Romo L, Ait-Yahia S, et al. (1996) T cell interleukin-17 induces stromal cells to produce proinflammatory and hematopoietic cytokines. J Exp Med 183:2593–2603.

44. Kawanokuchi J, Shimizu K, Nitta A, Yamada K, Mizuno T, et al. (2008) Production and functions of IL-17 in microglia. J Neuroimmunol 194:54–61.

45. Ma X, Reynolds SL, Baker BJ, Li X, Benveniste EN, et al. (2010) IL-17 enhancement of the IL-6 signaling cascade in astrocytes. J Immunol 184:4898.

46. Huppert J, Closhen D, Croxford A, White R, Kulig P, et al. (2010) Cellular Mechanisms of IL-17-Induced Blood-Brain Barrier Disruption. FASEB J 24:1023–1034.

47. Hofstetter HH, Ibrahim SM, Koczan D, Kruse N, Weishaupt A, et al. (2005) Therapeutic efficacy of IL-17 neutralization in murine experimental autoimmune encephalomyelitis. Cell Immunol 237:123–130.

48. Komiyama Y, Nakae S, Matsuki T, Nambu A, Ishigame H, et al. (2006) IL-17 plays an important role in the development of experimental autoimmune encephalomyelitis. J Immunol 177:566–573.

49. Uyttenhove C, Sommereyns C, Théate I, Michiels T, Van Snick J (2007) Anti-IL-17A Autovaccination Prevents Clinical and Histological Manifestations of Experimental Autoimmune Encephalomyelitis. Ann N Y Acad Sci 1110:330–336.

50. Kap YS, Jagessar SA, Driel N, Blezer E, Bauer J, et al. (2010) Effects of Early IL-17A Neutralization on Disease Induction in a Primate Model of Experimental Autoimmune Encephalomyelitis. J Neuroimmune Pharmacol 6:341–353.

51. Haak S, Croxford AL, Kreymborg K, Heppner FL, Pouly S, et al. (2009) IL-17A and IL-17F do not contribute vitally to autoimmune neuro-inflammation in mice. J Clin Invest 119:61–69.

52. Hu Y, Ota N, Peng I, Refino CJ, Danilenko DM, et al. (2010) IL-17RC Is Required for IL-17A– and IL-17F–Dependent Signaling and the Pathogenesis of Experimental Autoimmune Encephalomyelitis. J Immunol 184:4307–4316.

53. Kang Z, Altuntas CZ, Gulen MF, Liu C, Giltiay N, et al. (2010) Astrocyte-Restricted Ablation of Interleukin-17-Induced Act1-Mediated Signaling Ameliorates Autoimmune Encephalomyelitis. Immunity 32:414–425.

54. Hou W, Kang HS, Kim BS (2009) Th17 cells enhance viral persistence and inhibit T cell cytotoxicity in a model of chronic virus infection. J Exp Med 206:313–328.

55. Kelly MN, Kolls JK, Happel K, Schwartzman JD, Schwarzenberger P, et al. (2005) Interleukin-17/Interleukin-17 Receptor-Mediated Signaling Is Important for Generation of an Optimal Polymorphonuclear Response against Toxoplasma gondii Infection. Infect Immun 73:617–621.

56. Mucke L, Oldstone MB, Morris JC, Nerenberg MI (1991) Rapid activation of astrocyte-specific expression of GFAP-lacZ transgene by focal injury. New Biol 3:465–474.

57. Campbell IL, Abraham CR, Masliah E, Kemper P, Inglis JD, et al. (1993) Neurologic disease induced in transgenic mice by cerebral overexpression of interleukin 6. PNAS 90:10061.

58. Pagenstecher A, Lassmann S, Carson MJ, Kincaid CL, Stalder AK, et al. (2000) Astrocyte-targeted expression of IL-12 induces active cellular immune responses in the central nervous system and modulates experimental allergic encephalomyelitis. J Immunol 164:4481.

59. Boztug K, Carson MJ, Pham-Mitchell N, Asensio VC, DeMartino J, et al. (2002) Leukocyte infiltration, but not neurodegeneration, in the CNS of transgenic mice with astrocyte production of the CXC chemokine ligand 10. J Immunol 2002, 169:1505.

60. Giulian D, Baker T (1986) Characterization of ameboid microglia isolated from developing mammalian brain. J Neurosci 6:2163–2178.

61. Bancroft JD, Gamble M (2008) Theory and practice of histological techniques. Amsterdam, Elsevier Health Sciences.

62. Tamboli IY, Barth E, Christian L, Siepmann M, Kumar S, et al. (2010) Statins Promote the Degradation of Extracellular Amyloid β-Peptide by Microglia via Stimulation of Exosome-associated Insulin-degrading Enzyme (IDE) Secretion. J Biol Chem 285:37405–37414.

63. Getts DR, Terry RL, Getts MT, Müller M, Rana S, et al. (2008) Ly6c+ "inflammatory monocytes" are microglial precursors recruited in a pathogenic manner in West Nile virus encephalitis. J Exp Med 205:2319–2337.

64. de Haas AH, Boddeke HWGM, Biber K (2008) Region-specific expression of immunoregulatory proteins on microglia in the healthy CNS. Glia 56:888–894.

65. Krauthausen M, Ellis SL, Zimmermann J, Sarris M, Wakefield D, et al. (2011) Opposing roles for CXCR3 signaling in central nervous system versus ocular inflammation mediated by the astrocyte-targeted production of IL-12. Am J Pathol 179:2346–2359.

66. Saria A, Lundberg JM (1983) Evans blue fluorescence: quantitative and morphological evaluation of vascular permeability in animal tissues. J Neurosci Methods 8:41–49.

67. Ay I, Francis JW, Brown RH Jr (2008) VEGF increases blood-brain barrier permeability to Evans blue dye and tetanus toxin fragment C but not adeno-associated virus in ALS mice. Brain Research 1234:198–205.

68. Campbell IL, Hofer MJ, Pagenstecher A (2010) Transgenic models for cytokine-induced neurological disease. Biochimi Biophys Acta (BBA) - Molecular Basis of Disease 1802:903–917.

69. Stalder AK, Pagenstecher A, Yu NC, Kincaid C, Chiang CS, et al. (1997) Lipopolysaccharide-induced IL-12 expression in the central nervous system and cultured astrocytes and microglia. J Immunol 159:1344–1351.

70. Bohatschek M, Werner A, Raivich G (2001) Systemic LPS Injection Leads to Granulocyte Influx into Normal and Injured Brain: Effects of ICAM-1 Deficiency. Exp Neurol 172:137–152.

71. Cardona AE, Pioro EP, Sasse ME, Kostenko V, Cardona SM, et al. (2006) Control of microglial neurotoxicity by the fractalkine receptor. Nat Neurosci 9:917–924.

72. Chiang CS, Powell HC, Gold LH, Samimi A, Campbell IL (1996) Macrophage/microglial-mediated primary demyelination and motor disease induced by the central nervous system production of interleukin-3 in transgenic mice. J Clin Invest 97:1512–1524.

73. Akwa Y, Hassett DE, Eloranta M-L, Sandberg K, Masliah E, et al. (1998) Transgenic Expression of IFN-α in the Central Nervous System of Mice Protects Against Lethal Neurotropic Viral Infection but Induces Inflammation and Neurodegeneration. J Immunol 161:5016–5026.

74. Reiman R, Torres AC, Martin BK, Ting JP, Campbell IL, et al. (2005) Expression of C5a in the brain does not exacerbate experimental autoimmune encephalomyelitis. Neurosci Lett 390:134–138.

75. Yao Z, Fanslow WC, Seldin MF, Rousseau A-M, Painter SL, et al. (1995) Herpesvirus Saimiri encodes a new cytokine, IL-17, which binds to a novel cytokine receptor. Immunity 3:811–821.

76. Ge D, You Z (2008) Expression of interleukin-17RC protein in normal human tissues. Int Arch Med 1:19.

77. Martin AP, Coronel EC, Sano G-I, Chen S-C, Vassileva G, et al. (2004) A Novel Model for Lymphocytic Infiltration of the Thyroid Gland Generated by Transgenic Expression of the CC Chemokine CCL21. J Immunol 173:4791–4798.

78. Ploix CC, Noor S, Crane J, Masek K, Carter W, et al. (2011) CNS-derived CCL21 is both sufficient to drive homeostatic CD4+ T cell proliferation and necessary for efficient CD4+ T cell migration into the CNS parenchyma following Toxoplasma gondii infection. Brain Behav Immun 25:883–896.

79. Trajkovic V, Stosic-Grujicic S, Samardzic T, Markovic M, Miljkovic D, et al. (2001) Interleukin-17 stimulates inducible nitric oxide synthase activation in rodent astrocytes. J Neuroimmunol 119:183–191.

80. Joutel A, Monet-Leprêtre M, Gosele C, Baron-Menguy C, Hammes A, et al. (2010) Cerebrovascular dysfunction and microcirculation rarefaction precede white matter lesions in a mouse genetic model of cerebral ischemic small vessel disease. J Clin Invest 120:433–445.

81. Suzuki K, Masawa N, Sakata N, Takatama M (2003) Pathologic evidence of microvascular rarefaction in the brain of renal hypertensive rats. J Stroke Cerebrovasc Dis 12:8–16.

82. Jiang N, Li X, Qi T, Guo S, Liang F, et al. (2011) Susceptible gene single nucleotide polymorphism and hemorrhage risk in patients with brain arteriovenous malformation. J Clin Neurosci 18:1279–1281.

83. Erbel C, Dengler TJ, Wangler S, Lasitschka F, Bea F, et al. (2010) Expression of IL-17A in human atherosclerotic lesions is associated with increased inflammation and plaque vulnerability. Basic Res Cardiol 106:125–134.

84. Erbel C, Chen L, Bea F, Wangler S, Celik S, et al. (2009) Inhibition of IL-17A Attenuates Atherosclerotic Lesion Development in ApoE-Deficient Mice. J Immunol 183:8167–8175.

85. Smith E, Prasad KMR, Butcher M, Dobrian A, Kolls JK, et al. (2010) Blockade of Interleukin-17A Results in Reduced Atherosclerosis in Apolipoprotein E-Deficient Mice. Circulation 121:1746–1755.

86. Madhur MS, Funt SA, Li L, Vinh A, Chen W, et al. (2011) Role of Interleukin 17 in Inflammation, Atherosclerosis, and Vascular Function in Apolipoprotein E–Deficient Mice. Arterioscler Thromb Vasc Biol 31:1565–1572.

87. Kalueff A, Loseva E, Haapasalo H, Rantala I, Keranen J, et al. (2006) Thalamic calcification in vitamin D receptor knockout mice. Neuroreport 17:717–721.

88. Bruce D, Yu S, Ooi JH, Cantorna MT (2011) Converging pathways lead to overproduction of IL-17 in the absence of vitamin D signaling. Int Immunol 23:519–528.

89. Joshi S, Pantalena L-C, Liu XK, Gaffen SL, Liu H, et al. (2011) 1,25-Dihydroxyvitamin D3 Ameliorates Th17 Autoimmunity via Transcriptional Modulation of Interleukin-17A. Mol Cell Biol 31:3653–3669.

90. Wyss-Coray T, Feng L, Masliah E, Ruppe MD, Lee HS, et al. (1995) Increased central nervous system production of extracellular matrix components and development of hydrocephalus in transgenic mice overexpressing transforming growth factor-beta 1. Am J Pathol 147:53–67.

91. Wyss-Coray T, Lin C, Von Euw D, Masliah E, Mucke L, et al. (2000) Alzheimer's Disease-like Cerebrovascular Pathology in Transforming Growth Factor-β1 Transgenic Mice and Functional Metabolic Correlates. Ann N Y Acad Sci 903:317–323.

92. Wyss-Coray T, Lin C, Sanan DA, Mucke L, Masliah E (2000) Chronic Overproduction of Transforming Growth Factor-β1 by Astrocytes Promotes Alzheimer's Disease-Like Microvascular Degeneration in Transgenic Mice. Am J Pathol 156:139–150.

93. Audoy-Rémus J, Richard J-F, Soulet D, Zhou H, Kubes P, et al. (2008) Rod-Shaped Monocytes Patrol the Brain Vasculature and Give Rise to Perivascular Macrophages under the Influence of Proinflammatory Cytokines and Angiopoietin-2. J Neurosci 28:10187–10199.

94. Zhou H, Lapointe BM, Clark SR, Zbytnuik L, Kubes P (2006) A Requirement for Microglial TLR4 in Leukocyte Recruitment into Brain in Response to Lipopolysaccharide. J Immunol 177:8103–8110.

95. Paintlia MK, Paintlia AS, Singh AK, Singh I (2011) Synergistic activity of interleukin-17 and tumor necrosis factor-α enhances oxidative stress-mediated oligodendrocyte apoptosis. J Neurochem 116:508–521.

96. Chiricozzi A, Guttman-Yassky E, Suarez-Farinas M, Nograles KE, Tian S, et al. (2011) Integrative Responses to IL-17 and TNF-[alpha] in Human Keratinocytes Account for Key Inflammatory Pathogenic Circuits in Psoriasis. J Invest Dermatol 131:677–687.

97. Liu Y, Mei J, Gonzales L, Yang G, Dai N, et al. (2011) IL-17A and TNF-α Exert Synergistic Effects on Expression of CXCL5 by Alveolar Type II Cells In Vivo and In Vitro. J Immunol 186:3197–3205.

98. Hartupee J, Liu C, Novotny M, Li X, Hamilton T (2007) IL-17 Enhances Chemokine Gene Expression through mRNA Stabilization. J Immunol 179:v–4141.

2

Cartilage–Specific Over-Expression of CCN Family Member 2/Connective Tissue Growth Factor (CCN2/CTGF) Stimulates Insulin-Like Growth Factor Expression and Bone Growth

Nao Tomita[1,2], Takako Hattori[1]*, Shinsuke Itoh[1,2], Eriko Aoyama[3], Mayumi Yao[1], Takashi Yamashiro[2], Masaharu Takigawa[1,3]*

1 Department of Biochemistry and Molecular Dentistry, Okayama University Dental School, Okayama, Japan, 2 Department of Orthodontics, Okayama University Graduate School of Medicine, Dentistry, and Pharmaceutical Sciences, Okayama University Dental School, Okayama, Japan, 3 Biodental Research Center, Okayama University Dental School, Okayama, Japan

Abstract

Previously we showed that CCN family member 2/connective tissue growth factor (CCN2) promotes the proliferation, differentiation, and maturation of growth cartilage cells *in vitro*. To elucidate the specific role and molecular mechanism of CCN2 in cartilage development *in vivo*, in the present study we generated transgenic mice overexpressing CCN2 and analyzed them with respect to cartilage and bone development. Transgenic mice were generated expressing a *ccn2/lacZ* fusion gene in cartilage under the control of the 6 kb-*Col2a1*-enhancer/promoter. Changes in cartilage and bone development were analyzed histologically and immunohistologically and also by micro CT. Primary chondrocytes as well as limb bud mesenchymal cells were cultured and analyzed for changes in expression of cartilage–related genes, and non-transgenic chondrocytes were treated in culture with recombinant CCN2. Newborn transgenic mice showed extended length of their long bones, increased content of proteoglycans and collagen II accumulation. Micro-CT analysis of transgenic bones indicated increases in bone thickness and mineral density. Chondrocyte proliferation was enhanced in the transgenic cartilage. In *in vitro* short-term cultures of transgenic chondrocytes, the expression of *col2a1*, *aggrecan* and *ccn2 genes* was substantially enhanced; and in long-term cultures the expression levels of these genes were further enhanced. Also, *in vitro* chondrogenesis was strongly enhanced. *IGF-I* and *IGF-II* mRNA levels were elevated in transgenic chondrocytes, and treatment of non-transgenic chondrocytes with recombinant CCN2 stimulated the expression of these mRNA. The addition of CCN2 to non-transgenic chondrocytes induced the phosphorylation of IGFR, and *ccn2*-overexpressing chondrocytes showed enhanced phosphorylation of IGFR. Our data indicates that the observed effects of CCN2 may be mediated in part by CCN2-induced overexpression of IGF-I and IGF-II. These findings indicate that CCN2-overexpression in transgenic mice accelerated the endochondral ossification processes, resulting in increased length of their long bones. Our results also indicate the possible involvement of locally enhanced IGF-I or IGF-II in this extended bone growth.

Editor: Frank Beier, University of Western Ontario, Canada

Funding: This work was supported by the program Grants-in-Aid for Scientific Research (C) to TH and (S) to MT and Exploratory Research (to MT) from the Japan Society for the Promotion of Science and by internal grants from Okayama University (to TH) and by a grant from Senri Life Science Foundation (to TH). The funders had no role in study design, data collection and analysis, decision to publish, or preparation of the manuscript.

Competing Interests: The authors have declared that no competing interests exist.

* E-mail: hattorit@md.okayama-u.ac.jp (TH); takigawa@md.okayama-u.ac.jp (MT)

Introduction

CCN2(CCN family 2)/CTGF (connective tissue growth factor) is a member of the CCN family of secreted proteins, which also includes Cyr61/CCN1, NOV/CCN3, WISP1/CCN4, WISP2/CCN5, and WISP3/CCN6. CCN2 regulates diverse cell functions including mitosis, adhesion, apoptosis, extracellular matrix (ECM) production, growth arrest, and cellular migration [1,2]. The multimodular character of CCN factors allows multiple interactions between them and other growth factors such as TGFß, BMPs, IGFs or VEGF and networking between growth factors, extracellular matrix, and cell-surface receptors such as integrins [3]. Thus, it is not surprising that CCN factors are involved in a multiplicity of effects during development, differentiation, wound healing, and disease states, including tumorigenesis and fibrosis [2]. Most prominently, CCN2 has emerged as a major regulator of chondrogenesis, angiogenesis, and fibrogenesis [4]. CCN2 induces the migration of endothelial cells [5,6,7] and stimulates the synthesis of matrix proteins including collagens and fibronectin [8,9]. It is expressed in various tissues, with highest levels found in prehypertrophic chondrocytes and vascular tissues in developing embryos (for reviews, see refs [4,10]. Previously we demonstrated in a series of *in vitro* studies that CCN2 stimulates both the proliferation and synthesis of type II collagen and proteoglycans of growth-plate chondrocytes [11], human chondrosarcoma-derived chondrocytic cells [11,12], articular chondrocytes [13], and auricular chondrocytes [14]. Moreover, it induces hypertrophy and calcification of growth-plate chondrocytes, but not those of articular or auricular chondrocytes [11,14,15]. Also, osteoblast proliferation and maturation are stimulated by CCN2 [16]. These *in vitro* findings are consistent with studies on CCN2-deficient mice, which develop skeletal dysmorphisms including kinky bone and cartilage elements, due to impairment of chondrocyte proliferation and extracellular matrix deposition in the hypertrophic zone [17]. As a result of CCN2 deficiency, growth-plate angiogenesis and endochondral ossification are partially impaired, and CCN2-deficient mice die after birth because of respiratory failure caused by the skeletal defects [17]. Although multiple effects of CCN2 on differentiation, proliferation, and matrix synthesis of chondrocytes, fibroblasts, endothelial cells, and osteoblasts have been reported, the specific role of CCN2 synthesized by chondrocytes during cartilage and bone development *in vivo* remains unclear.

To elucidate the role of chondrocyte-derived CCN2, we generated CCN2-over-expressing mice with the gene expressed under the control of a 6 kb-*Col2a1* promoter that included a cartilage-specific enhancer element in the first intron of the *Col2a1* gene and obtained *in vivo* evidence for a key role of CCN2 in regulating chondrocyte gene expression and cartilage differentiation. Furthermore, our data suggest that CCN2 regulates the endochondral ossification process in long bones partially through increased expression of IGF-I and IGF-II.

Materials and Methods

Generation of Transgenic Mice

To express the *ccn2* as transgene in chondrocytes, we cloned the cDNA encoding a HA-tagged mouse *ccn2* gene into a vector containing 3 kb of the *Col2a1* promoter and 3.02 kb of the intron 1 sequence [18,19]. The *LacZ* gene preceded by an internal ribosomal entry site was placed downstream of the *ccn2* cDNA (Fig. 1A). This construct was microinjected into the pronuclei of fertilized C57BL/6CrSlc eggs to generate transgenic mice. Routine genotyping to identify the transgene was done by detecting the *LacZ* gene by performing a polymerase chain reaction (PCR) on genomic DNA. The primer sequences used were 5-GCATCGAGCTGGGTAATAAGCGTTGGCAAT-3′ and 5-GACACCAGACCAACTGGTAATGGTAGCGAC-3′.

All experimental procedures were performed in accordance with the Guidelines for Proper Conduct of Animal Experiments of the Science Council of Japan and approved by the Animal Research Control Committee of Okayama University (Approval No.: OKU-2012113).

LacZ Staining and Skeletal Preparation

LacZ activity was detected by staining with X-gal (5-bromo-4-chloro-3-indolyl-D-galactopyranoside; Roche) for 3–6 hours following fixation with glutaraldehyde and formaldehyde as described earlier [20]. For staining of embryos older than 15.5 days, the skin and internal organs were removed before fixation. LacZ-stained embryos were postfixed overnight in 4% formaldehyde, dehydrated, and embedded in paraffin. Sections were counterstained with eosin. Some LacZ-stained embryos were cleared with KOH –glycerol. Skeletal morphology was analyzed by alizarin red and alcian blue staining followed by clearing with 1% (w/v) KOH [21,22].

RNA Preparation and Northern Hybridization

RNA was prepared either directly from cartilage or from chondrocyte cultures. For the direct RNA preparation, rib cages of E18.5 or 19.5 embryos were separated from soft tissues, and single ribs were isolated. The isolated ribs were separated from bone, and the cartilage was soaked in Isogen (Nippon Gene) and homogenized until the tissue clumps had disappeared. The cartilage RNA were purified according to the Isogen instructions, and the purified RNA were further cleaned by using the RNeasy kit (Qiagen). For the RNA preparation from chondrocytes, the cells from rib cartilage were cultured as described below, harvested, and then subjected to RNA purification using the RNeasy kit. For Northern hybridization, 10 µg of RNA from costal cartilage was resolved on an agarose gel, transferred onto a nylon membrane (Bio-Rad), and hybridized with [^{32}P]-labeled *LacZ* or *ccn2* probes as described previously [23].

Western Blotting

Rib cartilage from E18.5 embryos was isolated as described above and homogenized with lysis buffer (50 mM Tris-HCl, pH 7.4, containing 150 mM NaCl, 1% Triton X-100, 0.1% SDS, and 1 mM PMSF). After centrifugation, the supernatant was collected; and 6 µg of protein per lane was loaded onto an SDS-PAGE gel. Western blotting was done as described previously [24] by using anti-HA (Covance), anti-actin (Sigma), anti-phospho IGF-1 receptor (Cell Signaling), and anti-IGF-1 receptor (Cell Signaling) antibodies.

Histological Examination

For histological analysis, tissues from E17.5 and E19.5 embryos and from 1- and 3- day postnatal mice were fixed with 10% formaldehyde/PBS, demineralized with 0.5 M EDTA, and embedded in paraffin. Then 7 µm-thick-sections were stained with hematoxylin, eosin, and safranin-O. Immunohistochemical staining was performed by using a peroxidase-conjugated polymer (Nichirei, Japan) and anti-type II collagen MoAb (CII D3, [25] or anti-type X MoAb (X53, kindly provided by Dr. K. von der Mark, Germany, [26,27]. For cell proliferation analysis, a PCNA staining kit (Zymed) was used. For detection of apoptotic cells, TUNEL analysis was performed by using an *In Situ* Cell Death Detection Kit, POD (Roche).

Cell Cultures

For preparation of primary cultures, chondrocytes were isolated from the rib cages of 18.5- or 19.5-day embryos and/or newborn mice as described previously [28]. Briefly, the rib cages were digested with collagenase (0.1% collagenase P, Roche, in F12/DMEM containing 10% fetal calf serum) after adhering connective tissue and muscle had been thoroughly removed by trypsin pretreatment. The cells were grown to confluence for 1 month to hypertrophy in α-modification of minimum essential medium (α-MEM) containing 10% fetal bovine serum (FBS) and supplemented with 50 µg/ml of ascorbic acid with or without recombinant CCN2, and then harvested for RNA extraction.

For preparation of CCN2 recombinant protein, human *ccn2* cDNA was amplified by PCR and subcloned into the pET-15b vector (Novagen), which harbors a His-tag; and *E. coli* BL21(DE3)pLysS Rosetta strain cells were subsequently transformed with this vector. Expressed His-tagged CCN2 protein was purified by the use of Ni-NTA agarose.

For inhibition of autophosphorylation of IGF-1 receptor, the IGF-1R inhibitor PPP (Calbiochem) was used, at a concentration of 60 nM. Anti-CCN2 monoclonal antibody (11H3, kindly provided by Dr. Seto, Nippn Flour Mills Co., LTD.), which had an inhibitory effect on the CCN2-mediated enhancement of aggrecan gene expression was also used to inhibit this autophosphorylation.

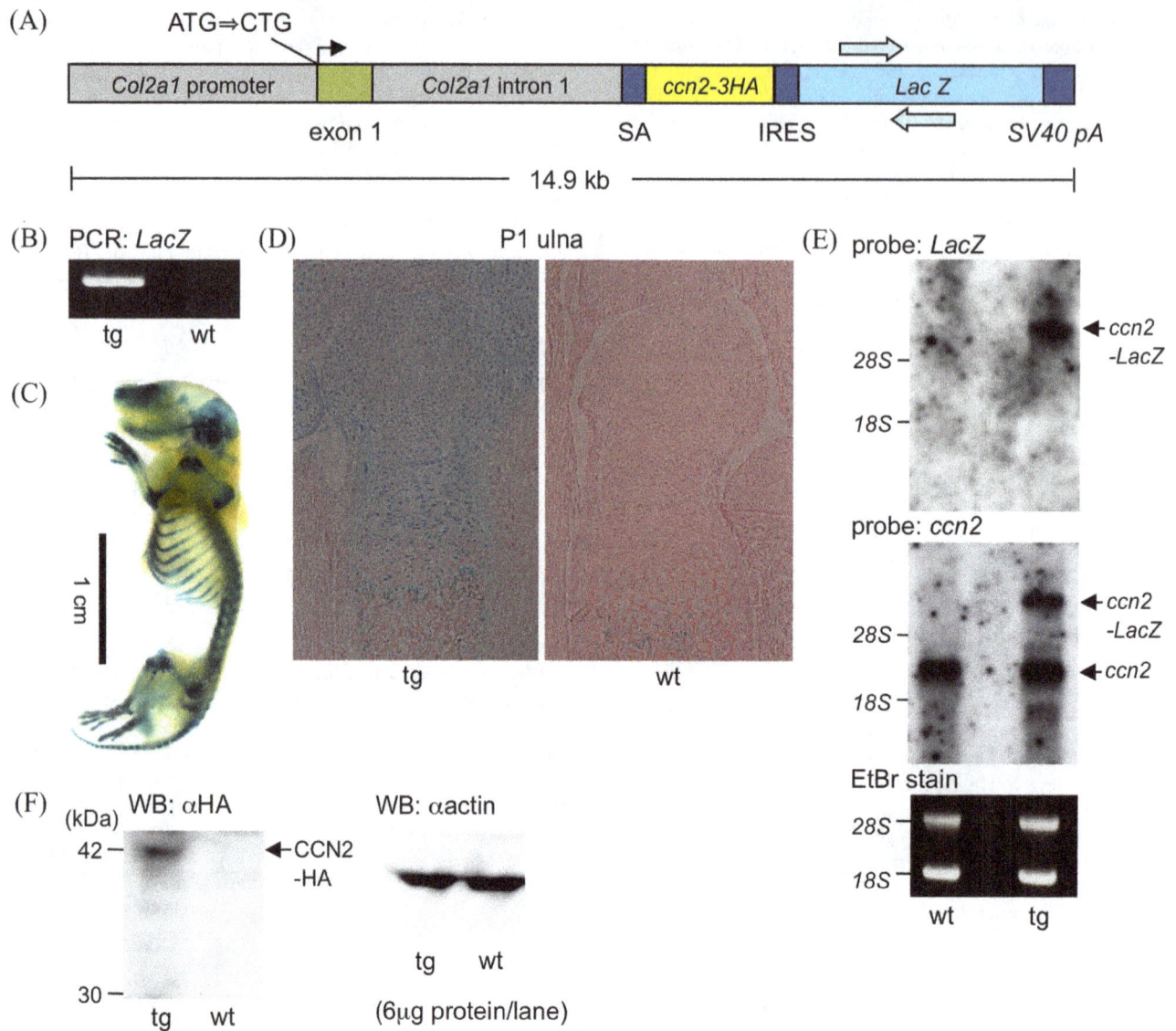

Figure 1. Generation of *Col2a1-ccn2* transgenic mice. (A) Schematic representation of the construct of the expression of HA-tagged CCN2 and IRES-LacZ in chondrocytes driven by the 6-kb *Col2a1* promoter-enhancer. The original initiation codon of *Col2a1* was mutated to CTG to facilitate translation from downstream cDNA. (B) Genotyping of transgenic mice (tg) by PCR to detect the transgene. wt, wild type. The location of the primers used are indicated in "A" by arrows. (C) Skeletal preparation of a newborn mouse after whole-mount X-gal staining, showing cartilage-specific expression of the transgene. (D) Sagittal sections of ulnae from wt and tg after whole-mount X-gal staining. All of the cartilaginous cells showed X-gal staining. The sections were counter-stained with Safranin-O. (E) Analysis of transgene expression by Northern hybridization using total RNA from tg and wt cartilage. *LacZ* (top) and *ccn2* (middle) probes were used to detect transgenic and endogenous *ccn2*, respectively. (F) Western blot (WB) analysis using cell lysates from tg and wt cartilage and anti-HA antibody recognizing only the CCN2-HA transgene products (left blot). The HA-tagged CCN2 was expressed in cartilage of tg mice. A Western blot of the same cell lysate reacted with anti-actin antibody as a loading control is also shown (right blot).

Quantitative real-time PCR

Reverse transcription (RT) was performed with 0.5 µg of total RNA as described above, and the resulting cDNA was amplified in triplicate by using the SYBR-Green PCR assay (TOYOBO SYBR Green PCR Master Mix; TOYOBO, Osaka, Japan), after which the products were detected with a LightCycler™ system (Roche, Basel, Switzerland). PCR reaction mixtures were incubated for 15 min at 95°C, followed by 50 amplification cycles of 30 s annealing at 60°C, 40 s extension at 72°C, and 30 s denaturation at 95°C. GAPDH was used to standardize the total amount of cDNA, as described previously [29].

The primers designed for real-time PCR were the following: *ccn2* (forward, 5'-GGTAAGGTCCGATTCCTACCAGG-3'; reverse, 5'-CTAGAAAGGTGCAAACATGTAAC-3'); *gapdh* (forward, 5'-GCCAAAAGGGTCATCATCTC-3'; reverse, 5'-GTCTTCTGGGTGGCAGTGAT-3'); *aggrecan* (forward, 5'-TCTTCAGTCCCGTTCTCCAC-3'; reverse, 5'-AACAT-CACTGAGGGCGAAGC-3'); *Col2a1* (forward, 5'-ATGACAA TCTGGCTCCCAACACTGC-3'; reverse, 5'-GACCGGCCC-TATGTCCACACCGAAT-3'); *Col10a1* (forward, 5'-CCCAGGGTTACCAGGACAAA-3'; reverse, 5'-GTTCACCTCTTGGACCTGCC-3'); *vegf* (forward, 5'-CCCAT-GAAGTGATCAAGTTC-3'; reverse, 5'-ACCCGCAT-

(A)

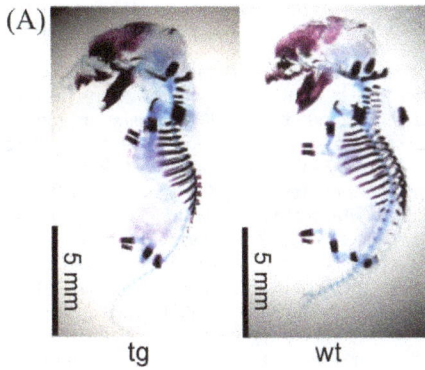

tg   wt

(B)

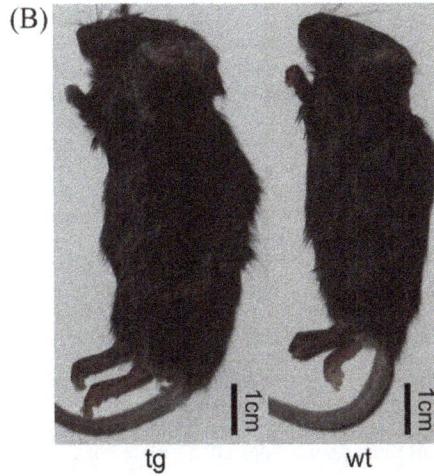

tg   wt

(C)

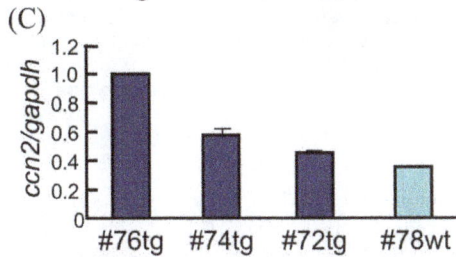

(D)

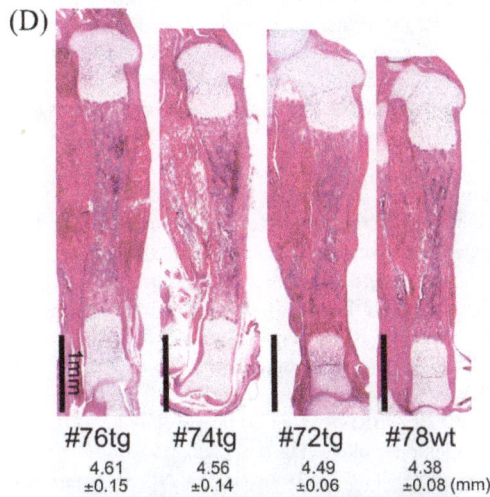

#76tg   #74tg   #72tg   #78wt
4.61   4.56   4.49   4.38
±0.15   ±0.14   ±0.06   ±0.08 (mm)

(E-1)   (E-2)

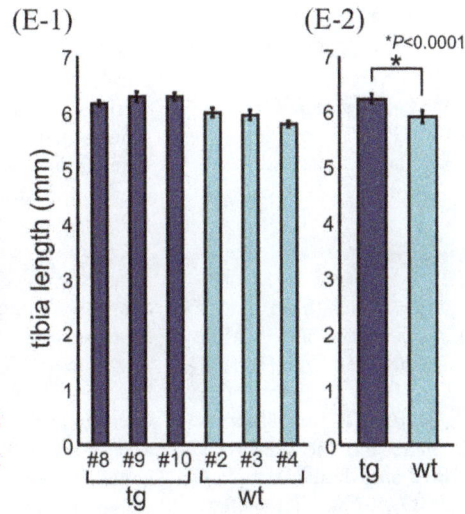

(F)

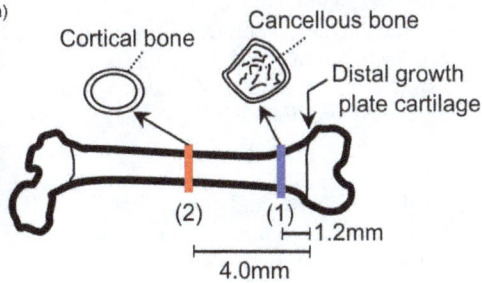

(G)

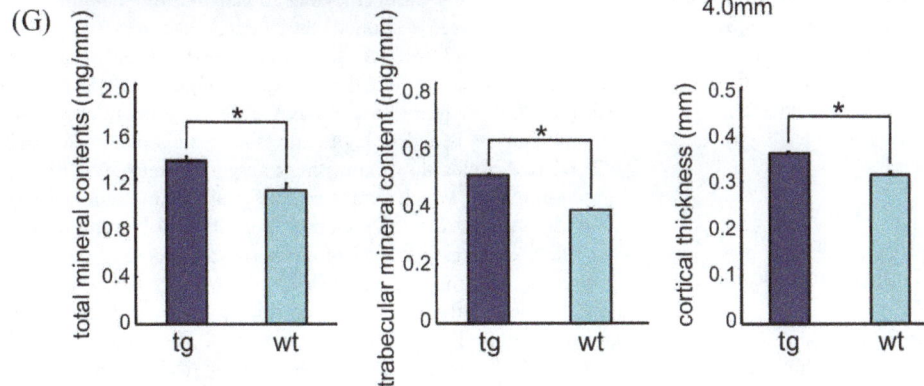

Figure 2. Skeletal analysis of *Col2a1-ccn2* transgenic mice. (A) Skeletal preparation of representative tg and wt littermates at E15.5 after alizarin red and Alcian blue staining. Skeletal development in tg mice appeared normal at this stage. (B) At 8 weeks the transgenic mice consistently showed an ~12% increase in body size. (C) Quantitative analysis by real-time PCR of *ccn2* mRNA levels in primary cultures of tg and wt rib chondrocytes revealed high-expressing transgenic mice (e.g., #76,#74) in each litter, besides low-expressing littermates (#tg72), the latter of which expressed *ccn2* at about the same level as the wt littermates (see also Fig. 2D). Real time-RCR analysis was repeated at least 2 times for each RNA preparation, and the 2 founder lines showed similar variations, but basically the same results. (D) Hematoxylin-eosin (HE) staining of transgenic and wild-type P1 tibiae from the same littermates as shown in Fig. 2C. Tibiae from transgenic mice showed a relatively extended length of the diaphyses in the high-expressing transgenic littermates. Tg and wt with a number indicate transgenic and non-transgenic littermates, respectively. Six litters from 2 different founder lines were investigated. (E-1) Diaphysis length of tibiae from transgenic and wild-type littermates of a P3 litter. Tibial diaphysis lengths of only pups that showed significantly enhanced levels of *ccn2* mRNA, measured in primary cultures of rib chondrocytes were measured. Serial sections (5–7 slides) were randomly selected every 3 slides from a single tibia, and stained with HE. The images were incorporated into a computer, and the length of diaphyses were measured. Bars indicate the mean length and standard deviations of diaphyses of tibia from wild-type and transgenic littermates (e.g. 2 wt, 8 tg). (E-2) Mean length of diaphyses of tibiae from the wild-type and transgenic mice indicated in E-1. *: p<0.0001. Two different founder lines with 3 litters each were analyzed and similar results were obtained. (F) Left: Representative micro-CT image (cross section) of femora of 8-week-old tg and wt littermates. Right: Positions of measurement in femur. (G) Peripheral quantitative computed tomography analysis of bone density and mineral content was made at 2 sites, one 1.2 mm (site #1, blue), and the other 4.0 mm (site #2, red), distal to the growth plate, as indicated in "F" (right). Bars represent the mean ±SD (n=9, males). In transgenic bones significant enhancement was seen in total mineral content (tg: 1.36±0.08 mg/mm *vs.* wt: 1.10±0.12 mg/mm), in trabecular mineral content (tg: 0.49±0.01 mg/mm *vs.* wt: 0.38±0.01 mg/mm), and in cortical thickness (tg: 0.060±0.013 mm *vs.* wt: 0.049±0.021 mm); but only in the femora at site #1 were the differences significant (*P<0.05).

GATCTGCATGG-3'); *mmp9* (forward, 5'-GGAACTCACAC-GACATCTTCCA-3'; reverse, 5'-GAAACTCACACGCCA-GAAGAATTT-3'; *IGF-I* (forward, 5'- GTGTGGACC-GAGGGGCTTTTACTTC-3'; reverse, 5'-GCTTCAGTGGGGCACAGTACATCTC-3'); and *IGF-II* (forward, 5'-GTGGCATCGTGGAAGAGTGC-3'; reverse, 5'-GGGGTGGGTAAGGAGAAACC-3'); lacZ (forward, 5'-GGTTACGATGCGCCCATCTA-3'; reverse, 5'-ACGGCG-GATTGACCGTAAT-3').

Micromass Culture

For preparation of micromass cultures, limbs from E11.5 embryos were digested in 0.05% trypsin for 1 hour on ice. After the cells had been suspended by pipeting, they were concentrated in 10% FCS-containing DMEM/F12 to 1×10^7 cells/ml. Ten microliters of cell suspension containing 1×10^5 cells was placed in the center of each well of a 24-well plate; and the cells were allowed to adhere to the bottom of the well for 1 h after the plate had been placed in an incubator (5% CO_2, 37°C). Thereafter, 1 ml of culture medium was added to each well; and the medium was replaced every 24 hours. Cell condensation in the cultures was visible after 1 or 2 days, and cartilage nodules appeared after 3 days. Some cells were stained with Alcian blue (pH 1) to visualize cartilage, and others were harvested for extraction of total RNA.

Analysis of Bone Mineralization

The femora from 8-week–old mice were removed, and the bones were scanned over the region from 1.2 mm to 4.0 mm from the distal epiphysial end by peripheral quantitative computed tomography (pQCT) analysis (XCT Research SA+[Stratec Medizintechnik GmbH, Pforzheim, Germany]). For the micro-computed tomography (micro-CT) analysis, the same position was scanned by using a Skyscan 1072 micro-CT machine (Skyscan, Aartselaar, Belgium).

Results

Cartilage-specific Over-expression of *ccn2* in Chondrocytes of Transgenic Mice Caused Increased Bone Size

For generation of transgenic mice over-expressing CCN2 in cartilage, HA-tagged *ccn2* cDNA was cloned into a vector containing 3 kb of the *Col2a1* promoter, 3.02 kb of the *Col2a1* intron 1 sequence, and *IRES-LacZ* (Fig. 1A). The purified vector

DNA was injected into oocytes, and 2 founders tested positive for the *ccn2–lacZ* transgene by PCR (Fig. 1B) and were kept to establish transgenic lines. X-gal staining of newborn transgenic mice showed intense, cartilage-specific lacZ expression in all cartilage elements (Fig. 1C). In tissue sections of newborns, all growth-plate and resting chondrocytes were positive after X-gal staining, indicating that the expression domains of the transgene overlapped with those of endogenous *ccn2* (Fig. 1D; and see also [30]).

Over-expression of the *ccn2* transgene in chondrocytes of the transgenic mice was confirmed by Northern and Western blot analyses. Northern blot hybridization of total RNA extracted from rib cage chondrocytes of E18.5 embryos with probes for *LacZ* and *ccn2* showed a reaction with the same 6-kb transcript in transgenic, but not wt, chondrocyte RNA (Fig. 1E, *LacZ* and *ccn2*). The intensity of the transgene signal obtained with the *ccn2* probe was about 75% of that of the endogenous *ccn2* mRNA (Fig. 1E, middle panel). Endogenous *ccn2* mRNA was also up-regulated (~110% of wild type) in transgenic cartilage (Fig. 1E, *ccn2*), possibly due to an autocrine mechanism. The HA-tagged CCN2 protein was detected in cell lysates from transgenic rib cartilage by Western blot analysis using an anti-HA antibody (Fig. 1F).

At day E15.5 of embryonic development, no major abnormalities in cartilage or bone development were detected in the transgenic animals (Fig. 2A). At 8 weeks, however, the majority of the transgenic mice were about 12% larger than their wild-type littermates (Fig. 2B).

For detailed analysis of the morphological alterations in the skeleton of postnatal transgenic mice, tibiae of transgene and wild-type newborns were sectioned, and their length was measured. The levels of *ccn2* mRNA in chondrocytes cultured from rib cartilage of the same animal were also monitored. Quantitative real-time PCR analysis of *ccn2* mRNA levels in rib chondrocytes in primary culture revealed high-expressing transgenic mice (e.g., #76,#74), as well as low-expressing transgenic littermates (#tg72) in the same litter, which expressed *ccn2* at about the same level as the wt littermates (Fig. 2C). Comparison of tibial length and the *ccn2* mRNA expression level of chondrocytes prepared from rib cartilage of the same animal showed a positive correlation (Fig. 2C and D). The length of diaphyses of tibiae from wt and transgenic littermates at the P3 stage was also measured. The expression level of *ccn2* mRNA in primary cultures of rib chondrocytes from littermates was monitored, and tibiae from pups with significantly higher levels of *ccn2* mRNA compared with wt levels were used for

(A)

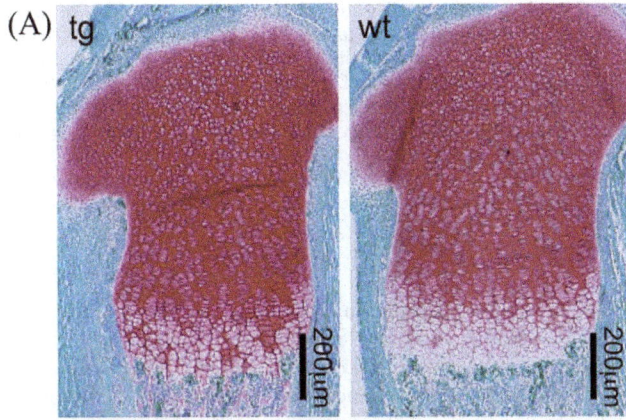

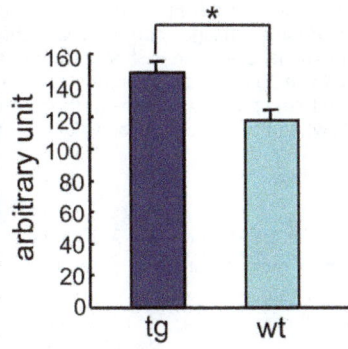

(B)

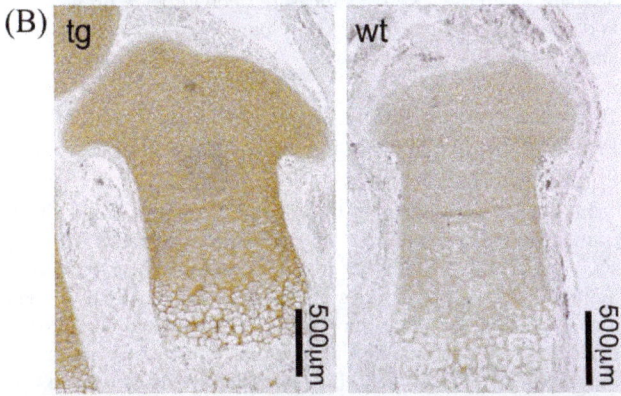

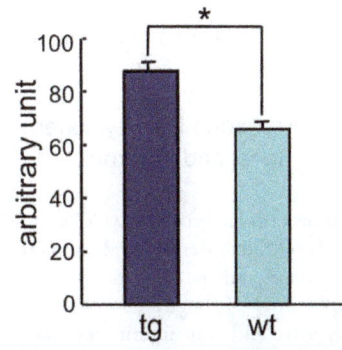

(C)

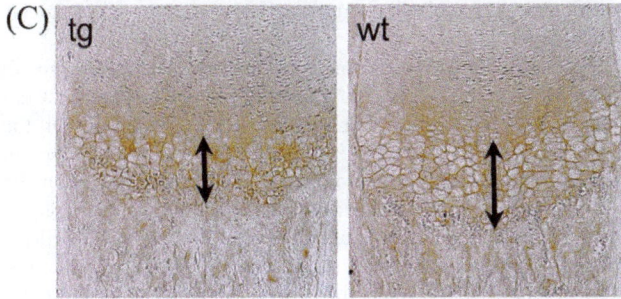

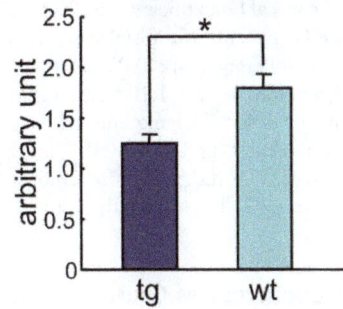

(D)

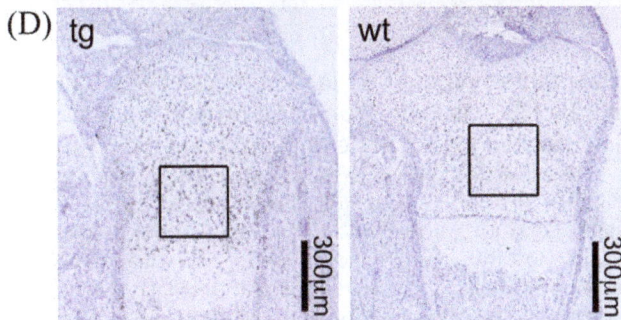

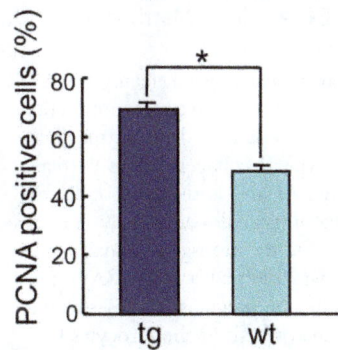

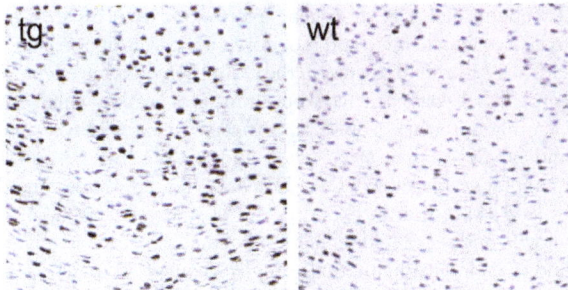

Figure 3. CCN2 overexpression causes enhanced type II collagen and proteoglycan deposition, enhanced chondrocyte proliferation and shortening of the hypertrophic cartilage zone. Tibiae from P1 littermates were stained with safranin-O for proteoglycans (A, left) and with anti-type II collagen antibody (B, left). Whole littermates were analyzed and the color intensity of 3 different wt or tg individuals was measured densitometrically; and the mean values are presented. (A, right; and B, right). *: p<0.005. Typical images from tg and wt littermates are shown. (C, left) Comparison of hypertrophic cartilage zone of CCN2 transgenic littermates. Tibiae were stained with type X collagen antibody. (C, right) The hypertrophic zone of tg cartilage appeared shorter compared with that of the wt cartilage. (D) Immunohistochemical analysis of proliferative cell nuclear antigen (PCNA) in tibiae of *ccn2* tg embryos at E19.5. Proliferative cells were observed in the whole epiphyseal cartilage of tg animals, whereas they were restricted to the proliferative zone of the wt littermates. The number of PCNA-positive cells inside of the boxed area was counted in 5 fields of 3 comparable wt and tg sections. Mean values indicate enhanced chondrocyte proliferation in the tg cartilage (graph at the lower right). *: p<0. 05.

comparison of length of diaphyses (Fig. 2E–1). Between 5–7 slides were randomly selected from serial sections of each tibia and stained with HE; and the length of diaphyses was measured by using an image analysis program. All of the transgenic tibiae with significantly enhanced expression levels of *ccn2* mRNA showed increased tibial length as compared with the wt tibiae (Fig. 2E–1). Comparison of mean length of diaphyses from 3 wt (5.897 ± 0.116 mm) and 3 transgenic mice (6.225 ± 0.080 mm) showed a significant difference (P<0.0001, Fig. 2E–2).

Over-expression of CCN2 Increased Bone Density, Extent of Mineralization of Cancellous Bone, and Thickness of Cortical Bone

Further evidence for a stimulation of bone growth by CCN2 in transgenic animals was obtained when bone density and mineral content of cancellous bone were monitored by using peripheral-quantitative computed-tomography (pQCT) analysis. For these studies, 8-week-old femora from 4 wt and 5 tg littermates were analyzed for mineral content and cortical bone thickness at 2 sites, one 1.2 mm, and the other more central 4.0 mm distal from the growth plate (Fig. 2F). Significant differences (p<0.05) in total mineral content (tg: 1.36 ± 0.08 mg/mm $vs.$ wt: 1.10 ± 0.12 mg/mm), trabecular mineral content (tg: 0.49 ± 0.01 mg/mm $vs.$ wt: 0.38 ± 0.01 mg/mm), and cortical thickness (tg: 0.060 ± 0.013 mm $vs.$ wt: 0.049 ± 0.021 mm) were observed for the part of the femora closer to the growth plate (Fig. 2F and 2G), but not for the central site (data not shown).

Over-expression of CCN2 in Chondrocytes Caused Enhanced Accumulation of Extracellular Matrix and Shortened Hypertrophic Zones

To examine the possibility that the extended skeletal growth of *ccn2* transgenic mice may have been due to enhanced production of cartilage matrix in the epiphysis, we analyzed the extracellular deposition of proteoglycans and type II collagen in the cartilage matrices by staining with safranin O and anti-type II collagen, respectively. Safranin-O staining indicated consistently an enhanced density of proteoglycans in the transgenic cartilage in comparison with cartilage of wt littermates (Fig. 3A). This observation is in accordance with our previous studies showing that CCN2 promotes proteoglycan synthesis in chondrocytes [11]. Also, the immunohistological analysis of type II collagen showed an enhanced reaction in resting chondrocytes and in the growth plate (Fig. 3B and Figure S1A). These results indicate that the over-expression of CCN2 enhanced the production and deposition of extracellular proteoglycans and type II collagen, which is in line with our previous *in vitro* findings. Surprisingly, however, the enhanced matrix deposition did not result in an increase in the size of the cartilaginous epiphysis; rather, the extended bone length was the result of an elongated bony shaft of the diaphysis.

Staining of the skeleton of transgenic embryos with type X collagen antibodies indicated that the hypertrophic zone was shorter in the transgenic embryos than in their wt littermates (Fig. 3C). This observation suggests an acceleration of chondrocyte proliferation and maturation, but possibly also accelerated cartilage resorption and chondrocyte apoptosis in these transgenic animals. Therefore, we next measured chondrocyte proliferation and apoptosis rates in the growing long bones of *ccn2* transgenic animals and their wt littermates.

Over-expression of CCN2 Resulted in Enhanced Cell Proliferation and Slightly Elevated Apoptosis of Epiphyseal Chondrocytes

In order to assess whether the enhanced bone growth of CCN2 transgenic animals was due to enhanced cell proliferation, we stained sections of E19.5-day transgenic and wt embryos with an antibody against proliferative cell nuclear antigen (PCNA). The data show that over-expression of CCN2 stimulated chondrocyte proliferation predominantly in the proliferative zone, but also in the resting zones (Fig. 3D). This observation is in accordance with previous *in vitro* studies showing that CCN2 promotes chondrocyte proliferation [11].

Curiously, however, staining for apoptotic cells in the growth plate of P3 by using the TUNEL assay revealed slightly, but not significantly, enhanced accumulation of apoptotic cells at the cartilage-bone interface and in the adjacent subchondral zone in the transgenic embryos as compared with their numbers in the wild-type (Fig. S1B). The length of the cartilaginous epiphyses seemed unaffected, since chondrocyte proliferation, cartilage matrix deposition, maturation, cartilage resorption, apoptosis, and assembly of trabecular bone were accelerated by the over-expressed CCN2.

Over-expression of CCN2 in Chondrocytes Resulted in Enhanced Gene Expression of *Col2a1* and *aggrecan*, and in Enhanced Chondrocyte Maturation *in vitro*

The increased accumulation of proteoglycan and type II collagen in the cartilage matrix of transgenic animals raised the question as to whether *ccn2* over-expression in chondrocytes stimulated cartilage and bone growth by enhancing cell proliferation, by stimulating the production of extracellular matrix or by accelerating the differentiation and maturation of chondrocytes. To obtain high *ccn2* transgene-expression, we crossed transgenic male and female mice and monitored the effects of over-expression of CCN2 in chondrocytes on the expression of extracellular matrix genes. RNA was extracted from short-term primary cultures of rib-cage chondrocytes from E18.5 transgenic or wild-type embryos and analyzed for *lacZ*, *ccn2*, and *Col2a1* mRNA levels by quantitative real-time PCR. The data showed about equal levels of lacZ expression in chondrocytes of transgenes #72,74, 76, 77 and 79, and a 2–3 fold higher level of the *lacZ* expression in tg #73 and #75, indicating that offspring #73 and #75 may bear double copies of transgene (Fig. 4A). Accordingly, the *ccn2* level in chondrocytes derived from those embryos (#73, 75 tg) was 2–3

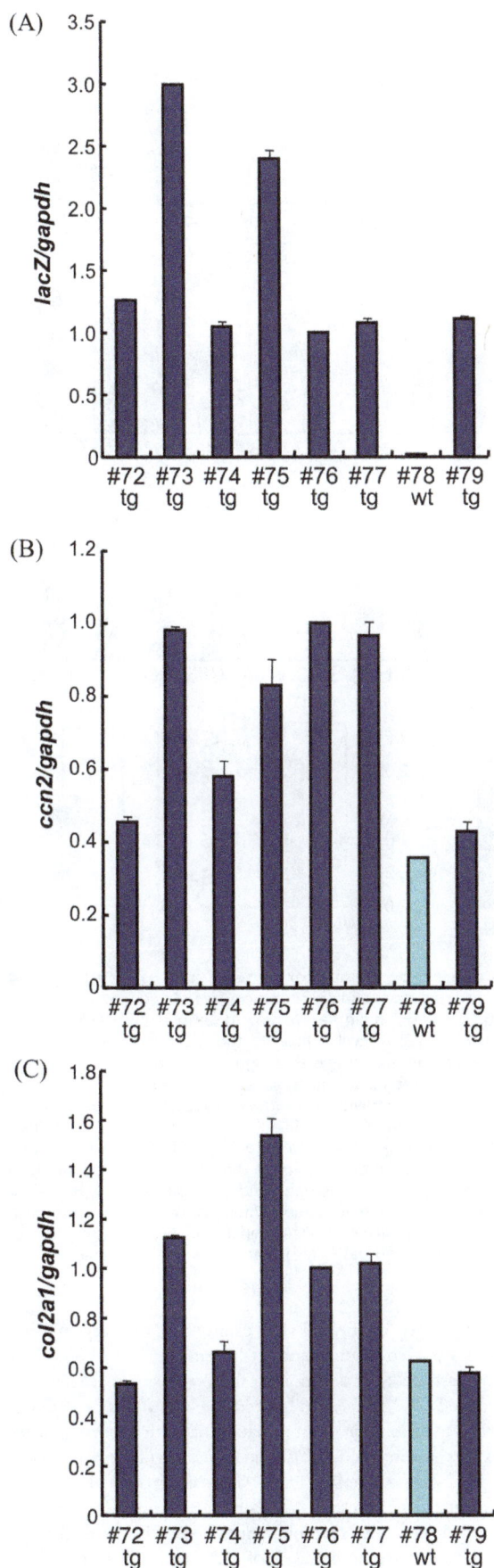

Figure 4. Gene expression analysis reveals enhanced *Col2a1* and *ccn2* in chondrocyte primary cultures of *Col2a1-ccn2* transgenic mice. To obtain high *ccn2* transgene-expressing littermates, we crossed transgenic male and female mice within same founder line; and expression of *LacZ*, *ccn2*, and *Col2a1* mRNA was measured by real-time PCR from 5 d chondrocyte cultures prepared from E18.5 wt and tg embryos. LacZ analysis revealed that high and low *lacZ*-expressing tg littermates and 1 wt were obtained (A). On average, *ccn2* expression levels in tg chondrocytes were significantly higher than those in wild-type littermates (B). *Col2a1* mRNA levels in tg chondrocytes were 2–3 fold higher than those in wt chondrocytes (C). Primary cultures of rib chondrocytes from individual littermates were prepared 3 times from each of the 2 founder lines, and total RNA were prepared. Real time-RCR analysis was repeated at least 2 times for each RNA preparation; and the 2 founder lines showed similar variations, but gave basically the same results. Primary-chondrocytes from *ccn2* tg and wt littermates were also pooled; and gene expression was analyzed as shown in figure S2.

fold enhanced as compared with the level for the wt chondrocytes (#78 wt, Fig. 4B). Also tg #76 and #77 showed enhanced levels of *ccn2* expression, whereas *ccn2* expression levels in tg #72, #74 and # 79 were not much higher than endogenous *ccn2* levels measured in the wt embryo #78, perhaps due to inactivation of the transgene. Tg chondrocytes with high over-expression of *ccn2* mRNA (#73, 75 tg), but also tg chondrocytes of #76 and 77 showed enhanced levels of *Col2a1* mRNA as compared with the wt level (#78 wt), as revealed by real-time PCR analysis (Fig. 4B and C), whereas tg cultures with low overexpression of *ccn2* (#72, 74, and 79) showed also low *col2a1* expression. To confirm the enhanced expression of *Col2a1* as well to estimate that of *aggrecan* mRNA in tg chondrocytes, we pooled primary–cultured chondrocytes from tg and wt littermates, and determined their *ccn2, col2a1, aggrecan* mRNA levels (figure S2). The levels of all 3 mRNA were greater in the tg than in the wt pooled cells.

The enhanced levels of *Col2a1* mRNA in the transgenic chondrocytes were also retained after 1 month in culture. During that time, chondrocytes ceased to proliferate and started to mature, but the *ccn2* transgene over-expression in tg cultures #86–88 remained at a high level compared with wt cultures #83–85 or low expressing tg cultures #80–82 (Fig. 5A). Primary cultures of chondrocytes with high levels of over-expressed *ccn2* mRNA continued to show strongly elevated *aggrecan* (Fig. 5B, 15-20,000 fold enhancement) and *Col2a1* (Fig. 5C, 100–1000 times enhancement) mRNA levels. The expression of *Col10a1*, a marker of hypertrophy, and that of *vegf* and of *mmp-9*, both vascular invasion factors expressed in the hypertrophic zone and boundary between cartilage and bone, were also enhanced; but not at the same extent as the enhancement of aggrecan and *col2a1* expression (Fig. 5D, 3–10 fold; 5E, 1.5–3 fold; and 5F, 1.5–3 fold enhancement). These results are in accordance with *in vitro* studies on the effect of *ccn2* on cultured chondrocytes [11] and are consistent with the notion that *ccn2* over-expression stimulated chondrocyte maturation.

Over-expression of *ccn2* Under the Control of the *col2a1* Promoter Accelerated Chondrogenesis

To investigate the effect of over-expression of *ccn2* on chondrogenic differentiation, we prepared micromass cultures of mesenchymal cells from 11.5-day embryonic transgenic and wt mouse limb buds. Mesenchymal cells from transgenic embryos started to develop Alcian blue-positive cartilaginous nodules after 2 days in culture (data not shown). After 3 days the cartilaginous nodule formation was significantly enhanced in cultures prepared from *ccn2*-overexpressing limb-buds cells as compared with that wild-type cells (Fig. 5G). The gene expression of *ccn2, Col2a1* and

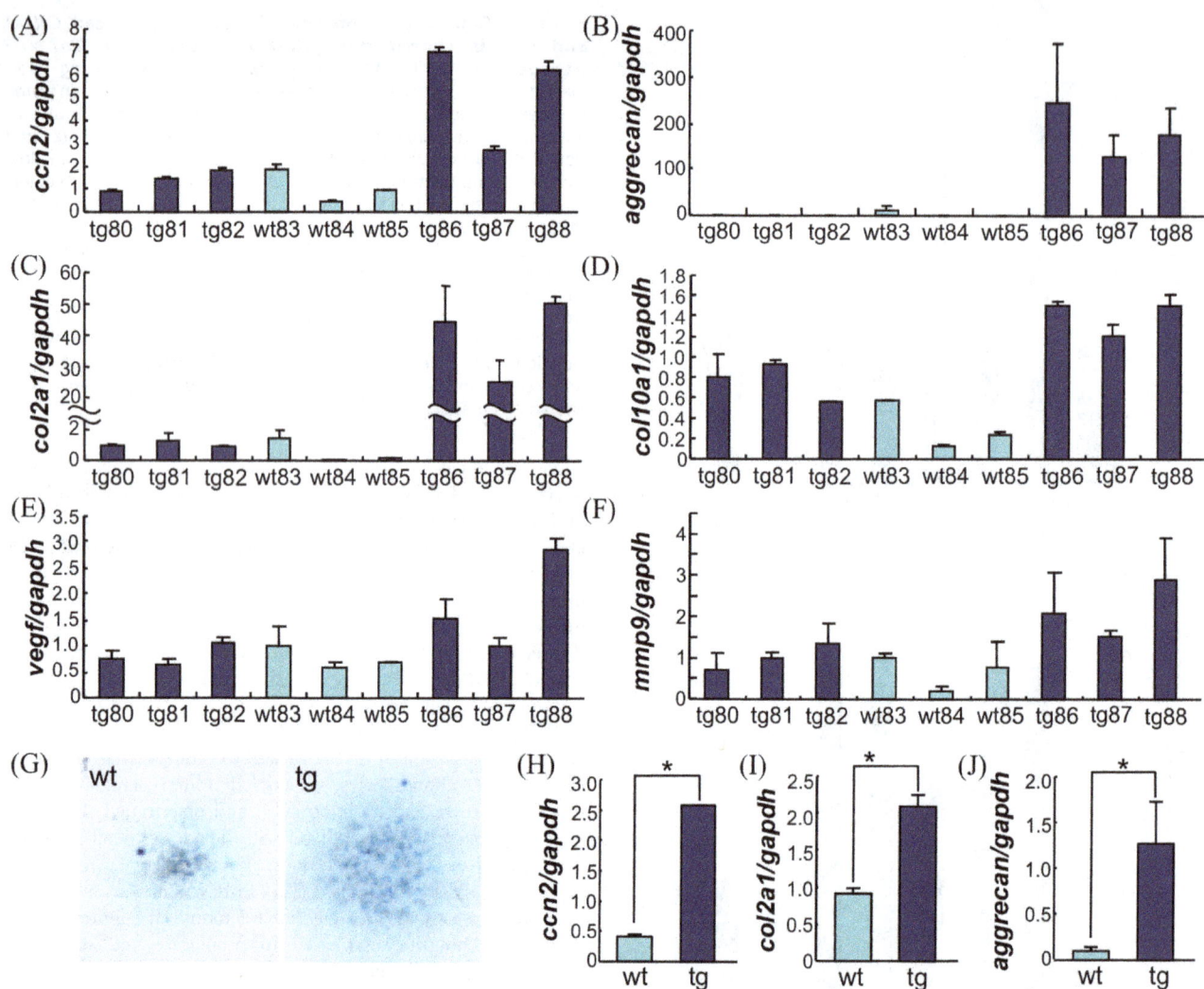

Figure 5. *Ccn2* overexpression on *Col2a1-ccn2* transgenic mice stimulates expression of marker gene of late hypertrophy and chondrogenesis of limb bud mesenchymal cells. For real-time PCR analysis of gene expression, primary chondrocytes isolated from ribs of *ccn2* tg and wt littermates were cultured for 1 month under differentiation-promoting conditions (A–F). In high-expressing tg samples, high levels of *ccn2* mRNA were retained during the entire culture time; whereas low expressers showed *ccn2* mRNA levels similar to those of wt chondrocytes (A). Expression of ECM components such as *aggrecan* (B) and *Col2a1* (C) was strongly up-regulated in the cultures that high levels of *ccn2* mRNA. Markers of late hypertrophic chondrocytes such as *Col10a1* (D), *vegf* (E), and *mmp9* (F) *were* also upregulated in those cultures. Expression levels of 1 representative litter out of 3 litters are shown. Primary 1 month cultures of rib chondrocytes from individual littermates were prepared twice from 2 founder lines; and total RNA was extracted. Real time-RCR analysis was repeated at least twice for the each RNA preparation. The 2 founder lines showed similar variations, but basically the same results. (G–J) Micromass cultures of mesenchymal cells derived from tg and wt E11.5 littermates. After 3 days in culture, nodule formation was accelerated in the cultures derived from ccn2-overexpressing mice as shown by Alcian blue (pH 1, G) and RNA was extracted for real-time PCR analysis. Cultures prepared from *ccn2* over-expressing mice showed enhanced expression of *ccn2* (H), *Col2a1* (I), and *aggrecan* (J) mRNA. *: $p < 0.005$. RNA of each littermates was individually analyzed and nodule formation among *ccn2* wt or tg was similar. Typical images of *ccn2* wt and tg are shown. Real time-RCR analysis was repeated at least 2 times for each RNA preparation. Micromass cultures of mesenchymal cells derived from E11.5 littermates were prepared 4 times from 2 founder lines, and basically similar results were obtained.

aggrecan was also up-regulated in *ccn2* transgenic micromass cultures as measured by quantitative RT-PCR (Fig. 5H, I and J, respectively).

Over-expression of CCN2 Enhanced Expression of *IGF-I* and *IGF-II*

In order to elucidate the mechanism of growth stimulation by the over-expressed CCN2, we analyzed changes in expression levels of growth factors known to be involved in skeletal growth. Remarkably, the RNA from tg chondrocytes contained clearly enhanced expression levels of *IGF-I* and *IGF-II* mRNA (Fig. 6A).

This finding was confirmed by examining primary-cultured chondrocytes pooled from tg and wt littermates (Figure S3). This finding suggests that, in addition to the possible direct effects of over-expressed CCN2, these enhanced levels of *IGF-I* or *II* might have been responsible for the stimulation of cortical bone growth, as well as for the enhanced *Col2a1* and *aggrecan* expression observed in the CCN2-over-expressing mice. To confirm this notion, we treated primary cultures of chondrocytes from 18.5-day wt embryos for 5 days with recombinant CCN2. The result showed a several-fold increase in the levels of *IGF-I* and *IGF-II* mRNA as well as a strong increase in endogenous *ccn2* expression (Fig. 6B).

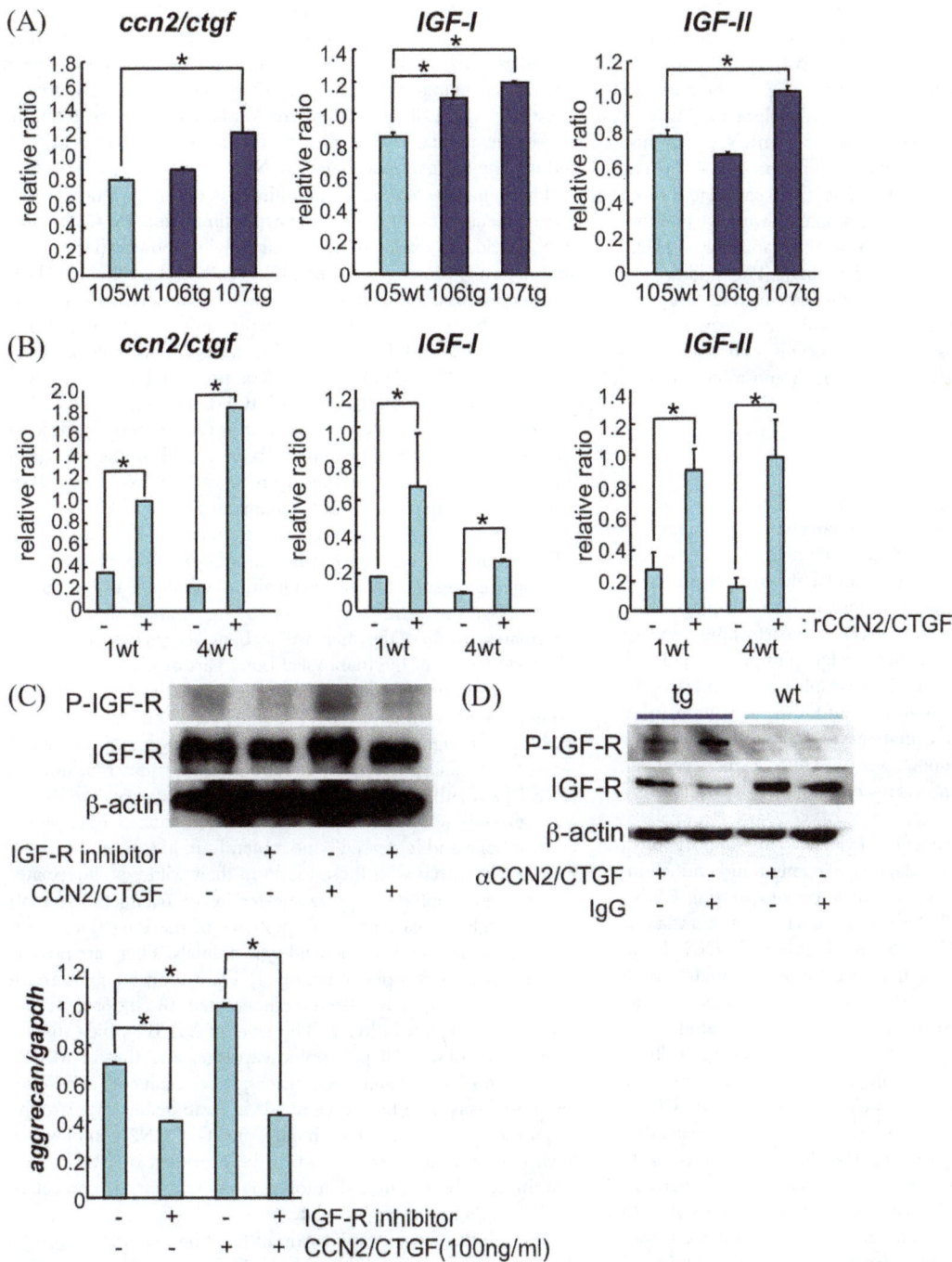

Figure 6. CCN2 stimulates IGF-IGFR pathway. Enhanced expression of IGF-I and IGF-II in primary cultures was found in primary cultures of chondrocytes prepared from the cartilage of ccn2-over-expressing mice, and in wt chondrocytes after treatment with recombinant CCN2. (A) Real-time PCR analysis of total RNA from tg cartilage which showed higher expression of ccn2 (107 tg) than wt cartilage (105 wt) also showed enhanced expression of IGF-I and II, whereas 106 tg with low ccn2 overexpression showed no enhanced IGF-II, but enhanced IGF-I expression. *:p<0.05. (B) Addition of recombinant CCN2 (50 ng/ml) to primary cultures of wt mouse rib chondrocytes stimulated IGF-I and II mRNA as well as ccn2 mRNA expression. Primary cultures of chondrocytes were prepared from wt E18.5 embryos; and the cells were seeded at 2×10^5 cells in 3.5-cm dishes with or without rCCN2 in the media, and incubated for 5 days. mRNA levels were standardized with gapdh; and all reactions were done in triplicate. Values for 1 wt and 4 wt are from 2 independently generated cultures. *:p<0.005. (C) Phosphorylation of IGF receptor induced by addition of rCCN2 (100 ng/ml) for 24 hours to primary cultures of wt rib chondrocytes, and inhibition of this phosphorylation of IGFR by PPP, an inhibitor of autophosphorylation of IGFR (upper panel). Aggrecan mRNA levels were measured (lower panel) and standardized to gapdh; and all reactions were done in triplicate. (D) Enhanced phosphorylation of IGFR in ccn2-overexpressing chondrocytes and inhibition of phosphorylation of IGFR by CCN2 antibody. Primary cultures of chondrocytes were pooled from P3 rib cages of ccn2 tg and wt littermates; and cells were seeded at 2×10^5 cells in 3.5-cm dishes and cultured for 2 days until the cells had reached the confluent state. CCN2 antibody or control IgG was added to the media, and the cultures were then incubated for 24 hours, after which the cells were collected with lysis buffer.

In order to elucidate whether CCN2 stimulated IGF-IGF receptor pathway, we examined the autophosphorylation of the IGF-1 receptor in response to the addition of CCN2. CCN2 enhanced the autophosphorylation of IGF-1 receptor (Fig. 6C and Figure S3), and the addition of PPP, IGFR inhibitor, abolished it (Fig. 6C). The CCN2-enhanced expression of *aggrecan* mRNA was also abolished by the addition of the IGFR inhibitor (Fig. 6C). The *ccn2*-overexpressing chondrocytes from *ccn2* tg rib cartilage showed enhanced phosphorylation of IGFR compared with wt chondrocytes; accordingly, the CCN2 neutralizing antibody, 11H3, repressed this autophosphorylation (Fig. 6D). The addition of 11H3 antibody down-regulated the expression of *ccn2*, *igf1*, and *igf2* mRNA (Figure S4). This finding of enhanced expression of IGF-I and -II in CCN2-transgenic chondrocytes is consistent with our finding of enhanced cortical bone growth and mineralization (see discussion).

Discussion

Previous *in vitro* studies on the response of rabbit growth-plate chondrocytes in primary culture and human chondrosarcoma cells HCS-2/8 to CCN2 demonstrated not only a significant stimulation of proliferation, differentiation, and enhanced synthesis of hyaline cartilage matrix components such as type II collagen and aggrecan, but also enhanced expression of hypertrophic cartilage proteins such as type X collagen and alkaline phosphatase [11,13]. Since CCN2 is expressed by prehypertrophic and hypertrophic chondrocytes, these findings indicate that CCN2 acts both in an autocrine and in a paracrine manner to promote chondrocyte proliferation and differentiation events. Thus, it may regulate cartilage matrix synthesis and turnover leading to endochondral ossification [4,11,15].

Here we provide experimental evidence in support of a significant role of CCN2 in cartilage development and endochondral ossification *in vivo* in transgenic mice over-expressing CCN2 driven by the cartilage-specific *Col2a1* promoter. Most remarkably, transgenic mice expressing high levels of transgenic CCN2 had greater bone length as compared with their wt littermates. By 8 weeks, some of the tg littermates had greater body mass (~12%), possibly caused by a better eating with tough skeleton. This morphological phenotype reflects several enhanced cellular activities observed in the transgenic cartilage: i) Chondrogenic differentiation of limb-bud mesenchymal cells from CCN2 transgenic animals was greatly enhanced as compared with that of their wild-type counterparts. ii) Histological analysis of tg cartilage revealed increased type II collagen and aggrecan deposition in the extracellular cartilage matrix, consistent with our *in vitro* data showing that chondrocytes isolated from transgenic animals had highly elevated levels of *Col2a1* and *aggrecan* mRNA shortly after isolation; iii) In long-term cultures, CCN2 transgenic rib chondrocytes also expressed higher levels of *Col10a1*, *vegf* and *mmp9* mRNA than wt chondrocytes, indicating accelerated maturation to hypertrophic chondrocytes. iv) PCNA staining revealed a significant increase in chondrocyte proliferation in resting and growth-plate cartilage of transgenic animals; and (v) CCN2 over-expression also caused slightly enhanced apoptosis of hypertrophic chondrocytes.

One explanation for these effects of the over-expressed CCN2 may be the enhanced levels of *IGF- I* and *IGF- II* mRNA in the transgenic chondrocytes. IGF-I and –II and IGF-binding proteins are known to be most potent regulators of cartilage and bone growth [31,32,33]. IGF-I and –II are well known to stimulate proliferation and proteoglycan synthesis in cultured chondrocytes [32,33,34]. Transgenic mice with an *IGF-I* gene under the control

of the *metallothionein-I* gene promoter weigh 1.3 times more than their non-transgenic littermates [35]. Furthermore, IGF-II is considered to be a fetal growth factor that promotes skeletal growth in young rats [36,37]. Therefore, it is likely that a substantial part, if not all, of the observed effects seen in the transgenic cartilage were due to the additional IGF-I and -II induced by the over-expressed CCN2.

These unexpected findings require revision of current views on the molecular mechanism of growth stimulation by CCN2 and may provide an explanation for our previous observations on the stimulation of proteoglycan and DNA synthesis by CCN2 in HCS-2/8 chondrosarcoma cells and rabbit chondrocytes [11,13]. Previous studies have shown an interaction between module 1 of CCN proteins and IGF, suggesting a regulatory role of CCN proteins on IGFs [38,39]. The data presented here, however, indicate that the up-regulation of IGF-I and -II by CCN2 in mouse chondrocyte cultures occurred at the transcriptional level. To which extent the stimulation of bone growth in the transgenic animals was caused by the up-regulated IGFs or by IGF-independent actions of CCN2 remains to be elucidated.

Surprisingly, the enhanced IGF-I and IGF-II levels in the CCN2 transgenic animals did not cause significant elongation of cartilaginous tissues, for the cartilaginous epiphyses of long bones were about the same size in transgenic animals and their wt littermates. Rather, the increase in bone length was due to an extended length of the diaphyseal bony part of the long bones. A paracrine stimulation of periosteal bone cells by CCN2 overexpressed by adjacent chondrocytes, or by IGFs induced by CCN2, is plausible in light of several *in vitro* studies showing stimulation of osteoblast proliferation and differentiation and mineralization by CCN2 [4,16,40]. A significant effect of over-expressed CCN2 on bone growth was also evident from the enhanced thickness of cortical bone and increased bone mineralization seen in transgenic mice as compared with those found in their wild-type littermates.

The hypertrophic zone was shorter in the transgenic animals, even though the cartilaginous epiphyses of the long bones were about the same size as in the wild-type animals. There are possible explanations for this phenomenon: 1) The enhanced chondrocyte proliferation may have been compensated by the increase in chondrocyte hypertrophy. 2) The level of VEGF, which induces vascular invasion of hypertrophic cartilage, and that of MMP9, which degrades cartilaginous matrices, were enhanced; in addition apoptosis was slightly accelerated in transgenic hypertrophic chondrocytes. On the other hand, VEGF- CCN2 complexes as formed *in vitro* have been shown to be degraded by MMPs [41]; and this may be an internal autoregulatory mechanism controlling CCN2 levels in the growth plate.

The results of our present gain-of-function experiment are for the most part in accordance with the findings of a loss-of-function study on CCN2-deficient mice [17], which develop skeletal dysmorphisms such as distorted cartilage and bone elements as a result of impaired chondrocyte proliferation and endochondral ossification. In line with the shortened hypertrophic zone observed in our CCN2 transgenic mice, the hypertrophic zone is extended in CCN2-deficient mice. Interestingly, however, CCN2 deficient mice do not show significant alterations in total bone size. Yet, this is in accordance with the notion that the major enhancing effect on bone growth in our CCN2 transgenic mice may have been caused by enhanced levels of IGFs. Thus, although the study on the CCN2-deficient mice confirmed the important role of CCN2 as a regulator of cartilage remodelling during endochondral ossification, the absence of more severe phenotypic alterations in these mice might have been due to redundant effects of other members of the CCN family [17].

In the CCN2 transgenic mouse lines presented here, the extent of bone elongation, as well as the extent of enhancement of *Col2a1* and *aggrecan* mRNA levels correlated with the extent of CCN2 over-expression in transgenic chondrocytes of both founder lines. Besides high-expressing chondrocytes, also transgenic chondrocytes showing low levels of CCN2 expression comparable to those of wt rib chondrocytes were seen in each litter, even when derived from the same founder. Enhanced bone size as well as reduced length of the hypertrophic zone was only observed in tg mice with high levels of CCN2 expression. This was probably due to unpredictable somatic inactivation of the transgene in some embryos and reflects the limitation of this technique, which relies on a random integration of the transgene into the genome. Our previous CCN2-transgenic mice under the control of the *Col9a1* promoter show dwarfism several months after birth and smaller testes, but not so much difference in body length [42]. The expression pattern and timing of *Col9a1* expression, however, differ to some extent from those of *Col2a1*, which may explain the difference in phenotype.

Elucidation of the exact molecular mechanisms involved in IGF-independent, CCN2-regulated chondrocyte responses is still hampered by the fact that currently no specific cell-surface signalling cellular receptor for CCN2 has been identified so far in chondrogenic or osteogenic cells; instead, CCN2 seems to control cellular events by complex interactions with numerous growth factors such as IGFs, and perhaps through integrins and their signalling pathways [2,43,44].

In conclusion, our study demonstrates that the use of the *Col2a1* promoter for specific over-expression of CCN2 or other members of the CCN family in chondrocytes may represent – together with *ccn2*-deficient chondrocytes - a powerful tool to provide further insight into the specific role of these growth factors in cartilage metabolism and skeletal development.

Supporting Information

Figure S1 Accumulation of type II collagen and slightly enhanced apoptosis in *ccn2*-overexpressing epiphyseal cartilage. (A) Comparison of accumulation of type II collagen in cartilage of *ccn2*-overexpressing and wt mice. Tibiae from P3 littermates were stained with anti-type II collagen antibody. The color intensity was measured densitometrically. Four wt and 5 *ccn2* tg littermates were analyzed. (B) TUNEL assay on tibiae from P3 littermates shows slightly enhanced apoptosis in the cartilage-bone transition zone in the tg mice.

Figure S2 Gene expression analysis in pooled primary chondrocytes from *ccn2* tg and wt littermates. Expression analysis of *ccn2*, *Col2a1*, and *Aggrean* mRNA of primary chondrocytes from pooled *ccn2* tg and wt littermates. Real time-RCR analysis was done in duplicate, *: p<0.005. The experiments were repeated 3 times and showed similar results.

Figure S3 Phosphorylation analysis of primary-cultured *ccn2* tg and wt chondrocytes pooled from different transgenic line from figure 6A. Results of Western blot analysis of IGF-1R and phospho-IGF-1R (upper photos) and those of gene expression analysis (graphs at bottom) of the same cells as used in Western blot analysis are shown. Real time-RCR analysis was done in duplicate and repeated 3 times, *: p<0.005.

Figure S4 Change in gene expression level of *ccn2*, *igf1*, and *igf2* mRNA by the addition of CCN2 antibody (11H3) to primary cultures of mouse rib chondrocytes from P3 littermates of *ccn2* tg mice. Cells from these cultures were seeded at 2×10^5 cells in 3.5-cm dishes and cultured for 2 days until the cells had reached to confluence. CCN2 antibody or control IgG was added to the media. The cells were incubated for 24 hours, and total RNA was then extracted from them. Real-time PCR demonstrated that CCN2 antibody repressed gene expression of *ccn2*, *igf1*, and *igf2* mRNA in the *ccn2*-overexpresssing chondrocytes. Real time-RCR analysis was done in duplicate, *: p<0.005. The experiments were repeated for 3 times and showed similar results.

Acknowledgments

We thank Drs. Shunichi Murakami and Benoit de Crombrugghe for their generous gifts of *Col2a1* promoter and lacZ constructs, as well as Drs. Satoshi Kubota and Takashi Nishida for their valuable discussion. We are also grateful to Dr. Hiroshi Ikegawa, Ms. Ayako Ogo, Ms. Yoshiko Miyake and Ms. Tomoko Yamamoto for technical assistance and to Ms. Eri Yashiro for secretarial assistance.

Author Contributions

Conceived and designed the experiments: NT TH MT. Performed the experiments: NT SI TH EA MY. Analyzed the data: NT SI TH MY MT. Contributed reagents/materials/analysis tools: TY. Wrote the paper: TH MT.

References

1. Brigstock DR (2003) The CCN family: a new stimulus package. J Endocrinol 178: 169–175.
2. Takigawa M, Nishida T, Kubota S (2005) Roles of CCN2/CTGF in the cntrol of growth and regeneration. In: Perbal B, Takigawa M, editors. CCN Proteins: A new family of cell growth and differentiation regulators. London: Imperial College Press. 19–59.
3. Kubota S, Takigawa M (2007) Role of CCN2/CTGF/Hcs24 in bone growth. Int Rev Cytol 257: 1–41.
4. Takigawa M, Nakanishi T, Kubota S, Nishida T (2003) Role of CTGF/HCS24/ecogenin in skeletal growth control. J Cell Physiol 194: 256–266.
5. Shimo T, Nakanishi T, Kimura Y, Nishida T, Ishizeki K, et al. (1998) Inhibition of endogenous expression of connective tissue growth factor by its antisense oligonucleotide and antisense RNA suppresses proliferation and migration of vascular endothelial cells. J Biochem 124: 130–140.
6. Shimo T, Nakanishi T, Nishida T, Asano M, Kanyama M, et al. (1999) Connective tissue growth factor induces the proliferation, migration, and tube formation of vascular endothelial cells in vitro, and angiogenesis in vivo. J Biochem 126: 137–145.
7. Babic AM, Chen CC, Lau LF (1999) Fisp12/mouse connective tissue growth factor mediates endothelial cell adhesion and migration through integrin alphavbeta3, promotes endothelial cell survival, and induces angiogenesis in vivo. Mol Cell Biol 19: 2958–2966.
8. Grotendorst GR, Duncan MR (2005) Individual domains of connective tissue growth factor regulate fibroblast proliferation and myofibroblast differentiation. Faseb J 19: 729–738.
9. Twigg SM, Cao Z, McLennan SV, Burns WC, Brammar G, et al. (2002) Renal connective tissue growth factor induction in experimental diabetes is prevented by aminoguanidine. Endocrinology 143: 4907–4915.
10. Kanaan RA, Aldwaik M, Al-Hanbali OA (2006) The role of connective tissue growth factor in skeletal growth and development. Med Sci Monit 12: RA277–281.
11. Nakanishi T, Nishida T, Shimo T, Kobayashi K, Kubo T, et al. (2000) Effects of CTGF/Hcs24, a product of a hypertrophic chondrocyte-specific gene, on the proliferation and differentiation of chondrocytes in culture. Endocrinology 141: 264–273.
12. Hattori T, Fujisawa T, Sasaki K, Yutani Y, Nakanishi T, et al. (1998) Isolation and characterization of a rheumatoid arthritis-specific antigen (RA-A47) from a human chondrocytic cell line (HCS-2/8). Biochem Biophys Res Commun 245: 679–683.
13. Nishida T, Kubota S, Nakanishi T, Kuboki T, Yosimichi G, et al. (2002) CTGF/Hcs24, a hypertrophic chondrocyte-specific gene product, stimulates

proliferation and differentiation, but not hypertrophy of cultured articular chondrocytes. J Cell Physiol 192: 55–63.

14. Fujisawa T, Hattori T, Ono M, Uehara J, Kubota S, et al. (2008) CCN family 2/connective tissue growth factor (CCN2/CTGF) stimulates proliferation and differentiation of auricular chondrocytes. Osteoarthritis Cartilage.

15. Nishida T, Kubota S, Fukunaga T, Kondo S, Yosimichi G, et al. (2003) CTGF/Hcs24, hypertrophic chondrocyte-specific gene product, interacts with perlecan in regulating the proliferation and differentiation of chondrocytes. J Cell Physiol 196: 265–275.

16. Nishida T, Nakanishi T, Asano M, Shimo T, Takigawa M (2000) Effects of CTGF/Hcs24, a hypertrophic chondrocyte-specific gene product, on the proliferation and differentiation of osteoblastic cells in vitro. J Cell Physiol 184: 197–206.

17. Ivkovic S, Yoon BS, Popoff SN, Safadi FF, Libuda DE, et al. (2003) Connective tissue growth factor coordinates chondrogenesis and angiogenesis during skeletal development. Development 130: 2779–2791.

18. Murakami S, Balmes G, McKinney S, Zhang Z, Givol D, et al. (2004) Constitutive activation of MEK1 in chondrocytes causes Stat1-independent achondroplasia-like dwarfism and rescues the *Fgfr3*-deficient mouse phenotype. Genes Dev 18: 290–305.

19. Zhang R, Murakami S, Coustry F, Wang Y, de Crombrugghe B (2006) Constitutive activation of MKK6 in chondrocytes of transgenic mice inhibits proliferation and delays endochondral bone formation. Proc Natl Acad Sci U S A 103: 365–370.

20. Bi W, Deng JM, Zhang Z, Behringer RR, de Crombrugghe B (1999) Sox9 is required for cartilage formation. Nat Genet 22: 85–89.

21. Otto F, Thornell AP, Crompton T, Denzel A, Gilmour KC, et al. (1997) *Cbfa1*, a candidate gene for cleidocranial dysplasia syndrome, is essential for osteoblast differentiation and bone development. Cell 89: 765–771.

22. Smits P, Li P, Mandel J, Zhang Z, Deng JM, et al. (2001) The transcription factors L-Sox5 and Sox6 are essential for cartilage formation. Dev Cell 1: 277–290.

23. Lefebvre V, Huang W, Harley VR, Goodfellow PN, de Crombrugghe B (1997) SOX9 is a potent activator of the chondrocyte-specific enhancer of the pro alpha1(II) collagen gene. Mol Cell Biol 17: 2336–2346.

24. Hattori T, Eberspaecher H, Lu J, Zhang R, Nishida T, et al. (2006) Interactions between PIAS proteins and SOX9 result in an increase in the cellular concentrations of SOX9. J Biol Chem 281: 14417–14428.

25. Holmdahl R, Rubin K, Klareskog L, Larsson E, Wigzell H (1986) Characterization of the antibody response in mice with type II collagen-induced arthritis, using monoclonal anti-type II collagen antibodies. Arthritis Rheum 29: 400–410.

26. Boos N, Nerlich AG, Wiest I, von der Mark K, Ganz R, et al. (1999) Immunohistochemical analysis of type-X-collagen expression in osteoarthritis of the hip joint. J Orthop Res 17: 495–502.

27. von der Mark K, Kirsch T, Nerlich A, Kuss A, Weseloh G, et al. (1992) Type X collagen synthesis in human osteoarthritic cartilage. Indication of chondrocyte hypertrophy. Arthritis Rheum 35: 806–811.

28. Lefebvre V, Garofalo S, Zhou G, Metsaranta M, Vuorio E, et al. (1994) Characterization of primary cultures of chondrocytes from type II collagen/beta-galactosidase transgenic mice. Matrix Biol 14: 329–335.

29. Hattori T, Coustry F, Stephens S, Eberspaecher H, Takigawa M, et al. (2008) Transcriptional regulation of chondrogenesis by coactivator Tip60 via chromatin association with Sox9 and Sox5. Nucleic Acids Res 36: 3011–3024.

30. Nakanishi T, Kimura Y, Tamura T, Ichikawa H, Yamaai Y, et al. (1997) Cloning of a mRNA preferentially expressed in chondrocytes by differential display-PCR from a human chondrocytic cell line that is identical with connective tissue growth factor (CTGF) mRNA. Biochem Biophys Res Commun 234: 206–210.

31. Hoeflich A, Gotz W, Lichanska AM, Bielohuby M, Tonshoff B, et al. (2007) Effects of insulin-like growth factor binding proteins in bone – a matter of cell and site. Arch Physiol Biochem 113: 142–153.

32. Schmid C (1995) Insulin-like growth factors. Cell Biol Int 19: 445–457.

33. Vetter U, Helbing G, Heit W, Pirsig W, Sterzig K, et al. (1985) Clonal proliferation and cell density of chondrocytes isolated from human fetal epiphyseal, human adult articular and nasal septal cartilage. Influence of hormones and growth factors. Growth 49: 229–245.

34. Takigawa M, Okawa T, Pan H, Aoki C, Takahashi K, et al. (1997) Insulin-like growth factors I and II are autocrine factors in stimulating proteoglycan synthesis, a marker of differentiated chondrocytes, acting through their respective receptors on a clonal human chondrosarcoma-derived chondrocyte cell line, HCS-2/8. Endocrinology 138: 4390–4400.

35. Quaife CJ, Mathews LS, Pinkert CA, Hammer RE, Brinster RL, et al. (1989) Histopathology associated with elevated levels of growth hormone and insulin-like growth factor I in transgenic mice. Endocrinology 124: 40–48.

36. Adams SO, Nissley SP, Handwerger S, Rechler MM (1983) Developmental patterns of insulin-like growth factor-I and -II synthesis and regulation in rat fibroblasts. Nature 302: 150–153.

37. Lund PK, Moats-Staats BM, Hynes MA, Simmons JG, Jansen M, et al. (1986) Somatomedin-C/insulin-like growth factor-I and insulin-like growth factor-II mRNAs in rat fetal and adult tissues. J Biol Chem 261: 14539–14544.

38. Bork P (1993) The modular architecture of a new family of growth regulators related to connective tissue growth factor. FEBS Lett 327: 125–130.

39. Burren CP, Wilson EM, Hwa V, Oh Y, Rosenfeld RG (1999) Binding properties and distribution of insulin-like growth factor binding protein-related protein 3 (IGFBP-rP3/NovH), an additional member of the IGFBP Superfamily. J Clin Endocrinol Metab 84: 1096–1103.

40. Safadi FF, Xu J, Smock SL, Kanaan RA, Selim AH, et al. (2003) Expression of connective tissue growth factor in bone: its role in osteoblast proliferation and differentiation in vitro and bone formation in vivo. J Cell Physiol 196: 51–62.

41. Hashimoto G, Inoki I, Fujii Y, Aoki T, Ikeda E, et al. (2002) Matrix metalloproteinases cleave connective tissue growth factor and reactivate angiogenic activity of vascular endothelial growth factor 165. J Biol Chem 277: 36288–36295.

42. Nakanishi T, Yamaai T, Asano M, Nawachi K, Suzuki M, et al. (2001) Overexpression of connective tissue growth factor/hypertrophic chondrocyte-specific gene product 24 decreases bone density in adult mice and induces dwarfism. Biochem Biophys Res Commun 281: 678–681.

43. Chen CC, Chen N, Lau LF (2001) The angiogenic factors Cyr61 and connective tissue growth factor induce adhesive signaling in primary human skin fibroblasts. J Biol Chem 276: 10443–10452.

44. Gao R, Brigstock DR (2004) Connective tissue growth factor (CCN2) induces adhesion of rat activated hepatic stellate cells by binding of its C-terminal domain to integrin alpha(v)beta(3) and heparan sulfate proteoglycan. J Biol Chem 279: 8848–8855.

3

Gain-of-Function of Stat5 Leads to Excessive Granulopoiesis and Lethal Extravasation of Granulocytes to the Lung

Wan-chi Lin[1][9]**, Jeffrey W. Schmidt**[1][9]**, Bradley A. Creamer**[1][¤]**, Aleata A. Triplett**[1]**, Kay-Uwe Wagner**[1,2]*****

1 Eppley Institute for Research in Cancer and Allied Diseases, University of Nebraska Medical Center, Omaha, Nebraska, United States of America, **2** Department of Pathology and Microbiology, University of Nebraska Medical Center, Omaha, Nebraska, United States of America

Abstract

The Signal Transducer and Activator of Transcription 5 (Stat5) plays a significant role in normal hematopoiesis and a variety of hematopoietic malignancies. Deficiency in Stat5 causes impaired cytokine-mediated proliferation and survival of progenitors and their differentiated descendants along major hematopoietic lineages such as erythroid, lymphoid, and myeloid cells. Overexpression and persistent activation of Stat5 are sufficient for neoplastic transformation and development of multi-lineage leukemia in a transplant model. Little is known, however, whether a continuous activation of this signal transducer is essential for the maintenance of hematopoietic malignancies. To address this issue, we developed transgenic mice that express a hyperactive mutant of Stat5 in hematopoietic progenitors and derived lineages in a ligand-controlled manner. In contrast to the transplant model, expression of mutant Stat5 did not adversely affect normal hematopoiesis in the presence of endogenous wildtype *Stat5* alleles. However, the gain-of-function of this signal transducer in mice that carry *Stat5a/b* hypomorphic alleles resulted in abnormally high numbers of circulating granulocytes that caused severe airway obstruction. Downregulation of hyperactive Stat5 in diseased animals restored normal granulopoiesis, which also resulted in a swift clearance of granulocytes from the lung. Moreover, we demonstrate that Stat5 promotes the initiation and maintenance of severe granulophilia in a cell autonomous manner. The results of this study show that the gain-of-function of Stat5 causes excessive granulopoiesis and prolonged survival of granulocytes in circulation. Collectively, our findings underline the critical importance of Stat5 in maintaining a normal balance between myeloid and lymphoid cells during hematopoiesis, and we provide direct evidence for a function of Stat5 in granulophilia–associated pulmonary dysfunction.

Editor: Kevin D. Bunting, Emory University, United States of America

Funding: This work was supported, in part, by the Public Health Service grants CA117930 (KUW) Additional financial support provided to KUW by the Nebraska Cancer and Smoking Disease Research Program (NE DHHS LB506 2012-44) was imperative to finance the maintenance of the Stat5 transgenic mice. WcL and BAC were supported through a research assistantship from the UNMC Graduate Studies Office. JWS received a graduate fellowship through the UNMC Cancer Research Training Program (CA009476), a Program of Excellence Graduate Assistantship from the UNMC Graduate Studies Office as well as a Breast Cancer Predoctoral Traineeship Award from the Department of Defense Congressionally Directed Medical Research Program (BC100147). The funders had no role in study design, data collection and analysis, decision to publish, or preparation of the manuscript.

Competing Interests: The authors have declared that no competing interests exist.

* E-mail: kuwagner@unmc.edu

[9] These authors contributed equally to this work.

[¤] Current address: Department of Biology, Missouri Southern State University, Joplin, Missouri, United States of America

Introduction

Signal Transducers and Activators of Transcription 5 (Stat5a and Stat5b) mediate extracellular signals from a variety of cytokine receptors and are therefore essential for the growth and differentiation of many cell types including those of hematopoietic lineages. Mice deficient in either Stat5a or Stat5b show defects in the prolactin-induced functional differentiation of the mammary gland [1] or in sexual dimorphism in the control of body size mediated by growth hormone [2]. The phenotypic examination of hypomorphic mutant mice that express low levels of truncated Stat5a and Stat5b (*Stat5$^{\Delta N/\Delta N}$*) revealed that both Stat5 isoforms have redundant functions. *Stat5$^{\Delta N/\Delta N}$* double mutant mice exhibit abnormalities during erythropoiesis and reduced proliferation of peripheral T cells [3–5]. The Cre-mediated ablation of the entire *Stat5* locus from the murine genome caused much more severe phenotypes and resulted in perinatal lethality due to anemia and other defects [6]. Subsequent studies using Stat5a/Stat5b conditional knockout mice also showed that the combined functions of these evolutionarily conserved transcription factors are critical for the homeostasis and differentiation of hematopoietic stem cells and derived descendants along the lymphoid lineage [7–11]. Moreover, Stat5 is required for granulocyte macrophage colony-stimulating factor receptor (GM-CSF) signaling and controls granulopoiesis by promoting the generation of granulocytes from granulocyte-macrophage progenitors (GMPs) as well as the survival of mature neutrophils [12,13].

The phenotypes associated with a knockout of Stat5 in mice provided guidance to the identification of the first germline mutations in the coding region of the *STAT5B* gene in patients who were insensitive to growth hormone (GH) and who did not carry any mutations in the GH receptor [14–16]. Interestingly, the majority of STAT5B deficient cases in humans were associated with symptoms of severe infection, autoimmune diathesis, and lymphocytic interstitial pneumonitis. These patients also exhibited a reduction in the numbers of regulatory T cells, suggesting that loss of STAT5B in humans appears to be sufficient for the initiation of certain immune phenotypes as well as chronic lung disease [17].

Both STAT5 isoforms are frequently overexpressed and activated in a broad range of human cancers and hematologic malignancies. Cytokine-independent cell growth and survival, which is a hallmark of neoplastic transformation, can be caused by aberrant autocrine signaling as well as genetic and epigenetic changes in intracellular signal networks that involve tyrosine kinases and negative regulators [18]. Chromosomal translocations that lead to the formation of hyper-active JAK2 fusion proteins such as TEL-JAK2, BCR-JAK2, and PCM1-JAK2 signal through STAT5 and are frequently detected in various leukemia subtypes [for references see reviews by Valentino and Pierre (2006) and Ghoreschi et al. (2009) [19,20]. Additionally, missense mutations in the *JAK2* gene (e.g. JAK2V617F) have been shown to be associated with many myeloproliferative disorders [21–24]. Besides JAK2, STAT5 can be persistently activated in leukemias by the BCR/ABL tyrosine kinase and mutations in c-KIT or FLT-3 [25–28]. Using retroviral gene transfer of a hyperactive mutant of Stat5 (caStat5a-S710F) into hematopoietic stem/progenitor cells and their transplantation into recipient mice, Moriggl and colleagues [29] demonstrated that a persistent activation of this signal transducer was sufficient to cause multi-lineage leukemia in mice. Then again, expression of the same mutant under regulation of the lymphoid-specific Eμ enhancer was weakly oncogenic, and only 6 out of 50 transgenic mice developed B-cell malignancies after a long latency [30]. This suggests that secondary mutations are required for disease onset in this transgenic line. Regardless of whether active Stat5 acts as a weak or strong oncogene depending on the experimental model, it still remains to be elucidated whether a Stat5-induced hematopoietic malignancy requires a continuous expression of the mutant signal transducer for disease maintenance.

In this report, we describe the development and analysis of transgenic lines that express the S710F mutant of Stat5a in hematopoietic progenitors and derived lineages in a ligand-controlled manner to assess whether a sustained activation of Stat5 is required for the growth and survival of preneoplastic or neoplastic hematopoietic cells following disease initiation. We show here that, depending on the levels of endogenous Stat5, the persistent activation of this signal transducer causes excessive granulopoiesis and prolonged survival of granulocytes in circulation. Upon extravasation, these cells can cause severe pulmonary obstruction and lethality. Stat5 is clearly important for maintaining a normal balance between lymphoid and myeloid cells during hematopoiesis, and all of these major phenotypes are reversible through downregulation of hyperactive Stat5.

Results

Generation of transgenic mice that express hyperactive Stat5 in a temporally and spatially controlled manner

We developed transgenic mice in which the expression of exogenous Stat5 can be targeted to hematopoietic stem cells and derived lineages in a ligand regulatable manner. Upon expression of a tetracycline-controlled transactivator, the TetO-Stat5 responder transgene (Fig. 1A) permits the co-expression of Stat5 and

a luciferase reporter for *in vivo* bioluminescence imaging. To facilitate a gain-of-function of this signal transducer, we utilized a Flag-tagged S710F mutant of Stat5a, which has been reported to exhibit a persistent tyrosine phosphorylation in the absence of cytokines [29]. Following the generation of 15 founders by pronuclear injection, the correct temporally controlled expression of the TetO-Stat5 transgene was examined in detail in two independent founder lines (11651 and 11676). For this purpose, we derived primary mouse embryonic fibroblasts from both lines and infected them with a retrovirus expressing the reverse transactivator (pBabe-rtTA, Tet-ON). Expression of luciferase and Stat5 was only detected when these cells were treated with doxycycline (Dox), and the levels declined within 48 hours following the withdrawal of the ligand (Fig. 1B and 1C). To target the expression of hyperactive Stat5 to hematopoietic cells, we crossed TetO-Stat5 mice with a transgenic strain that expresses the tetracycline-controlled transactivator (tTA, Tet-OFF) under control of the MMTV-LTR [31]. In contrast to the rtTA (Tet-ON), which we used in the cell culture experiments (Fig. 1B and 1C), the tTA (Tet-OFF) transactivator is better suited for a constitutive expression of TetO-driven transgenes *in vivo* for a prolonged period since a continuous administration of Dox is not required. Besides expression in various secretory organs, the MMTV-tTA construct is active in hematopoietic cells and has been successfully used in the past to establish models of B-cell leukemia and lymphoma [32–34]. Expression of the luciferase reporter was only observed in MMTV-tTA TetO-Stat5 double transgenic animals using *in vivo* imaging, and the TetO-Stat5 transgene alone does not exhibit any background activation in the absence of the transactivator (Fig. 1D, left panel). Upon administration of Dox, the tTA-mediated transactivation and expression of the TetO-Stat5 construct can be rapidly and completely silenced (Fig. 1D, right panel). The immunoblot analysis revealed that the MMTV-tTA-mediated transactivation of the TetO-Stat5 transgene occurred in multiple secretory organs, and overexpression of exogenous Stat5 was readily detectable in the salivary gland and in the prostate (Fig. 1E). In organs that express high levels of endogenous Stat5 (spleen, thymus, and seminal vesicle), the transgenic protein was proportionally less abundant. However, we observed a significant increase in the steady-state levels of active Stat5 in splenocytes, which was confirmed using immunoprecipitation (IP) and western blotting (Fig. 1F). In summary, while the expression of exogenous Stat5 did not lead to a significant elevation in total Stat5, there was a noticeable increase in the activation of Stat5 in hematopoietic cells.

To assess whether the MMTV-tTA-mediated expression of TetO-driven responder genes is restricted to lymphoid cells, we examined in more detail the expression profile of the MMTV-tTA transgene in major hematopoietic cell types. For this purpose, we generated MMTV-tTA TetO-H2B-GFP double transgenic mice and performed a flow cytometric analysis for GFP expression in bone marrow cells that were labeled with various markers for long-term (LT-HSC) and short-term (ST-HSC) hematopoietic stem cells (Fig. 2A; Fig. S1) as well as various lineage-specific markers (Fig. 2B; Fig. S2). This study revealed that, in addition to lymphoid cells, the MMTV-tTA is active in a significant subset of hematopoietic stem cells (HSCs) under nonselective conditions (e.g., without expression of an oncogene). Within differentiated hematopoietic lineages, the tTA is expressed in erythroid and B cells and to a lesser extend in T cells and granulocytes.

Next, we assessed possible effects caused by the expression of hyperactive Stat5 on the number of hematopoietic stem and progenitor cells. The flow cytometric analysis revealed that gain-

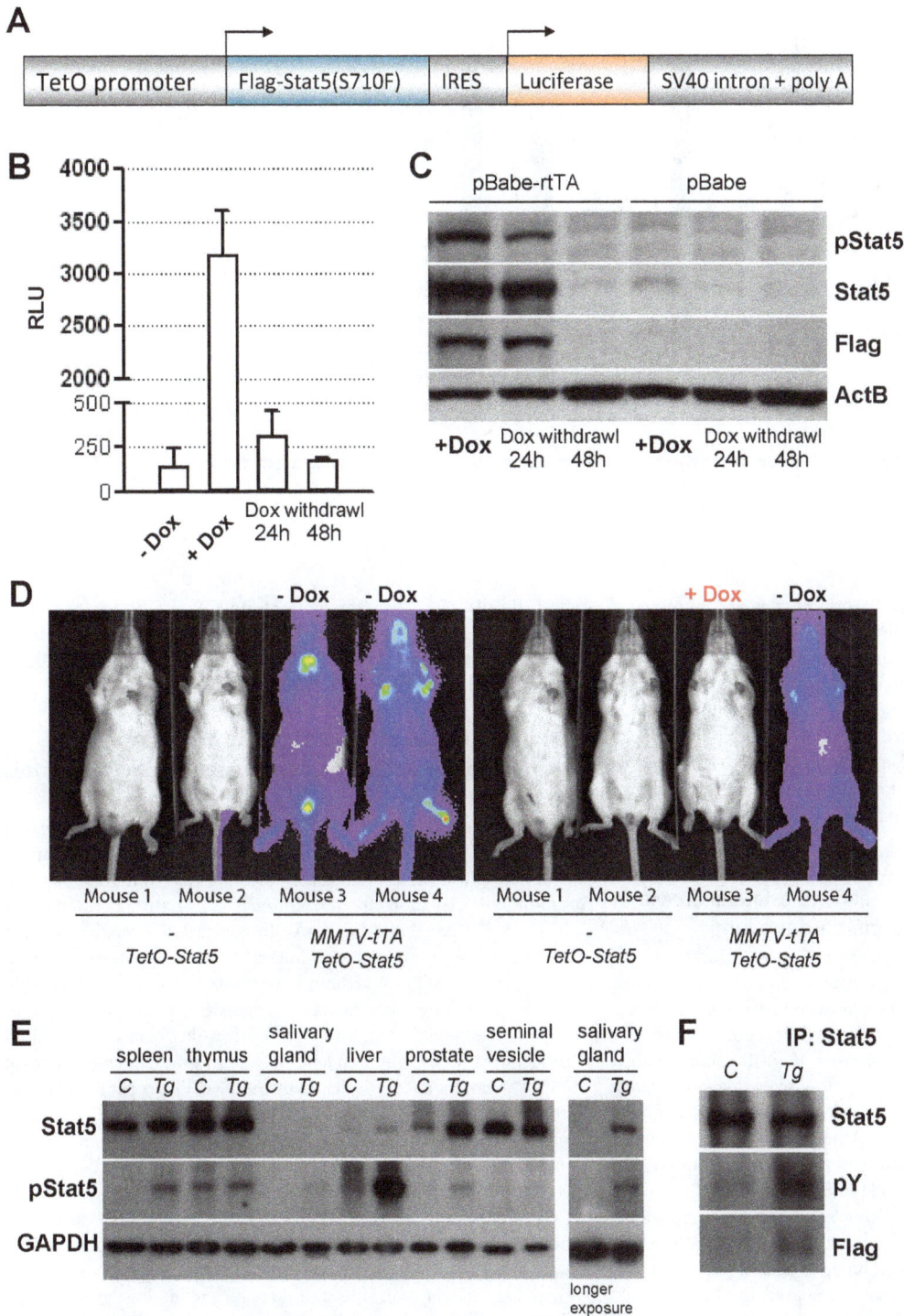

Figure 1. Generation and analysis of transgenic mice that co-express a Flag-tagged hyperactive mutant of Stat5a (S710F) and luciferase under control of the tetracycline-controlled operator (TetO). A. Schematic outline of the transgene. B. Conventional luciferase assay on mouse embryonic fibroblasts (MEFs) derived from TetO-Stat5 mice that were infected with a pBabe-rtTA-puro retroviral construct and maintained in the absence or presence of doxycycline (± Dox). C. Western blot analysis to verify the expression of Flag-tagged, exogenous Stat5 (tyrosine phosphorylated and total) in MEFs derived from TetO-Stat5 mice that were infected with the pBabe-rtTA-puro or pBabe-puro control vector. Cells were maintained in the presence of Dox, and the withdrawal of this ligand led to a significant decline in transgenic Stat5 expression within 48 hours in cells expressing the reverse transactivator protein (rtTA). D. In vivo bioluminescence imaging of two MMTV-tTA TetO-Stat5 double transgenic mice (mouse 3 and 4) and their TetO-Stat5 single transgenic controls (mouse 1 and 2) prior to administration of Dox (left panel). The right panel shows the same animals after mouse 3 was treated for 48 hours with Dox. E. Western blot to assess the expression of total and tyrosine phosphorylated Stat5 in a panel of tissues from double transgenic mice (Tg) and their TetO-Stat5 single transgenic controls (C). F. IP/Western blot for total and active Stat5 on spleen tissues of a control mouse (C) and a double transgenic animal expressing exogenous Stat5a (Tg).

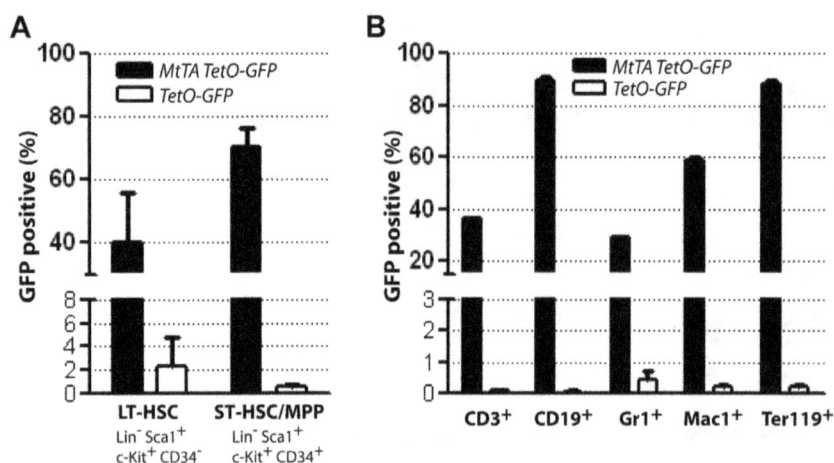

Figure 2. MMTV-tTA-mediated transactivation of TetO-responder transgenes in hematopoietic stem cells and derived lineages. Flow cytometric analysis of the presence of the GFP reporter in long-term and short-term (LT-HSC, ST-HSC) hematopoietic stem cells (**A.**) as well as lymphoid, myeloid, and erythroid cells (**B.**) in the bone marrow of MMTV-tTA TetO-GFP double transgenic mice and their TetO-GFP single transgenic controls.

of-function of Stat5 did not significantly change the number of particular stem/progenitor cells in the bone marrow (Fig. 3A, 3B, and Figures S3 and S4). In addition, the MMTV-tTA TetO-Stat5 double transgenic animals exhibited blood counts that were similar to single transgenic controls (Fig. 3C). Based on the previous report that expression of Stat5 from a retroviral vector leads to a rapid onset of multi-lineage leukemias [29], we were surprised to note that MMTV-tTA TetO-Stat5 double transgenic mice did not develop any hematopoietic malignancies within 18 months of age.

Expression of hyperactive Stat5 in mutant mice with Stat5a/b hypomorphic alleles leads to a fatal increase in circulating granulocytes that infiltrate the lung

Recent evidence suggests that a disease phenotype induced by a hyperactive Jak2/Stat5 signaling pathway is influenced by the level of signaling and the ratio of mutant to wildtype protein [35,36]. Since the level of exogenous Stat5 under regulation of the MMTV-tTA is not substantially higher than the endogenous Stat5, we next examined whether the loss of endogenous Stat5 might lead to differences in the functionality of the exogenous mutant protein. For this purpose, we generated mice that carry the MMTV-tTA and TetO-Stat5 transgenes in a Stat5 hypomorphic mutant background ($Stat5^{\Delta N/\Delta N}$). Homozygous $Stat5^{\Delta N/\Delta N}$ mice are known to express very low levels of an N-terminal truncated Stat5 protein. Overexpression of transgenic, hyperactive Stat5 in the Stat5 hypomorphic background had a profound impact on the disease-free lifespan of these animals compared to their controls (Fig. 4A). Starting at two months of age, a significant subset of MMTV-tTA TetO-Stat5 $Stat5^{\Delta N/\Delta N}$ mice started to exhibit signs of severe airway obstruction and had to be euthanized. The histopathological analysis revealed a substantial extravasation of immune cells into the lung (Fig. 4B). We also observed hematopoietic cells, albeit to a much lesser extent, in the liver and kidney (not shown). Initially, young animals exhibited a granulocyte count that was similar to controls. However, prior to or during disease onset, we observed a sharp increase in the relative number of granulocytes and a significant elevation in the total number of granulocytes in circulation (Fig. 4C). Hence, although the MMTV-tTA is only expressed in approximately one third of all Gr1-positive cells in comparison to 90% of erythroid

and B-cells (Fig. 2B), the gain-of-function of Stat5 lead to a selective elevation in the absolute number of granulocytes in circulation whereas the number of all other white and red blood cells as well as platelets remained unaltered.

Downregulation of Stat5 swiftly restores normal granulopoiesis and reverts extravasation of granulocytes to the lung

The flow cytometric analysis of splenocytes from diseased animals and their controls showed that the gain-of-function of Stat5 resulted in a moderate increase in immature and a substantial elevation in the number of mature granulocytes (Fig. 5A, left panel and Fig. S5A). In contrast, the relative amount of myeloblasts remained unchanged. While hypomorphic Stat5-deficient mice with or without exogenous Stat5 have comparable numbers in long-term hematopoietic stem cells, expression of hyperactive Stat5 led to a reduction in common myeloid progenitors (CMPs) as well as a decline in granulocyte/monocyte (GMP) and megakaryocyte/erythroid (MEP) progenitors in the bone marrow (Fig. 5A, right panel and Fig. S5B).

To assess whether the survival of circulating granulocytes, including those that extravasated into the lung, depended on the continuous expression of hyperactive Stat5, we treated MMTV-tTA TetO-Stat5 $Stat5^{\Delta N/\Delta N}$ mice with Dox after they developed high granulocyte counts and exhibited signs of airway obstruction. Within only two to seven days, the administration of Dox led to a significant decline in the number of circulating granulocytes (Fig. 5B), and after 2 weeks of treatment, the granulocyte count was normalized to the level of healthy control mice (compare to Fig. 4C). Notably, the downregulation of exogenous Stat5 also resulted in the restoration of the drastically depleted pool of CMPs and MEPs in the bone marrow (Fig. 5C and Fig. S6). Following treatment with Dox, the mice were also visibly healthier, and the histological analysis of the lungs revealed that, unlike the untreated controls, the administration of Dox resulted in a complete clearance of granulocytes from the airway system (Fig. 5D). Since treatment with another antibiotic (i.e., ampicillin) did not alleviate the severity of this condition, the reversal of the phenotype was likely a specific result of the downregulation of exogenous Stat5 and not a broader effect of the antibiotic properties of doxycycline.

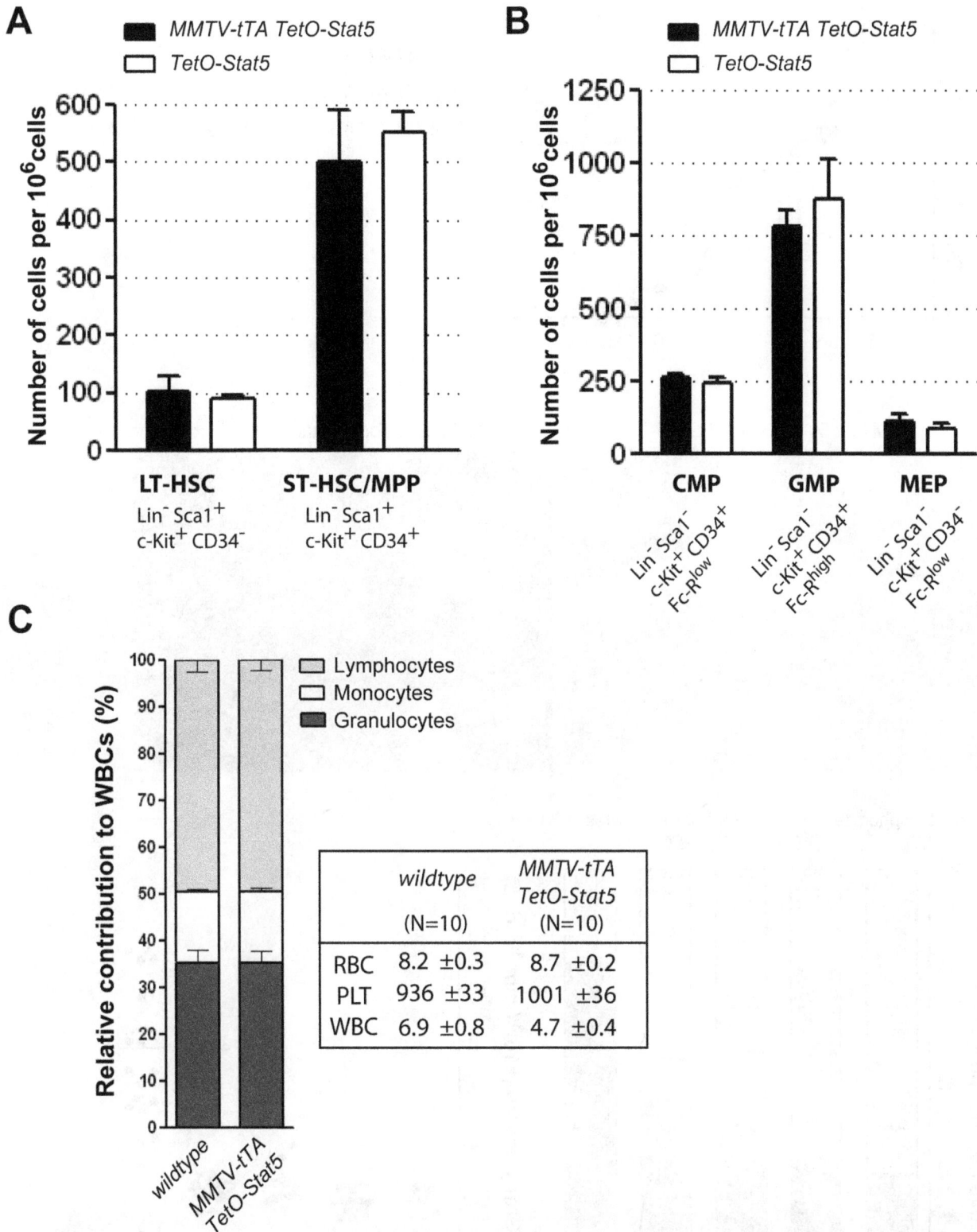

Figure 3. Mice expressing hyperactive Stat5 under control of the MMTV-tTA have normal blood cell counts. Relative numbers of long-term and short-term (LT-HSC, ST-HSC) hematopoietic stem cells (**A.**) as well as common myeloid (CMP), granulocyte-macrophage (GMP), and megakaryocyte-erythrocyte (MEP) progenitors (**B.**) in the bone marrow of MMTV-tTA TetO-Stat5 double transgenic mice and their TetO-Stat5 single transgenic controls. **C.** Absolute numbers of peripheral blood cells ($10^3/mm^3$) and relative contribution of lymphocytes, monocytes, and granulocytes to the total number of white blood cells. RBC, red blood cells; PLT, platelets; WBC, white blood cells.

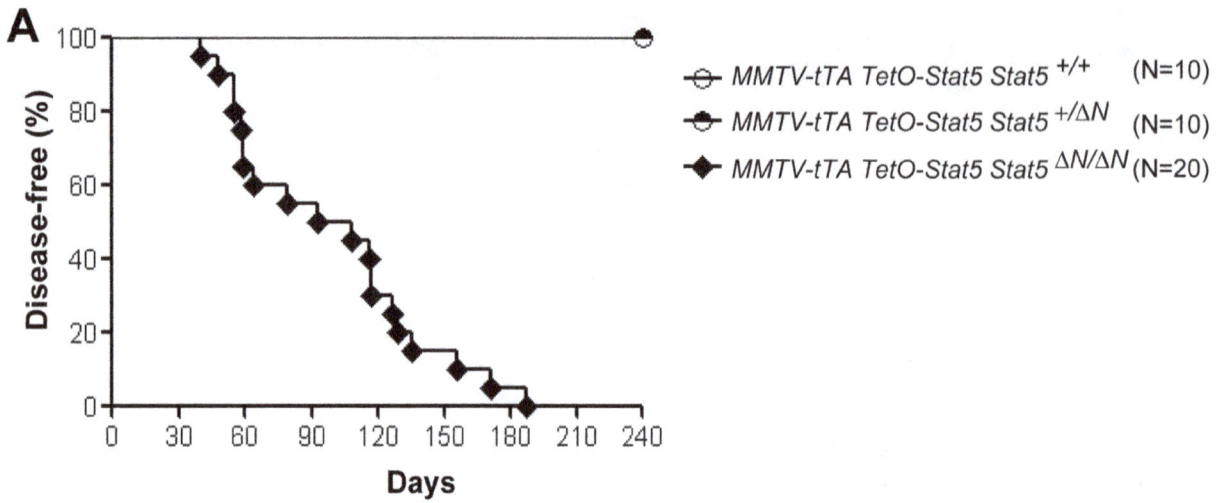

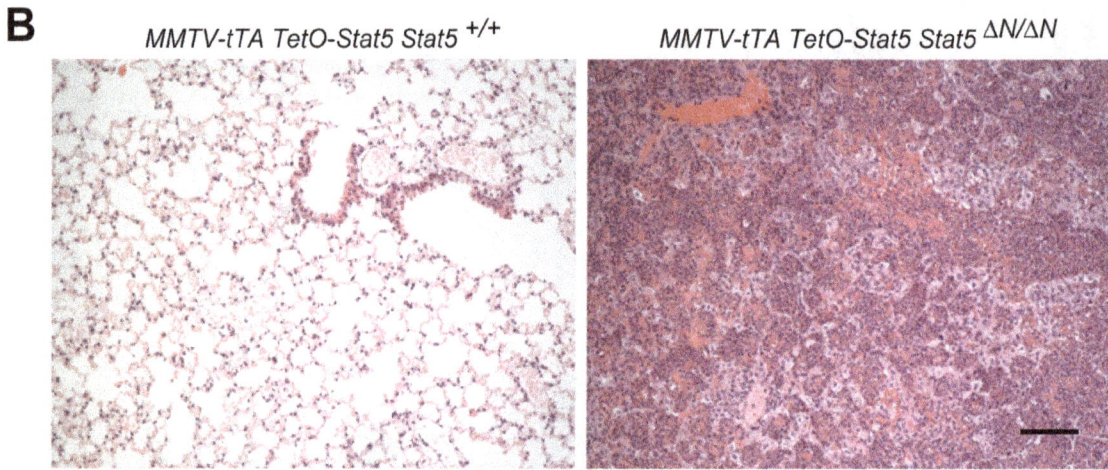

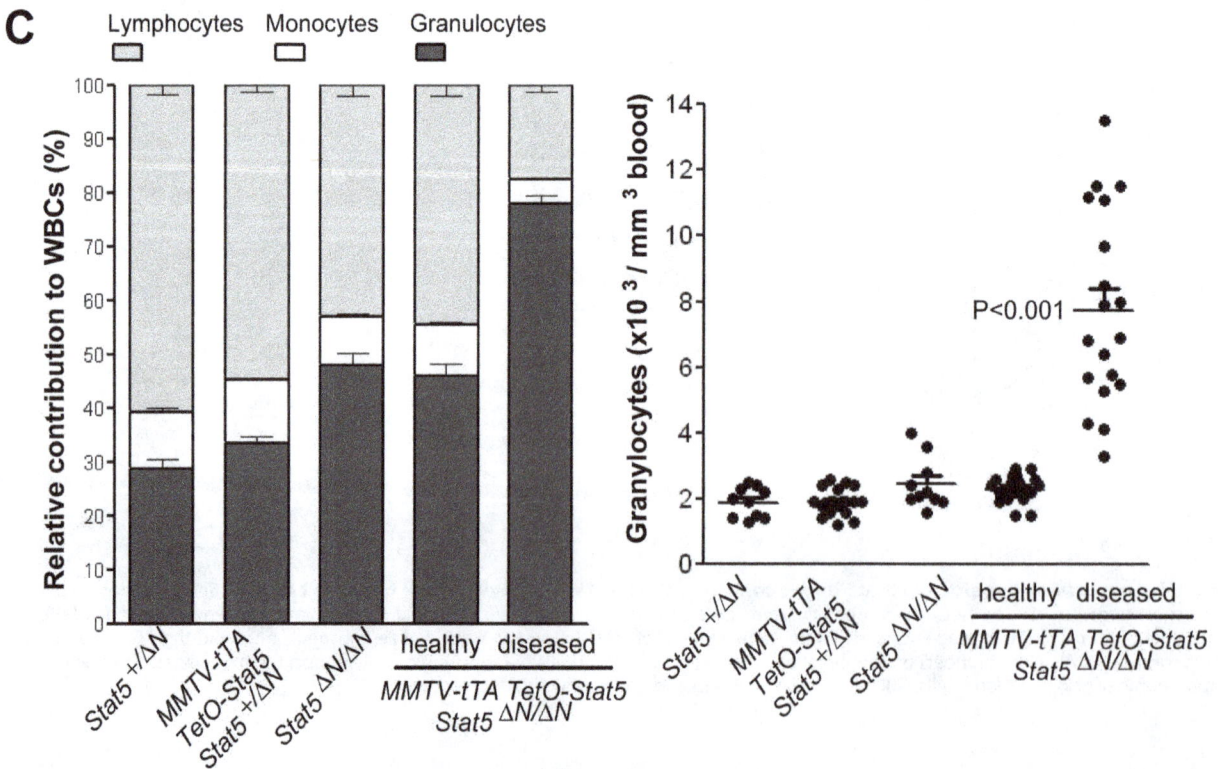

Figure 4. Gain-of-function of Stat5a in mice with hypomorphic _Stat5a/b_ alleles causes a fatal increase in circulating granulocytes. A. Kaplan–Meier curve illustrating the disease-free survival of mice expressing hyperactive Stat5 (MMTV-tTA TetO-Stat5) in the presence of two wildtype endogenous _Stat5a/b_ alleles (_Stat5$^{+/+}$_) or in animals that carry _Stat5a/b_ hypomorphic alleles in a heterozygous or homozygous configuration (_Stat5$^{+/\Delta N}$_ and _Stat5$^{\Delta N/\Delta N}$_). B. Sections of lung tissues from diseased MMTV-tTA TetO-Stat5 _Stat5$^{\Delta N/\Delta N}$_ mice and healthy MMTV-tTA TetO-Stat5 controls. Bar represents 100 μm. C. Relative contribution of lymphocytes, monocytes, and granulocytes to the total number of white blood cells (WBCs) (left panel) and total granulocyte counts (right panel) in healthy and diseased MMTV-tTA TetO-Stat5 _Stat5$^{\Delta N/\Delta N}$_ mice and controls (_P_ value, _t_ test).

Animals receiving Dox remained disease free during a prolonged treatment period. However, the withdrawal of this ligand led to a recurrence of elevated granulocyte counts within two weeks (not shown). This suggests that the downregulation of hyperactive Stat5 did not cause a selective elimination of precursors that are the cellular basis for accelerated disease recurrence upon re-expression of active Stat5.

Hyperactive Stat5 promotes excessive granulopoiesis in a cell autonomous manner

It has been reported that Stat5 hypomorphic mutant mice (_Stat5$^{\Delta N/\Delta N}$_) mice develop mild neutrophilia through a cell extrinsic mechanism, i.e. an overproduction of granulocyte colony-stimulating factor in liver endothelial cells [37] While the relative amount of granulocytes among WBCs was increased on average by 10–15%, only two of these mutant animals actually exhibited a marginal increase in the total number of granulocytes (Fig. 4C, right panel). Severe granulophilia was only observed in Stat5 hypomorphic mutant mice when they also express hyperactive Stat5. To assess whether the disease onset in these mice was due to the proposed extrinsic production of cytokines or a Stat5-mediated intrinsic mechanism, we transplanted bone marrow cells from MMTV-tTA TetO-Stat5 _Stat5$^{\Delta N/\Delta N}$_ mice as well as MMTV-tTA TetO-Stat5 _Stat5$^{\Delta N/wt}$_ controls into conditioned wildtype mice. Recipient animals expressing hyperactive Stat5 in the hypomorphic mutant background developed high relative and absolute granulocyte counts that were similar to donor mice (Fig. 6A and 6B). This suggested that the excessive granulopoiesis was likely a direct effect of the gain-of-function of Stat5 within myeloid progenitors and their differentiated descendants. This cell intrinsic mechanism does not require an increased production of G-CSF in extra-hematopoietic cell types. This assumption is supported by the fact that the treatment of diseased animals with Dox resulted in the clearance of circulating granulocytes and restoration of normal granulopoiesis that is indistinguishable from control animals (Fig. 6C). Although wildtype recipient mice expressing mutant Stat5 exhibited some abnormalities in their lungs (Fig. 6D), the extravasation of granulocytes to the airway system was less widespread compared to the donors.

It has been reported previously that Stat5 is essential for GM-CSF signaling, and Stat5a/b conditional double knockout mice show decreased numbers of mature neutrophils [12]. Since we observed significantly elevated numbers of circulating granulocytes that also accumulated over time in the lungs of our Stat5 overexpression model, we assessed next whether the gain-of-function of Stat5 promoted a prolonged survival in addition to the increased production and release of these particular hematopoietic cells into circulation. For this purpose, we isolated peripheral white blood cells (WBCs) through retro-orbital blood draw from diseased MMTV-tTA TetO-Stat5 _Stat5$^{\Delta N/\Delta N}$_ mice. These WBCs had a relative contribution of more than 90% granulocytes and were maintained _ex vivo_ for up to 96 hours in the presence and absence of Dox and GM-CSF (Fig. 7). After 96 hours in culture, less than 10% of seeded cells lacking Stat5 (<2,000 of 20,000 plated cells per well) were still viable (Fig. 7 B, open bars). Even when treated with GM-CSF, these cells cannot be effectively maintained _in vitro_.

In contrast, more than twice the amount of cells that express Stat5 were alive during that period, and we observed a clear synergistic effect of Stat5 expression and growth factor treatment, which yielded a significant increase in cell survival (Fig. 7 B, solid bars). In a separate experiment, we isolated hematopoietic cells from the lung, and we were able to maintain them on top of lung-derived fibroblasts for at least 12 weeks. This suggested that many extravasated cells were alive and a subset may have acquired an immortal state _in vivo_ or after a short term in culture. The flow cytometric analysis revealed that the majority of these proliferating cells were immature granulocytes, and some cells underwent differentiation into mature granulocytes in culture (Fig. 8). The notion that myeloid precursors expressing hyperactive Stat5 show accelerated proliferation and an increase in survival might be supported by the fact that mice expressing exogenous Stat5 exhibited a significant increase in the activation of the _Cyclin D1_ gene, i.e. a direct transcriptional target of Stat5 in hematopoietic cells [38], and elevated levels of _Bcl2_ (Fig. 9).

Discussion

As a signal transducer and transcription factor downstream of many cytokines, Stat5 acts as a mitogenic factor in most cell types, including hematopoietic progenitors. It has been recently suggested that activation of Stat5 maintains the quiescence of hemato-poietic stem cells (HSCs) during steady-state hematopoiesis [39]. The conditional deletion of both Stat5 isoforms in HSCs using Mx1-Cre in adult mice led to increased stem cell cycling and depletion of the long-term HSC pool. We did not observe any significant changes in the numbers of LT-HSCs and ST-HSCs in MMTV-tTA TetO-Stat5 transgenic mice. In part, this might be the result of a mosaic expression pattern of the MMTV-tTA, which appears to be active in only 40% of LT-HSCs as opposed to 70% of ST-HSCs in this transgenic line. Despite expression of hyperactive Stat5 in over 90% of erythroid and B cells as well as the majority of macrophages, MMTV-tTA TetO-Stat5 transgenic mice did not develop multi-lineage leukemia similar to the retroviral gene transfer and transplantation model generated by Moriggl and colleagues [29]. We initially assumed that the transactivation of the TetO-Stat5 transgene by the MMTV-tTA might not have been efficient enough, and we, therefore, crossed both TetO-Stat5 responder lines with Eμ-tTA and Tal1-tTA transgenics that have been utilized to generate multiple leukemia and lymphoma models [40,41]. None of the resulting double transgenics developed hematopoietic malignancies within 12 months of age. We also used retroviral and lentiviral gene transfer to express the tTA in HSCs and differentiated lineages in bone marrow cells derived from TetO-Stat5 transgenics to induce a cytokine storm that is generally associated with the transplantation of infected cells into conditioned wildtype recipient mice. Despite verification of the permanent engraftment of the infected bone marrow cells, recipient animals did not develop leukemia or lymphoma after a prolonged latency. Additionally, we examined a possible effect of the genetic background on Stat5-induced neoplastic transformation. For this purpose, we transferred the MMTV-tTA and one of the TetO-Stat5 transgenes (line 11676) into the C57/Bl6 strain through eight or more backcrosses, but

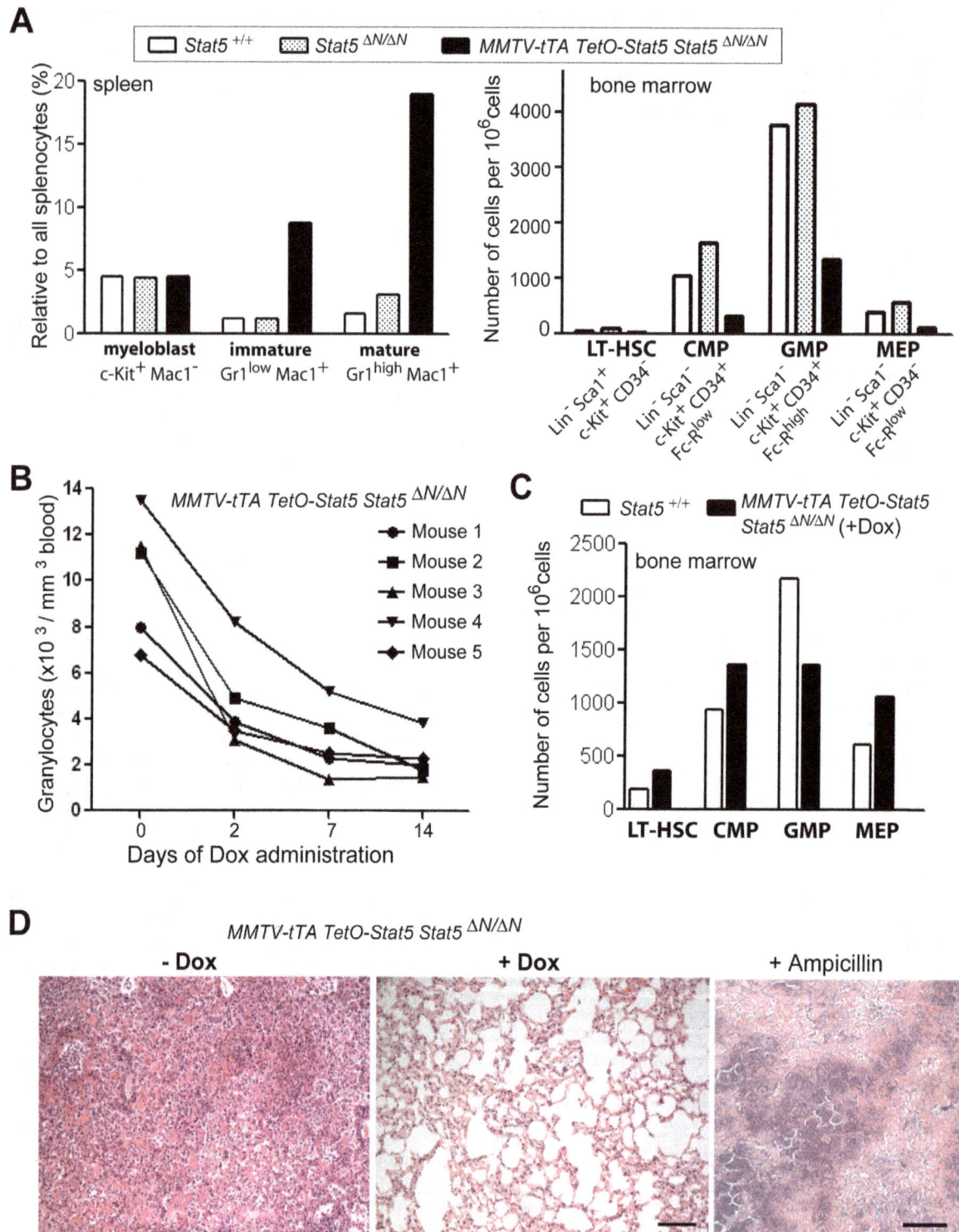

Figure 5. Downregulation of exogenous Stat5a leads to clearance of excessive granulocytes in circulation and in the lung as well as restoration of a normal progenitor pool. A. Relative contribution of myeloblasts and immature and mature granulocytes among the total number of splenocytes (left panel) and numbers of long-term hematopoietic stem cells (LT-HSC) as well as common myeloid (CMP), granulocyte-macrophage (GMP), and megakaryocyte-erythrocyte (MEP) progenitors in the bone marrows of diseased MMTV-tTA TetO-Stat5 $Stat5^{\Delta N/\Delta N}$ mice and their controls. B. Changes in the number of granulocytes in five diseased animals that were treated for 14 days with Dox. C. Numbers of hematopoietic stem cells and progenitors in a wildtype control and a diseased MMTV-tTA TetO-Stat5 $Stat5^{\Delta N/\Delta N}$ mouse following Dox administration. D. Sections of lung tissues from diseased MMTV-tTA TetO-Stat5 $Stat5^{\Delta N/\Delta N}$ mice prior to and following downregulation of Stat5 (±Dox). The right panel shows the lung of a mouse treated with ampicillin to control for nonspecific effects of Dox as an antibiotic as opposed to its direct control of Stat5 expression. Bars represent 100 μm (left, middle) and 200 μm (right).

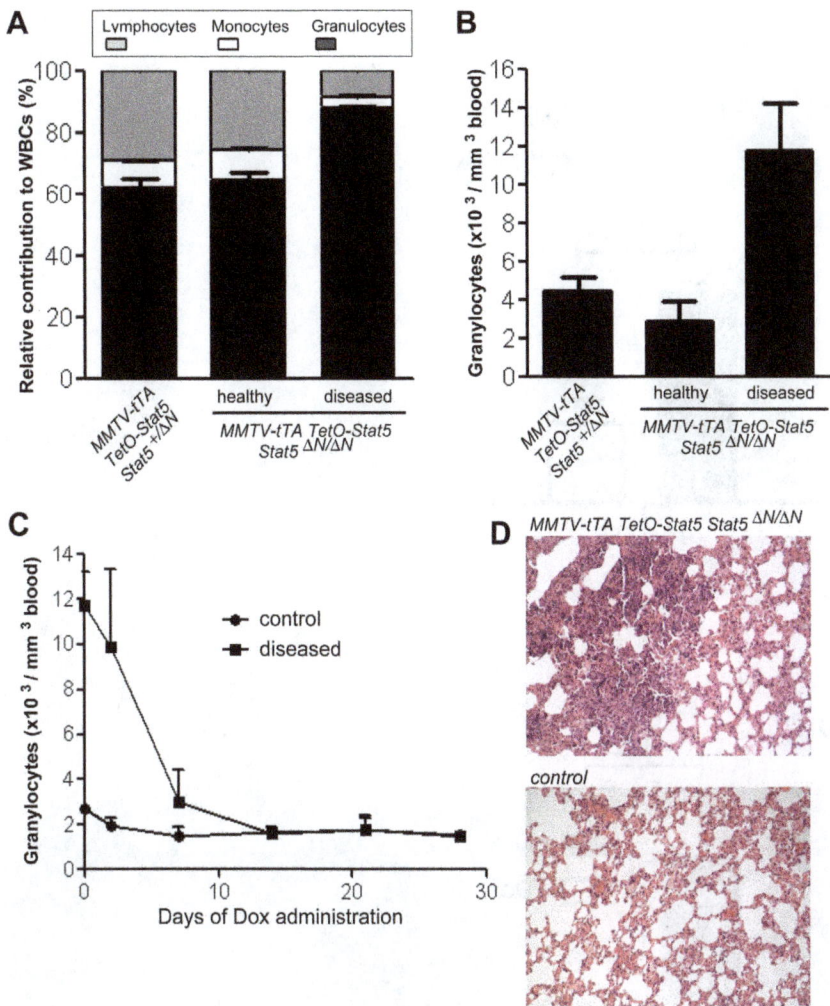

Figure 6. Excessive granulopoiesis is controlled by hyperactive Stat5 in a cell autonomous manner. A. Relative contribution of lymphocytes, monocytes, and granulocytes to the total number of white blood cells (WBCs) and **B.** total granulocyte counts in healthy and diseased recipients that were engrafted with bone marrow cells from MMTV-tTA TetO-Stat5 $Stat5^{\Delta N/\Delta N}$ mice and MMTV-tTA TetO-Stat5 $Stat5^{+/\Delta N}$ controls. **C.** Changes in the number of granulocytes in three diseased recipient mice and controls that were treated for 28 days with Dox. **D.** Corresponding sections of lung tissues from diseased recipient mice and their normal controls.

none of the double transgenic mice developed hematopoietic abnormalities within 12 months of age. The results of our combined efforts to generate a reversible, Stat5-induced leukemia model indicate that this signal transducer appears to be a rather weak oncogene that requires additional, cancer-initiating mutations to trigger a malignant phenotype. Hence, our observations are similar to Eμ-Stat5 mice that exhibit a very low penetrance in cancer formation [30]. This is perhaps not unique to the hematopoietic system. Evidently, the expression of wildtype or constitutively active Stat5 in the mammary epithelium is only weakly oncogenic [42–44]. Moreover, despite expression of mutant Stat5 in other secretory tissues such as the prostate and salivary gland in our model (see Fig. 1E), we also did not see any consistent malignant changes in these organs in aging MMTV-tTA TetO-Stat5 transgenics.

While expression of mutant Stat5 in the presence of the wildtype, endogenous protein did not seem to cause any major phenotypic abnormalities, the transfer of the MMTV-tTA and TetO-Stat5 transgenes into the Stat5 hypomorphic mutant background ($Stat5^{\Delta N/\Delta N}$) had a profound effect on the differenti-

ation of hematopoietic cells and the wellbeing of the mice. The results of our study indicate that expression of hyperactive Stat5a in the presence of hypomorphic $Stat5a/b$ alleles promotes the numeric expansion of immature and mature granulocytes and a synchronous depletion of progenitors (i.e., CMPs and MEPs). In addition, the gain-of-function of Stat5 leads to an increase in the survival of peripheral granulocytes and their lethal extravasation to multiple organs, most prominently, the lung. All of these major phenotypes are reversible through downregulation of the mutant Stat5. Our observations confirm previous reports that highlight the importance of Stat5 in maintaining a normal balance between lymphoid and myeloid cells during hematopoiesis as well as essential functions of this transcription factor downstream of GM-CSF signaling [7,12,13,37]. It might be possible that the low-level expression of N-terminally truncated, endogenous Stat5 plays some role in the initiation of the phenotypes. We confirmed that $Stat5^{\Delta N/\Delta N}$ mice have a slight increase in the relative number of granulocytes. This was suggested to be a consequence of emergency granulopoiesis that might overcompensate for the intrinsic survival defect in this hematopoietic lineage [37].

A

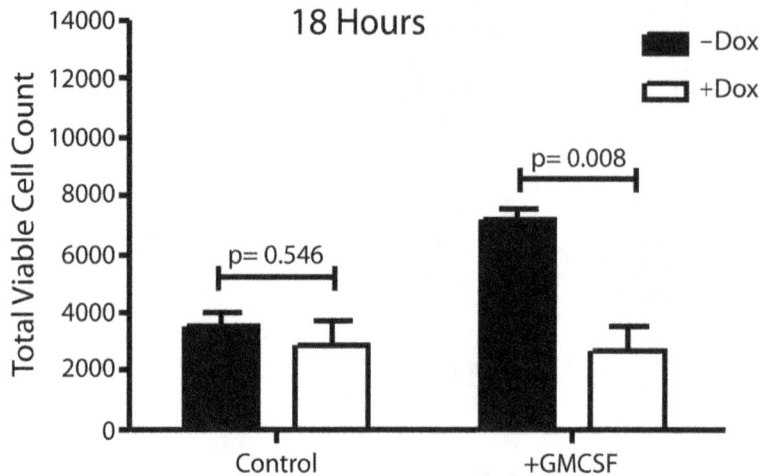

B

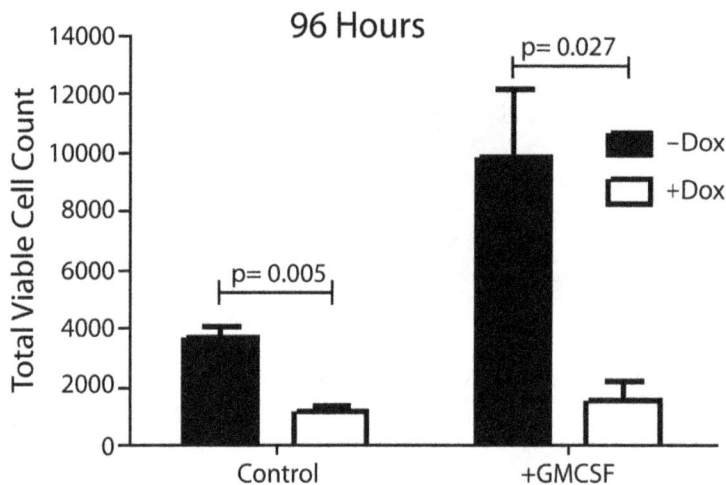

Figure 7. Gain-of-function of Stat5 promotes survival of circulating granulocytes ex vivo. Viable cell counts of peripheral WBCs that consisted of more than 90% granulocytes from diseased MMTV-tTA TetO-Stat5 $Stat5^{\Delta N/\Delta N}$ mice that were maintained in culture for 18 (**A.**) or 96 hours (**B.**) in the presence and absence of Dox and GM-CSF (*P* value, *t* test).

Expression of hyperactive Stat5 might act in a synergistic manner and delay the turnover of circulating granulocytes and thereby promote the severity of the phenotype. But this process might not depend exclusively on the elevated secretion of G-CSF from liver endothelial cells in $Stat5^{\Delta N/\Delta N}$ mutants as previously proposed [37]. The transplant experiments in this study suggest that hyperactive Stat5 is still able to promote excessive granulopoiesis in a cell autonomous manner. This notion is supported by the reversibility of the severe consequences of Stat5 overexpression in the MMTV-tTA TetO-Stat5 $Stat5^{\Delta N/\Delta N}$ transgenics and their derived hematopoietic transplants in wildtype recipients.

The increase in granulocyte count and extravasation into the lung were typically observed after two to six months. This might imply that additional genetic or epigenetic changes were required to promote disease manifestation. In contrast to the initiation of the disease, the reactivation of the transgene expressing mutant

Stat5 following complete disease remission led to a re-accumulation of granulocytes in circulation after just 14 days. This suggests that precursors, which are highly responsive to the expression of mutant Stat5 and that might carry secondary hits, can survive the downregulation of the transgene. The subsequent reactivation of exogenous Stat5 then leads to an accelerated expansion of immature and mature granulocytes. The implication of this observation is that Stat5 needs to be effectively and permanently inactivated to prevent rapid disease recurrence.

Despite clear evidence that exogenous Stat5 promotes extended survival of peripheral granulocytes, the underlying cause for their extravasation into the lung remains unknown. Since the lungs of diseased mice resemble necrotizing pneumonitis on the histopathological level, we initially assumed the presence of foreign organisms as disease initiating agents. A gram, Ziehl–Neelsen, and silver staining was performed and evaluated by Dr. Robert

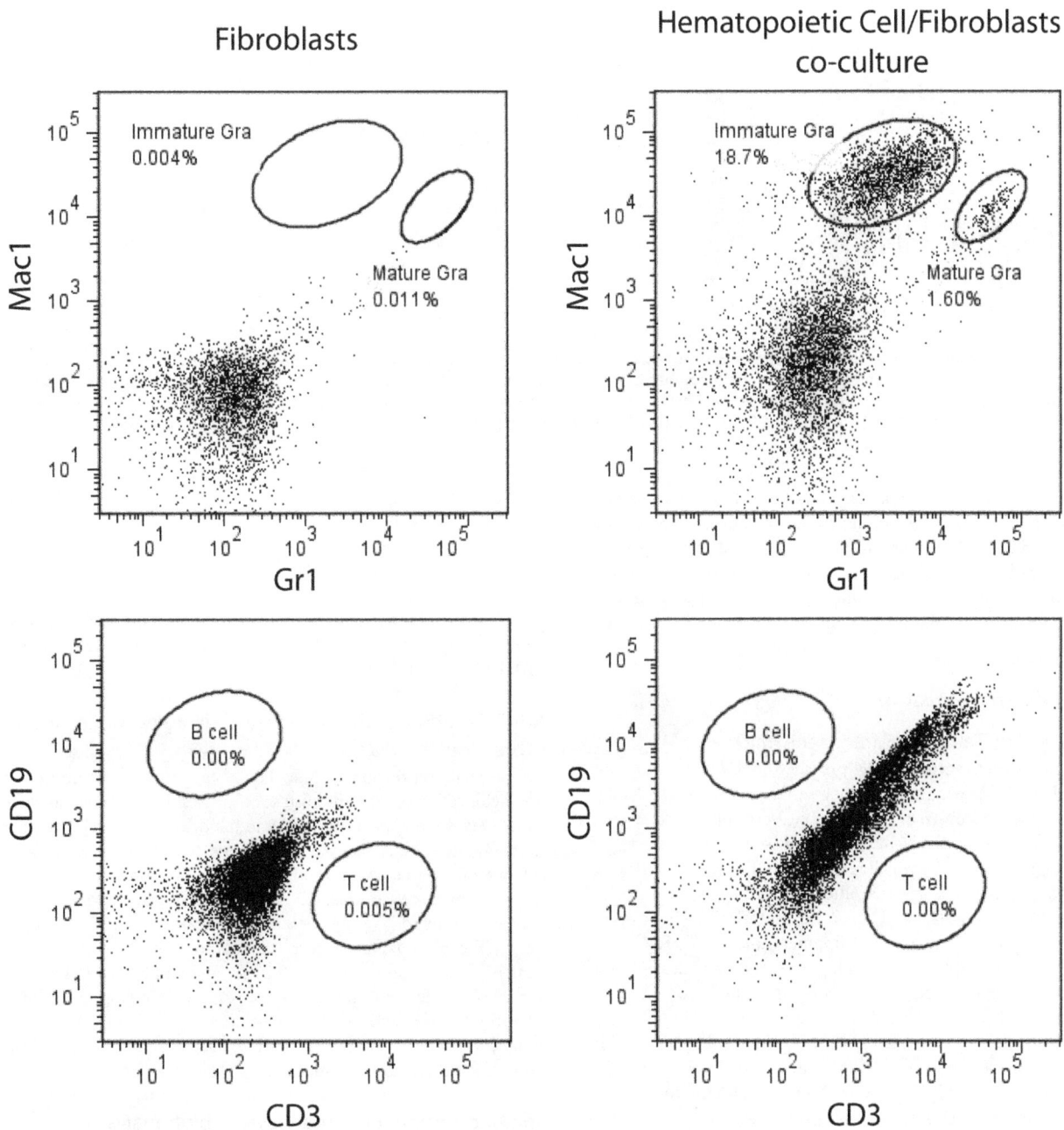

Figure 8. The majority of lung-derived hematopoietic cells expressing exogenous Stat5 that can be maintained for a prolonged period in culture are immature granulocytes. Flow cytometric analysis of various lineage markers on lung-derived hematopoietic cells that were maintained *ex vivo* together with pulmonary fibroblasts for 12 weeks.

Cardiff (UC Davis), and all tests were negative, suggesting that various types of bacteria, pseudomonas, legionella, and fungi like Pneumocystis and Candida did not seem to play a significant role in disease manifestation. At this point, however, we cannot exclude that the extravasation is trigged by a variety of environmental toxins. In addition to pathogens, we considered that a deregulated expression of Stat5 in the lung epithelium might fuel a local production of cytokines that provoke excessive extravasation. However, a quantitative RT-PCR assay to assess the expression of various cytokines did not reveal any significant differences among disease-free and diseased animals with or without expression of exogenous Stat5. Regardless of the underlying events that trigger

the massive accumulation of immune cells in the lungs of mice expressing hyperactive Stat5a in a Stat5a/b deficient background, our observations might potentially have clinical implications in humans. As mentioned in the *Introduction* section, chronic lung disease and immune dysfunction appear to be common among patients that are deficient in STAT5B [17]. It is very likely that the remaining STAT5A plays a significant role in this process. Unfortunately, the expression and activation of STAT5A has not been specifically assessed in these patients, but this particular STAT5 isoform appears to be the predominant target for GM-CSF signaling [45]. More importantly, STAT5 has been shown to be an essential mediator for GM-CSF signaling and survival of

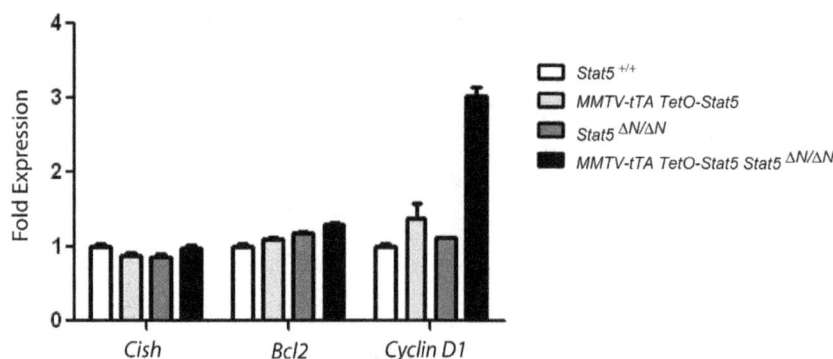

Figure 9. Hematopoietic cells from mice with a gain-of-function of Stat5 show a significant increase in the transcriptional activation of *Cyclin D1*. Quantitative real-time RT-PCR to assess the transcriptional activation of *Cyclin D1*, *Bcl2*, and *Cish* in bone marrow derived hematopoietic cells of experimental mice that express hyperactive Stat5 in the absence or presence of *Stat5a/b* hypomorphic alleles (MMTV-tTA TetO-Stat5 *Stat5*[+/+] and MMTV-tTA TetO-Stat5 *Stat5*[ΔN/ΔN]). Wildtype littermates (*Stat5*[+/+] and Stat5 hypomorphic mutants (*Stat5*[ΔN/ΔN]) served as controls.

lung-derived granulocytes in a veterinary animal model for asthma [46]. If lack of STAT5B results in a hyperactivation of the remaining STAT5A in the myeloid lineage, targeting JAK2/STAT5 locally might alleviate some of the severe symptoms in STAT5B deficient patients for whom there are currently limited treatment options available to improve the complex immune defects.

Materials and Methods

Generation of TetO-Stat5 transgenic mice

The pTet-Splice vector (kindly provided by Dr. Jun-Lin Guan, Cornell University) was modified by inserting of a short DNA oligo into the *Xho*1 site upstream of the TetO-promoter sequence to introduce additional *Swa*I/*Not*I restriction sites. The Flag-tagged mutant of *Stat5a* (S710F), which is a gift from Dr. Richard Moriggl (LBI-CR, Vienna), was inserted into the *Sma*I site in front of an IRES-Luciferase construct. The entire Stat5-IRES-Luc cassette was cloned as a blunted *Not*I/*Sac*I fragment into the *Eco*RV site of the modified pTet-Splice vector. The transgene was released from the vector backbone using *Not*I, gel purified, and injected into FVB zygotes by the UNMC Mouse Genome Engineering Core Facility. Fifteen transgenic founder lines were obtained following pronuclear injection, and two lines (11651 and 11676) were used for experiments described here. The Mouse Genome Informatics (MGI) nomenclature for both of these strains is Tg(tetO-Stat5a*S710F,-luc)11651Kuw and Tg(tetO-Stat5a*S710F,-luc)11676Kuw. Offspring were screened by genomic PCR with primers recognizing the junction (approximately 345 bps) between the TetO/CMVmin promoter and the *Stat5a* cDNA (primer 1978: 5′-CCG TCA GAT CGC CTG GAG ACG-3′; primer 1977: 5′-GCC TGA ATC CAG CCC GCC ATG-3′).

Other genetically modified mouse strains

MMTV-tTA [129/B6.Cg-Tg(MMTVtTA)1Mam] transgenic mice [31] were kindly provided by Dr. Lothar Hennighausen (NIH). In addition to the experiments described here, we recently performed a more detailed analysis of the expression profile of this strain in extrahematopoietic tissues [47]. TetO-H2B-GFP [Tg(tetO-HIST1H2BJ/GFP)47Efu/J] transgenic mice [48] were obtained from the Jackson Laboratory. Genetically modified mice that carry both targeted alleles for *Stat5a* and *Stat5b* [Stat5a[tm1Jni]/Stat5b[tm2Jni]] on chromosome 11 [3], designated here as *Stat5*[ΔN/ΔN], were kindly provided by Dr. James N. Ihle (St. Jude Children's

Research Hospital). The MMTV-tTA transgene and *Stat5a/b* targeted alleles were transferred into an FVB background (n = 8). This study was carried out in strict accordance with the recommendations in the Guide for the Care and Use of Laboratory Animals of the National Institutes of Health. The protocol was approved by the Institutional Animal Care and Use Committee of the University of Nebraska Medical Center (IACUC#: 06-082-03).

Administration of Dox and in vivo imaging of luciferase expression

Complete repression of the TetO-Stat5-IRES-Luc transgene was achieved through administration of freshly prepared doxycycline (Dox; Sigma, St. Louis, MO) in the drinking water (2 mg/ml supplemented with 50 mg/ml sucrose). The expression and activity of the luciferase reporter gene was monitored using *in vivo* bioluminescence imaging machine (IVIS200, Caliper Life Sciences, Alameda, CA) as described previously [49,50]. According to the manufacture's recommendations, luciferin (1 mg D-luciferin potassium salt in 0.21ml 1× PBS) was injected intraperitoneally ten minutes prior to the imaging procedure. The mice were kept under anesthesia (isoflurane) during the acquisition of the images that were collected at intervals ranging from ten seconds to four minutes.

Immunoprecipitation and western blot analysis

The preparation of whole-cell extracts of clarified cell lysates and tissues homogenates as well as the experimental procedures for immunoprecipitation (IP) and western blot analysis were described in detail elsewhere [51]. The following antibodies were used: anti-Stat5 antibodies E289 (Abcam, Cambridge, MA) and N-20 as well as anti-β-actin (I-19) (Santa Cruz Biotechnology, Santa Cruz, CA); anti-pStat5 antibody (Cell Signaling Technology, Danvers, MA); anti-Flag antibody M2 (Sigma, St. Louis, MO); anti-pY antibody 4G10 (Millipore, Billerica, MA); and anti-GAPDH antibody D16H11 (Cell Signaling Technology, Danvers, MA).

Histology

Lung tissues from healthy and diseased mice were fixed overnight at 4°C in buffered formalin, repeatedly washed in 1× PBS, and paraffin embedded. Histological sections were prepared and stained with hematoxylin and eosin (H&E) at the UNMC Tissue Sciences Core Facility. Images of histological slides were taken on a Zeiss AxioImager microscope (Carl Zeiss, Inc., Germany) equipped with a SPOT FLEX camera (Diagnostic Instruments, Inc., Sterling Heights, MI).

Peripheral blood cell counts

Approximately 50 µl of blood was collected from the retro-orbital sinus of mice into a tube containing EDTA using capillary tubes. The blood cell counts were measured on a Scil Vet ABC hematology analyzer (scil animal care company Ltd., Gurnee, IL).

Flow cytometry analysis

All hematopoietic cells used for flow cytometric analysis were prepared freshly. Blood was drawn from the retro-orbital venous of anesthetized mice. Bone marrow cells were flushed from tibias and femurs. The cells of spleen or thymus tissues were squeezed through a 70 µm cell strainer (Becton Dickinson, Franklin Lakes, NJ) with the plunger of a syringe. Red blood cells were removed from these samples using the erythrocyte lysing kit WBL1000 VitaLyse ® from BioE (St. Paul, MN) according to the manufacturer's protocol. The remaining nucleated cells were then washed once with 1× PBS, resuspended in 1× PBS/1% FBS and stained with saturating amounts of antibodies against cell surface markers of various hematopoietic cells. For the cell linage analysis, the following antibodies were used: CD3-Alexa 647, CD19-Percp-Cy5.5, Ly6G (Gr-1)-APC-Cy7, CD11b (Mac-1)-Alexa 700, and Ter119-PE-Cy7. All the antibodies were purchased from BD Biosciences or eBioscience. For the analysis of stem or progenitor cells, the bone marrow cells were first purified by lineage depletion, using a magnetically labeled cocktail of biotinylated antibodies against lineage antigens as well as anti-biotin microbeads and a magnetic separator (Miltenyi Biotec, Auburn, CA). Subsequently, cells were labeled with APC-conjugated streptavidin (Invitrogen, Molecular Probe, Eugene, OR), and stained with Ly-6A/E (Sca-1)-Percp-Cy5.5, CD117 (c-kit)-PE-Cy7 as well as additional antibodies to define hematopoietic stem cells and progenitors as described [52–55]. The flow cytometry was performed on an LSRII and analyzed with the DIVA software (Becton Dickinson, Franklin Lakes, NJ).

Transplantation of bone marrow cells into wildtype mice

Bone marrow cells were harvested from the femurs and tibias of transgenic mice and resuspended in 1× PBS. 5×10^6 cells per 0.2 ml were injected into the tail veins of sublethally irradiated (1,300 rads) FVB recipient mice. Mice were housed in micro-isolator cages under sterile conditions.

Ex vivo cell cultures and survival assay

Peripheral white blood cells (WBCs) were isolated through retro-orbital blood draw from diseased transgenic mice, and erythrocytes were subsequently lysed using the VitaLyse® kit. Purified WBCs were comprised of 95% granulocytes according to hematological analysis (Scil Vet ABC hematology analyzer). 20,000 viable WBCs were plated in triplicates into 96-well plates, and cells were maintained in RPMI 1640 medium in the presence or absence of 40 ng/ml GM-CSF (PeproTech, Rocky Hill, NJ) and/or 10 µg/ml doxycycline. Cell viability was then determined after 18 and 96 hours using the trypan blue exclusion assay.

Quantitative real-time RT-PCR

Total RNA was extracted from bone marrow cells using standard guanidinium thiocyanate-phenol-chloroform extraction. A Superscript II kit from Invitrogen with Oligo(dT) primers was utilized to perform the fist strand synthesis. Quantification of *Bcl2*, *Cyclin D1*, *Cish* and *Gapdh* expression was performed using iQ

SYBR green Supermix (Bio-Rad) and gene-specific primer sets (*Bcl2*: 5′-CTC GTC GCT ACC GTC GTG ACT TCG-3′ and 5′-CAG ATG CCG GTT CAG GTA CTC AGT C-3′; *Cyclin D1*: 5′- CAG ACG GCC GCG CCA TGG AA-3′ and 5′- AGG AAG TTG TTG GGG CTG CC-3′; *Cish*: 5′- GCA GAG AAT GAA CCG AAG GTG C-3′ and 5′- GGA AGC TAG AAT CGG CGT ACT C-3′; *Gapdh*: 5′- GTG TCC GTC GTG GAT CTG ACG-3′ and 5′- CAA CCT GGT CCT CAG TGT AGC-3′). The quantitative PCRs (qPCRs) were carried out in triplicate in a CFX96 real-time PCR detection system (Bio-Rad, Hercules, CA). The expression values obtained were normalized against *Gapdh*, and standard curves were generated from the same samples.

Statistical analysis

All graphic illustrations and statistics were performed with Prism 5 software (GraphPad Software, Inc., La Jolla, CA). Data are expressed as mean ±SD unless otherwise indicated and were compared using an unpaired Student t test. A P of less than .05 was considered significant.

Supporting Information

Figure S1 **Representative flow cytometric blots showing the presence of GFP in hematopoietic stem cells of MMTV-tTA TetO-GFP double transgenic mice and their TetO-GFP single transgenic controls.** These blots correspond to the bar graph shown in Fig. 2A.

Figure S2 **Representative flow cytometric blots showing the presence of GFP in differentiated hematopoietic lineages of MMTV-tTA TetO-GFP double transgenic mice and their TetO-GFP single transgenic controls.** These blots correspond to the bar graph shown in Fig. 2B.

Figure S3 **Representative flow cytometric blots of the relative numbers of long-term and short-term (LT-HSC, ST-HSC) hematopoietic stem cells in the bone marrow of MMTV-tTA TetO-Stat5 double transgenic mice and their TetO-Stat5 single transgenic controls.** These blots correspond to the quantitative analysis shown in Fig. 3A.

Figure S4 **Representative flow cytometric blots of common myeloid (CMP), granulocyte-macrophage (GMP), and megakaryocyte-erythrocyte (MEP) progenitors in the bone marrow of MMTV-tTA TetO-Stat5 double transgenic mice and their TetO-Stat5 single transgenic controls.** These blots correspond to the quantitative analysis shown in Fig. 3B.

Figure S5 **Representative flow cytometric blots of the contribution of myeloblasts and immature and mature granulocytes among the total number of splenocytes (A) and numbers of long-term hematopoietic stem cells (LT-HSC) as well as common myeloid (CMP), granulocyte-macrophage (GMP), and megakaryocyte-erythrocyte (MEP) progenitors (B) in the bone marrows of diseased MMTV-tTA TetO-Stat5 $Stat5^{\Delta N/\Delta N}$ mice and their controls.** The blots shown in panel A correspond to the bar graph illustrated in the left panel of Fig 5A. The blots from the backgating shown in panel B correspond to the bar graph shown in the right panel of Fig 5A.

Figure S6   Representative flow cytometric blots of long-term hematopoietic stem cells (LT-HSC) as well as common myeloid (CMP), granulocyte-macrophage (GMP), and megakaryocyte-erythrocyte (MEP) progenitors in the bone marrows of a wildtype control and a diseased MMTV-tTA TetO-Stat5 $Stat5^{\Delta N/\Delta N}$ mouse following Dox administration. These backgated blots against the total number of bone marrow cells correspond to the bar graph shown in Fig 5C.

Author Contributions

Conceived and designed the experiments: WcL JWS KUW. Performed the experiments: WcL JWS BAC AAT. Analyzed the data: WcL JWS KUW. Wrote the paper: K.-U.W.

References

1. Liu X, Robinson GW, Wagner KU, Garrett L, Wynshaw-Boris A et al. (1997) Stat5a is mandatory for adult mammary gland development and lactogenesis. Genes Dev 11: 179-186.

2. Udy GB, Towers RP, Snell RG, Wilkins RJ, Park SH et al. (1997) Requirement of STAT5b for sexual dimorphism of body growth rates and liver gene expression. Proc Natl Acad Sci U S A 94: 7239-7244.

3. Teglund S, McKay C, Schuetz E, van Deursen JM, Stravopodis D et al. (1998) Stat5a and Stat5b proteins have essential and nonessential, or redundant, roles in cytokine responses. Cell 93: 841-850.

4. Socolovsky M, Fallon AE, Wang S, Brugnara C, Lodish HF (1999) Fetal anemia and apoptosis of red cell progenitors in Stat5a-/-5b-/- mice: a direct role for Stat5 in Bcl-X(L) induction. Cell 98: 181-191.

5. Moriggl R, Topham DJ, Teglund S, Sexl V, McKay C et al. (1999) Stat5 is required for IL-2-induced cell cycle progression of peripheral T cells. Immunity 10: 249-259.

6. Cui Y, Riedlinger G, Miyoshi K, Tang W, Li C et al. (2004) Inactivation of Stat5 in mouse mammary epithelium during pregnancy reveals distinct functions in cell proliferation, survival, and differentiation. Mol Cell Biol 24: 8037-8047.

7. Li G, Wang Z, Zhang Y, Kang Z, Haviernikova E et al. (2007) STAT5 requires the N-domain to maintain hematopoietic stem cell repopulating function and appropriate lymphoid-myeloid lineage output. Exp Hematol 35: 1684-1694. S0301-472X(07)00535-8 [pii];10.1016/j.exphem.2007.08.026 [doi].

8. Kimura A, Martin C, Robinson GW, Simone JM, Chen W et al. (2010) The gene encoding the hematopoietic stem cell regulator CCN3/NOV is under direct cytokine control through the transcription factors STAT5A/B. J Biol Chem .285: 32704-.32709 M110.141804 [pii];10.1074/jbc.M110.141804 [doi]

9. Yao Z, Cui Y, Watford WT, Bream JH, Yamaoka K et al. (2006) Stat5a/b are essential for normal lymphoid development and differentiation. Proc Natl Acad Sci U S A 103: 1000-1005. 0507350103 [pii];10.1073/pnas.0507350103 [doi].

10. Hoelbl A, Kovacic B, Kerenyi MA, Simma O, Warsch W et al. (2006) Clarifying the role of Stat5 in lymphoid development and Abelson-induced transformation. Blood 107: 4898-4906. 2005-09-3596 [pii];10.1182/blood-2005-09-3596 [doi].

11. Dai X, Chen Y, Di L, Podd A, Li G et al. (2007) Stat5 is essential for early B cell development but not for B cell maturation and function. J Immunol 179: 1068-1079. 179/2/1068 [pii].

12. Kimura A, Rieger MA, Simone JM, Chen W, Wickre MC et al. (2009) The transcription factors STAT5A/B regulate GM-CSF-mediated granulopoiesis. Blood 114: 4721-4728.blood-2009-04-216390 [pii];10.1182/blood-2009-04-216390 [doi].

13. Kieslinger M, Woldman I, Moriggl R, Hofmann J, Marine JC et al. (2000) Antiapoptotic activity of Stat5 required during terminal stages of myeloid differentiation. Genes Dev 14: 232-244.

14. Kofoed EM, Hwa V, Little B, Woods KA, Buckway CK et al. (2003) Growth hormone insensitivity associated with a STAT5b mutation. N Engl J Med 349: 1139-1147.

15. Hwa V, Little B, Adiyaman P, Kofoed EM, Pratt KL et al. (2005) Severe growth hormone insensitivity resulting from total absence of signal transducer and activator of transcription 5b. J Clin Endocrinol Metab90: 4260-4266.jc.2005-0515 [pii];10.1210/jc.2005-0515 [doi].

16. Bernasconi A, Marino R, Ribas A, Rossi J, Ciaccio M et al. (2006) Characterization of immunodeficiency in a patient with growth hormone insensitivity secondary to a novel STAT5b gene mutation. Pediatrics 118: e1584-e1592.peds.2005-2882 [pii];10.1542/peds.2005-2882 [doi].

17. Nadeau K, Hwa V, Rosenfeld RG (2011) STAT5b deficiency: an unsuspected cause of growth failure, immunodeficiency, and severe pulmonary disease. J Pediatr 158: 701-708.S0022-3476(10)01154-6 [pii];10.1016/j.jpeds.2010.12.042 [doi].

18. Schmidt JW, Wagner KU (2012) Activation of Janus Kinases during Tumorigenesis. Chapter 15 259-288.

19. Valentino L, Pierre J (2006) JAK/STAT signal transduction: regulators and implication in hematological malignancies. Biochem Pharmacol 71: 713-721.

20. Ghoreschi K, Laurence A, O'Shea JJ (2009) Janus kinases in immune cell signaling. Immunol Rev 228: 273-287.

21. Baxter EJ, Scott LM, Campbell PJ, East C, Fourouclas N et al. (2005) Acquired mutation of the tyrosine kinase JAK2 in human myeloproliferative disorders. Lancet 365: 1054-1061.S0140-6736(05)71142-9 [pii];10.1016/S0140-6736(05)71142-9 [doi].

22. James C, Ugo V, Le Couedic JP, Staerk J, Delhommeau F et al. (2005) A unique clonal JAK2 mutation leading to constitutive signalling causes polycythaemia vera. Nature 434: 1144-1148.nature03546 [pii];10.1038/nature03546 [doi].

23. Levine RL, Wadleigh M, Cools J, Ebert BL, Wernig G et al. (2005) Activating mutation in the tyrosine kinase JAK2 in polycythemia vera, essential thrombocythemia, and myeloid metaplasia with myelofibrosis. Cancer Cell 7: 387-397.S1535-6108(05)00094-2 [pii];10.1016/j.ccr.2005.03.023 [doi].

24. Kralovics R, Passamonti F, Buser AS, Teo SS, Tiedt R et al. (2005) A gain-of-function mutation of JAK2 in myeloproliferative disorders. N Engl J Med 352: 1779-1790.352/17/1779 [pii];10.1056/NEJMoa051113 [doi].

25. Buettner R, Mora LB, Jove R (2002) Activated STAT signaling in human tumors provides novel molecular targets for therapeutic intervention. Clin Cancer Res 8: 945-954.

26. Brizzi MF, Dentelli P, Rosso A, Yarden Y, Pegoraro L (1999) STAT protein recruitment and activation in c-Kit deletion mutants. J Biol Chem 274: 16965-16972.

27. Mizuki M, Schwable J, Steur C, Choudhary C, Agrawal S et al. (2003) Suppression of myeloid transcription factors and induction of STAT response genes by AML-specific Flt3 mutations. Blood 101: 3164-3173.10.1182/blood-2002-06-1677 [doi];2002-06-1677 [pii].

28. Taketani T, Taki T, Sugita K, Furuichi Y, Ishii E et al. (2004) FLT3 mutations in the activation loop of tyrosine kinase domain are frequently found in infant ALL with MLL rearrangements and pediatric ALL with hyperdiploidy. Blood 103: 1085-1088.10.1182/blood-2003-02-0418 [doi];2003-02-0418 [pii].

29. Moriggl R, Sexl V, Kenner L, Duntsch C, Stangl K et al. (2005) Stat5 tetramer formation is associated with leukemogenesis. Cancer Cell 7: 87-99.

30. Joliot V, Cormier F, Medyouf H, Alcalde H, Ghysdael J (2006) Constitutive STAT5 activation specifically cooperates with the loss of p53 function in B-cell lymphomagenesis. Oncogene 25: 4573-4584.1209480 [pii];10.1038/sj.onc.1209480 [doi].

31. Hennighausen L, Wall RJ, Tillmann U, Li M, Furth PA (1995) Conditional gene expression in secretory tissues and skin of transgenic mice using the MMTV-LTR and the tetracycline responsive system. J Cell Biochem 59: 463-472.

32. Huettner CS, Zhang P, Van Etten RA, Tenen DG (2000) Reversibility of acute B-cell leukaemia induced by BCR-ABL1. Nat Genet 24: 57-60.

33. Letai A, Sorcinelli MD, Beard C, Korsmeyer SJ (2004) Antiapoptotic BCL-2 is required for maintenance of a model leukemia. Cancer Cell 6: 241-249.

34. Refaeli Y, Young RM, Turner BC, Duda J, Field KA et al. (2008) The B cell antigen receptor and overexpression of MYC can cooperate in the genesis of B cell lymphomas. PLoS Biol 6: e152.

35. Tiedt R, Hao-Shen H, Sobas MA, Looser R, Dirnhofer S et al. (2008) Ratio of mutant JAK2-V617F to wild-type Jak2 determines the MPD phenotypes in transgenic mice. Blood 111: 3931-3940.

36. Li J, Spensberger D, Ahn JS, Anand S, Beer PA et al. (2010) JAK2 V617F impairs hematopoietic stem cell function in a conditional knock-in mouse model of JAK2 V617F-positive essential thrombocythemia. Blood 116: 1528-1538.

37. Fievez L, Desmet C, Henry E, Pajak B, Hegenbarth S et al. (2007) STAT5 is an ambivalent regulator of neutrophil homeostasis. PLoS One 2: e727.

38. Matsumura I, Kitamura T, Wakao H, Tanaka H, Hashimoto K et al. (1999) Transcriptional regulation of the cyclin D1 promoter by STAT5: its involvement in cytokine-dependent growth of hematopoietic cells. EMBO J 18: 1367-1377.

39. Wang Z, Li G, Tse W, Bunting KD (2009) Conditional deletion of STAT5 in adult mouse hematopoietic stem cells causes loss of quiescence and permits efficient nonablative stem cell replacement. Blood 113: 4856-4865.blood-2008-09-181107 [pii];10.1182/blood-2008-09-181107 [doi].

40. Felsher DW, Bishop JM (1999) Reversible tumorigenesis by MYC in hematopoietic lineages. Mol Cell 4: 199-207.

41. Koschmieder S, Gottgens B, Zhang P, Iwasaki-Arai J, Akashi K et al. (2005) Inducible chronic phase of myeloid leukemia with expansion of hematopoietic stem cells in a transgenic model of BCR-ABL leukemogenesis. Blood 105: 324-334.10.1182/blood-2003-12-4369 [doi];2003-12-4369 [pii].

42. Iavnilovitch E, Groner B, Barash I (2002) Overexpression and forced activation of stat5 in mammary gland of transgenic mice promotes cellular proliferation, enhances differentiation, and delays postlactational apoptosis. Mol Cancer Res 1: 32-47.

43. Iavnilovitch E, Cardiff RD, Groner B, Barash I (2004) Deregulation of Stat5 expression and activation causes mammary tumors in transgenic mice. Int J Cancer 112: 607-619.

44. Wagner KU, Rui H (2008) Jak2/Stat5 signaling in mammogenesis, breast cancer initiation and progression. J Mammary Gland Biol Neoplasia 13: 93-103.

45. Feldman GM, Rosenthal LA, Liu X, Hayes MP, Wynshaw-Boris A et al. (1997) STAT5A-deficient mice demonstrate a defect in granulocyte-macrophage colony-stimulating factor-induced proliferation and gene expression. Blood 90: 1768-1776.

46. Turlej RK, Fievez L, Sandersen CF, Dogne S, Kirschvink N et al. (2001) Enhanced survival of lung granulocytes in an animal model of asthma: evidence for a role of GM-CSF activated STAT5 signalling pathway. Thorax 56: 696-702.

47. Sakamoto K, Schmidt JW, Wagner KU (2012) Generation of a novel MMTV-tTA transgenic mouse strain for the targeted expression of genes in the embryonic and postnatal mammary gland. PLoS One 7: e43778.10.1371/journal.pone.0043778 [doi];PONE-D-12-16928 [pii].

48. Tumbar T, Guasch G, Greco V, Blanpain C, Lowry WE et al. (2004) Defining the epithelial stem cell niche in skin. Science 303: 359-363.

49. Creamer BA, Triplett AA, Wagner KU (2009) Longitudinal analysis of mammogenesis using a novel tetracycline-inducible mouse model and in vivo imaging. Genesis 47: 234-245.

50. Zhang Q, Triplett AA, Harms DW, Lin WC, Creamer BA et al. (2010) Temporally and spatially controlled expression of transgenes in embryonic and adult tissues. Transgenic Res 19: 499-509.

51. Sakamoto K, Creamer BA, Triplett AA, Wagner KU (2007) The Janus kinase 2 is required for expression and nuclear accumulation of cyclin D1 in proliferating mammary epithelial cells. Mol Endocrinol 21: 1877-1892.

52. Akashi K, Traver D, Miyamoto T, Weissman IL (2000) A clonogenic common myeloid progenitor that gives rise to all myeloid lineages. Nature 404: 193-197.10.1038/35004599 [doi].

53. Kalaitzidis D, Neel BG (2008) Flow-cytometric phosphoprotein analysis reveals agonist and temporal differences in responses of murine hematopoietic stem/progenitor cells. PLoS One3: e3776. 10.1371/journal.pone.0003776 [doi].

54. Nishida S, Hosen N, Shirakata T, Kanato K, Yanagihara M et al. (2006) AML1-ETO rapidly induces acute myeloblastic leukemia in cooperation with the Wilms tumor gene, WT1. Blood 107: 3303-3312.2005-04-1656 [pii];10.1182/blood-2005-04-1656 [doi].

55. Zhang Y, Diaz-Flores E, Li G, Wang Z, Kang Z et al. (2007) Abnormal hematopoiesis in Gab2 mutant mice. Blood 110: 116-124.blood-2006-11-060707 [pii];10.1182/blood-2006-11-060707 [doi].

4

Ectopic Expression of Human *BBS4* Can Rescue Bardet-Biedl Syndrome Phenotypes in *Bbs4* Null Mice

Xitiz Chamling[1], **Seongjin Seo**[2], **Kevin Bugge**[1,3], **Charles Searby**[1,3], **Deng F. Guo**[4,5], **Arlene V. Drack**[2], **Kamal Rahmouni**[4,5], **Val C. Sheffield**[1,2,3]*

1 Department of Pediatrics, University of Iowa Interdisciplinary Program of Genetics, Iowa City, Iowa, United States of America, **2** Department of Ophthalmology and Visual Sciences, University of Iowa Carver College of Medicine, Iowa City, Iowa, United States of America, **3** Howard Hughes Medical Institute, Chevy Chase, Maryland, United States of America, **4** Department of Internal Medicine, University of Iowa Carver College of Medicine, Iowa City, Iowa, United States of America, **5** Department of Pharmacology, University of Iowa Carver College of Medicine, Iowa City, Iowa, United States of America

Abstract

Bardet-Biedl syndrome (BBS) is a genetically heterogeneous autosomal recessive disorder characterized by obesity, retinal degeneration, polydactyly, hypogenitalism and renal defects. Recent findings have associated the etiology of the disease with cilia, and BBS proteins have been implicated in trafficking various ciliary cargo proteins. To date, 17 different genes have been reported for BBS among which *BBS1* is the most common cause of the disease followed by *BBS10*, and *BBS4*. A murine model of *Bbs4* is known to phenocopy most of the human BBS phenotypes, and it is being used as a BBS disease model. To better understand the in vivo localization, cellular function, and interaction of BBS4 with other proteins, we generated a transgenic *BBS4* mouse expressing the human *BBS4* gene under control of the beta actin promoter. The transgene is expressed in various tissues including brain, eye, testis, heart, kidney, and adipose tissue. These mice were further bred to express the transgene in *Bbs4* null mice, and their phenotype was characterized. Here we report that despite tissue specific variable expression of the transgene, human *BBS4* was able to complement the deficiency of *Bbs4* and rescue all the BBS phenotypes in the *Bbs4* null mice. These results provide an encouraging prospective for gene therapy for BBS related phenotypes and potentially for other ciliopathies.

Editor: Yann Herault, IGBMC/ICS, France

Funding: This work was supported by National Institute of Health Grants R01EY022616 (to S.S.), RO1EY017168 and RO1EY110298 (to V.C.S.). V.C.S. is a Howard Hughes Medical Institute investigator. The funders had no role in study design, data collection and analysis, decision to publish, or preparation of the manuscript.

Competing Interests: The authors have declared that no competing interests exist.

* E-mail: val-sheffield@uiowa.edu

Background

Bardet-Biedl syndrome (BBS) is a pleiotropic, autosomal recessive disorder with the primary clinical features of obesity, retinopathy, polydactyly, learning disabilities, hypogenitalism, and renal defects [1,2,3]. BBS is rare in the general population: The prevalence in most of North America and Europe is 1 in 140,000 to 1 in 160,000 newborns [4]. However, components of the BBS phenotype are common, and debilitating including retinopathy, renal failure and obesity.

BBS is a genetically heterogeneous disease, and 17 different causative genes have been reported to date. Among these 17 genes, *BBS1* and *BBS10* have been reported to be the most common cause of the disease [5]. The data collected from our lab shows that *BBS4* is the third most common cause of this disorder. *BBS4* along with *BBS5* and *BBS8* are more common in patients of Middle Eastern and North African descent [6]. *BBS4* was the third BBS genes to be identified, after *BBS6* and *BBS2*, and its localization and interaction was analyzed soon after [7]. To better understand the function of BBS4, a Knockout murine model was generated and shown to phenocopy most of the human BBS phenotypes [8]. Indeed, *Bbs4−/ −* mice have obesity, retinal degeneration, primary cilia dyskenisia, and lack spermatozoa flagella [8]. *Bbs4$^{-/-}$* mice

are also hypertensive, leptin resistant and have renal defects [9]. Due to these significant phenotypes, *Bbs4$^{-/-}$* mice are commonly used as a BBS disease model.

Structurally, BBS4 is comprised almost entirely of tetratricopeptide repeats (TPR) that are predicted to fold into rod-shaped alpha solenoids [10]. Human *BBS4* shares 89% similarity with mouse and 81% with zebrafish. BBS4 has been shown to form a complex known as the BBSome with six other BBS proteins [1]. In addition, BBS4 has been shown to interact with centrosomal protein PCM1 and potentially with p150^{glued} [7]. Although the function of the BBSome complex has been studied, the role of BBS4 as a part of the PCM1 complex is not well understood. Moreover, due to the abundance of proteins in and around cilia where BBS4 localizes, it is likely that BBS4 interacts with other proteins that remains to be indentified. To better understand the in vivo localization and cellular function of BBS4, and to indentify novel interacters of BBS4, we generated a transgenic *BBS4* (*BBS4^{tg}*) mouse expressing the human *BBS4* gene under control of the ß-actin promoter. Here we report that despite tissue specific variable expression of the transgene, human *BBS4* was able to complement the deficiency of *Bbs4* and rescue all BBS phenotypes when crossed onto the *Bbs4* null genetic background.

Results

Generation of Transgenic BBS4

We generated transgenic mice carrying human *BBS4* tagged with a lap tag (localization and purification tag) under the control of β-actin promoter (fig. S1A), which is known to drive consistently strong, and ubiquitous expression. A lap tag consists of GFP and S tag flanking a TEV cleavage site. We used a pHβApr expression vector to clone the N-terminally tagged *BBS4*. Expression of the construct was confirmed by transient transfection of the plasmid in 293T cells followed by western blotting as well as immunoflourescence (data not shown). Transgenic mice were generated using pronuclear injection [11] by the Transgenic Animal Facility at the University of Iowa. Several founder mice were obtained, and a transgenic line that most broadly expressed the highest level of LAP-BBS4 was selected for further study. We crossed mice carrying the human *BBS4* transgene (*BBS4^tg^*) with *Bbs4^+/−^* mice to generate (*BBS4^tg^*/*Bbs4^+/−^*) mice. These mice were crossed to generate *BBS4^tg^*/*Bbs4^−/−^* mice and other combinations of genotypes for the study (table 1). In the following sections *Bbs4^+/+^* are referred to as WT, *Bbs4^−/−^* mice are referred to as KO, and (*BBS4^tg^*/*Bbs4^−/−^*) are referred to as TG.

Expression of the Transgene (LAP-BBS4)

Although a strong and ubiquitous promoter was used, expression of the LAP-BBS4 was limited to the few tissues, where endogenous Bbs4 is highly expressed. We analyzed the expression of the transgene via western blot. Comparing the expression of endogenous Bbs4 to the LAP-BBS4 using western blotting was not feasible since it was hard to get clean blot from a tissue lysate except in eye and testis (fig. 1B). When equal amount of lysate was loaded, lower expression of the transgene in the eye and higher expression in the testis, compared to the endogenous Bbs4, was observed (fig. 1B). For the other tissues, we used S-agarose beads to pull down and test which tissues express the LAP-BBS4 (fig. 1A). Highest expression of the transgene was found in testis, followed by brain, adipose, eye, heart, and very slight expression in trachea, lungs and bone; no expression in the liver and skin was observed.

We also performed rtqPCR to compare the amount of *LAP-BBS4* RNA expressed in transgenic mice compared to

endogenous *Bbs4* expressed in WT mice (fig. S1C). Only tissues with robust observed based on western blotting were used for rtqPCR. Since immunohistochemisty (IHC) was not able to detect the transgene in the brain, different brain regions were also included in rtqPCR to better understand which brain region has the transgene expression. The level of *LAP-BBS4* RNA compared to endogenous *Bbs4* was higher in testis, and in regions within the brain including hypothalamus and cerebellum, but lower in other regions of the brain and also other tissues including eye, kidney, and adipose tissue (fig. S1C). Due to low expression of BBS proteins and lack of a good functioning antibody, it is difficult to stain tissues with BBS4. Therefore, to confirm that the transgenic protein product (LAP-BBS4) localizes similarly to the endogenous Bbs4 in our transgenic animals, we performed immunoflourescence, on MEF cells prepared from the transgenic animal; as we expected, the localization of LAP-BBS4 is similar to endogenous BBS4 such that it localizes to the centrosome and cilia (fig. 1C).

Expression in Testis Rescues Male Infertility

Loss of sperm flagellum leading to male infertility is one of the most evident phenotypes in the *Bbs4^−/−^* mice [8]. To test if the transgene can act to rescue the deletion of the endogenous *Bbs4* and reverse the male infertility, we crossed male transgenic mice, *BBS4^tg^* to female *Bbs4^+/−^* mice. Resulting transgenic heterozygous males (*BBS4^tg^*/*Bbs4^+/−^*) were further crossed with heterozygous females (*Bbs4^+/−^*). We selected the *Bbs4^−/−^* animals expressing the transgene (*BBS4^tg^*/*Bbs4^−/−^*) and crossed them again. The *BBS4^tg^*/*Bbs4^−/−^* male and female were able to breed normally and produce similar litter size and number of pups as the control mating (Fig. 2B, Table 1); the pups produced were either *BBS4^tg^*/*Bbs4^−/−^* or *Bbs4^wt^*/*Bbs4^−/−^*. Female *Bbs4^−/−^* mice are fertile, and the reason of infertility in the *Bbs4^−/−^* male mice is lack of sperm flagella, which are completely missing or severely decrease in size in the *Bbs4^−/−^* mice (Fig. 2C–D). We also analyzed the structure of the sperm flagellum, and histology of testis in these *BBS4^tg^*/*Bbs4^−/−^* males using immunoflouresence and H&E staining respectively, and we could see elongated, healthy, and normal sperm flagella, similar to the WT (fig. 2C–D).

Table 1. A table showing combination of male and female of various genotypes crossed, their litter size and total number of pups produced.

Dam ID	Dam genotype	Sire ID	Sire genotype	# Litters	Total Pups	# Born Dead
* (1291-4)-53	Bbs4 ^wt^Bbs4^+/+^	(B4T78-2)-33	Bbs4 ^wt^Bbs4^+/+^	7	37	**0**
(B4T78-4)-17	Bbs4 ^wt^ Bbs4^+/−^	(B4T78-4)-14	Bbs4 ^tg^Bbs4^+/+^	6	54	0
(B4T78-5)-20	Bbs4 ^wt^ Bbs4^+/−^	(B4T78-5)-17	Bbs4 ^tg^Bbs4^+/+^	3	21	0
(1294-16)-3	Bbs4 ^wt^ Bbs4^+/−^	(B4T78-1)-23	Bbs4 ^tg^Bbs4^+/+^	4	21	0
(1294B4T-1)-1	Bbs4T ^wt^ Bbs4^+/−^	(1294B4T-1)-6	Bbs4 ^tg^Bbs4^+/−^	7	49	0
Ψ (1294B4T-2)-5	Bbs4^tg^ Bbs4^−/−^	(1294B4T-1)-7	Bbs4 ^tg^Bbs4^−/−^	6	47	0
Ψ (1294B4T-2)-19	Bbs4^tg^ Bbs4^−/−^	(1294B4T-2)-9	Bbs4 ^tg^Bbs4^−/−^	10	84	0
(1294B4T-2)-18	Bbs4^tg^ Bbs4^+/−^	(1294B4T-2)-12	Bbs4 ^tg^Bbs4^+/−^	4	41	0
(1294B4T-3)-14	Bbs4^tg^ Bbs4^+/−^	(1294B4T-5)-20	Bbs4 ^wt^Bbs4^+/−^	7	38	0
(1294B4T-6)-28	Bbs4^tg^ Bbs4^+/−^	(1294B4T-6)-22	Bbs4 ^tg^Bbs4^+/−^	3	21	0
Ψ (1294B4T-4)-51	Bbs4^wt^ Bbs4^−/−^	(1294B4T-4)-46B	Bbs4 ^tg^Bbs4^−/−^	3	15	0

Rows indicated by (Ψ) are the matings between male *BBS4^tg^/Bbs4^−/−^* and female *BBS4^tg^/Bbs4^−/−^* or *Bbs4^−/−^* animals; their litter sizes and the pups produced are also shown. No abnormality was found in litters produced or number of pups produced compared to the control matings (*). Genotyping was performed for *LAP-BBS4* (*BBS4^tg^*) as well as endogenous *Bbs4* as shown in (fig. S1B).

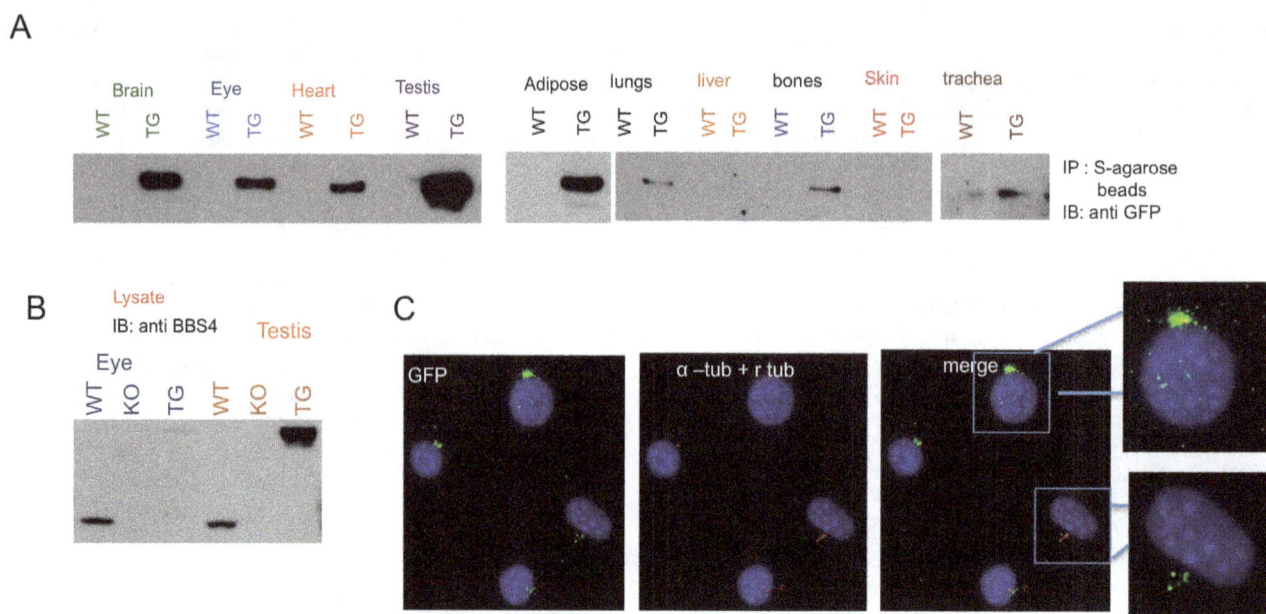

Figure 1. LAP-BBS4 is expressed in multiple different tissues. Due to multiple non-specific bands, comparing the expression of LAP-BBS4 to endogenous Bbs4 using protein lysates was not feasible except in eye and testis. **A)** For the other tissues, we used S-agarose beads to pull down the LAP-BBS4 from the lysates and immunoblotted with GFP antibody. The bands indicate the presence of LAP-BBS4, which is missing in the WT mice. **B)** Protein lysates from eye and testis of WT, $Bbs4^{-/-}$, and $BBS4^{tg}$ animals were run on SDS gels and immunoblotted with BBS4 antibody. The upper band shows the amount of LAP-BBS4, and the lower band is the endogenous Bbs4. **C)** Localization of LAP-BBS4 was studied using immunoflourescence in MEF cells prepared using transgenic mice. Cilia were stained with acetylated α-tubulin and LAP-BBS4 was stained using GFP. As expected the centrosomal and ciliary localization of the LAP-BBS4 was observed.

Low Expression of the BBS4 Transgene Rescues Retinopathy

In the $Bbs4^{-/-}$ mouse model, the photoreceptors degenerate progressively with a loss of nuclei and shortening of the inner and outer segments leading to complete loss of photoreceptor cells overtime [12] (fig. 3B). Our transgenic mice have low transgenic expression in the eye (fig. 1, S1C). Within retina, expression of transgene was observed mostly in the inner segment of an eye with very low expression in the RPE layer (Fig. 3A). The qPCR data show that expression of the transgene is approximately 30 fold lower compared to endogenous $Bbs4$ expression (fig. S1C). We performed electroretinography (ERG) on 2,4 and 6 months old mice to test whether the observed amount of transgene expression was able to rescue the blindness in $BBS4^{tg}$ mice. In WT eye, the average b-wave amplitude is approximately 800 uV at 2 months of age, whereas b-wave amplitude is drastically reduced to 200 uV in $Bbs4^{-/-}$ mice (fig. 3C). In $Bbs4^{tg}$ mice compared to the $Bbs4^{-/-}$ mice, the b- wave is significantly increased to approximately 550 uV. Although the ERG b-wave in $Bbs4^{tg}$ mice remained significantly different compared to the WT mice, the difference becomes less with age, which indicates prevention of retinal degeneration by the transgene. By the end of 6 months, the b-wave amplitude difference between the $Bbs4^{-/-}$ and $BBS4^{tg}$ mice was almost 5 fold (fig. 3C). We also performed histology on mouse eyes of different ages. The histology data from 2 months old $BBS4^{tg}$ mice (fig. 3B) show that the outer segment and ONL were preserved.

Rescue of Obesity

Obesity is one of the cardinal features of BBS [5]. Mouse models of BBS including $Bbs4^{-/-}$ mice become obese [8]. Since the $BBS4$ transgene is expressed in the brain as well as in adipose

tissues (fig. 1A), we expected obesity to be rescued in $Bbs4^{-/-}$ mice carrying the transgene. To test this hypothesis we weighed $Bbs4^{-/-}$, WT, and $BBS4^{tg}$ mice every month to six months of age (fig. 4A, S2). $Bbs4^{-/-}$ mice were significantly heavier than WT and $BBS4^{tg}$ mice (P<0.01) by two months of age (fig. 4A, S2A). The weight difference increased with time. At six months of age the $Bbs4^{-/-}$ mice were more than 20(+/−3) grams heavier than WT mice. The weight gain is prevented in $BBS4^{tg}$ mice with no significant difference between these and WT mice (fig. 4A, S2A). Interestingly, we observed a difference in weight and onset of obesity in $Bbs4^{-/-}$ mice is also dependent on genetic background. Our $Bbs4^{-/-}$ mice on the 129/SvJ background become obese earlier and to a greater extent than $Bbs4^{-/-}$ mice on the C57 and 129 mixed background reported by Mykyten et al. [8].

Leptin Resistance and Renal Sympathetic Nerve Activity (SNA)

We previously compared the effect of exogenous leptin in BBS $Bbs4^{-/-}$ and WT mice by administrating leptin and tracking weight loss as well as food intake [9]. Compared to WT mice, $Bbs4^{-/-}$ mice were resistant to leptin with no significant change in body weight or food intake with leptin administration. Since there is expression of the transgene in the brain and hypothalamus, we decided to test whether the transgene can rescue the leptin resistance phenotype seen in $Bbs4^{-/-}$ mice. We injected 1 ug of leptin or vehicle (PBS) per gram of body weight twice a day, and monitored the weight change as well as food intake (Fig. 4B, S2B). There was no significant weight loss upon vehicle injection in any of the mice (fig. 4b ii). Upon leptin injection, there was less than 1% body weight loss in the $Bbs4^{-/-}$ mice. Compared to the $Bbs4^{-/-}$ mice, WT mice lost significant weight (p<0.01) starting the day following first leptin injection. They progressively lost weight each day, and 7% of body weight was lost by the end of day

A

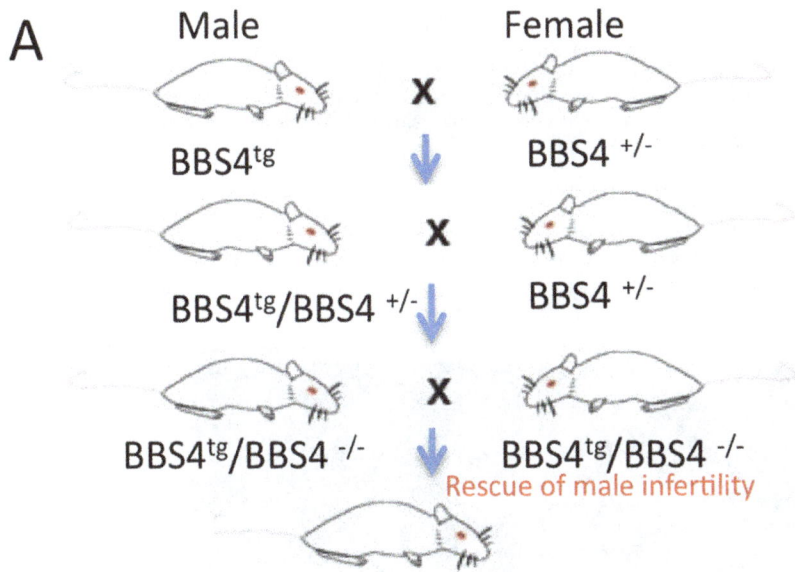

Male   Female

BBS4^tg   X   BBS4 ^+/-

BBS4^tg/BBS4 ^+/-   X   BBS4 ^+/-

BBS4^tg/BBS4 ^-/-   X   BBS4^tg/BBS4 ^-/-

Rescue of male infertility

B

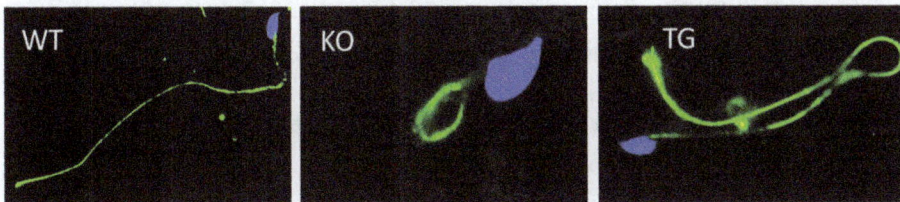

C   WT   KO   TG

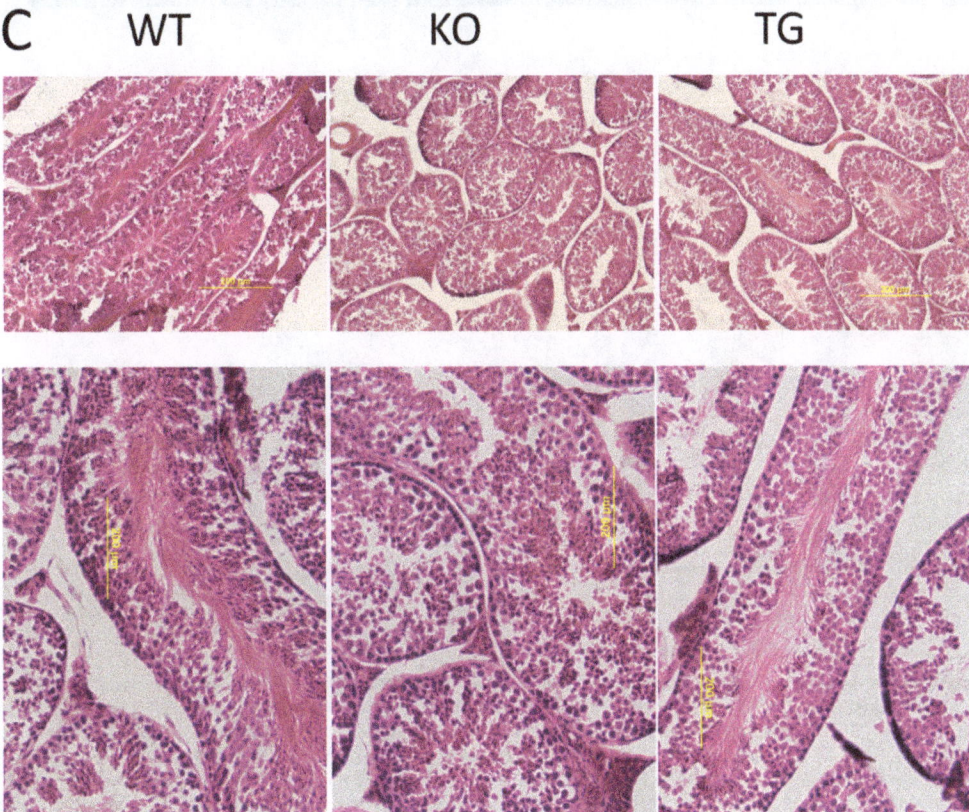

Figure 2. Infertility in *Bbs4*^{−/−} mice is rescued by LAP-BBS4. A) We started with male *Bbs4*^{tg} mice crossed with female *Bbs4*^{−/−} mice. In the F3 generation, we selected male transgenic animals lacking endogenous *Bbs4* (*BBS4*^{tg}/*Bbs4*^{−/−}) and crossed them with female *Bbs4*^{−/−}. Male *BBS4*^{tg}/ *Bbs4*^{−/−} mice were able to produce pups indicating rescue of the infertility. **B)** Sperm flagella are normal in *BBS4*^{tg} mice. Sperm heads are stained with DAPI, while the sperm tails are stained with acetylated α tubulin. *Bbs4*^{−/−} male mice produce spermatozoa that have either no flagella or in a few cases short, truncated flagella. Sperm tails in *BBS4*^{tg} animals are normal and full length. **C)** H&E staining show the testis histology demonstrating diminished sperm production in *Bbs4*^{−/−} testis compared to WT and the *BBS4*^{tg} animals.

5. Although the change in body weight was not as drastic after the first day of injection, progressive loss was apparent in the transgenic mice as well. Compared to the *Bbs4*^{−/−} mice, *BBS4*^{tg} mice lost significant weight (P<0.01) by the third day after leptin injection was started. By the end of day 5, 6% (^{+/−}1%) of the body weight was lost in *BBS4*^{tg} mice (fig. 4B ii), which indicates that

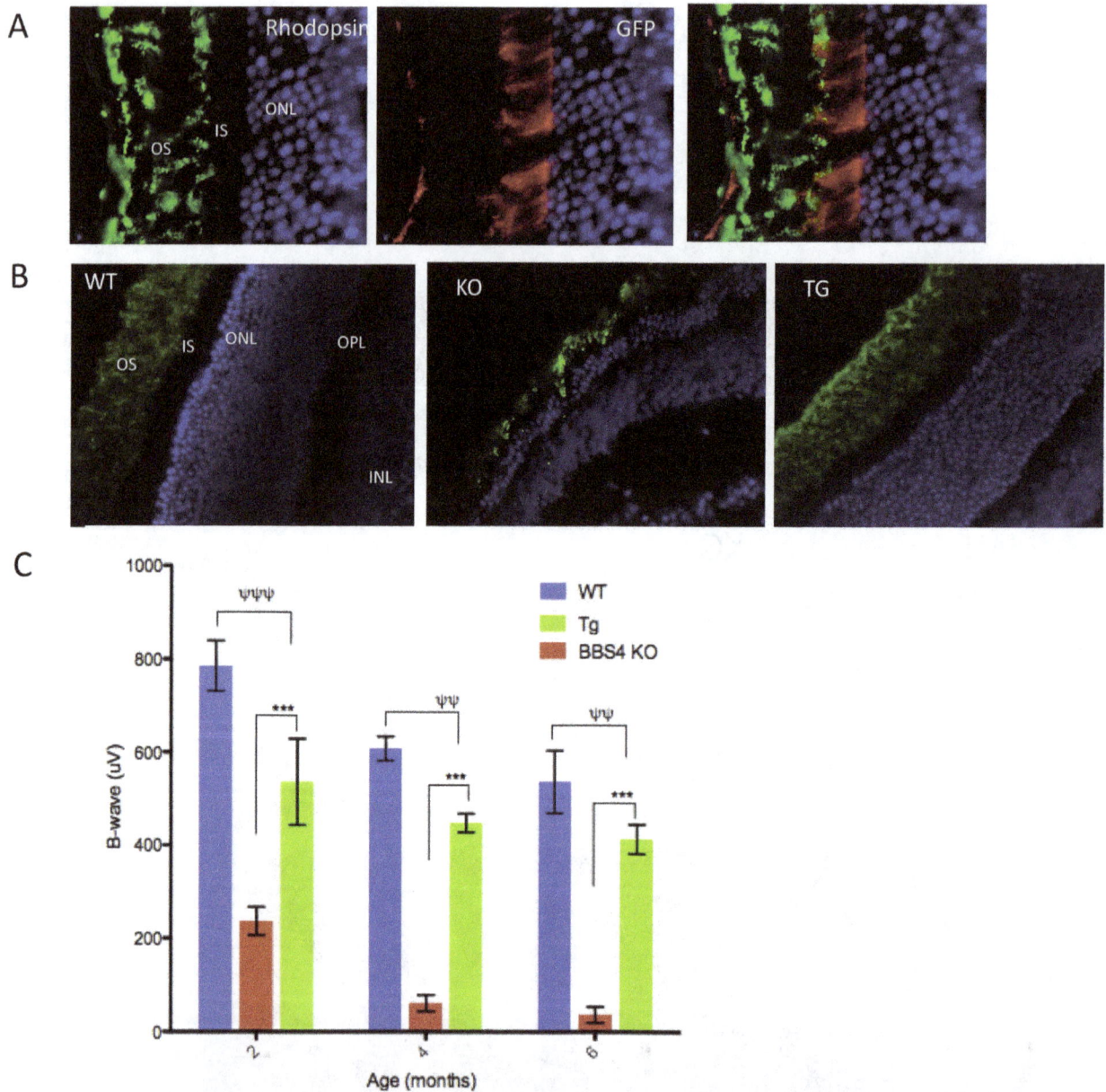

Figure 3. *LAP-BBS4* can significantly improve retinal degeneration in *Bbs4*^{−/−} animals. A) LAP BBS4 is expressed in the inner segment of a retina (red) as observed using antibody against GFP. Green is outer segment rhodopsin staining and blue is nuclear staining with DAPI; slight LAP-BBS4 expression in the RPE layer is also seen. **B)** The histology of retina observed using rhodopsin (green) and DAPI (blue) staining in 2 months old mice show that *Bbs4*^{tg} mice have normal retinas. **C)** ERG measured at 4 different time points in WT, *Bbs4*^{tg}, and *Bbs4*^{−/−} mice. Data shown are mean +/− SEM with n = 4 for each genotype. B-wave value shows that the ERG remains significantly different in the *BBS4*^{tg} mice when compared to the WT (Ψ Ψ Ψ P<0.001, Ψ Ψ P<0.01), however, it is drastically and significantly improved in the *Bbs4*^{tg} mice compared to the *Bbs4*^{−/−} mice (***P<0.001). The difference between *Bbs4*^{tg} and *Bbs4*^{−/−} mice becomes more significant as the mice get older.

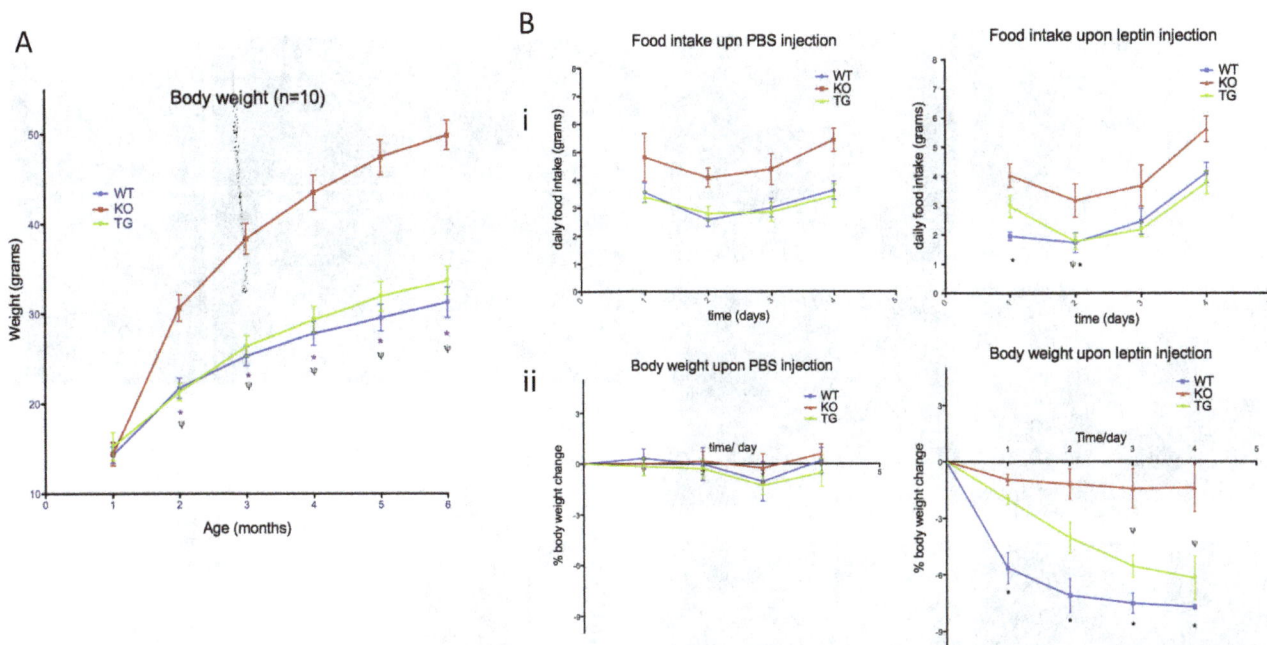

Figure 4. Obesity and leptin resistance in Bbs4$^{-/-}$ mice are rescued by LAP-BBS4. A) 10 mice per group (5 male and 5 female) mice were weighed every month to 6 months of age. Bbs4$^{-/-}$ mice (red line) become significantly heavier than WT mice (*P<0.01) and the Bbs4^{tg} mice (Ψ P<0.01) beginning at 2 months of age. Bbs4^{tg} mice (blue line) are not significantly heavier than WT mice (green line) at any age. Data are presented as mean +/− SEM, n = 10 per genotype. B) Weight change and food intake was measured in the WT (blue), BBS4^{tg} (green), and Bbs4$^{-/-}$ (red) mice upon vehicle (PBS) or leptin injection. i) When compared to vehicle injection, food intake of WT mice was significantly reduced on day 1 and 2 after beginning leptin injection (*P<0.05), and the food intake of BBS4^{tg} mice was decreased on day 2 after beginning leptin injection (Ψ P<0.05). T-test was used to calculate significance ii) Leptin injection had no significant effect on body weight or food intake in Bbs4$^{-/-}$ mice. When compared to the Bbs4$^{-/-}$ mice, WT mice lost significant amount of weight after day 1 of leptin injection (*P<0.01) and they progressively lost up to 7% of total body weight by the end of day 5. Bbs4^{tg} mice when compared to Bbs4$^{-/-}$ mice, lost significant weight after day 3 of injection (Ψ P<0.01) and progressively lost up to 6% of body weight by the end of day 5. Four mice (2 males and 2 females) of each genotype were used for the study.

these mice respond to leptin, similar to WT mice. We also measured the food intake in mice upon vehicle or leptin injection (fig. 4B i). On average, WT and BBS4^{tg} mice eat 3.5 grams/day, while Bbs4$^{-/-}$ mice eat 5 grams/day (fig. 4C). Upon leptin injection, there was no significant change in the food intake of Bbs4$^{-/-}$ mice. However, the food intake was reduced in WT after the first day of injection, and food intake slowly progressed back to normal (3.5 grams/day) by the end of day 5. Although the change in Bbs4^{tg} mice was not significant on the first day, the food intake subsequently followed the same trend as in WT mice (fig. 4B i). When compared to food intake upon vehicle injection, food intake of WT mice significantly (P<0.05) decreased beginning on day 1 of leptin injection. In BBS4^{tg} mice a significant reduction in food intake (P<0.05) was observed beginning on day 2 after leptin injection was started. In general, food intake and the response to leptin in BBS4^{tg} mice were significantly different than the food intake and response to leptin of Bbs4$^{-/-}$ mice.

Previously, we observed that renal sympathetic nerve activity (SNA) was significantly greater in Bbs4$^{-/-}$ animals compared to the WT [9]. Average renal SNA in a WT mouse is 65+7 spikes/sec. Consistent with our previous finding, in Bbs4$^{-/-}$ mice, renal SNA is increase (79±19 spikes/sec). The renal SNA in the BBS4^{tg} mice is 66±9 spikes/sec, which is not significantly different from WT mice (fig. S3A).

Rescue of Hydrocephalus and Motile Cilia

Hydrocephalus is observed in all BBS murine models, including Bbs4$^{-/-}$ mice. There is evidence linking ventricular motile cilia abnormalities to congenital hydrocephalus [13,14]. Along with

hydrocephalus, our BBS murine models have motile cilia defects in the ependymal cells lining the cerebral ventricles, as well as in airway epithelia [15,16]. We performed scanning electron microscopy (SEM) to examine the cilia in the lateral ventricles of WT, Bbs4$^{-/-}$, and BBS4^{tg} mice (fig. 5A, S3B). Bbs4$^{-/-}$ mice have very few cilia and most of them are grossly deformed with large bulges at the tip when compared to WT mice. BBS4^{tg} mice have normal numbers of cilia and the cilia are morphologically normal (fig. 5A, S3). The expression of the transgene in the brain is also capable of rescuing the hydrocephalus in BBS4^{tg} mice. Coronal and horizontal MRI sections were taken in 3 months old WT, BBS4^{tg} and Bbs4$^{-/-}$ mice (fig. 5C, S3B). The MRI images show enlarged lateral ventricles (LV) in the Bbs4$^{-/-}$ brain, but not in WT and BBS4^{tg} brains (fig. 5B, S3C). Relative LV volume of WT and BBS4^{tg} brain is significantly lower compared to Bbs4$^{-/-}$ (P<0.001) (fig. 5C).

Discussion

This study was designed to better understand the localization, function, and interactions of BBS4, and to determine whether ectopic expression can functionally complement endogenous Bbs4. In our study, we used human BBS4 with a LAP tag at the N-terminus of BBS4 and ectopically expressed the protein in mice lacking endogenous Bbs4. Previously; we generated similar transgenic mice with the BBS4 gene controlled by the CMV promoter [17]. While we saw significant improvement in body weight and fertility, the transgene did not appear to be ubiquitously expressed [17]. In the mouse generated in this

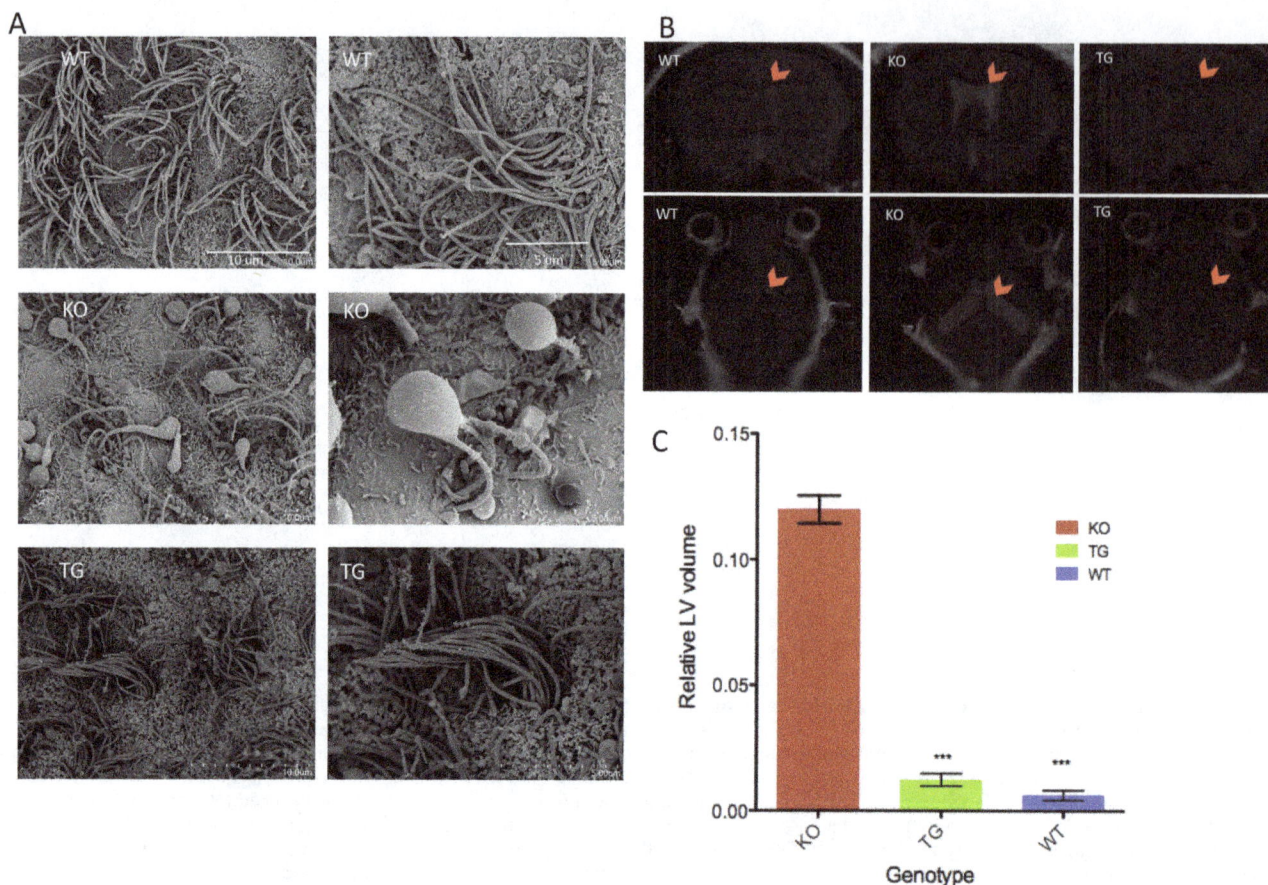

Figure 5. LAPBBS4 can rescue motile cilia defects and hydrocephalus. A) SEM image of cilia from the lateral ventricles of the brain in 6 months old WT, $Bbs4^{-/-}$ and $BBS4^{tg}$ mice. Motile cilia appear normal and healthy in $BBS4^{tg}$ mice. Right panel shows the motile cilia in higher magnification. **B)** $Bbs4^{-/-}$ mice have hydrocephalus (enlarged ventricles, arrowheads). The top panel shows an MRI of 3-month-old mouse brains in coronal section, and the bottom panel shows the horizontal sections of the same mice. The ventricles are slightly visible in the WT mice and in the $Bbs4^{tg}$ mice. **C)** Relative volume of lateral ventricle is significantly low (***P<0.001) in WT and $BBS4^{tg}$ mice compared to the $Bbs4^{-/-}$ mice (data shown as mean +/− SEM, n = 3 per genotype).

study, the transgene appeared to be more broadly expressed, although the level of expression varied among tissues. Despite an apparent low level of expression in some tissues, the human *BBS4* transgene was able to rescue almost all of the BBS phenotypes observed in *Bbs4* null mice.

Human and mouse BBS4 share 89% similarity, and both, almost entirely, consist of TRP repeats flanked by short N-and C-terminal regions [7,18]. Each of the proteins contains 8 TPR repeats spanning from 67 amino acids to 371 amino acids. Therefore, we expected the human homologue of BBS4 to replicate the function of mouse BBS4. In fact, BBS4 interacting proteins including PCM1 and BBSome components are the same in human as well as mouse. BBS null mice demonstrate most of the human BBS phenotypes. However, there are some differences in the protein sequence including a 14 aa low complexity region at the N-terminus of the mouse protein that is absent in the human BBS4 protein (SMART database). Such differences could lead to differences in interacting partners and could account for some variation in phenotypes between mice and human. For example, polydactyly is a major feature of BBS in human patients, but it is not found in *Bbs* null mice. Furthermore, hydrocephalus is not highly penetrant in human patients, but is frequently observed in *Bbs* null mice. Of note, all observed mouse BBS phenotypes,

including hydrocephalus were rescued in Bbs4 null mice by ectopic expression of human BBS4.

An important function of BBS proteins is ciliary proteins trafficking. In sperm, BBS proteins localize to the annulus and are thought to traffic essential proteins to form elongated sperm tails. In the retina, the BBS proteins localize to the ciliary body and assist in protein trafficking for maintenance of the outer segment of photoreceptors. We observed retinal degeneration, as well as either completely missing or only partially formed sperm flagella in $Bbs4^{-/-}$ mice. In addition, recent findings have implicated leptin resistance and defective leptin receptor b trafficking in the hypothalamus to obesity in $Bbs4^{-/-}$ mice (2, 9). The role of BBS proteins and the BBSome in protein trafficking has been widely studied in primary cilia using cell culture, but in-vivo confirmation is lacking. Using our transgenic mice we have now shown that ectopic expression of BBS protein can effectively restore sperm flagella formation, rescue obesity, and successfully restore rhodopsin trafficking in photoreceptor cells preventing retinal degeneration. By using this in-vivo model we have further confirmed the role of BBS proteins in ciliary trafficking.

The role of BBS proteins in motile cilia has not been as widely explored compared to the role in primary cilia. Shah et al [16] noted few deformed motile cilia in the airway epithelial cells in *Bbs* null mice. Deformed motile cilia at the ependymal lining of the ventricles are also seen in $Bbs3^{-/-}$ mice [15]), which could be the cause of hydrocephalus in these mice. $Bbs4$ null mice also have these phenotypes. The ectopic expression of $BBS4$ was able to correct the motile cilia defect in $Bbs4$ null mice. The results indicate an important role of BBS proteins in protein trafficking in motile, as well as primary cilia.

In $Bbs4^{tg}$ mice, expression of LAP-BBS4 was highly variable among tissue types. High expression of LAP-BBS4 compared to endogenous mouse Bbs4 was seen in testis, while low expression was observed in the eye (fig. 1A, S1B). Within the brain, the cortex, hippocampus, and amygdala had lower expression levels compared to the hypothalamus and cerebellum. Expression of the transgene in the hypothalamus appeared higher than endogenous $Bbs4$ expression (fig. S1B). Despite the variable and often weak expression of the transgene, there was rescue of BBS phenotypes, which suggests that the amount of BBS4 expression is less important than its presence. Collectively, our results suggest that the rescue of BBS phenotypes is not dose dependent, and almost all the known BBS phenotypes can be corrected by supplementing the gene.

Simons et al [19] reported that sub retinal injection of AAV with $BBS4$ with coverage factor of as low as 5% was able to rescue the mislocalization of rhodopsin in $Bbs4$ null mice and preserve the rod outer segment. In agreement with their data, we were able to show that a greater than 20-fold lower level of expression of transgene compared to endogenous Bbs4 expression was adequate to improve the ERG, rhodopsin localization and overall histology of the retina. Similarly, obesity in BBS has been associated with hyperleptinemia, which occurs due to leptin receptor mislocalization in hypothalamic neurons in the absence of BBS4 [9]. Our data shows that ectopic expression of human $BBS4$ was able to reduce hyperleptinemia and rescue obesity in $Bbs4$ deficient mice. The expression of the transgene in the brain was able to preserve motile cilia in the ventricles of the brain, and rescue hydrocephalus. Our results support the concept that syndromic obesity, retinopathy, hydrocephalus and other ciliary phenotypes are candidates for gene therapy.

Methods

Ethics statement: The University Animal Care and Use Committee at the University of Iowa approved all animal work in this study (animal protocol number: 1003062). Every person involved in handing mice was properly trained to the standard proposed by the committee.

Generation of Transgenic Mice

We generated transgenic mice carrying human $BBS4$ tagged with the lap tag under the control of the β-actin promoter. We used a pHβApr expression vector [20,21] to clone the N-terminally tagged $BBS4$. Transgenic mouse lines were generated using pronuclear injection [11] by the transgenic animal facility in University of Iowa. B6SJL (C57BL/6J X SJL/J: Jackson Laboratory) embryos were used for the pronuclear injection. Transgenic animals were maintained on 129/SvJ background. To generate $BBS4^{tg}/Bbs4^{-/-}$ and $Bbs4^{-/-}$ mice used in this study, $BBS4^{tg}/Bbs4^{+/-}$ lines were crossed with $Bbs4^{+/-}$ animals on the 129/SvJ background. Genotypes for the lap-$BBS4$ transgene were determined using the following primers: GTCCTGCTGGAGTTCGTGAC and GGCGAAATAT-CAATGCTTGG). For $Bbs4^{-/-}$ genotyping, the following primers were used: GCTACCCGTGATATTGCTGAA and TTGGGTGCTCTATTCTGCTG.

Antibodies and Reagents

BBS4 antibody was kindly provided by Dr Maxence Nachury (Stanford University). Anti-acetylated tubulin (6-11B-1) was purchased from Sigma (St. Louis, MO, USA), GFP antibody (A11120), Alexa 488-conjugated and Alexa 568-conjugated secondary antibodies were purchased from Invitrogen (Carlsbad, CA, USA). S-agarose beads for immunoprecipitation were purchased from (novagen).

Quantitative Real-time PCR, MEF Cells and Immunoflourescence Microscopy

For qPCR, RNA from different tissues was extracted using IBI RNA extraction kit (cat# IBI47302) following their protocol. cDNA was prepared from 2 ug of total RNA using SperScript III reverse transcriptase (Invitrogen). qPCR was performed using SYBR Green PCR Master Mix (Applied Biosystems # 4309155) in the Mx3000p unit (Stratagene). GAPDH was used to normalize the mRNA level and ΔΔCt [20,22] method was used to calculate the expression of LAP-BBS4 compared to endogenous Bbs4. MEF cells from the transgenic mice was prepared from embryonic day 12.5 embryos. Pregnant mice were sacrificed by cervical dislocation following IACUC guidelines. Uterine horns of sacrificed mice were dissected out, and cultured in Dulbeco's modified Eagle medium with 10% serum following the protocol by Jozefczuk et al. [23]. Immunoflourescence was performed as previously described (22).

Tissue Sectioning, Staining, and SEM

For tissue sectioning, animals were perfused with 4% PFA in PBS (2.5 ml/min, 50 ml) following a standard protocol. Prior to perfusion, animals were fully anesthetized by injecting Ketamine (0.1 ml/20 g BW). Eyes were post-dissection fixed with fixative for 30 more minutes and frozen in OCT. Sectioning was performed on the frozen section using CryoStar NX70 (thermo Scientific). Mouse sperms were extracted and stained as describe [17]. For brain, the tissue was freshly harvested without perfusion and snap frozen in liquid nitrogen. Frozen tissues were used for sectioning with the CryoStar. For SEM, animals were perfused as described above. Brain was harvested, and dissected to remove the hippocampus and expose the ventricle. The brains were then fixed with 2.5% glutaraldehyde for 24–48 hrs. Standard SEM protocol was followed to prepare the sample and a Hitachi S-4800 electron microscope was used for imaging.

Obesity, Leptin Resistance, Food Intake and Body Weight Determination

Feeding and body weight response to leptin was compared between WT, $Bbs4^{-/-}$, and $BBS4^{tg}$ animals as mentioned previously [9]. Mice were housed separately a week before the injection, and mock injections were performed as training for a week to eliminate stress and difference in environment as a cause of weight loss and differential food intake. One micro liter (1 mg/ml) of leptin per gram of body weight was given to each mouse via i.p injection twice a day for 4 days. Weight and food intake was measured every morning. For the obesity study, 10 mice (5 male and 5 female) from each genotype (WT, $Bbs4^{-/-}$ and $BBS4^{tg}$) were weighed each month for up to 6 months. Littermates from 3 different parents were used for the study.

ERG, and Renal SNA

For ERG measurement, mice were dark-adapted overnight prior to anesthetizing by ip injection with ketamine (87.5 mg/kg) and xylazine (2.5 mg/kg). Gold ring electrodes referenced to a needle electrode used to record ERG from corneal surface of each eye simultaneously following the description in [24]. To measure renal SNA, mice were anesthetized as mentioned above and a nerve fascicle to the left kidney was isolated and a bipolar platinum-iridium electrode was suspended as described in [9]. Baseline renal SNA was recorded for 10 min and an average of two separate measurements was used as the baseline value for each animal.

Statistics: Results are expressed as mean $+/-$ SEM. Data was analyzed using 1-way ANOVA and TUKEY unless specified otherwise. $P<0.05$ is considered significant.

Supporting Information

Figure S1 Genotypes and relative expression of transgene varies among tissues. A) Expression cassette of *LAP-BBS4*, where a LAP-tag is at the N-terminus of the *BBS4* gene. *LAP-BBS4* was cloned in phßAPr expression vector. **B)** Example of genotyping performed on the animals. Each animal was tested for the presence of the transgene as well as the endogenous *Bbs4* gene. Left side represents genotype for endogenous *Bbs4*, and the right side is the genotype for the transgene. **C)** Graph showing relative expression of the transgene (*LAP-BBS4*) compared to the endogenous *Bbs4*. The Y-axis is the relative fold change (*LAP-BBS4/msBbs4*), and the X-axis shows tissues that were used. Horizontal dotted line at 1 in Y-axis represents the point where equal expression of the transgene and endogenous *Bbs4* are observed. Above the line represents higher expression of *LAP-BBS4* compared to endogenous *Bbs4*, and below that reference line shows higher expression of endogenous *Bbs4* than transgene. For example, retina has relatively higher expression of endogenous *Bbs4* than *LAP-BBS4*, and higher amount of *LAP-BBS4* than the endogenous *Bbs4* is expressed in testis.

Figure S2 Obesity and leptin resistance in *Bbs4*^{tg} mice compared to *Bbs4*^{−/−} mice. A) 5 male and 5 female mice were weighed every month for 6 months. *Bbs4*^{−/−} mice (red line) are significantly heavier than WT (*P<0.01)(green) and *BBS4*^{tg} mice (Ψ P<0.01) beginning at 2 months of age. There is no significant

difference in weight between *Bbs4*^{tg} and WT mice in either male or female. **B)** Food intake and body weight were monitored in 4 mice (2 male 2 female each, 9–12 weeks old) of all three genotypes. Mice were kept in separate cages for a week followed by a training session of mock injection to adapt them to the stress caused during injection. Compared to the vehicle injection, body weight as well as food intake in WT and *BBS4*^{tg} mice are decreased upon leptin injection. Although body weight decreased in *BBS4*^{tg} and WT mice but not in *Bbs4*^{−/−} mice upon leptin injection, the change was not significant due to low N, and higher variation in mouse weight. However, when compared to vehicle injection, a significant reduction in food intake in WT (*P<0.05) and *BBS4*^{tg} (Ψ <0.05) mice were observed after leptin injection.

Figure S3 Baseline renal sympathetic nerve activity (RSNA), hydrocephalus and motile cilia are improved in transgenic mice. A) Three mice (9–12 weeks) per genotype group were used to measure the baseline renal RSNA. Baseline RSNA is increased in the *Bbs4*^{−/−} mice compared to the WT mice. *BBS4*^{tg} mice have normal renal RSNA. **B)** SEM image of lateral ventricles in WT, *Bbs4*^{−/−}, and *BBS4*^{tg} mice. Compared to WT and *BBS4*^{tg}, *Bbs4*^{−/−} mouse brain has very few motile cilia in the lateral ventricle. C) MRI image showing enlarged LV (red arrowhead) in the sagittal section of *Bbs4*^{−/−} compared to WT and *BBS4*^{tg} mouse brain.

Acknowledgments

We thank Ruth Swiderski and Valerie Buffard for help with genotyping, Sajag Bhattarai for ERG measurements, Donald A. Morgan for assisting with SNA experiments, Jean Ross for help with SEM, and Daniel Thedens for his help with MRI.

Author Contributions

Conceived and designed the experiments: XC SS AVD KR VCS. Performed the experiments: XC KB CS DFG. Analyzed the data: XC SS KR. Wrote the paper: XC VCS.

References

1. Nachury MV, Loktev AV, Zhang Q, Westlake CJ, Peranen J, et al. (2007) A core complex of BBS proteins cooperates with the GTPase Rab8 to promote ciliary membrane biogenesis. Cell 129: 1201–1213.

2. Seo S, Guo DF, Bugge K, Morgan DA, Rahmouni K, et al. (2009) Requirement of Bardet-Biedl syndrome proteins for leptin receptor signaling. Hum Mol Genet 18: 1323–1331.

3. Seo S, Baye LM, Schulz NP, Beck JS, Zhang Q, et al. (2010) BBS6, BBS10, and BBS12 form a complex with CCT/TRiC family chaperonins and mediate BBSome assembly. Proc Natl Acad Sci U S A 107: 1488–1493.

4. Pereiro I, Valverde D, Pineiro-Gallego T, Baiget M, Borrego S, et al. (2010) New mutations in BBS genes in small consanguineous families with Bardet-Biedl syndrome: detection of candidate regions by homozygosity mapping. Mol Vis 16: 137–143.

5. Forsythe E, Beales PL (2012) Bardet-Biedl syndrome. Eur J Hum Genet.

6. Billingsley G, Deveault C, Heon E (2011) BBS mutational analysis: a strategic approach. Ophthalmic Genet 32: 181–187.

7. Kim JC, Badano JL, Sibold S, Esmail MA, Hill J, et al. (2004) The Bardet-Biedl protein BBS4 targets cargo to the pericentriolar region and is required for microtubule anchoring and cell cycle progression. Nat Genet 36: 462–470.

8. Mykytyn K, Mullins RF, Andrews M, Chiang AP, Swiderski RE, et al. (2004) Bardet-Biedl syndrome type 4 (BBS4)-null mice implicate Bbs4 in flagella formation but not global cilia assembly. Proc Natl Acad Sci U S A 101: 8664–8669.

9. Rahmouni K, Fath MA, Seo S, Thedens DR, Berry CJ, et al. (2008) Leptin resistance contributes to obesity and hypertension in mouse models of Bardet-Biedl syndrome. J Clin Invest 118: 1458–1467.

10. Loktev AV, Zhang Q, Beck JS, Searby CC, Scheetz TE, et al. (2008) A BBSome subunit links ciliogenesis, microtubule stability, and acetylation. Dev Cell 15: 854–865.

11. Ittner LM, Gotz J (2007) Pronuclear injection for the production of transgenic mice. Nat Protoc 2: 1206–1215.

12. Eichers ER, Abd-El-Barr MM, Paylor R, Lewis RA, Bi W, et al. (2006) Phenotypic characterization of Bbs4 null mice reveals age-dependent penetrance and variable expressivity. Hum Genet 120: 211–226.

13. Lechtreck KF, Delmotte P, Robinson ML, Sanderson MJ, Witman GB (2008) Mutations in Hydin impair ciliary motility in mice. J Cell Biol 180: 633–643.

14. Tissir F, Qu Y, Montcouquiol M, Zhou L, Komatsu K, et al. (2010) Lack of cadherins Celsr2 and Celsr3 impairs ependymal ciliogenesis, leading to fatal hydrocephalus. Nat Neurosci 13: 700–707.

15. Zhang Q, Nishimura D, Seo S, Vogel T, Morgan DA, et al. (2011) Bardet-Biedl syndrome 3 (Bbs3) knockout mouse model reveals common BBS-associated phenotypes and Bbs3 unique phenotypes. Proc Natl Acad Sci U S A 108: 20678–20683.

16. Shah AS, Farmen SL, Moninger TO, Businga TR, Andrews MP, et al. (2008) Loss of Bardet-Biedl syndrome proteins alters the morphology and function of motile cilia in airway epithelia. Proc Natl Acad Sci U S A 105: 3380–3385.

17. Seo S, Zhang Q, Bugge K, Breslow DK, Searby CC, et al. (2011) A novel protein LZTFL1 regulates ciliary trafficking of the BBSome and Smoothened. PLoS Genet 7: e1002358.

18. Jin H, White SR, Shida T, Schulz S, Aguiar M, et al. (2010) The conserved Bardet-Biedl syndrome proteins assemble a coat that traffics membrane proteins to cilia. Cell 141: 1208–1219.

19. Simons DL, Boye SL, Hauswirth WW, Wu SM (2011) Gene therapy prevents photoreceptor death and preserves retinal function in a Bardet-Biedl syndrome mouse model. Proc Natl Acad Sci U S A 108: 6276–6281.

20. Livak KJ, Schmittgen TD (2001) Analysis of relative gene expression data using real-time quantitative PCR and the 2(-Delta Delta C(T)) Method. Methods 25: 402–408.

21. Gunning P, Leavitt J, Muscat G, Ng SY, Kedes L (1987) A human beta-actin expression vector system directs high-level accumulation of antisense transcripts. Proc Natl Acad Sci U S A 84: 4831–4835.

22. Zhang Q, Seo S, Bugge K, Stone EM, Sheffield VC (2012) BBS proteins interact genetically with the IFT pathway to influence SHH-related phenotypes. Hum Mol Genet 21: 1945–1953.

23. Jozefczuk J, Drews K, Adjaye J (2012) Preparation of mouse embryonic fibroblast cells suitable for culturing human embryonic and induced pluripotent stem cells. J Vis Exp.

24. Drack AV, Dumitrescu AV, Bhattarai S, Gratie D, Stone EM, et al. (2012) TUDCA slows retinal degeneration in two different mouse models of retinitis pigmentosa and prevents obesity in Bardet-Biedl syndrome type 1 mice. Invest Ophthalmol Vis Sci 53: 100–106.

5

A Constitutively Active Gαi3 Protein Corrects the Abnormal Retinal Pigment Epithelium Phenotype of Oa1−/− mice

Alejandra Young[1,2], Ying Wang[4], Novruz B. Ahmedli[1], Meisheng Jiang[4]*, Debora B. Farber[1,2,3]*

1 Jules Stein Eye Institute, University of California Los Angeles, Los Angeles, California, United States of America, 2 Molecular Biology Institute, University of California Los Angeles, Los Angeles, California, United States of America, 3 Brain Research Institute, University of California Los Angeles, Los Angeles, California, United States of America, 4 Department of Molecular and Medical Pharmacology, University of California Los Angeles, Los Angeles, California, United States of America

Abstract

Purpose: Ocular Albinism type 1 (OA1) is a disease caused by mutations in the OA1 gene and characterized by the presence of macromelanosomes in the retinal pigment epithelium (RPE) as well as abnormal crossing of the optic axons at the optic chiasm. We showed in our previous studies in mice that Oa1 activates specifically Gαi3 in its signaling pathway and thus, hypothesized that a constitutively active Gαi3 in the RPE of Oa1−/− mice might keep on the Oa1 signaling cascade and prevent the formation of macromelanosomes. To test this hypothesis, we have generated transgenic mice that carry the constitutively active Gαi3 (Q204L) protein in the RPE of Oa1−/− mice and are now reporting the effects that the transgene produced on the Oa1−/− RPE phenotype.

Methods: Transgenic mice carrying RPE-specific expression of the constitutively active Gαi3 (Q204L) were generated by injecting fertilized eggs of Oa1−/− females with a lentivirus containing the Gαi3 (Q204L) cDNA. PCR, Southern blots, Western blots and confocal microscopy were used to confirm the presence of the transgene in the RPE of positive transgenic mice. Morphometrical analyses were performed using electron microscopy to compare the size and number of melanosomes per RPE area in putative Oa1−/−, Gαi3 (Q204L) transgenic mice with those of wild-type NCrl and Oa1−/− mice.

Results: We found a correlation between the presence of the constitutively active Gαi3 (Q204L) transgene and the rescue of the normal phenotype of RPE melanosomes in Oa1−/−, Gαi3 (Q204L) mice. These mice have higher density of melanosomes per RPE area and a larger number of small melanosomes than Oa1−/− mice, and their RPE phenotype is similar to that of wild-type mice.

Conclusions: Our results show that a constitutively active Gαi3 protein can by-pass the lack of Oa1 protein in Oa1−/− mice and consequently rescue the RPE melanosomal phenotype.

Editor: Tiansen Li, National Eye Institute, United States of America

Funding: Financial support from the Vision of Children Foundation is gratefully acknowledged. The funders had no role in study design, data collection and analysis, decision to publish, or preparation of the manuscript.

Competing Interests: The authors have declared that no competing interests exist.

* E-mail: jm@ucla.edu (MJ); farber@jsei.ucla.edu (DBF)

Introduction

X-linked ocular albinism is a disorder of melanosome biogenesis leading to congenital visual impairment in males [1]. Affected individuals exhibit nystagmus, reduced visual acuity, hypopigmentation of the iris and retinal pigment epithelium (RPE), foveal hypoplasia, ocular misrouting, reduced or absent binocular functions, photoaversion, strabismus and giant melanosomes in the RPE and skin melanocytes [2]. In addition to macromelanosomes, OA1 patients (as well as Oa1 knockout mice) have a reduction of ipsilateral retinal ganglion axons at the optic chiasm. These are the two main phenotypic characteristic of ocular albinism [3]. Mutations in the OA1 gene [4], also known as the GPR143 gene, are responsible for this disease. More than 60 mutations (missense, nonsense, frameshift or splice-site mutations)

have been identified in affected individuals [4,5]. These changes result in a nonfunctional OA1 protein and often prevent it from reaching its normal location at melanosomal membranes [6,7,8] or from interacting with other molecules of its signaling pathway [9]. Without a functional GPR143 protein, melanosomes in the RPE and melanocytes of the skin become abnormally large, but it is unclear how these macromelanosomes are related to vision abnormalities in patients with ocular albinism.

The OA1 protein is an intracellular G protein-coupled receptor localized in the RPE. We previously identified the inhibitory GTP-binding protein alpha subunit polypeptide 3 (Gαi3) as the specific downstream component of the Oa1 signaling cascade of mice. In addition, we showed that Oa1 and Gαi3 knockout mice present similar abnormal macromelanosomes in the RPE and misrouting of optic axons at the optic chiasm [10]. These findings strongly

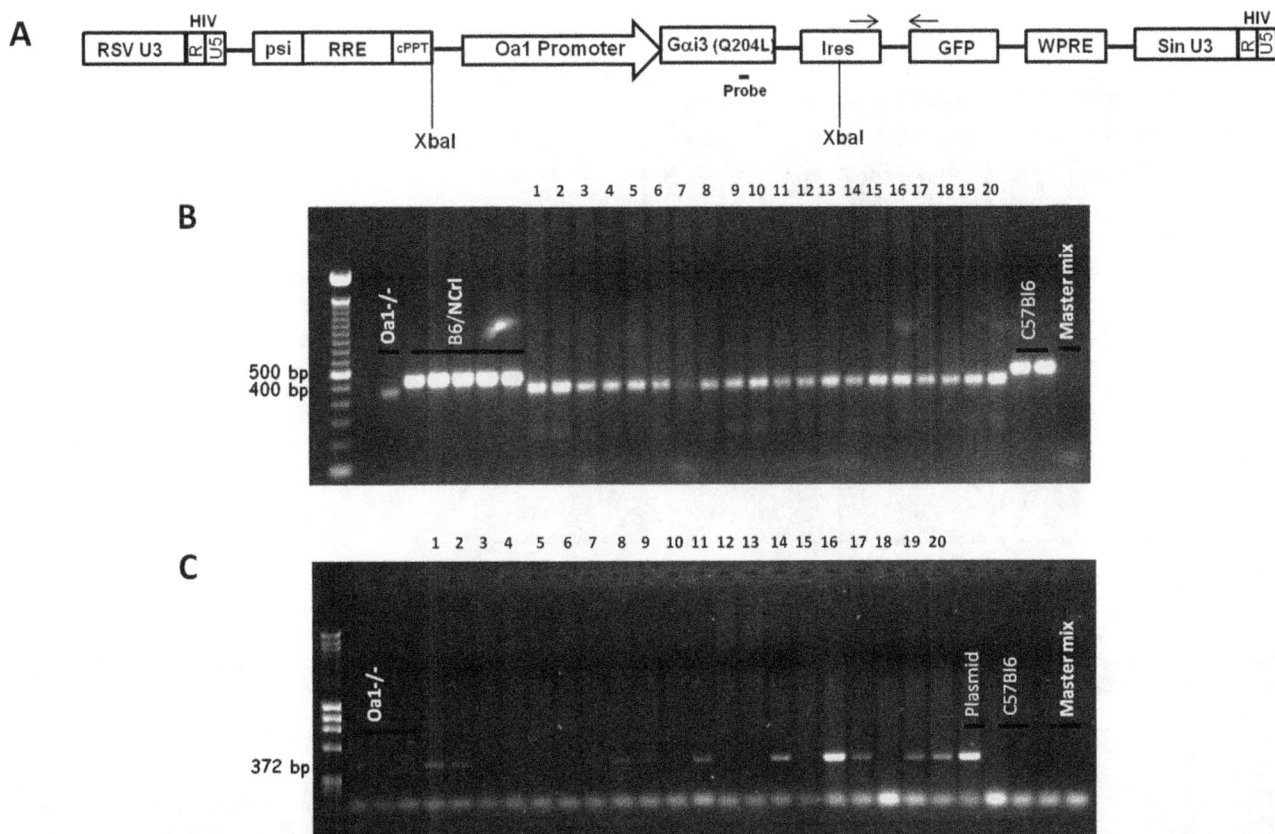

Figure 1. A: Identification of Oa1−/−, Gαi3 (Q204L) transgenic mice. The lentivirus construct used to generate Oa1−/−, Gαi3 (Q204L) transgenic mice contains the constitutively active Gαi3 (Q204L) cDNA driven by the Oa1 promoter and ires-GFP. Abbreviations: RSV, Rous sarcoma virus promoter; U3-HIV-R-U5, 5′ and 3′ detailed long terminal repeats (LTRs); psi, packaging signal for viral RNA into virus capsids to continue the infection of HIV in its host; RRE, Rev-responsive element; cPPT, central polypurine tract; Oa1 (ocular albinism type 1) promoter; Gαi3 (Q204L), constitutively active Gαi3 cDNA; ires, internal ribosome entry site; GFP, green fluorescence protein; WPRE, Woodchuck hepatitis virus post-transcriptional regulatory element; sin U3, self-inactivating element that relies on the introduction of a deletion in the U3 region of the 3′ (LTR). XbaI, restriction enzyme site. Two short thin black arrows indicate the forward and reverse ires-GFP primers used for PCR amplification. The probe used for Southern blots binds to Gαi3 (Q204L) cDNA at the position indicated in the figure. **B:** 1.8% agarose gel showing PCR analysis of the Oa1 gene in putative founder transgenic mice. Specific primer sets (HPRT and Oa1) were used to amplify a 400 bp band of the HPRT cassette and a 500 bp of the endogenous Oa1 gene, respectively. The 20 putative transgenic founders analyzed on the gel had the 400 bp band, indicating that all animals were generated in the Oa1−/− background. **C:** Identification of positive transgenic founders by PCR. 1.8% agarose gel identifying positive Gαi3 (Q204L) transgenic mice. Genotyping was done using specific primers that amplify a 372 bp fragment between the ires and GFP regions of the transgenic construct. Controls used: Oa1−/−, NCrl, C57Bl/6 genomic DNA samples and transgenic plasmid. The Master mix was loaded as a control for contamination.

suggested a common Oa1-Gαi3 signaling pathway, supporting a previously unsuspected role for Gαi3 in the events directly or indirectly related with melanosomal biogenesis.

Gαi3 has been shown to regulate multiple pre- and post-Golgi trafficking steps, suggesting that it may function at variable sites across the Golgi stacks of different cells. In renal cells, Gαi3 is exclusively located on Golgi membranes [11,12]. In exocrine pancreatic cells Gαi3 is found not only in cis- and trans-Golgi membranes, but also on vesicles at both sides of the Golgi stack [13]. In addition, Gαi3 has been demonstrated to regulate protein trafficking in a variety of pathways, i.e., in human intestinal cells, human colon cancer cell line HT-29, human epithelial cells LLC-PK1/NRK and murine erythroleukemia (MEL) cells [12,14,15,16,17,18]. Moreover, Gαi3 has been suggested to have

Table1. Primer sequences (5′ to 3′) used for PCR and Southern blot genotyping.

Primer set	Forward	Reverse	Product Size (bp)
HPRT	taagttctttgctgacctgct	ggctttgtatttgccttttcc	400
Oa1	acatgacgcccaatctccctc	tagactaccctctgagtccag	500
ires-GFP	tgaccctgaagttcatctgca	ttcttctgcttgtcggcggtg	372
Gαi3 (Q204L)Exon 6-7	gtttgatgtaggtggcctaag	ggataacagatagttaacgga	272

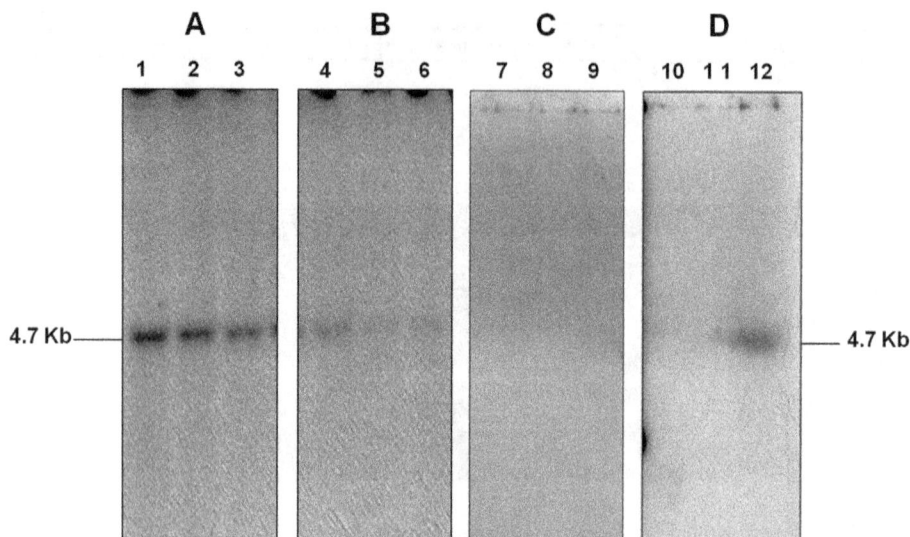

Figure 2. Southern blot analysis of transgenic progeny. Positive transgenic mice were identified by the presence of the expected 4.7 kb band. Each lane contained 10 µg of mouse genomic DNA. **A.** Positive transgenic mice presenting a strong signal on the Southern blot: Lane 1: line 142. Lane 2: line 131. Lane 3: line 157. **B:** Positive transgenic mice with a weak signal: Lane 4: line 377. Lane 5: 374, Lane 6: line 396. **C:** Mice without integrated transgene show no radioactive signal: Lane 7: line 223. Lane 8: Line 275. Lane 9: line 276. **D.** Negative controls NCrl and line 13 (Lanes 10 and 11) and positive control transgenic mouse line 16 (Lane 12).

a role in macroautophagy in the colon carcinoma cell line HT-29 [17] and a role in the insulin-mediated regulation of autophagy [19].

While the precise function of Gαi3 in RPE melanosome biogenesis remains to be delineated, our studies suggest that this protein plays an important role in the control of the size and density of RPE melanosomes and therefore in the regulation of RPE pigmentation [10,20]. We hypothesize that a constitutively active Gαi3 protein could by-pass the lack of Oa1 in $Oa1-/-$ mice and keep the Oa1 signaling cascade going leading to the normalization of their RPE pigmentation. In this study we show that, indeed, introducing the constitutively active Gαi3 (Q204L) protein as a transgene in $Oa1-/-$ mice rescues in them the RPE melanosomal phenotype characteristic of wild-type mice.

Materials and Methods

Transgenic Construct

The expression vector containing the active mutant Gαi3 (Q204L) under the control of the *Oa1* promoter was constructed by subcloning a 3.5 kb genomic DNA fragment of the mouse *Oa1* promoter region, a 1.1 kb human Gαi3 cDNA fragment encoding the Q204L mutation, and a 0.5 kb polyadenylation signal sequence into the pKS pBluescript II KS+ plasmid (Clontech, Palo Alto, CA). The ~5.1 kb *NotI-XbaI Oa1*-Gαi3 (Q204L) fragment was cloned into the *BamHI* and *PstI* sites of a VSVG pseudotyped lentivirus vector. The vector also contained an ires-GFP sequence downstream of the *Oa1*-Gαi3 (Q204L) cassette. Both ends of the insert were sequenced to ensure that the joining was correct. The final lentivirus expression vector (Figure 1A) was packed in 293T cells. The titer of the virus was determined by measuring the viral capsid p24 protein. There are approximately 1×10^4 particles of lentivirus for every pg of p24 antigen, which works out to be 100 transducing units of virus for every pg of p24 antigen present (Sigma Aldridge protocol). The virus packing and production were done at the UCLA Vector Core Facility.

Ethics Statement

All experiments involving mice were carried out using protocols approved by the UCLA Animal Research Committee, and in accordance with the ARVO Statement for Use of Animals in Ophthalmic and Vision Research.

Generation of transgenic mice carrying the constitutively active Gαi3

Wild-type C57Bl/6NCrl (hereafter NCrl) mice and congenic *Oa1* knock-out mice ($Oa1-/-$) were obtained from The Charles River Labs, USA and Italy, respectively. Mice were housed and bred in conventional cages and environmental conditions at the UCLA animal facility. Transgenic animals were generated by microinjection of lentivirus to embryos at the UCLA Transgenic Core Facility. Briefly, fertilized eggs at the pronuclear stage from $Oa1-/-$ mice were microinjected with the lentivirus (~1 picoliter, equivalent to the size of nuclei) containing the constitutively active human Gαi3 (Q204L) cDNA. The injected eggs were then implanted into surrogate mothers to complete the gestation. These putative $Oa1-/-$, Gαi3 (Q204L)-ires-GFP transgenic mice were tagged, genotyped by PCR and used to expand the colony.

Genotyping of transgenic mice

Genomic DNAs were extracted from tail biopsies with Protease K digestion and phenol/chloroform extraction. Mouse genotyping was carried out by PCR and used to confirm the $Oa1-/-$ background and to determine the presence of the Gαi3 (Q204L) transgene in their genome. To corroborate the absence of the *Oa1* gene, the *Oa1* primer set (Table 1) was used to produce a 500 bp fragment representative of the wild-type *Oa1+/+* allele. In addition, the HPRT primer set was used to produce a 400 bp product (Table 1). The HPRT band is a distinct feature of the knockout *Oa1* $-/-$ mice. PCR reactions were performed in a total volume of 25 µl [2.5 µl of 10X PCR buffer, 2 µl of dNTPs (each 2.5 mM), 1 µl of each primer (10 pmol/µl each), 0.2 µl of Taq polymerase (1U), and ddH$_2$O to 25 µl], for a total of 30 cycles

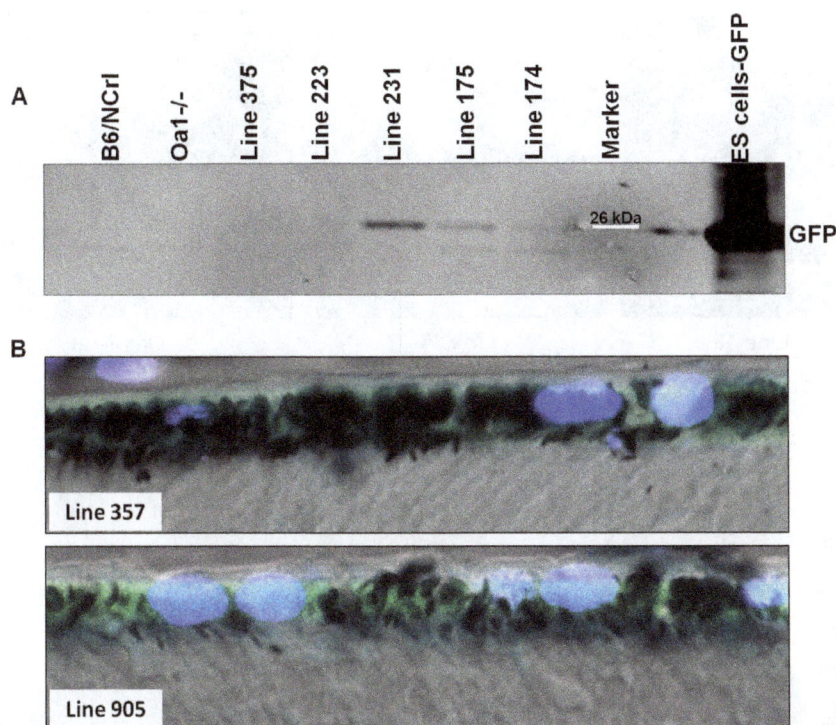

Figure 3. GFP expression in the RPE varies in different transgenic mice, suggesting a possible correlation between level of GFP expression and transgene copy number. A: Western blot using RPE/choroid protein extracts and GFP antibody. Positive transgenic mouse lines 231,175 and 174 expressed GFP in the RPE. Controls NCrl and *Oa1−/−*, as well as mice without integrated transgene (lines 375 and 223) did not express GFP. B: Confocal images (magnification: 174X) of the RPE from positive transgenic mice show expression of GFP.

using the following conditions: denaturing at 94°C for 30 sec., annealing at 55°C for 45 sec. and extension at 72°C for 45 sec. To determine the presence of the Gαi3 (Q204L) transgene in the mice genome, the ires-GFP region of the lentivirus construct was amplified. The ires-GFP primer set was used to produce a 372 bp DNA fragment. The PCR reaction was set up as above and run using a total of 33 cycles of denaturing at 94°C for 30 sec., annealing at 62°C for 1 min. and extension at 72°C for 2 min. All samples were run on 1.8% agarose gels.

Southern blot

Genomic DNA from each mouse (10 µg) was digested with *XbaI* separated by electrophoresis on a 1% agarose gel and transferred onto Hybond-N+ membrane (Amersham Biosciences, NJ). The hybridization probe was generated by PCR using the Gαi3 (Q204L) primer set designed to amplify a sequence of the Gαi3 (Q204L) cDNA that corresponds to the end of exon 6 and beginning of exon 7 of the *Gαi3* gene. The gel purified-PCR product was then labeled with [α-^{32}P] dCTP using Ready-To-Go DNA labeling beads (Amersham Biosciences). Presence of the transgene was confirmed by an expected 4.7 kb radioactive band.

Morphometrical Analysis

3-month-old *Oa1−/−*, Gαi3 (Q204L)-ires-GFP positive and negative transgenic mice, NCrl and *Oa1−/−* mice were deeply anesthetized by Isoflurane (30% concentration) inhalation. The left eyes were enucleated for protein extraction before intravenous perfusion with 2% formaldehyde and 2.5% glutaraldehyde in 0.1 M sodium phosphate buffer, pH 7.4. After perfusion, the right eyes were enucleated, washed in 0.1 M sodium phosphate buffer, fixed with 1% osmium tetroxide, washed with double distilled cold

water, dehydrated in ethanol solutions of increasing concentrations, and embedded in Epon/Araldite mixed resin.

Electron microscopy. The nasal regions of all eyes were sectioned into superior and inferior quadrants. Eye quadrants were rinsed in 0.1 M sodium phosphate buffer, post-fixed with 1% buffered osmium tetroxide, dehydrated in ethanol solutions of increasing concentrations, and embedded in araldite 502 (Electron Microscopy Sciences, PA). Sections (60–70 nm) were cut on a Leica Ultracut UCT and collected on 200 mesh uncoated copper grids. For the ultrastructural analysis, stained sections (5% uranyl acetate and 0.4% lead citrate) were analyzed with a 910 Zeiss electron microscope and the RPE fields were photographed using a Keenview^{TM} digital camera. For quantification of melanosomes and determination of their area in the RPE, we analyzed 12 micrographs corresponding to the nasal region of the eye at 16,000X magnification using the analysis^{TM} software for LEO 900 TEM, version 3.2. (Soft Imaging System, Lakewood, CO). To determine the size and number of melanosomes, each micrograph was loaded into the Soft Imaging System analySIS software platform and the area of individual melanosomes was selected and measured with the magic wand tool. In addition, the total RPE area containing the melanosomes was selected and measured using the pencil tool. The program produced two spreadsheets per micrograph containing the melanosomal count, the individual areas, and their mean size. We analyzed the differences in the melanosomal size in the RPEs from the different eyes and the percent frequency of melanosomes in the 100 nm^2 to more than 15,000 nm^2 size range. We also calculated the number of melanosomes per RPE area in the transgenic mice and the controls NCrl and *Oa1−/−* mice. Comparisons among all groups

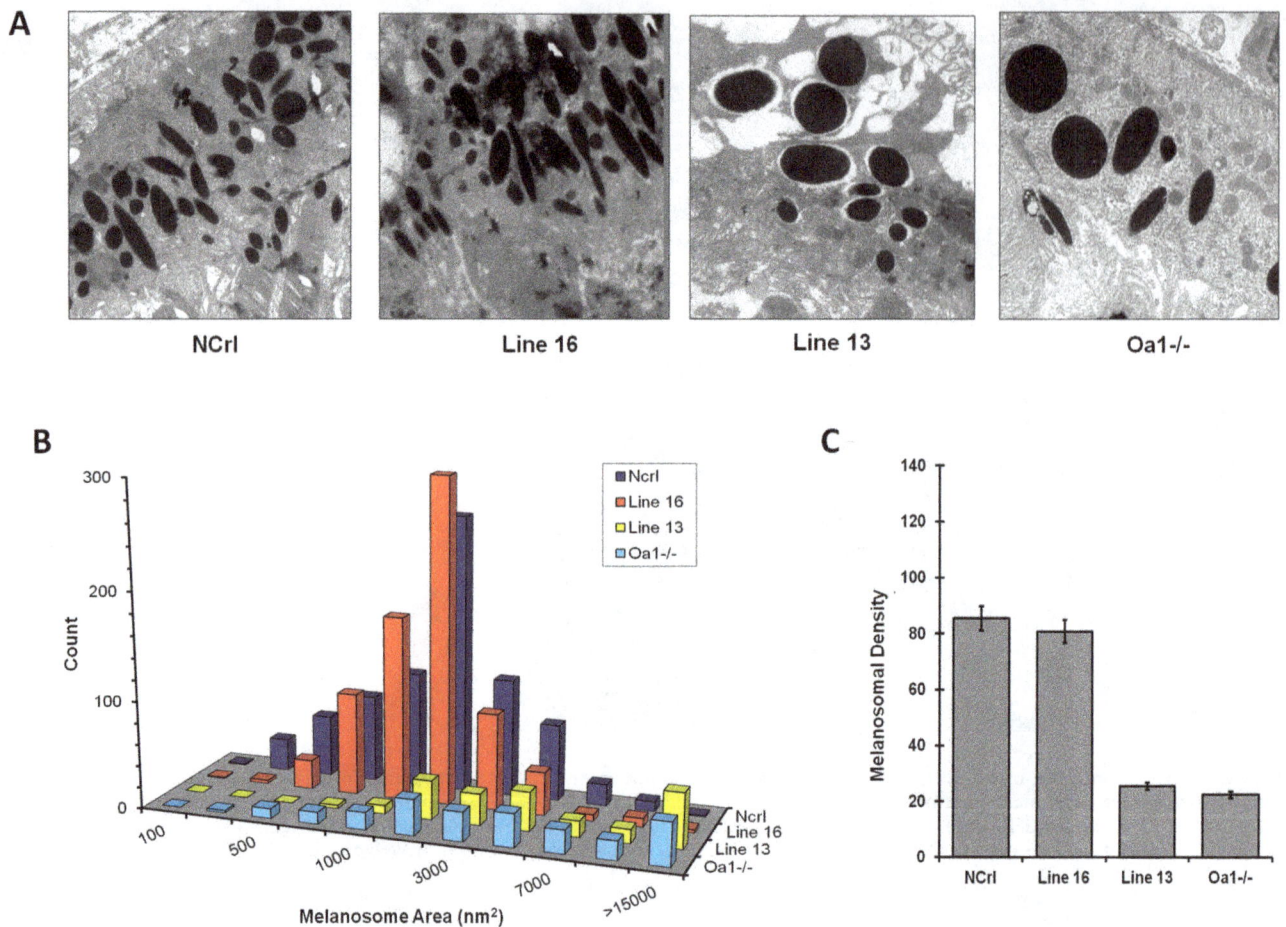

Figure 4. Electron microscopy and densitometry analysis of putative transgenic founder mice. A: Electron micrographs comparing RPEs from positive and negative transgenic founder mouse lines 16 and 13 to wild-type NCrl and $Oa1-/-$ RPEs, respectively. Panel **A** shows electron micrographs of RPEs from: NCrl, transgenic line 16 and line 13 without integrated transgene and $Oa1-/-$ mice. **B-C:** Morphometrical analysis of RPE melanosomes from lines 16 and 13. **B:** Histogram representing the percent distribution of melanosomes by size. RPEs from $Oa1-/-$ mice and line 13 have highest percentage of melanosomes larger than 15,000 nm^2 (19.1% and 26.7% respectively), while NCrl, and line 16 RPEs have higher frequency of smaller and medium size melanosomes (500–3000 nm^2), 67.6% and 81.6%, respectively. **C.** $Oa1-/-$ and line 13 RPEs have very low mean melanosomal densities, 22.3 and 25.3 melanosomes per RPE area, respectively. NCrl and line 16 RPEs have mean densities of 85.5 and 80.7 melanosomes per RPE area, respectively.

were done using a one-way analysis of variance (ANOVA). Standard errors of the mean are shown in the statistical analysis.

Western blot analysis

The GFP protein expression in RPE-Choroid protein extracts from one eye of each NCrl and $Oa1-/-$ control mice and from five putative transgenic mice was determined using an anti-GFP antibody (Cell Signaling technology, Danvers, MA). Protein extracts (15 µg per lane) were separated by SDS-PAGE using a 4–12% gradient Tris-gel (Life Technologies, NY) and blotted onto nitrocellulose. The membranes were blocked in Odyssey's blocking buffer (Li-Cor Biotechnology, NE) and probed overnight at 4°C with primary polyclonal rabbit anti-GFP antibody diluted 1:2,000 (Santa Cruz Biotechnology, Santa Cruz, CA), followed by incubation for 45 min at RT with secondary Alexa Fluor 680-conjugated goat anti-rabbit IgG diluted 1:5,000. Antibody recognition was detected with an Odissey scanner according to the manufacturer's instructions (Licor Biotech, NE).

Confocal Microscopy

Eyes from 6 month-old mice positive for the transgene (lines 357 and 905) were enucleated, rinsed in 0.1 M sodium phosphate buffer, post-fixed with 4% paraformaldehyde, and embedded in OCT compound prior to cryostat sectioning. Cryostat sections were cut at 8 µm, mounted on histological slides, and air dried at room temperature for direct observation of GFP fluorescence and nuclei stained with DAPI. Images were taken using a 60X objective lens and a Zoom of 2.9 for a total magnification of 174X.

Results

Generation of $Oa1-/-$, Gαi3 (Q204L) transgenic mice

We generated transgenic mice using a VSVG pseudotyped lentivirus vector carrying the constitutively active Gαi3 (Q204L) protein expression cassette. Appropriate expression of this constitutively active Gαi3 in the RPE of $Oa1-/-$ mice was obtained using the RPE-specific promoter of the $Oa1$ gene (Figure 1A). It has been shown that transgenic mice can be generated

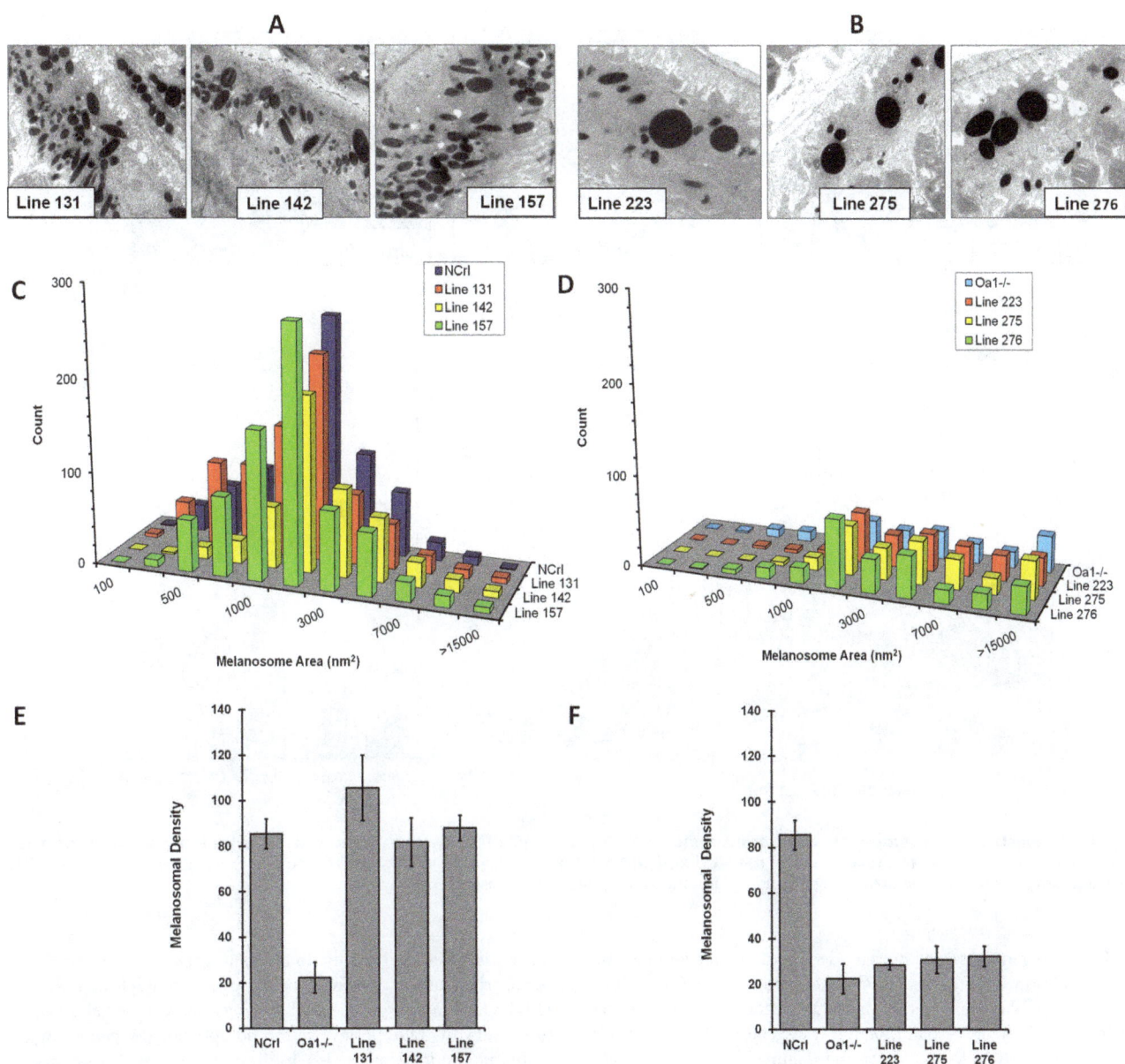

Figure 5. Electron microscopy and densitometry analysis of the progeny of transgenic founders, comparing the RPEs of strong positive transgenic Gαi3 (Q204L) animals and those of mice without integrated transgene. A. Electron micrographs of RPEs from positive-strong transgenic mice lines 131, 142, and 157. B. Electron micrographs of RPEs from mice without integrated transgene lines 223, 275, and 276. C and D. Histograms representing the percent of RPE melanosomal size distribution in transgenic mice and animals without integrated transgene, respectively. E and F. Melanosomal density analysis of transgenic and non-transgenic progeny, respectively.

efficiently by microinjection of lentivirus into the perivitelline space of single-cell mouse embryos, and that injecting a high dose of virus can produce near 100% transgenesis with variable copies of transgene in transgenic lines [21]. In this study, we intended to generate transgenic lines with one or few copies of the transgene by lowering the lentivirus input, aiming for a 40~70% transgenesis rate. The titer of lentivirus used was 18.9 µg/ml of p24 protein, which was estimated to be 1.9×10^9 transducing units/ml. ~0.5–1pL of the stock virus, which is equivalent to ~1–2 transducing units, were injected into the perivitelline space of single-cell mouse embryos. Embryos were implanted into pseudopregnant females and were carried to full term yielding 20 putative transgenic founder pups (tagged with IDs 1-20).

Initial genotyping verified that these pups were *Oa1* knockouts. The *Oa1*−/− mice used for these experiments had been generated by introducing an HPRT cassette to delete exon 1 of the *Oa1* gene [22]. PCR amplification of mouse genomic DNA using the HPRT cassette primer set (Table 1) produces a 400 bp band characteristic of the *Oa1*−/− allele. The *Oa1* wild-type primer set (Table 1), which amplifies a 500 bp fragment of the *Oa1* gene exon 1 was used to identify the *Oa1* wild-type allele. The DNA samples from all founder transgenic mice showed only the targeted allele 400 bp amplicon and lacked the wild-type 500 bp PCR band (Figure 1B), indicating that indeed, all 20 pups were derived from *Oa1*−/− mice.

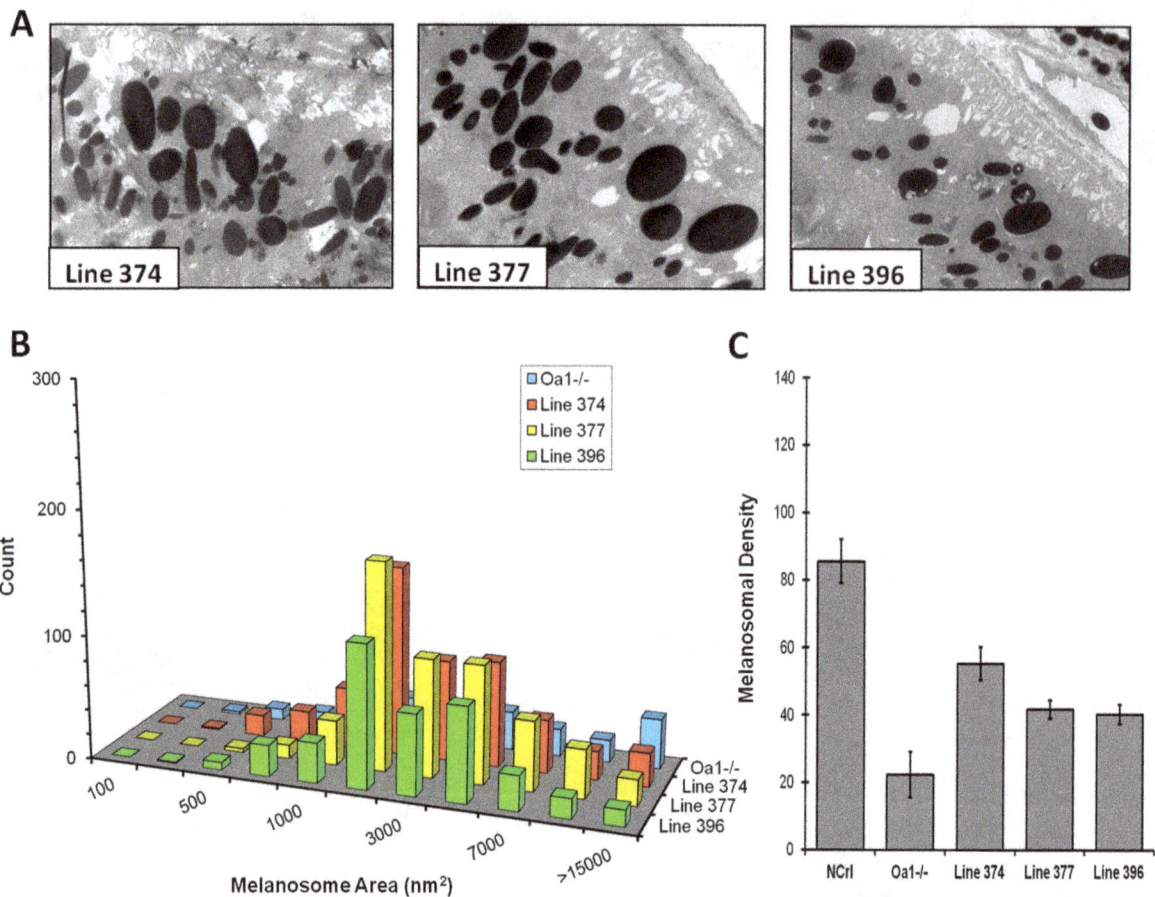

Figure 6. Electron microscopy of weak transgenic *Oa1*−/−, Gαi3 (Q204L) mouse RPEs and densitometry analysis of their melanosomes compared to those of wild-type NCrl and Oa1 −/− mice. A. Electron micrograph of mouse RPEs from lines 374, 377, and 396. B. Histogram representing the percent distribution of melanosomal sizes. C. Melanosomal density analysis.

Then, we genotyped the pups to identify transgenic founders by PCR analysis using the ires-GFP primer set (Table 1). Our results showed that 11 pups out of the 20 generated presented the expected 372 bp PCR band for the transgene while the other 9 pups were negative for the transgene (Figure 1C). Since we only injected few transducing units per each embryo, the transgenesis rate was expected to be 40~70%, and positive transgenic founders were anticipated to contain one or two copies of the transgene. The PCR results showed the generation of 60% transgenic founders, which falls into the ballpark of our strategy. In addition, there was a variation in the intensity of the transgene PCR bands among the positive transgenic lines, which might reflect the different copy number. We grouped founder lines according to the intensity of the transgene PCR band normalized to the corresponding HPRT cassette amplicon intensity in the pups. The progeny group with lower intensity more likely represents that with a single copy of the transgene, while samples with strong intensity perhaps had two or more copies.

Germline transmission of the Gαi3 (Q204L) transgene

The innate ability of lentiviral vectors to integrate into the host genome with high efficiency is limited by the fact that each lentiviral integration site contains a single copy of the transgene, rather than the head to tail arrays of multiple transgenes that occur when using DNA pronuclear injection. This means that multiple copies of the transgene will be segregated in the progeny.

Therefore, we interbred the founders to generate a diverse group of offspring, having negative, single copy and multiple copies of the Gαi3 (Q204L) transgene. This breeding scheme could also help us to identify any false positive restoration phenotypes caused by the lentivirus random integration leading to the host genes inactivation. The resulting progeny and subsequent generations of mice were tagged with ID numbers 100–1000, without parental identification. Southern blots were used to determine the relative transgene expression level in fifty mice derived from the interbreeding of transgenic lines. Results of the Southern blots indicate that the transgenic mice progeny could be classified into three different groups (negative, low and high copy number) according to the relative hybridization intensity (Figure 2), similar to the different intensities of the PCR bands obtain from the founder transgenic mice, showed in Figure 1C. Among the progeny, the first group represents positive transgenic mice with a strong signal: lines 142, 131, 157, (Figure 2A, lanes 1–3, respectively). The second group corresponds to positive transgenic mice with weak radioactive band intensity: lines 377, 374, and 396 (Figure 2B, lanes 4–6, respectively). The last group shows negative transgenic mice with absence of radioactive band: lines 223, 275, and 276 (Figure 3C, lanes 7–9, respectively). We conclude that the transgene can effectively be transmitted to offspring and that we were able to obtain lines with different copy numbers.

Specific RPE expression of GFP in transgenic lines

Since no highly specific anti-Gαi3 antibody is available, we designed the viral expression cassette with ires-GFP to monitor the expression of the transgene indirectly. We performed Western blot analysis to detect GFP expression in RPE cells from a single eye using anti-GFP antibodies. The 26-kDa GFP band was detected in the positive transgenic lines and absent in non-transgenic lines (Figure 3A). Variable intensity of the GFP signal was also observed among the positive lines. Line 231 showed strong staining intensity, while lines 174 and 175, had a weak signal. Lines 223 and 375 were negative for GFP. In addition, confocal microscopy of fixed, OCT embedded and mounted retina sections showed expression of GFP in the RPE of positive transgenic mice from lines 357 and 905 (Figure 3B). Thus, Western and confocal analyses of GFP expression in the RPEs of transgenic mice demonstrated that the transgene is stably integrated and transmitted through the germline and not silenced. Genotyping analyses (PCR and Southern) demonstrated a correlation of transgene copy number with the transgene protein expressed (Western).

Rescue of the RPE phenotype in Oa1−/− mice by expression of the Gαi3 (Q204L) transgene

Lentiviral microinjection at high percentage of transgenesis rate allowed us to study directly the effect of RPE phenotype restoration by the constitutively active Gαi3 (Q204L) transgene in transgenic founders. To test whether expression of the active Gαi3 mutant in the RPE of Oa1−/− mice can bypass the requirement of a functional Oa1 receptor, we performed morphometrical analysis of melanosomal size and density in the RPEs of transgenic mice using electron microscopy. We analyzed and compared the transgene line 16, which has several copies of the transgene, with the negative control line 13. The RPE of the positive transgenic founder line 16 did not present macromelanosomes, which are characteristic of the parental Oa1−/− mice; instead it had a homogeneous, smaller in size, melanosomal population similar to that observed in wild-type NCrl RPE (Figure 4, A and B). On the other hand, the RPE of negative transgenic mouse line 13 exhibited a similar Oa1−/−, abnormal, macromelanosomal phenotype (Figure 4, C and D). Accordingly, melanosomes 2,000 nm^2 in size were present in high numbers in NCrl and line 16 RPEs (34.6% and 45.7% respectively) but were found at a lower frequency in Oa1−/− and negative transgenic line 13 RPEs (15.8% and 18.8%, respectively) (Figure 4B). In addition, our results indicate that the presence of the transgene in mice of the Oa1−/− background correlates with a high number of melanosomes per RPE area, similar to what is observed in wild-type NCrl RPE: i.e., positive transgenic mouse line 16 had a mean density of 80.7±12.7 melanosomes per RPE area, and wild-type NCrl of 85.5±6.5 (Figure 4C). Thus, these results suggest that the presence of the transgene could restore the normal melanosomal size and density in the Oa1−/− RPE.

Genotyping and protein expression studies of the progeny of founder transgenic mice showed that transgenic lines also could be divided into two expressions groups, high and low. Therefore, we analyzed the RPE morphology of the progeny with high transgene expression to determine whether the RPE melanosomal phenotype was restored similarly to that in founder line 16. Electron microscopy analysis showed, that the population and distribution of melanosomes in the high transgene expression group are completely different from their parental Oa1−/−, and rather similar to Oa1+/+ NCrl control. On the other hand, as expected, the negative transgenic mice retained the Oa1−/− RPE

phenotype, showing the presence of macromelanosomes (Figure 5 A and B).

Furthermore, morphometrical analysis of RPE melanosomes in these groups showed that presence of the transgene correlated with a high number of medium and small melanosomes. Mice identified as lines 131, 142, and 157, showed highest frequency of melanosomes 2,000 nm^2 in size (29.1%, 37.3% and 35.3%, respectively) and had very low percentages of melanosomes larger than 15,000 nm^2: 0.8%, 1.4% and 0.8%, respectively (Figure 5C). Conversely, the group without an integrated transgene had the highest percentage of macromelanosomes and lowest number of small and medium melanosomes. For example, mouse lines 223, 275, and 276 had 13.0%, 16.9%, and 11.8% melanosomes larger than 15,000 nm^2, respectively (Figure 5D and Table S1). The mean RPE melanosomal density in high expression lines 131, 142 and 157 was 105.6±14.3, 81.9±10.6 and 87.9±5.6, respectively (Figure 5E). On the contrary, negative transgenic lines 223, 275, and 276 had a reduced number of melanosomes per RPE area, 28.3±2.1, 30.6±6.0 and 32.2±4.4 respectively, which resembles the characteristic Oa1−/− RPE melanosomal phenotype (Figure 5F). Thus, our results demonstrated that transgenic animals with high transgene expression showed complete rescue of the melanosomal phenotype.

We also analyzed the transgenic lines with low transgene expression, which likely carry a single copy of the Gαi3 (Q204L) transgene. In these mice, a partial rescue of the RPE melanosomal phenotype was observed, with the presence of few macromelanosomes among many more small melanosomes (Figure 6A). Lines 374, 377, and 396 are examples of this type of mice that have 30.2%, 31.9%, and 30.5% melanosomes of 2,000 nm^2, respectively, and 5.4%, 4.0%, and 3.4% melanosomes of 15,000 nm^2, respectively (Figure 6B). They also showed a partial rescue of the melanosomes per RPE area and presented a mean density of 55.3±4.9, 41.7±2.8, and 40.3±2.9, respectively (Figure 6C). This suggests there is a Gαi3 dose-dependent restoration for RPE melanogenesis.

In summary, our results demonstrate that the constitutively active Gαi3 protein indeed can rescue the abnormal RPE phenotype observed in Oa1−/− mice, by restoring the size and number of its melanosomes. This further corroborates the involvement of Gαi3 in RPE melanogenesis.

Discussion

We have previously demonstrated by in-vitro and in-vivo studies on mice that Gαi3 physically interacts with the Oa1 protein in the RPE and that lack of Gαi3 from the mouse genome leads to the presence of RPE macromelanosomes and a reduced melanosomal density, suggesting that Gαi3 is the first downstream component in the Oa1 signaling pathway [10,20]. This work was designed to test the hypothesis that Gαi3 is the major transducer for Oa1-mediated melanogenesis in RPE cells. We introduced a constitutively active Gαi3 protein in Oa1−/− RPE cells, to test in-vivo if the activation of Gαi3 and its downstream signaling could bypass the lack of Oa1 protein and keep the cascade going, leading to normal pigmentation and melanosomal morphology of the RPE.

Transgenic mice were efficiently generated by infecting Oa1−/− fertilized eggs with a lentiviral vector carrying the constitutively active Gαi3 protein driven by the RPE-specific promoter of the Oa1 gene. Substitution of a conserved glutamine for a leucine residue (Q204L) within the GTPase domain of Gαi3, results in a constitutively active form of this protein. A lentiviral vector was chosen to generate the transgenic animals, since it is able to transduce efficiently both dividing and non dividing cells [23]. Moreover, it has been shown that the transgenes delivered by lentiviral vectors are capable of escaping gene silencing and expressing stably *in-vivo* [24,25]. The lentiviral construct contained not only the Gαi3 (Q204L) cDNA, but also the ires-GFP cDNA, which allowed us to determine the expression pattern and levels of the transgene.

After generating *Oa1*−/−, Gαi3 (Q204L) transgenic positive founder mice, we demonstrated the germline transmission of the transgene among their progeny. Our results showed that the presence of the constitutively active Gαi3 in the *Oa1*−/− RPE rescued the size distribution and density of its melanosomes. Most interestingly, the data showed that there is a strong correlation between the degree of rescue observed and the relative transgene expression level. Positive mice with a high expression level of the transgene presented an RPE with an average number of melanosomes (727±8.1) similar to that of the wild-type NCrl control mice (743±2.6), indicating that a high level of expression of the constitutively active Gαi3 (Q204L) is most likely sufficient to activate a normal melanogenesis signaling pathway. In contrast, transgenic mice with weak Gαi3 (Q204L) expression showed an increased average number of melanosomes (506±4.1) from that of Oa1−/− RPE (209±1.8) but still presented some macromelanosomes. This corroborated our hypothesis that there might be a necessary threshold of activated Gαi3 protein needed for proper melanosome biogenesis. The precise mechanism for regulation of the number of melanosomes in the RPE remains to be determined. It is noteworthy that transgenic founder mice and their progeny from interbreeding that have multiple copies of the transgene exhibit full restoration of the RPE phenotype while those transgenic founders or their offspring with a single copy showed partial restoration. This suggests that the melanosomal rescue observed in individual transgenic lines correlates simply to copy number rather than to the integration site in the genome of the transgenic founders.

Our data once again confirmed that Gαi3 is indeed the major transducer of Oa1 in this G-protein coupled receptor signaling cascade as well as its importance in controlling the regulation of RPE pigmentation. Furthermore, it depicted Gαi3 as the lead factor in melanogenesis, since its constitutive activation when the Oa1 receptor is not functional results in a normal signaling pathway. Thus, Gαi3 (Q204L) ensures the proper recruitment of melanosome biogenesis factors needed for the normal regulation of the size and number of RPE melanosomes. We previously proposed a model suggesting that Gαi3 controls the size of melanosomes in the RPE through the inhibition of vesicle trafficking from the trans-Golgi network (TGN) to the melanosomes [20]. This could explain the changes in RPE phenotype observed in ocular albinism when a non-functional OA1 protein (that cannot activate Gαi3) leads to a continuous vesicular traffic of

membrane proteins to melanosomes resulting in the formation of macromelanosomes. This further supports the notion that giant melanosomes are formed by overgrowth of single melanosomes rather than by fusion of several small melanosomes. None of the electronmicrographs that we analyzed had melanosomes in the process of merging or dividing. Therefore, we conclude that there is no fusion or fission of these organelles during melanogenesis.

It has been shown that both α and βγ subunits of heterotrimeric G proteins are involved in the receptor signaling regulating many biological processes [26]. Interestingly, our data suggest that melanogenesis signaling is mediated mainly by Gαi3 alone, and that the Gi βγ moiety plays little or no role at all.

RPE abnormal pigmentation has been shown to be closely related to the abnormal routing of the optic axons [27,28,29]. An important question that remains to be address is whether the introduction of the constitutively active Gαi3 protein into the RPE rescues the misrouting of RGC axons at the *Oa1*−/− transgenic mice optic chiasm. If that were the case, we would be able to conclude that Gαi3, helping to maintain a healthy melanized RPE, is needed for proper development of the optic axon projections.

In summary, we have proved that a constitutively active Gαi3 protein circumvents the need for a functional Oa1 receptor to regulate RPE melanogenesis, and have confirmed that Gαi3 is essential in the Oa1 signaling cascade. The discovery of other members of this pathway will increase our understanding of ocular albinism and might open possibilities to develop therapeutic strategies for this disease.

Supporting Information

Table S1 Percentage of melanosomes by size distribution in transgenic lines, control wild-type, and *Oa1*−/− RPEs. Each percentage was calculated from the total number of melanosomes in 12 electronmicrographs analyzed/sample, as described in Materials and Methods.

Acknowledgments

We thank Dr. Fei Yu for his assistance with the statistical analysis and Shannan Eddington and Marcia Lloyd for their valuable help and guidance with electron microscopy.

Author Contributions

Conceived and designed the experiments: DBF MJ AY. Performed the experiments: AY YW. Analyzed the data: AY NBA DBF. Contributed reagents/materials/analysis tools: DBF MJ. Wrote the paper: AY DBF MJ.

References

1. Lang GE, Rott HD, Pfeiffer RA (1990) X-linked ocular albinism. Characteristic pattern of affection in female carriers. Ophthalmic Paediatr Genet 11: 265–271.
2. King RA, Summers CG (1988) Albinism. Dermatol Clin 6: 217–228.
3. Shen B, Samaraweera P, Rosenberg B, Orlow SJ (2001) Ocular albinism type 1: more than meets the eye. Pigment Cell Res 14: 243–248.
4. Schiaffino MV, Bassi MT, Galli L, Renieri A, Bruttini M, et al. (1995) Analysis of the OA1 gene reveals mutations in only one-third of patients with X-linked ocular albinism. Hum Mol Genet 4: 2319–2325.
5. Schnur RE, Gao M, Wick PA, Keller M, Benke PJ, et al. (1998) OA1 mutations and deletions in X-linked ocular albinism. Am J Hum Genet 62: 800–809.
6. Newton JM, Orlow SJ, Barsh GS (1996) Isolation and characterization of a mouse homolog of the X-linked ocular albinism (OA1) gene. Genomics 37: 219–225.
7. Bassi MT, Incerti B, Easty DJ, Sviderskaya EV, Ballabio A (1996) Cloning of the murine homolog of the ocular albinism type 1 (OA1) gene: sequence, genomic structure, and expression analysis in pigment cells. Genome Res 6: 880–885.
8. Schiaffino MV, Baschirotto C, Pellegrini G, Montalti S, Tacchetti C, et al. (1996) The ocular albinism type 1 gene product is a membrane glycoprotein localized to melanosomes. Proc Natl Acad Sci U S A 93: 9055–9060.
9. d'Addio M, Pizzigoni A, Bassi MT, Baschirotto C, Valetti C, et al. (2000) Defective intracellular transport and processing of OA1 is a major cause of ocular albinism type 1. Hum Mol Genet 9: 3011–3018.
10. Young A, Powelson EB, Whitney IE, Raven MA, Nusinowitz S, et al. (2008) Involvement of OA1, an intracellular GPCR, and G alpha i3, its binding protein, in melanosomal biogenesis and optic pathway formation. Invest Ophthalmol Vis Sci 49: 3245–3252.
11. Ercolani L, Stow JL, Boyle JF, Holtzman EJ, Lin H, et al. (1990) Membrane localization of the pertussis toxin-sensitive G-protein subunits alpha i-2 and alpha i-3 and expression of a metallothionein-alpha i-2 fusion gene in LLC-PK1 cells. Proc Natl Acad Sci U S A 87: 4635–4639.
12. Stow JL, de Almeida JB, Narula N, Holtzman EJ, Ercolani L, et al. (1991) A heterotrimeric G protein, G alpha i-3, on Golgi membranes regulates the

secretion of a heparan sulfate proteoglycan in LLC-PK1 epithelial cells. J Cell Biol 114: 1113–1124.

13. Denker SP, McCaffery JM, Palade GE, Insel PA, Farquhar MG (1996) Differential distribution of alpha subunits and beta gamma subunits of heterotrimeric G proteins on Golgi membranes of the exocrine pancreas. J Cell Biol 133: 1027–1040.

14. Ogier-Denis E, Petiot A, Bauvy C, Codogno P (1997) Control of the expression and activity of the Galpha-interacting protein (GAIP) in human intestinal cells. J Biol Chem 272: 24599–24603.

15. Brand SH, Holtzman EJ, Scher DA, Ausiello DA, Stow JL (1996) Role of myristoylation in membrane attachment and function of G alpha i-3 on Golgi membranes. Am J Physiol 270: C1362–1369.

16. Leyte A, Barr FA, Kehlenbach RH, Huttner WB (1992) Multiple trimeric G-proteins on the trans-Golgi network exert stimulatory and inhibitory effects on secretory vesicle formation. EMBO J 11: 4795–4804.

17. Ogier-Denis E, Couvineau A, Maoret JJ, Houri JJ, Bauvy C, et al. (1995) A heterotrimeric Gi3-protein controls autophagic sequestration in the human colon cancer cell line HT-29. J Biol Chem 270: 13–16.

18. Wilson BS, Palade GE, Farquhar MG (1993) Endoplasmic reticulum-through-Golgi transport assay based on O-glycosylation of native glycophorin in permeabilized erythroleukemia cells: role for Gi3. Proc Natl Acad Sci U S A 90: 1681–1685.

19. Gohla A, Klement K, Nurnberg B (2007) The heterotrimeric G protein G(i3) regulates hepatic autophagy downstream of the insulin receptor. Autophagy 3: 393–395.

20. Young A, Jiang M, Wang Y, Ahmedli NB, Ramirez J, et al. (2011) Specific interaction of Galphai3 with the Oa1 G-protein coupled receptor controls the size and density of melanosomes in retinal pigment epithelium. PLoS One 6: e24376.

21. Lois C, Hong EJ, Pease S, Brown EJ, Baltimore D (2002) Germline transmission and tissue-specific expression of transgenes delivered by lentiviral vectors. Science 295: 868–872.

22. Incerti B, Cortese K, Pizzigoni A, Surace EM, Varani S, et al. (2000) Oa1 knock-out: new insights on the pathogenesis of ocular albinism type 1. Hum Mol Genet 9: 2781–2788.

23. Naldini L, Blomer U, Gallay P, Ory D, Mulligan R, et al. (1996) In vivo gene delivery and stable transduction of nondividing cells by a lentiviral vector. Science 272: 263–267.

24. Hamaguchi I, Woods NB, Panagopoulos I, Andersson E, Mikkola H, et al. (2000) Lentivirus vector gene expression during ES cell-derived hematopoietic development in vitro. J Virol 74: 10778–10784.

25. Somia N, Verma IM (2000) Gene therapy: trials and tribulations. Nat Rev Genet 1: 91–99.

26. Birnbaumer L (1992) Receptor-to-effector signaling through G proteins: roles for beta gamma dimers as well as alpha subunits. Cell 71: 1069–1072.

27. Guillery RW, Okoro AN, Witkop CJ Jr (1975) Abnormal visual pathways in the brain of a human albino. Brain Res 96: 373–377.

28. Creel DJ, Summers CG, King RA (1990) Visual anomalies associated with albinism. Ophthalmic Paediatr Genet 11: 193–200.

29. Rachel RA, Mason CA, Beermann F (2002) Influence of tyrosinase levels on pigment accumulation in the retinal pigment epithelium and on the uncrossed retinal projection. Pigment Cell Res 15: 273–281.

p53 Selectively Regulates Developmental Apoptosis of Rod Photoreceptors

Linda Vuong[1], Daniel E. Brobst[1], Ivana Ivanovic[1], David M. Sherry[1,2,3], Muayyad R. Al-Ubaidi[1,2]*

1 Department of Cell Biology, University of Oklahoma Health Sciences Center, Oklahoma City, Oklahoma, United States of America, **2** Oklahoma Center for Neurosciences, University of Oklahoma Health Sciences Center, Oklahoma City, Oklahoma, United States of America, **3** Department of Pharmaceutical Sciences, University of Oklahoma Health Sciences Center, Oklahoma City, Oklahoma, United States of America

Abstract

Retinal cells become post-mitotic early during post-natal development. It is likely that p53, a well-known cell cycle regulator, is involved in regulating the genesis, differentiation and death of retinal cells. Furthermore, retinal cells are under constant oxidative stress that can result in DNA damage, due to the extremely high level of metabolic activity associated with phototransduction. If not repaired, this damage may result in p53-dependent cell death and ensuing vision loss. In this study, the role of p53 during retinal development and in the post-mitotic retina is investigated. A previously described super p53 transgenic mouse that expresses an extra copy of the mouse p53 gene driven by its endogenous promoter is utilized. Another transgenic mouse (HIP) that expresses the p53 gene in rod and cone photoreceptors driven by the human interphotoreceptor retinoid binding protein promoter was generated. The electroretinogram (ERG) of the super p53 mouse exhibited reduced rod-driven scotopic a and b wave and cone-driven photopic b wave responses. This deficit resulted from a reduced number of rod photoreceptors and inner nuclear layer cells. However, the reduced photopic signal arose only from lost inner retinal neurons, as cone numbers did not change. Furthermore, cell loss was non-progressive and resulted from increased apoptosis during retinal developmental as determined by TUNEL staining. In contrast, the continuous and specific expression of p53 in rod and cone photoreceptors in the mature retinas of HIP mice led to the selective loss of both rods and cones. These findings strongly support a role for p53 in regulating developmental apoptosis in the retina and suggest a potential role, either direct or indirect, for p53 in the degenerative photoreceptor loss associated with human blinding disorders.

Editor: Michael E. Boulton, University of Florida, United States of America

Funding: This research was partially supported by the Foundation Fighting Blindness (MRA), the National Center For Research Resources P20RR017703, and the National Eye Institute P30EY12190 and R01EY018137 (MRA). The funders had no role in study design, data collection and analysis, decision to publish, or preparation of the manuscript. The content is solely the responsibility of the authors and does not necessarily represent the official views of National Institutes of Health or any of its institutes.

Competing Interests: The authors have declared that no competing interests exist.

* E-mail: muayyad-al-ubaidi@ouhsc.edu

Introduction

p53 is a tumor suppressor that is activated in response to cellular stressors such as DNA damage, oncogene activation, and loss of contact between cells (for review [1]). Its main functions include cell cycle arrest in response to cell stress and facilitating the repair of damaged DNA. If the damage cannot be repaired, p53 initiates apoptosis through mitochondrial membrane permeabilization and the caspase cascade [2].

Although p53 is known to be expressed in different ocular tissues [3,4], the absence of p53 in C57BL×CBA [5] and 129/Sv×C57BL/6 [6] mice does not lead to any ocular abnormalities, implying either that other p53 family members compensate for its absence or that p53 may not be essential for eye development. However, severe ocular abnormalities arise in the p53 null mouse in the C57BL/6 and BALB/c OlaHsd backgrounds, suggesting that alleles from the C57BL/6 genetic background contribute to the observed phenotypes in the absence of p53 [7]. This implies that p53, or the pathway in which it functions, is important for normal development and/or maintenance of the eye [7].

During early embryogenesis in the mouse, p53 is expressed at high levels but as cells exit the cell cycle and terminally differentiate, p53 transcript and protein levels decline [8]. Similarly, the steady state levels of p53 in the developing mouse eye are highest at embryonic days (E) 17 and 18, drop precipitously to very low levels and then remain at those low levels throughout adulthood [9]. Although this finding suggests a role for p53 in early retinal development, it is not clear what role p53 plays beyond E18, the peak of differentiation of retinal cells [10], during postnatal retinal development, or in the mature retina. Furthermore, p53 may have important roles in the retina during stress or disease although these potential roles remain unclear. Although p53 may be dispensable for light- or chemical stress-induced apoptosis and in certain animal models of retinitis pigmentosa (RP), p53 has been linked to retinal responses to irradiation, oxidative stress, and the development of retinoblastoma ([11]for review).

To better understand the role of p53 in the developing retina and the significance of its downregulation in post-mitotic retinal cells, we studied the developing and adult retina in the super p53 mouse, a p53-overexpressing transgenic mouse model that has been previously characterized [12], and in a newly generated transgenic mouse model that overexpresses p53 specifically in retinal photoreceptors from mid-embryonic stages into adulthood (HIP, **H**uman **I**nterphotoreceptor retinoid binding protein promoter-**P**53).

We demonstrate that the super p53 mouse exhibits increased developmental retinal apoptosis, supporting an important role for p53 in retinal development. This p53-induced developmental apoptosis decreased the total number of rod photoreceptors and postreceptoral neurons in the inner nuclear layer (INL), but did not induce loss of cone photoreceptors or any further progressive degeneration of retinal cells in the mature retina. These changes were also reflected in the functional responses recorded from the super p53 mouse. To specifically assess the effects of p53 expression in adult photoreceptors, and to test whether the site of transgene integration was important to p53 overexpression effects, retinal structure and function also were examined in HIP transgenic mice, and corroborated the findings from the super p53 mouse.

Materials and Methods

Animals

All animal experiments were approved by the University of Oklahoma Health Sciences Center Institutional Animal Care and Use Committee (IACUC) and conformed to the National Institute of Health Guide for the Care and Use of Laboratory Animals and the Association for Research in Vision and Ophthalmology Resolution on the Use of Animals in Research. Animals were maintained at 30 to 50 lux on a 12-hour day/light cycle unless otherwise specified. Food and water were available ad libitum. Details for each mouse line are provided below.

Super p53 Mice

Super p53 transgenic mice were a gracious gift from Dr. Manuel Serrano (Spanish National Center of Biotechnology, Madrid, Spain) [12]. These transgenic mice were created using a 130 kb segment of genomic p53 that is under endogenous control regardless of the site of integration [13]. In functionality tests, the transgene was able to rescue p53 deficiency in mouse embryonic fibroblasts and in p53 null mice [12]. Super p53 mice showed enhanced responses to DNA damage and were more tumor resistant than wild-type controls but maintain normal aging processes [12].

Transgenic HIP Mice

A high-fidelity PCR cloning strategy was used to assemble a construct in which the human interphotoreceptor binding protein (hIRBP) promoter drives the expression of *Trp53*, the murine p53 gene, specifically in rod and cone photoreceptors of the retina from E10 onward. Briefly, a 1.4 kb hIRBP promoter was amplified from the plasmid using a primer set in which the forward primer (5′-CCACGTCCCTGAGACCACCTTCTC-CAGTCGACGCTGCCTACTGAGGCACACA-3′) corresponds to the hIRBP sequence and the reverse primer (5′-GAGAGAA-GAGATTGTGTACTGTATGGATGCTGGTGGACA-GAAGGTCTGGGGCTAAAC-3′) was chimeric with the hIRBP sequence followed by an overhang encompassing the 5′-end of the *Trp53* sequence. Likewise, a 5.1 kb fragment encompassing exons 2–11 of the *Trp53* gene [14] was amplified using a forward primer (5′-GTTTAGCCCCAGACCTTCTGTCCACCAGCATCCA-TACAGTACACAATCTCTTCTCTC-3′) that was complementary to the reverse primer used above to amplify the hIRBP sequence. The corresponding reverse primer (5′-TCTCAAA-

GAGGCTTAGTCGACTGACTCCAACAGACTGCCTG-GAC-3′) was complementary to the 3′-end of the *Trp53* sequence. The transgenic vector was sequenced to ensure the absence of any mutations prior to injection into mouse embryos. The general strategy for the generation of the construct and the transgenic mice was previously published [15].

Potential transgenic founders (F_0) and their offspring (F_1 and later) were identified by PCR screening. Briefly, DNA was extracted from ear punches and used as the template for an amplification reaction in the presence of an hIRBP-specific primer (5′-AGACCTTCTGTCCACCAGCA-3′) and a transgene-specific primer (5′-GATCGTCCATGCAGTGAGGTG-3′). To ensure the transgene was passed through the germ line, potential founder mice were mated to in-house inbred normal (wild-type) mice [16].

Southern blot analysis was used to determine both the pattern of transgene integration and copy number in the transgenic lines. Comparative quantitative densitometry measurements of the transgene and endogenous specific bands on the blots were used to determine the transgene copy numbers.

All mice were either in the C57BL/6 background or in our in-house inbred breeder strain [16].

Due to a malfunction in the HVAC system of the mouse facility, the ambient temperature increased leading to the accidental death of many mouse lines including all the HIP mice.

The Retinal Degeneration (rd/rd) Mouse

Stocks of the *rd/rd* mouse were obtained from Jackson Laboratories (Bar Harbor, ME) and raised under conditions as outlined above. These mice were crossed to the HIP mice and then backcrossed to establish *rd/rd* mice that express HIP.

Immunoblot Analysis

Sample preparation and immunoblotting were performed as previously described [15]. Briefly, primary antibodies were diluted in blocking solution and blots were incubated overnight at 4°C or 1 hour at room temperature. Appropriate secondary antibodies were applied for 1 hour at room temperature, and detection was performed using an ECL-detection kit according to the manufacturer's instructions (Pierce, Rockford, IL) and a Kodak Image Station 4000R (Eastman Kodak, Rochester, NY). Details of the antibodies used for immunoblot analysis are provided in Table 1.

Light and Electron Microscopy

Eyes were collected, processed for Spurr's resin embedment and microtomy, and examined by light and electron microscopy as previously described [17]. For light microscopy, 0.75 μm to 1 μm thick tissue sections were stained with toluidine blue and examined using an Olympus BH-2 photomicroscope (Olympus America, Center Valley, PA) and photographed using a Nikon DXM-1200 digital camera system (Nikon, Inc., Tokyo, Japan). For electron microscopy, silver-gold tissue sections were examined with a JEOL 100 EX transmission electron microscope (JEOL, Peabody, MA).

Immunohistochemistry and Lectin Cytochemistry

Samples for immunohistochemistry (IHC) and lectin labeling were processed as previously described [18]. Details of the antibodies used for fluorescent immunolabeling and lectins are provided in Table 2.

Frozen sections (10–15 μm thick) of eyecups fixed in 4% paraformaldehyde were fluorescently labeled and imaged as described previously [18]. Cell-specific markers included the 65 kDa isoform of glutamic acid decarboxylase (GAD-65), $G_0\alpha$, glutamine synthetase, calbindin, and the α isoform of protein

Table 1. Antibodies used for immunoblot analysis.

Antibody	Dilution	Source	Catalog Number
Monoclonal p53	1:1000	Cell Signaling Technology	2524
Polyclonal p53	1:2000	Santa Cruz	sc-6243
p63	1:1000	Santa Cruz	sc-8343
p73	1:1000	Santa Cruz	sc-7957
Mdm2	1:1000	EMD Biosciences	OP46T
Mdm4 [48]	1:2000	Dr. Steven J. Berberich, Boonshoft School of Medicine, Wright State University, Dayton, OH	
Actin	1:1000	Abcam	ab6276
Anti-rabbit	1:10000	KPL	0751–1506
Anti-mouse	1:10000	Amersham ECL	NA931

kinase C (PKC). Wheat germ agglutinin (WGA) conjugated to AlexaFluor-488 and peanut agglutinin (PNA) conjugated to AlexaFluor-568 were used to assess the rod and cone-specific domains of the interphotoreceptor matrix. Goat anti-mouse and goat anti-rabbit secondary antibodies conjugated to AlexaFluor-488 or -568 (Invitrogen-Molecular Probes, Carlsbad, CA) were used at a dilution of 1:200 to visualize primary antibody binding.

All antibodies and lectins were diluted in blocking agent consisting of 10% normal goat serum +5% BSA +1% fish gelatin +0.5% triton X-100 in Hank's buffered saline solution. Images were captured using an Olympus IX70 fluorescence microscope fitted with a QiCAM camera controlled by QCapture software (QImaging, Surrey, British Columbia, Canada). Image scale was set and images were imported into Photoshop software (Adobe Systems, Mountain View, CA) for preparation of figures.

Electroretinography

ERGs were recorded from the corneal surface of mice using an LKC UTAS-E 3000 Ganzfeld (LKC Technologies, Inc., Gaithersburg, MD) and analyzed with EM for Windows software (LKC Technologies, Inc.) as previously described [16].

Assessment of Cell Numbers

Image montages derived from semithin resin sections spanning the vertical meridian of the retina and passing through the optic nerve head were captured at 20× magnification using a Zeiss microscope (Carl Zeiss Meditec, Dublin, CA) and AxioVision software (Carl Zeiss Microscopy, Thornwood, NY). Using Adobe Photoshop CS (Adobe, Mountain View, CA), regions of 25.4 μm by 33.9 μm in the outer nuclear layer (ONL) were selected for cell counting at intervals of every 84.7 μm from the optic nerve head to the periphery of the retina. All nuclei fully contained within each area were counted. Cell numbers in the INL were assessed similarly.

In situ TUNEL Analysis

Eyes collected from P1 pups were fixed in 4% paraformaldehyde for two hours, processed, embedded in paraffin blocks, and cut into 5 μm sections. Apoptosis in the neuroblastic layer (NBL) was determined by TUNEL labeling using the In Situ Cell Death Detection kit (Roche, Indianapolis, IN) with fluorescein as the fluorophore. Labeled sections were mounted in ProLong Gold Antifade Reagent mounting medium containing DAPI (Invitrogen, Carlsbad, CA). For each mouse, the total number of

Table 2. Antibodies and lectins used for fluorescence IHC labeling.

Antibody	Host	Dilution	Source	Catalog Number
Monoclonal p53	mouse	1:10	Oncogene Research Products	OP29
Monoclonal p53	mouse	1:2000	Cell Signaling Technology	2524
Polyclonal p53	rabbit	1:50–1:500	Santa Cruz	sc-6243
Mdm2	mouse	1:20–1:100	Oncogene Research Products	OP46 IF2
Mdm4 [48]	rabbit		Dr. Steven J. Berberich, Boonshoft School of Medicine, Wright State University, Dayton, OH	
Calbindin	Mouse	1:300	Sigma-Aldrich, St. Louis, MO	C9848 (clone CB955)
G$_o$α	Mouse	1:500–1:1,000	Millipore, Bellerica, MA	MAB3073 (clone 2A)
Glutamic Acid Decarboxylase, 65 kDa (GAD-65)	Mouse	1:500–1:1000	Developmental Studies Hybridoma Bank, University of Iowa, Iowa City, IA	GAD-6 (clone GAD-6)
Glutamine synthetase	Mouse	1:1000	Millipore, Bellerica, MA	MAB302 (clone GS6)
Peanut Agglutinin (PNA)	–	1:20	Invitrogen-Molecular Probes, Carlsbad, CA	L21409
Protein Kinase C (PKC)	Rabbit	1:1000–1:2000	Sigma-Aldrich, St. Louis, MO	P-4334
Wheat germ agglutinin (WGA)	–	1:50	Invitrogen-Molecular Probes, Carlsbad, CA	W11261

TUNEL-positive nuclei in the entire NBL from three to four different sections was averaged for analysis. Sections from four non-transgenic (wild type, referred to as wt henceforth) and four super p53 mice were analyzed.

Statistical Analysis

Unless otherwise stated, statistical significance was determined using one-way analysis of variance (ANOVA) with Bonferroni *post hoc* multiple pairwise comparison tests (Prism, GraphPad Software, San Diego, CA). Statistical significance was accepted if $p<0.05$. Data are presented as mean $\pm$ SEM.

Results

Super p53 Mice Exhibit Reduced Photoreceptor Function

To determine whether increased steady state levels of p53 affected retinal function, scotopic and photopic ERGs were performed on age-matched super p53 mice and wt controls. Figure 1 presents dark-adapted ERG responses recorded from wt and super p53 mice to flash stimuli spanning a range of intensity of several log units. This range includes light intensities that elicit purely rod-driven (scotopic) responses, mixed rod- and cone-driven responses, and purely cone-driven (photopic) responses (Figure 1A).

The ERG response of wt mice at the lowest light intensity was dominated by the positive-going b wave, which reflects primarily the activity of rod bipolar cells. At increasing intensities, the scotopic b wave was preceded by the negative-going a wave, which represents the collective response of the rods. In super p53 mice, both scotopic a and b wave amplitudes were reduced and mostly fell outside the normal range observed for wt mice at all light intensities (Figure 1B&C). However, there were no changes in implicit time ($\sim$6 milliseconds) of the a wave between super p53 and wt mice (data not shown). Developmentally, the scotopic a and b wave amplitudes of super p53 mice started significantly below the wt amplitudes at P30 and remained lower than wt for all ages examined (Figure 1D&E) but did not show any additional progressive decline over time.

The reduction in ERG response was not limited to rod-driven responses. Cone-driven photopic ERG responses also were reduced in super p53 mice compared to wt controls (Figure 1F). The reduction ($\sim$41%) in photopic b wave amplitude in super p53 mice at P30 was proportionally much larger than the reduction in either the scotopic a wave ($\sim$13%) or the b wave ($\sim$17%) amplitude. To determine whether the reduction in the photopic response from super p53 mice resulted from a selective functional deficit in either green or blue cones, spectral ERGs were performed. The responses elicited from both cone types in post-natal day (P) 90 super p53 mice were reduced (Figure 1G), suggesting that the functional deficit was not cone type-specific. The age-dependent reduction in scotopic and photopic ERG responses from super p53 retinas paralleled that of wt mice, suggesting that the functional deficit initially arose during retinal development and was not followed by any additional progressive functional deficits that were p53-dependent.

The observed reductions in scotopic and photopic ERG responses from super p53 mice potentially could arise from degenerative changes in photoreceptors and/or higher-order neurons in the inner retina. Alternatively, the reduced ERG responses might result from functional changes unrelated to cell loss. Examination of P30 wt and super p53 retinas at the light microscopy level showed normal retinal lamination of the p53 mouse retina (Figure 2A–D) although the number of cells in the ONL and in the INL were reduced (Figure 2E&F). Electron microscopy showed preservation of normal ultrastructure (Figure S1).

Quantification of the difference in the number of photoreceptor nuclei at P30 in wt and super p53 mice along the vertical meridian of the retina showed that photoreceptors were lost equally across the retina of the super p53 mouse retina (Figure 2E&F). The loss of photoreceptors from the super p53 retina is consistent with the reduced ERG responses in super p53 mice. To determine whether photoreceptor loss in the super p53 retina affected rod and cone photoreceptors differentially, we assessed cone numbers in the retinas of super p53 and age-matched wt control mice by counting cone photoreceptors identified by M- and S-opsin labeling in vertical sections (Figure S2) and in retinal whole mounts (data not shown). Cone numbers in the super p53 and wt retina were similar in both superior and inferior retinal regions. Consistent with these findings, western blot analysis showed no difference in cone opsin levels between the super p53 and wt retinas (data not shown). The rod and cone-specific domains of the interphotoreceptor matrix (IPM), labeled using WGA and PNA, also appeared unaffected in the super p53 retina (data not shown). Together, these data indicate that constitutive overexpression of p53 reduced the total number of photoreceptors but did not cause cone loss. Thus, reduced cell numbers in the ONL of the super p53 retina must reflect a selective reduction in the numbers of rods.

The reduced b wave amplitudes observed in the super p53 mouse ERG, particularly at low light intensities that evoke rod-driven responses, might arise from the reduced number of rods. However, a reduction in the number of inner retinal neurons, particularly bipolar cells, also might contribute to reduced b wave amplitude, particularly for light intensities that drive cone-mediated responses since cone number in the super p53 retina is normal. Therefore, we similarly quantified cell numbers in the INL along the vertical meridian of the retina. These analyses showed that the INL of the super p53 mouse had reduced numbers of cells compared to age-matched wt control mice (Figure 2F). Thus, the reduced numbers of cells in the INL may contribute to the reduced amplitude of the scotopic b wave in the super p53 mouse. Furthermore, the reduced numbers of cells in the INL, rather than defects in cones, may be responsible for the reduced photopic b wave amplitude observed in super p53 mice.

Immunolabeling with markers for specific populations of cells in the INL showed that all cell types examined were present in the super p53 retina and showed morphology comparable to the same cell types in the wt retina (summarized in Table 3). Cell types examined included rod bipolar cells (PKC, shown in Figure S3), ON-bipolar cells ($G_o\alpha$), horizontal cells (calbindin), GABAergic amacrine cells (GAD-65), and Müller glial cells (glutamine synthetase) (data not shown). These results suggest that although p53 over-expression reduced the total number of cells in the INL, the cell populations present in the INL were appropriate and showed their normal cell-specific morphological characteristics.

These studies showed that overexpression of p53 led to functional deficits and reduced cell numbers in the ONL and INL in the super p53 mouse retina in the absence of further progressive p53-dependent functional decline or cell loss, but did not establish how these defects originated. The deficits observed in the super p53 retina could arise from either increased developmental apoptosis or from altered cell production during retinogenesis.

To test whether transgenic p53 expression led to increased developmental apoptosis, TUNEL staining was performed on retinal sections from P1 retinas from both super p53 and wt mice. The transgenic expression of p53 increased numbers of TUNEL-positive cells in the super p53 retina compared to the wt retina

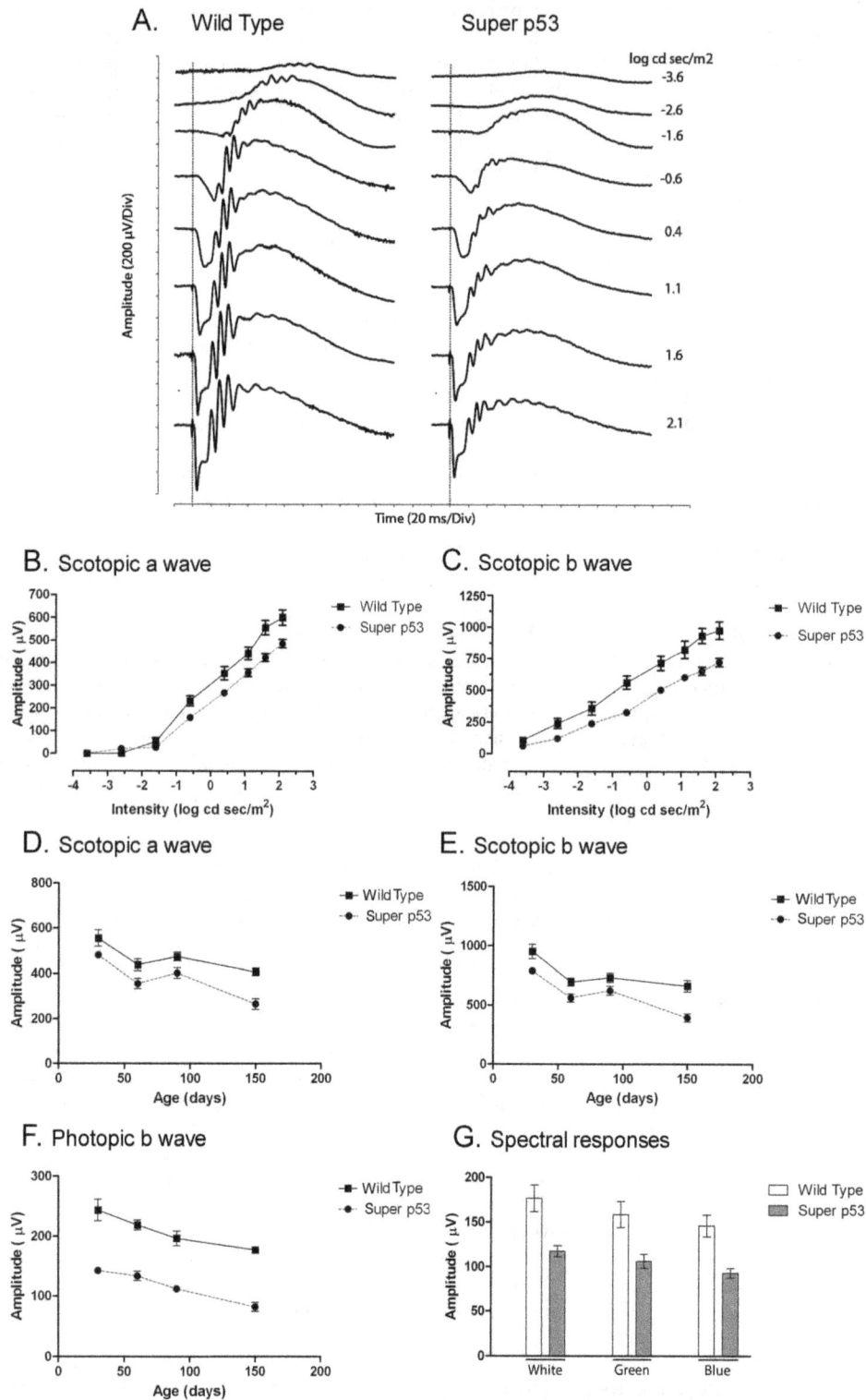

Figure 1. Electroretinographic (ERG) analysis of the super p53 mouse. (A) ERG responses from P30 wild type (left) and super p53 (right) mice to a series of stimulus intensities spanning a several log unit range presented to the dark-adapted mouse eye. Dashed vertical lines represent the time of presentation of stimulus. Amplitude of the dark-adapted, scotopic a wave (**B**) and b wave (**C**) plotted as a function of stimulus intensity. (**D**) Amplitude of the scotopic a wave plotted as a function of mouse age. (**E**) Amplitude of the scotopic b wave plotted as a function of mouse age. (**F**) Amplitude of the light-adapted photopic b wave plotted as a function of mouse age. Solid squares represent responses from wild type mice; filled circles represent the super p53 mouse responses. Symbols represent the average obtained from testing 6 to 12 transgenic animals; error bars represent SEM. (G) Spectral electroretinography. Recordings of maximal b wave responses to either a 400 nm flash (blue) or to a 500 nm flash (green) made from P90 wt (filled bars) or super p53 (open bars) mice. Symbols represent the average obtained from testing 6 transgenic animals; error bars represent SEM. The responses marked by "white' represent the full photopic b wave response.

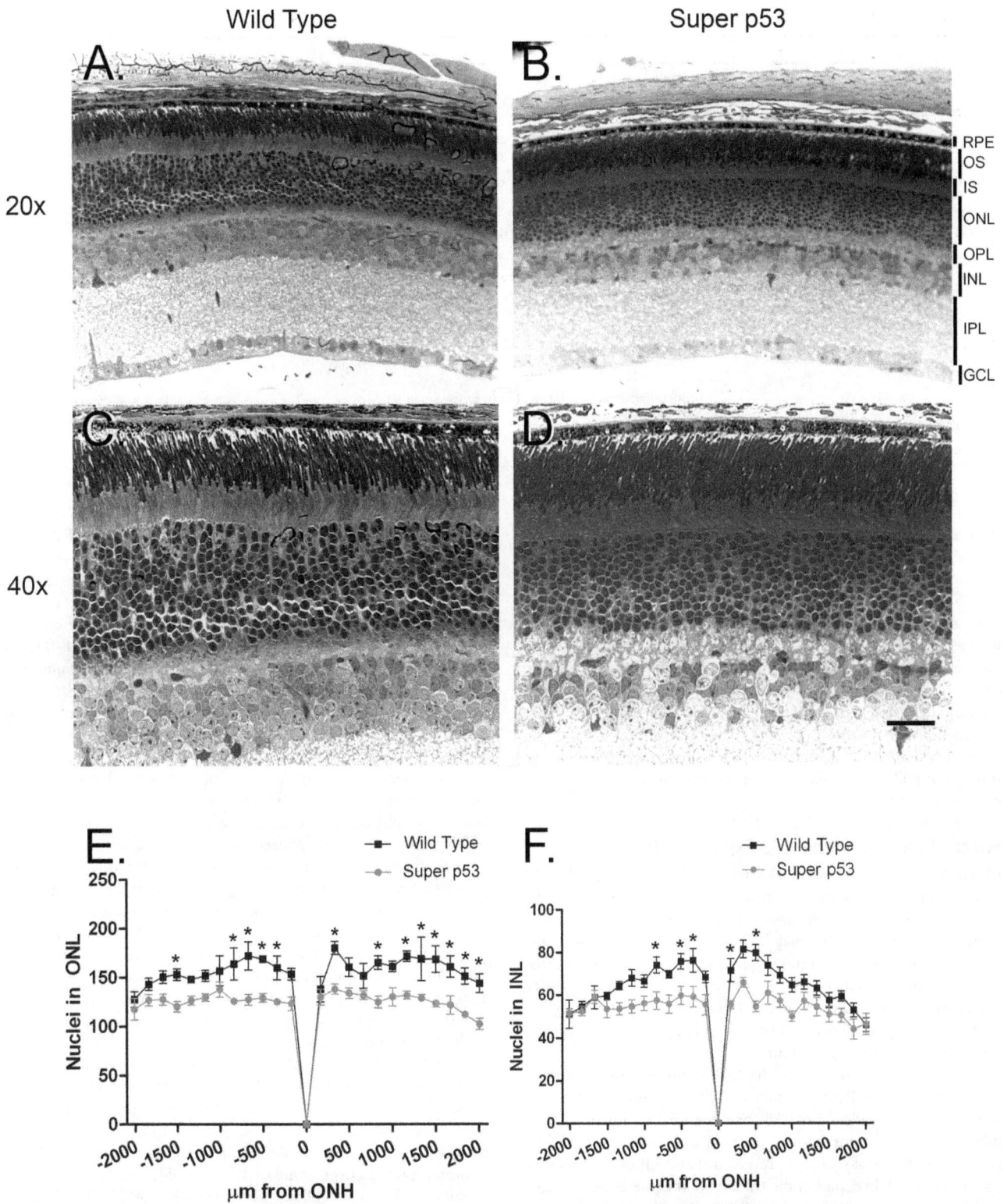

Figure 2. Histologic examination of super p53 retina. Light micrographs of retinal cross-sections taken from Wild Type (**A&C**) and Super p53 (**B&D**) mice. **A&B** present images obtained at 20X and **C&D** present images obtained at 40X. RPE: pigment epithelium; OS: outer segments; IS inner segments; ONL: outer nuclear layer; OPL: outer plexiform layer; INL: inner nuclear layer; IPL: inner plexiform layer; GCL: ganglion cell layer. Bar is equal to 50 μm. **E**. The number of nuclei in the ONL was counted along the vertical meridian at each indicated distance from the optic nerve head (ONH, 0) from inferior to superior retinal margin (n7–8 mice). **F**. The number of nuclei in the INL was counted along the vertical meridian at each indicated distance from the optic nerve head (0) from inferior to superior retinal margin (n7–8 mice). P values for points marked by an asterisk ranged between <0.001 and <0.05.

Table 3. Normal cell-specific morphology or distribution of retinal cells in the super p53 retina.

Cell- or synapse-specific marker	Wild type	Super p53
Photoreceptors and IPM		
PNA	IPM surrounding cone outer segments; flat contacts with OFF-cone bipolar cells at cone terminals	Comparable to wild type
WGA	IPM surrounding rod and cone outer segments	Comparable to wild type
Bipolar Cells		
PKC-α	Rod bipolar cell dendrites, cell bodies, axons and their terminals in the innermost IPL	Comparable to wild type
Goα	All ON-type bipolar cell bodies and dendrites. Other, non-bipolar cell processes in IPL	Comparable to wild type
Horizontal cells		
Calbindin	Horizontal cell bodies, dendrites and axons in the OPL	Comparable to wild type
Amacrine cells		
GAD-65	GABAergic amacrine cells and their processes in IPL	Comparable to wild type
Glial cells		
Glutamate synthetase	Müller cells and astrocytes	Comparable to wild type

(Figure 3). The increase in TUNEL positive cells in the super p53 retina suggests that the super p53 retina produces a normal number of cells but undergoes a higher degree of developmental cell death, ultimately leading to a reduced final number of retinal cells and reduced functional competence. To confirm that the increased apoptosis in the super p53 retina is associated with increased levels of p53, immunoblots were performed on retinal extracts at early retinal developmental stages. As shown in Figure 3D, the levels of p53 are considerably higher in super p53 at both P1 and P3. However, the levels of p53 drop at P3, suggesting that due to the inclusion of its endogenous promoter the transgenic p53 mimics the endogenous p53 in its developmental pattern of expression.

Effect of HIP Transgenic p53 Expression in Photoreceptors Mimics that Observed in Super p53 Mice

The super p53 mouse clearly exhibited a developmental phenotype. However, because it is a transgenic mouse, it is possible that the transgene may have integrated into a locus in the genome that modulates retinal development. Furthermore, the transgene in the super p53 mouse is regulated by its native promoter, which shows dramatic transcript downregulation by E18 [9], making it difficult to assess the consequences of continued p53 expression in the adult retina.

To address these questions, the HIP construct was generated and injected into E1 mouse embryos. The injection yielded four potential HIP founder mice. All four potential founders successfully passed the transgene to their offspring in a Mendelian fashion. Southern blot analysis revealed that each of the founders contained multiple copies of the transgene integrated into a single site (data not shown) but showed varying levels of transgene expression.

Immunoblot analysis of retinal lysates collected at P30 showed that the HIP mouse lines generated from F_093, F_089, F_044, and F_041 expressed increasing amounts of p53 protein (Figure 4A). Compared to 661W cells, a continuously dividing cone photoreceptor cell line [19], founder F_093 expressed $0.2\times$ the amount of p53, F_089 expressed $0.9\times$, F_044 expressed $2.9\times$, and F_041 expressed $23\times$. Due to their moderate levels of expression of

transgenic p53, the HIP mouse lines generated from F_089 and F_044 were chosen for subsequent studies.

To confirm that expression of transgenic p53 driven by the hIRBP promoter was restricted to photoreceptors, IHC was performed on P30 HIP and wt retinal sections. As expected, transgenic p53 was expressed in the retina of both HIP F_089 and HIP F_044 at levels reflective of that observed on immunoblots (Figures 4 and 5). Transgenic p53 expression was localized to the inner segment (IS), which contains the cytosolic fractions of the rod and cone photoreceptors, and the ONL, which consists of the rod and cone photoreceptor cell bodies.

In addition, p53 signal was also detected in the distal INL near the OPL (asterisks, Figure 5C–F), a region that normally contains horizontal and bipolar cells but not photoreceptors. Because the hIRBP promoter was utilized in the generation of the HIP construct, expression was expected to be limited to the photoreceptors of both rod and cones [20]. During normal retinogenesis, dividing photoreceptor progenitor cells migrate within the NBL [21]. As these progenitors permanently exit the cell cycle, they migrate toward the outer region of the retina where they differentiate into photoreceptors that later establish the ONL. Photoreceptor cells that fail to migrate into the nascent ONL and/ or are unsuccessful in establishing proper connections normally undergo apoptosis [22].

In order to determine whether the expression of transgenic p53 in the INL was due to ectopic expression in INL cells or to the mislocalization of p53-overexpressing photoreceptors, the HIP transgenic mice were bred with the *rd* mouse, which has a mutation in the β-subunit of phosphodiesterase that causes all rod photoreceptors to degenerate by P21 [23–25], leaving a single row of cone nuclei in the ONL. The retinas of F_089 and F_044 HIP mice bred onto the *rd/rd* background (HIP$^{rd/rd}$) showed degeneration of all rods in the ONL and the elimination of all p53-expressing cells from the INL (Figure S4). The specific loss of p53 labeling from the INL in the retina of HIP$^{rd/rd}$ mice indicates that the p53 labeling in the INL of HIP mice on the wt background was due to mislocalized photoreceptors rather than ectopic expression of transgenic p53 in horizontal or bipolar cells.

The majority of transgenic p53 was observed in the cytoplasm and the perinuclear region of photoreceptors (Figure 5). Overex-

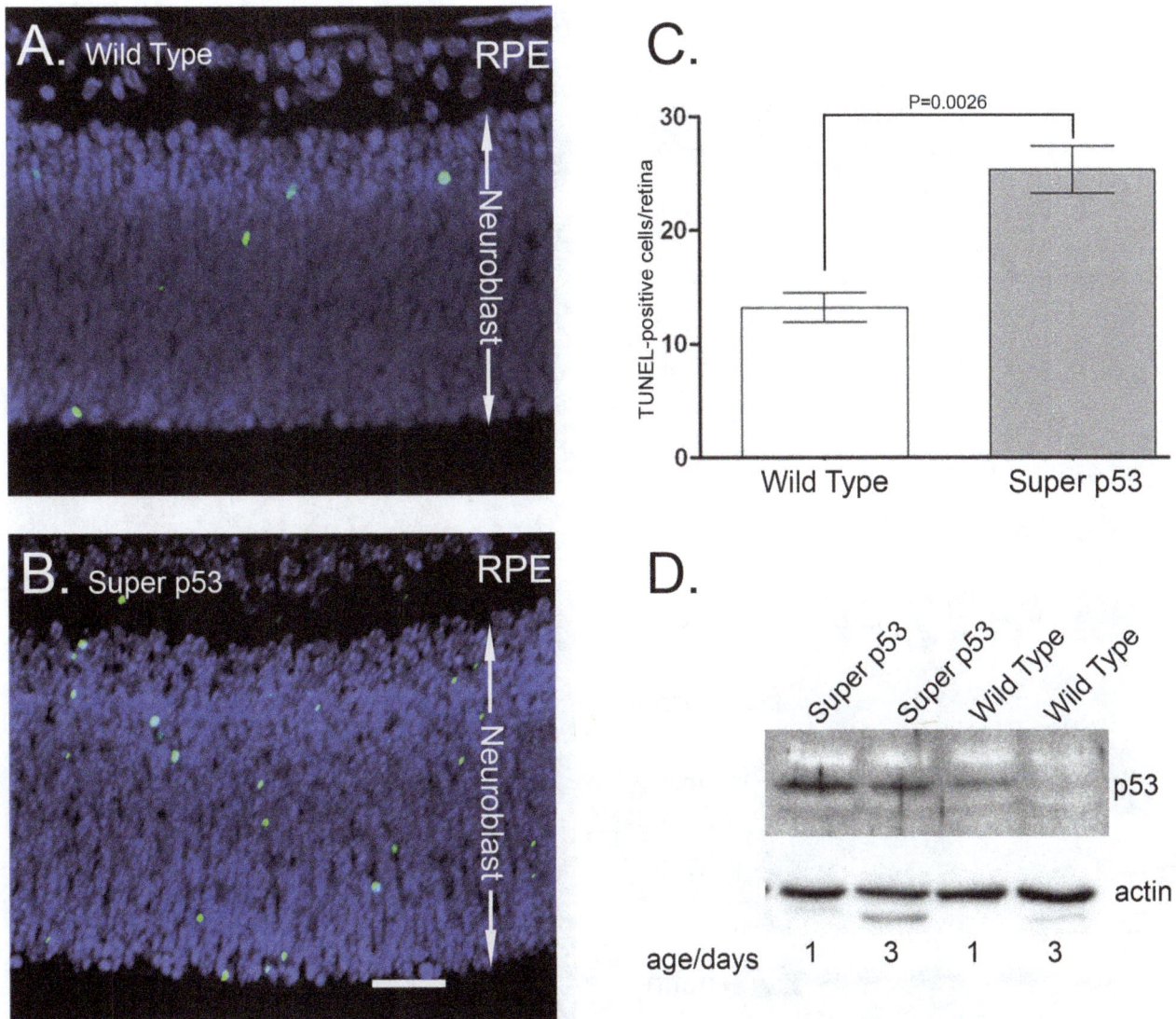

Figure 3. TUNEL labeling of cross sections of P1 retina. TUNEL labeling showed increased apoptotic death in the neuroblastic layer (NBL) of P1 super p53 mice (**B**) when compared to wt retinas (**A**). TUNEL-positive cells are green; nuclei (blue) were stained with DAPI. **C.** Bar graphs showing the increase in the total number of apoptotic cells in a cross section of the retina of the super p53 mouse at P1. N3–4 sections from 4 mice of each genotype. Bars represent SEM. Scale bar50 μm. D. Immunoblot analysis of retinal extracts from super p53 and wild type mice at P1& P3. Upper blot was probed with anti-p53 antibody then stripped and reprobed with anti-actin antibody.

pression of cytoplasmic p53 has been observed in some cancers [26–33] and during apoptotic signaling [34]. Mouse double minute 2 (Mdm2) or Mdm4, two p53 regulators, can promote p53 nuclear export [35]. If cytoplasmic sequestration of p53 was Mdm2- or Mdm4-dependent, the levels of Mdm2 and/or Mdm4 should have changed. However, Mdm4 did not show altered levels by immunoblot analysis in HIP $F_0 44$ or HIP $F_0 89$ retinas (Figure 4B), suggesting that Mdm4 does not play a role in the cytoplasmic sequestration of transgenic p53. We did observe reduced levels of Mdm2 in the HIP $F_0 89$ retina and little, if any, Mdm2 expression in the HIP F_{044} retina (Figure 4B). This might arise from downregulation of Mdm2 by increased levels of p53 or the integration of the transgene into the Mdm2 locus. However, homozygous HIP $F_0 44$ and homozygous HIP $F_0 89$ mice are viable (data not shown) while homozygous Mdm2 knockout mice are not [36], suggesting that Mdm2 downregulation arose from transgenic p53 expression rather than integration of the transgene into the

Mdm2 locus. This is further supported by the fact that the dividing 661W cone photoreceptor cell line expressed substantial levels of p53 in the absence of detectable Mdm2 (Figure 4B).

The p53 family of proteins also includes p63 and p73, which are known to play essential roles in development [37–42]. However, increased levels of p53 in the retinas of HIP transgenic mice caused no observable changes in levels of either p63 or p73 in comparison to wt (Figure 4B), suggesting that overexpression of p53 did not lead to compensatory changes in the expression of other members of the p53 family.

To determine the effects of increased expression of p53 on its target genes, the levels of p21 were assessed on immunoblots of retinal extracts from P6–7 mice. As shown in Figure 4C, the levels of p21 are increased in HIP retina ($\sim$3X) when compared to their non-transgenic counterparts, indicating that elevated p53 levels do in fact alter the expression of its target genes.

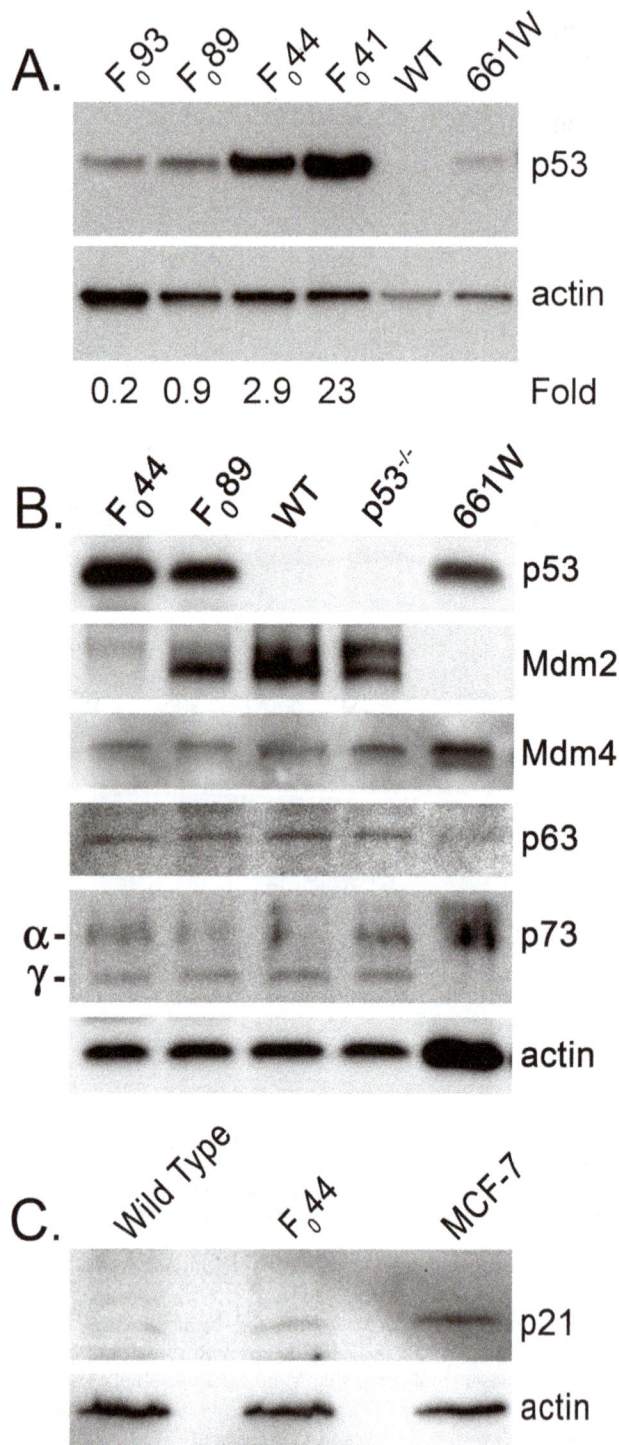

Figure 4. Pattern of expression of p53, its family members, and its regulators in HIP retinas. (A) Western blot analysis showing the levels of expression of p53 in four transgenic HIP families relative to its levels in 661W cells, a continuously dividing cone photoreceptor cell line. **(B)** Levels of Mdm2, Mdm4, p63 and p73 were determined in two of the HIP families. Actin served as a loading control. **(C)** Immunoblot demonstrating the increased levels of p21 in HIP retinas at P6–7.

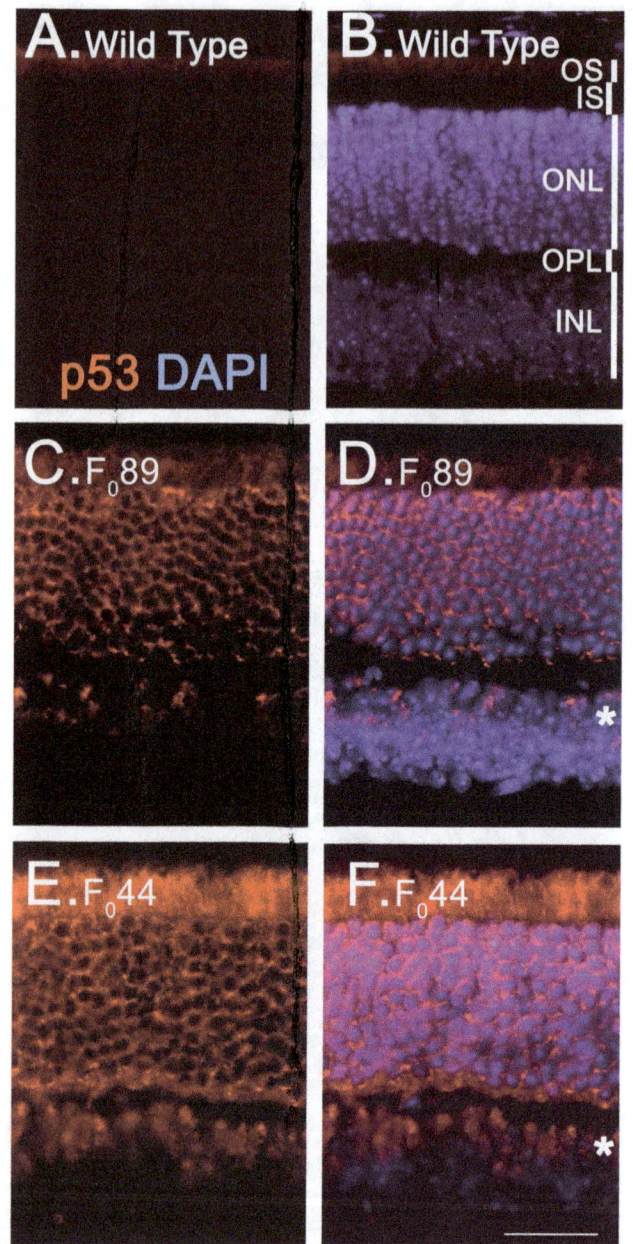

Figure 5. Immunohistochemical localization of p53 in two of the HIP mouse lines. Retinal sections from P30 wild type (**A&B**), $F_0 89$ (**C&D**) and $F_0 44$ (**E&F**) mice were labeled using an anti-p53 antibody (red). Asterisks in **D&F** point to p53-expressing cells in the INL of both HIP families. OS outer segments; IS inner segments; ONL outer nuclear layer; OPL, outer plexiform layer; INL, inner nuclear layer. Scale bar 20 μm.

To determine whether overexpression of p53 in the retina of HIP mice led to functional deficits similar to those observed in the super p53 mouse, ERGs were performed on the HIP $F_0 44$ line (Figure 6A). Scotopic ERG recordings show that ERG responses started at lower levels than those recorded from wt, reached almost wt levels between P60 and P120, and then declined as the mice aged. Furthermore, the age-related decreases in the scotopic a and b waves of HIP $F_0 44$ mice paralleled one another. Cone-driven function, as determined by the photopic b wave, also showed a reduced response at P30 that then improved somewhat by P60, reaching about 80% of wt levels. However, after P60 the photopic b wave then declined steadily with age.

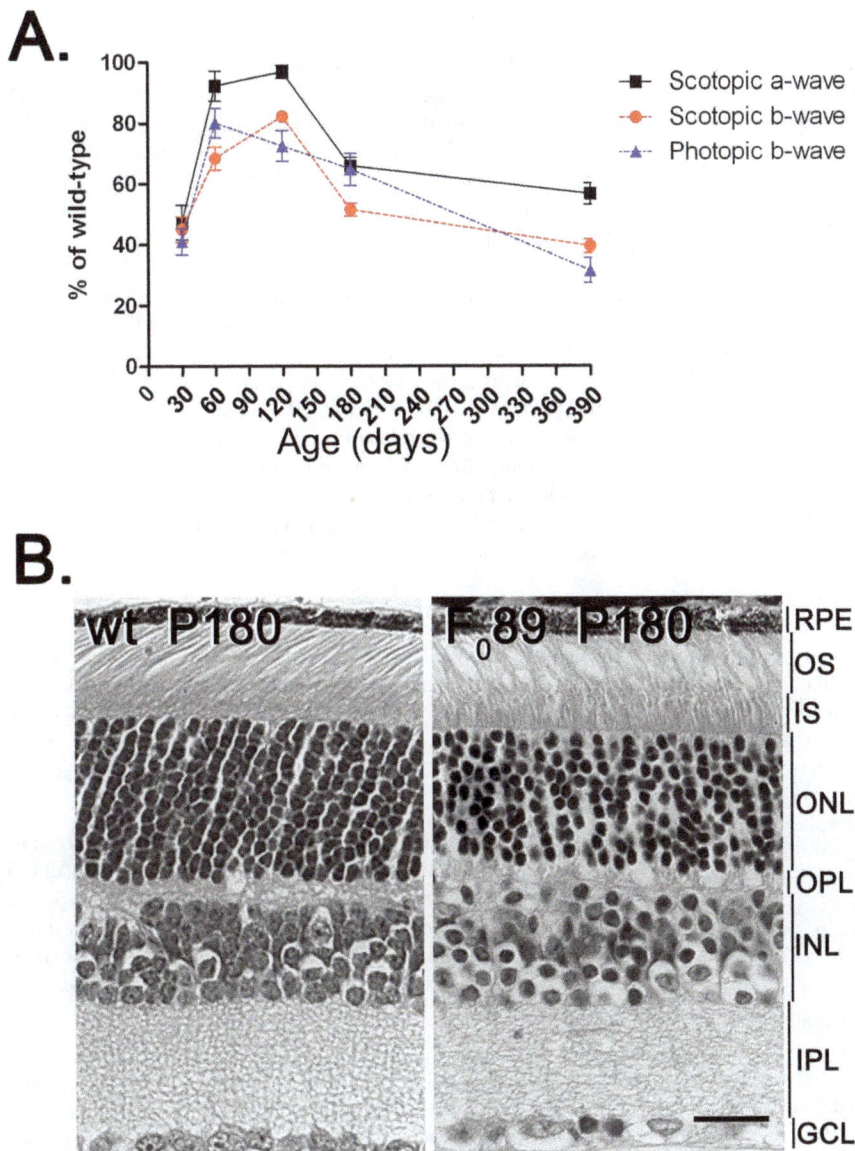

Figure 6. Functional and histologic analyses of the HIP $F_0$89 retina. A. Developmental electroretinographic analysis of scotopic a and b waves responses and the photopic b wave responses from the HIP $F_0$89 retinas presented as percent of wt. **B.** Histologic cross sections of wild type (wt) and HIP $F_0$89 retinas. Scale bar50 μm.

The early reduced responses might arise from either developmental delays or secondary effects on retinal function. The apparently normal retinal structure of the HIP transgenic mice (Figure 6B) suggests that secondary functional effects may be a more likely cause than developmental delay. However, histologic examination of retinal sections from HIP mice revealed degenerative changes, evidenced by a reduced number of nuclei in the ONL (Figure 6B), suggesting that the progressive decline in ERG responses is likely to reflect degenerative changes.

Discussion

To determine the functional role of p53 in the retina, the super p53 mouse, a line that globally overexpresses p53 under the control of the endogenous p53 promoter, and transgenic HIP mouse lines, which overexpress p53 specifically in photoreceptor cells under the control of the hIRBP promoter, were studied. The studies performed in the super p53 mouse indicate that developmental overexpression of p53 in the retina leads to the selective loss of rod photoreceptors, but leaves the cone photoreceptor population apparently intact. TUNEL labeling studies indicate that the reduced rod population in the super p53 retina arises from increased apoptotic rod cell death rather than from failure of rod generation. The selective loss of rods also is consistent with the functional deficits shown by the diminished a and b waves of the rod-driven ERG in the super p53 mouse. The reduced numbers of cells in the INL of the super p53 mouse also might contribute to the reduced scotopic b wave amplitude, which reflects the activity of cells in the inner retina, particularly rod bipolar cells. The respective contributions of the reduced rod populations and the reduced number of cells in the INL to the decline in scotopic b wave decline are not clear at present.

One possibility for the selective effects of p53 on rods is the fact that p53 protein expression peaks at E18 [9], after most, if not all, cones have already differentiated [43]. In addition, the lack of any further progressive degeneration in the super p53 retina suggests that the transgenic p53 gene is regulated in a manner similar to that of the endogenous p53 gene and that the introduction of the extra copy only leads to increased p53 levels without affecting the onset of expression or the timing of downregulation. This also supports the notion of the absence of co-regulation between p53 allelic genes.

The developmental overexpression of p53 also compromised cone function, as shown by the reduced amplitude of the photopic b wave, even though the cone population of the super p53 mouse retina appeared to be intact. The reduced photopic b wave in the super p53 mouse suggests that the loss of cells from the INL compromises cone-driven retina function, although secondary effects related to the reduced numbers of rods in the super p53 mouse retina also could indirectly contribute to the compromised cone-driven responses.

Developmental overexpression of p53 led to a decrease in the numbers of cells residing in the INL, but precisely how this decrease in cell number occurs is not currently clear. One possibility is that developmental overexpression of p53 might cause a small, generalized increase in cell death across multiple types of inner retinal cells. It is also conceivable that developmental overexpression of p53 might selectively reduce some specific subset of retinal neurons. Although the IHC studies presented here were not exhaustive, it is clear that developmental overexpression of p53 did not compromise the ability of the retina to generate the different classes of inner retina cells, as all classes were present. Furthermore, all specific subtypes of inner retinal cells that were examined showed their normal cell-specific morphological characteristics. Together, these studies suggest that p53 is not critical to the generation and differentiation of the various retinal cell types, but rather is more likely involved in regulating developmental apoptosis and determining the final number of cells in the mature retina.

Results from the HIP studies show that continued expression of p53 into adulthood is deleterious and seems to equally affect both rods and cones, leading to a degenerative phenotype.

Two pieces of information can be gleaned from the above studies: The levels of p53 are tightly regulated, and the timely downregulation of p53 during retinal development is necessary to maintain a healthy number of cells in the retina. Any perturbation in p53 levels can lead to disastrous consequences for the retina.

The continued expression of p53 in HIP retinas beyond E18 caused a delay in the development of retinal function, emphasizing the significance of the downregulation of p53 during retinal development. Alternatively, the developmental functional delay may result from the early expression of p53 in the HIP retina because the onset of expression by the human IRBP promoter is between E10–13 [44].

A surprising finding in the HIP mice is lack of co-regulation of Mdm2 as a result of expression of higher levels of p53. This may reflect a behavior limited to the retina since it has been shown that Mdm2 pattern of expression does not follow that of p53 during early ocular development while that of Mdm4 seems to mimic the pattern of expression of p53 [9]. This is supported by the presence of only Mdm4 in the 661W cells.

Another interesting observation in HIP retinas is the presence of the majority of p53 perinuclearly and in the cytoplasm rather than the anticipated nuclear localization. This may be due to the higher levels of p53 expressed and the relatively unchanged levels of its regulators, Mdm2 and As a result under these conditions p53 may end up interacting with novel binding partners or organelles, altering its signaling and leading to cell death.

Because p53 null mice showed no ocular phenotype in the C57BL×CBA and 129/Sv×C57BL/6 backgrounds, investigators concluded that p53 lacks any biological role in retinal development and homeostasis [6,45]. Additionally, it has been shown that cell death in the N-methyl-N-nitrosourea-induced model of retinitis pigmentosa is p53-independent because the pattern of photoreceptor degeneration was similar whether on a wt or p53 null background [39]. In both of these cases, the patterns of expression of the p53 family members p63 and 73 were not examined to determine whether these other members of the p53 family may have compensated for the loss of p53. Both p63 and 73 can regulate the cell cycle and apoptotic cell death in a fashion similar to that of p53 [46,47]. Although the conclusions made were correct that in certain cases retinal degeneration is p53-independent, it is yet to be determined definitively whether it also independent of all members of the p53 family.

In summary, this report demonstrates that p53 has a major role in developmental apoptosis in the retina and may play an as yet unidentified role in human blinding diseases associated with increased levels of p53 resulting from mutations in the p53 gene.

Supporting Information

Figure S1 Expression of p53 does not disrupt general retinal ultrastructure. Montages show cross sections that span the entire retina of both Wild Type and super p53 mice. All retinal layers are preserved and the subcellular organization is intact. Scale bar2 μm.

Figure S2 Cone numbers in wild type and super p53 retinas. Whole eyes were sectioned along the superior-inferior axis. All OS structures labeled by IHC using M-opsin or S-opsin-specific antibodies within the first 324.6 μm on either side of the optic nerve head were counted. The graph shows the number of cones per 100 μm. N2 sections from 4–5 mice of each genotype. Bars represent SEM.

Figure S3 Expression of p53 does not specifically alter retinal cell morphology or distribution. Rod bipolar cells immunolabeled for PKC shown. Rod bipolar cells in the wild type (**A**) and super p53 expressing retina (**B**) show comparable morphology and distribution, similar to other cell types tested (see text for details). ONL outer nuclear layer; OPL, outer plexiform layer; INL, inner nuclear layer; IPL, inner plexiform layer; GCL, ganglion cell layer. Scale bars20 μm.

Figure S4 Cross sectional analysis of retinas of wt and super p53 in the *rd/rd* background. Sections were immunolabeled for p53 (red) and rhodopsin (green). Nuclei (blue) are stained with DAPI. Mice from $F_0$89 (**A**) and $F_0$44 (**C**) were bred to *rd/rd* mice and then backcrossed to generate *rd/rd* mice expressing the p53 transgene from $F_0$89 (**B**) and $F_0$44 (**D**). Retinal sections from wt (**E**), *rd/rd* (**F**) and p53$^{-/-}$ (**G**) mice served as controls.

Acknowledgments

The authors thank Dr. Steven J. Berberich for his generous gift of Mdm4 antibody and Leann Garrett, Barbra Nagel, and Eileen Parks for excellent technical assistance.

Author Contributions

Conceived and designed the experiments: MRA DMS. Performed the experiments: LV DEB II DMS. Analyzed the data: MRA DMS LV. Wrote the paper: MRA LV.

References

1. Levine AJ, Oren M (2009) The first 30 years of p53: growing ever more complex. Nat Rev Cancer 9: 749–758.
2. Schuler M, Bossy-Wetzel E, Goldstein JC, Fitzgerald P, Green DR (2000) p53 induces apoptosis by caspase activation through mitochondrial cytochrome c release. J Biol Chem 275: 7337–7342.
3. Tendler Y, Weisinger G, Coleman R, Diamond E, Lischinsky S, et al (1999) Tissue-specific p53 expression in the nervous system. Brain Res Mol Brain Res 72: 40–46.
4. Shin DH, Lee HY, Lee HW, Kim HJ, Lee E, et al (1999) In situ localization of p53, bcl-2 and bax mRNAs in rat ocular tissue. Neuroreport 10: 2165–2167.
5. Armstrong JF, Kaufman MH, Harrison DJ, Clarke AR (1995) High-frequency developmental abnormalities in p53-deficient mice. Curr Biol 5: 931–936.
6. Jacks T, Remington L, Williams BO, Schmitt EM, Halachmi S, et al (1994) Tumor spectrum analysis in p53-mutant mice. Curr Biol 4: 1–7.
7. Ikeda S, Hawes NL, Chang B, Avery CS, Smith RS, et al (1999) Severe ocular abnormalities in C57BL/6 but not in 129/Sv p53-deficient mice. Invest Ophthalmol Vis Sci 40: 1874–1878.
8. Schmid P, Lorenz A, Hameister H, Montenarh M (1991) Expression of p53 during mouse embryogenesis. Development 113: 857–865.
9. Vuong L, Brobst DE, Saadi A, Ivanovic I, Al-Ubaidi MR (2012) Pattern of expression of p53, its family members, and regulators during early ocular development and in the post-mitotic retina. Invest Ophthalmol Vis Sci 53: 4821–4831.
10. Young RW (1985) Cell differentiation in the retina of the mouse. Anat Rec 212: 199–205.
11. Vuong L, Conley SM, Al-Ubaidi MR (2012) Expression and Role of p53 in the Retina. Invest Ophthalmol Vis Sci 53: 1362–1371.
12. Garcia-Cao I, Garcia-Cao M, Martin-Caballero J, Criado LM, Klatt P, et al (2002) "Super p53" mice exhibit enhanced DNA damage response, are tumor resistant and age normally. EMBO J 21: 6225–6235.
13. Giraldo P, Montoliu L (2001) Size matters: use of YACs, BACs and PACs in transgenic animals. Transgenic Res 10: 83–103.
14. Bienz B, Zakut-Houri R, Givol D, Oren M (1984) Analysis of the gene coding for the murine cellular tumour antigen p53. EMBO J 3: 2179–2183.
15. Tan E, Wang Q, Quiambao AB, Xu X, Qtaishat NM, et al (2001) The relationship between opsin overexpression and photoreceptor degeneration. Invest Ophthalmol Vis Sci 42: 589–600.
16. Xu X, Quiambao AB, Roveri L, Pardue MT, Marx JL (2000) Degeneration of cone photoreceptors induced by expression of the Mas1 protooncogene. Exp Neurol 163: 207–219.
17. Stricker HM, Ding XQ, Quiambao A, Fliesler SJ, Naash MI (2005) The Cys214->Ser mutation in peripherin/rds causes a loss-of-function phenotype in transgenic mice. Biochem J 388: 605–613.
18. Sherry DM, Murray AR, Kanan Y, Arbogast KL, Hamilton RA, et al (2010) Lack of protein-tyrosine sulfation disrupts photoreceptor outer segment morphogenesis, retinal function and retinal anatomy. Eur J Neurosci 32: 1461–1472.
19. Tan E, Ding XQ, Saadi A, Agarwal N, Naash MI, et al (2004) Expression of cone-photoreceptor-specific antigens in a cell line derived from retinal tumors in transgenic mice. Invest Ophthalmol Vis Sci 45: 764–768.
20. Liou GI, Geng L, Al-Ubaidi MR, Matragoon S, Hanten G, et al (1990) Tissue-specific expression in transgenic mice directed by the 5′-flanking sequences of the human gene encoding interphotoreceptor retinoid-binding protein. J Biol Chem 265: 8373–8376.
21. Baye LM, Link BA (2008) Nuclear migration during retinal development. Brain Res 1192: 29–36.
22. Linden R, Rehen SK, Chiarini LB (1999) Apoptosis in developing retinal tissue. Prog Retin Eye Res 18: 133–165.
23. Keeler C (1966) Retinal degeneration in the mouse is rodless retina. J Hered 57: 47–50.
24. Pittler SJ, Keeler CE, Sidman RL, Baehr W (1993) PCR analysis of DNA from 70-year-old sections of rodless retina demonstrates identity with the mouse rd defect. Proc Natl Acad Sci U S A 90: 9616–9619.
25. Drager UC, Hubel DH (1978) Studies of visual function and its decay in mice with hereditary retinal degeneration. J Comp Neurol 180: 85–114.
26. Ali IU, Schweitzer JB, Ikejiri B, Saxena A, Robertson JT, et al (1994) Heterogeneity of subcellular localization of p53 protein in human glioblastomas. Cancer Res 54: 1–5.
27. Moll UM, LaQuaglia M, Benard J, Riou G (1995) Wild-type p53 protein undergoes cytoplasmic sequestration in undifferentiated neuroblastomas but not in differentiated tumors. Proc Natl Acad Sci U S A 92: 4407–4411.
28. Weiss J, Schwechheimer K, Cavenee WK, Herlyn M, Arden KC (1993) Mutation and expression of the p53 gene in malignant melanoma cell lines. Int J Cancer 54: 693–699.
29. Pezzella F, Morrison H, Jones M, Gatter KC, Lane D, et al (1993) Immunohistochemical detection of p53 and bcl-2 proteins in non-Hodgkin's lymphoma. Histopathology 22: 39–44.
30. Van Veldhuizen PJ, Sadasivan R, Cherian R, Dwyer T, Stephens RL (1993) p53 expression in incidental prostatic cancer. Am J Med Sci 305: 275–279.
31. Moll UM, Riou G, Levine AJ (1992) Two distinct mechanisms alter p53 in breast cancer: mutation and nuclear exclusion. Proc Natl Acad Sci U S A 89: 7262–7266.
32. Sun XF, Carstensen JM, Zhang H, Stal O, Wingren S, et al (1992) Prognostic significance of cytoplasmic p53 oncoprotein in colorectal adenocarcinoma. Lancet 340: 1369–1373.
33. Sun XF, Carstensen JM, Stal O, Zhang H, Nilsson E, et al (1993) Prognostic significance of p53 expression in relation to DNA ploidy in colorectal adenocarcinoma. Virchows Arch A Pathol Anat Histopathol 423: 443–448.
34. Green DR, Kroemer G (2009) Cytoplasmic functions of the tumour suppressor p53. Nature 458: 1127–1130.
35. Lohrum MA, Woods DB, Ludwig RL, Balint E, Vousden KH (2001) C-terminal ubiquitination of p53 contributes to nuclear export. Mol Cell Biol 21: 8521–8532.
36. Montes de Oca LR, Wagner DS, Lozano G (1995) Rescue of early embryonic lethality in mdm2-deficient mice by deletion of p53. Nature 378: 203–206.
37. Mills AA, Zheng B, Wang XJ, Vogel H, Roop DR, et al (1999) p63 is a p53 homologue required for limb and epidermal morphogenesis. Nature 398: 708–713.
38. Yang A, Schweitzer R, Sun D, Kaghad M, Walker N, et al (1999) p63 is essential for regenerative proliferation in limb, craniofacial and epithelial development. Nature 398: 714–718.
39. Bakkers J, Hild M, Kramer C, Furutani-Seiki M, Hammerschmidt M (2002) Zebrafish DeltaNp63 is a direct target of Bmp signaling and encodes a transcriptional repressor blocking neural specification in the ventral ectoderm. Dev Cell 2: 617–627.
40. Lee H, Kimelman D (2002) A dominant-negative form of p63 is required for epidermal proliferation in zebrafish. Dev Cell 2: 607–616.
41. Yang A, Walker N, Bronson R, Kaghad M, Oosterwegel M, et al (2000) p73-deficient mice have neurological, pheromonal and inflammatory defects but lack spontaneous tumours. Nature 404: 99–103.
42. Rentzsch F, Kramer C, Hammerschmidt M (2003) Specific and conserved roles of TAp73 during zebrafish development. Gene 323: 19–30.
43. Young RW (1985) Cell differentiation in the retina of the mouse. Anat Rec 212: 199–205.
44. Berberich T, Kusano T (1997) Cycloheximide induces a subset of low temperature-inducible genes in maize. Mol Gen Genet 254: 275–283.
45. Donehower LA, Harvey M, Slagle BL, McArthur MJ, Montgomery CA, Jr., (1992) Mice deficient for p53 are developmentally normal but susceptible to spontaneous tumours. Nature 356: 215–221.
46. Murray-Zmijewski F, Lane DP, Bourdon JC (2006) p53/p63/p73 isoforms: an orchestra of isoforms to harmonise cell differentiation and response to stress. Cell Death Differ 13: 962–972.
47. Allocati N, Di IC, De L, V (2012) p63/p73 in the control of cell cycle and cell death. Exp Cell Res 318: 1285–1290.
48. Jackson MW, Berberich SJ (2000) MdmX protects p53 from Mdm2-mediated degradation. Mol Cell Biol 20: 1001–1007.

7

Anti-Bacterial Activity of Recombinant Human β-Defensin-3 Secreted in the Milk of Transgenic Goats Produced by Somatic Cell Nuclear Transfer

Jun Liu[◑], Yan Luo[◑], Hengtao Ge[◑], Chengquan Han, Hui Zhang, Yongsheng Wang, Jianmin Su, Fusheng Quan, Mingqing Gao, Yong Zhang*

College of Veterinary Medicine, Northwest A&F University, Key Laboratory of Animal Biotechnology of the Ministry of Agriculture, Yangling, Shaanxi, China

Abstract

The present study was conducted to determine whether recombinant human β-defensin-3 (rHBD3) in the milk of transgenic goats has an anti-bacterial activity against *Escherichia coli (E. coli)*, *Staphylococcus aureus (S. aureus)* and *Streptococcus agalactiae (S. agalactiae)* that could cause mastitis. A HBD3 mammary-specific expression vector was transfected by electroporation into goat fetal fibroblasts which were used to produce fourteen healthy transgenic goats by somatic cell nuclear transfer. The expression level of rHBD3 in the milk of the six transgenic goats ranged from 98 to 121 μg/ml at 15 days of lactation, and was maintained at 90–111 μg/ml during the following 2 months. Milk samples from transgenic goats showed an obvious inhibitory activity against *E. coli*, *S. aureus* and *S. agalactiae in vitro*. The minimal inhibitory concentrations of rHBD3 in milk against *E. coli*, *S. aureus* and *S. agalactiae* were 9.5–10.5, 21.8–23.0 and 17.3–18.5 μg/mL, respectively, which was similar to those of the HBD3 standard ($P > 0.05$). The *in vivo* anti-bacterial activities of rHBD3 in milk were examined by intramammary infusion of viable bacterial inoculums. We observed that 9/10 and 8/10 glands of non-transgenic goats infused with *S. aureus* and *E. coli* became infected. The mean numbers of viable bacteria went up to 2.9×10^3 and 95.4×10^3 CFU/ml at 48 h after infusion, respectively; the mean somatic cell counts (SCC) in infected glands reached up to 260.4×10^5 and 622.2×10^5 cells/ml, which were significantly higher than the SCC in uninfected goat glands. In contrast, no bacteria was presented in glands of transgenic goats and PBS-infused controls, and the SSC did not significantly change throughout the period. Moreover, the compositions and protein profiles of milk from transgenic and non-transgenic goats were identical. The present study demonstrated that HBD3 were an effective anti-bacterial protein to enhance the mastitis resistance of dairy animals.

Editor: Vladimir N. Uversky, University of South Florida College of Medicine, United States of America

Funding: This work was supported by the National High Technology Research and Development Program of China (863 Program) (No. 2011AA100303) (http://www.863.gov.cn/). The funders had no role in study design, data collection and analysis, decision to publish, or preparation of the manuscript.

Competing Interests: The authors have declared that no competing interests exist.

* E-mail: zhangylab@yahoo.com.cn

◑ These authors contributed equally to this work.

Introduction

Mastitis is inflammation of the mammary gland, which is usually caused by microbial infection [1]. Bovine mastitis is highly prevalent and the most costly disease in the dairy industry worldwide [2]. Traditional antibiotic treatments for clinical mastitis have resulted in antibiotic-resistant bacterial strains and antibiotic residues in milk [3]. In addition, there is no effective vaccine for the management of mastitis [4]. Recently, several research teams have demonstrated that the expression of anti-bacterial proteins in the milk of transgenic dairy animals can inhibit the bacterial pathogens that cause mastitis [5–8]. Therefore, the production of mastitis-resistant animals by genetic engineering technology has been proposed as an alternative approach to enhance mastitis resistance [9].

It has been reported that some kinds of anti-bacterial peptides (AMPs) play an important role in the innate immunity of the bovine mammary gland [10]. Beta-defensins, a kind of AMPs, are expressed in response to mastitis, and protect the body from bacterial invasion [11,12]. Scientists have extensively studied human β-defensin-3 (HBD3). HBD3 is widely expressed in many tissues [13,14], has a broad-spectrum anti-bacterial activity against bacteria, fungi and enveloped viruses, and plays important roles in immunity [15]. Therefore, HBD3 may be a candidate gene to enhance mastitis resistance. In addition, recombinant HBD3 (rHBD3) purified from milk may be useful for pharmacological and therapeutic applications.

The current study was designed to determine whether rHBD3 secreted in the milk of cloned transgenic goats has an anti-bacterial activity against bacteria that could cause mastitis. We transfected a HBD3 mammary-specific expression vector into goat fetal fibroblast cells (GFFs) by electroporation, and produced transgenic goats by somatic cell nuclear transfer (SCNT). In addition, we examined the *in vitro* and *in vivo* anti-bacterial activity of rHBD3 in the milk of cloned transgenic goats. This study demonstrated that HBD3 could be an effective anti-bacterial protein to enhance the mastitis resistance of dairy animals.

Materials and Methods

Ethics Statement

All experiments were approved by Care and Use of Animals Center, Northwest A&F University. This study was carried out in strict accordance with the Guidelines for the Care and Use of Animals of Northwest A&F University. Goat's ovaries were collected from Tumen abattoir, a local slaughterhouse of Xi'An, P.R. China. Fetuses and recipient goats were obtained from Yangling Keyuan Cloning Co., Ltd. Every effort was made to minimize animal pain, suffering and distress and to reduce the number of animal used, and all surgery was performed under anesthesia created by intravenous injection of Sumianxing (Veterinary Research Institute, Jilin, China).

Chemicals

Unless otherwise indicated, all chemicals and reagents were purchased from Sigma Chemical Company (St. Louis, MO), and the culture medium and fetal bovine serum (FBS) used for preparation of donor cells were obtained from Gibco (Grand Island, NY).

Construction of the HBD3 Mammary-specific Expression Vector pEBB

Construction and assessment strategies of the mammary-specific expression vector pEBB were performed as described previously [16]. Genomic DNA was extracted from the blood of Holstein cattle using a TIANamp Genomic DNA Kit (Tiangen Biotech, Beijing, China). The 2.2 kb promoter region (BBC5, including the 1.7 kb 5'-flanking sequence, exon 1 and part of intron 1) and 0.6 kb 3'-untranslated region (BBC3, including part of the last intron and exon) of the bovine β-casein gene (GenBank: X14711) were amplified using rTaq DNA polymerase (Takara, Dalian, China). We obtained the 1156 bp HBD3 DNA sequence (GenBank: 55894) from human genomic DNA by PCR. Then, the pEBB vector was constructed by inserting 2.2 kb BBC5, 0.6 kb BBC3, and 1156 bp HBD3 DNA sequences into a pEGFP-C1 plasmid (Clontech, Mountain View, CA, USA) (Fig. 1A). The vector was linearized with *ApaL*I, purified by a Wizard DNA Clean-Up System (Promega, Madison, USA), diluted with double distilled water, quantified, and then used for transfection.

Preparation of HBD3 Transgenic Donor Cells

GFFs were isolated as described previously [17]. An established cell line, GFF1, was seeded on a 60-mm culture dish, and cultured in 4 ml Dulbecco's modified Eagle's medium/Ham's F-12 (DMEM/F12) supplemented with 10% FBS. At 70–80% confluence, cells were harvested and transfected by electroporation [18]. Briefly, the cell concentration was adjusted to 5×10^6 cells/ml in transfection medium containing 20 μg/ml plasmid DNA. Then, the mixture was transferred to a 4-mm gap electroporation cuvette and pulsed by a BTX ECM 2001 (500 V, 1 ms, three pulses). After the pulses, cells were seeded on a 100-mm culture dish containing 10 ml DMEM/F12 supplemented with 10% FBS. After 24 h of culture, the cells were selected by 800 μg/ml geneticin (G418) for 7 days, and then 400 μg/ml G418 was used to obtain cell colonies. Cell colonies were counted under an inverted fluorescence microscope (Nikon, Tokyo, Japan). Enhanced green fluorescent protein (EGFP)-positive colonies were picked up and expanded in culture. A portion of the cell culture was harvested for cryopreservation, and the remaining cells were used for PCR screening. The number of chromosomes in transgenic clonal cells was determined as described previously [19]. More than twenty metaphase chromosome spreads from

each sample were examined, and cells with 60 chromosomes were classified as normal cells. Each cell line was analyzed three times.

Detection of Transfected Cells by PCR

G418-resistant clonal cells were lysed in lysis buffer, and then analyzed by PCR to confirm transgene integration. The upstream primer (5'-GGTCATTAGTTCATAGCCCATATATG-GAGTTC-3') is located in the CMV promoter, and the downstream primer (5'-CGTCGACTTTCCACAGCTCTTTT-TAACATC3-') is located in BBC3 (Fig. 1A, P1/P2). LATaq DNA Polymerase (Takara, Dalian, China) was used to amplify a 5100 bp fragment containing CMV promoter, EGFP, BBC5 and BBC3 sequences. PCR amplification conditions were 94°C for 5 min, followed by 30 cycles of 94°C for 30 sec, 58°C for 30 sec and 72°C for 4 min, and then 72°C for 10 min. Following amplification, PCR products were analyzed on a 0.8% agarose gel containing ethidium bromide. PCR products isolated from the gel were cloned into a pMD19-T vector using a TA Cloning Kit (Takara, Dalian, China), and then sequenced. PCR analysis was also performed on the primary cultured fibroblasts (GFF1) and pEBB plasmid as negative and positive controls, respectively.

Somatic Cell Nuclear Transfer

Transgenic cells were seeded on a 48-well plate, and allowed to reach confluence for 2–3 d prior to SCNT. The SCNT procedures were performed as described previously by our laboratory [20,21]. Briefly, oocytes with the first polar body were selected for enucleation after 22–24 h of *in vitro* maturation. Both the polar body and metaphase plate were removed, and then a single round donor cell was injected into the perivitelline space of the enucleated oocyte. Karyoplast-cytoplast couplets were fused by electrofusion. Couplets were incubated for 2–3 h in TCM-199 supplemented with 10% FBS and 7.5 μg/ml cytochalasin B. Fused embryos were activated by treatment with 5 μM ionomycin for 5 min and then 2 mM 6-dimethylaminopurine for 4 h. Following activation, the embryos were washed extensively and cultured in 200 μl mSOF supplemented with 10% FBS while covered with mineral oil at 38.5°C in a humified atmosphere with 5% CO_2. Embryos were cultured for 7 days to evaluate the *in vitro* developmental rate. One- to two-cell stage embryos cultured for 20–24 h were transferred into the oviducts of synchronized recipients on day 1 of estrus (day 0 = estrus, 19–28 embryos per recipient). Pregnancy was determined by ultrasonography.

Identification of Transgenic Goats by PCR and Southern Blotting

Genomic DNA from cloned goats was extracted, and PCR was performed as described above. Genomic DNA from wild-type goats and transgenic cells was used as negative and positive controls, respectively.

Genomic DNA isolated from cloned goats was analyzed by Southern blotting using a DIG-High Prime DNA Labeling and Detection Starter Kit II (Roche Molecular Biochemicals, Mannheim, Germany). Genomic DNA (20 μg) was thoroughly digested with restriction endonucleases *Sac*I and *Sal*I (New England Biolabs, Beverly, MA). Digested DNA was separated by 1.0% agarose gel electrophoresis, and then transferred to nylon membranes (Amersham, Piscataway, NJ). A digoxigenin-labeled 1156 bp HBD3 DNA probe (Fig. 1A, Probe) was prepared following the manufacturer's recommended procedure. The membranes were subsequently hybridized with the probe overnight at 68°C in a hybridization solution. The bands were detected using a chemiluminescent substrate system.

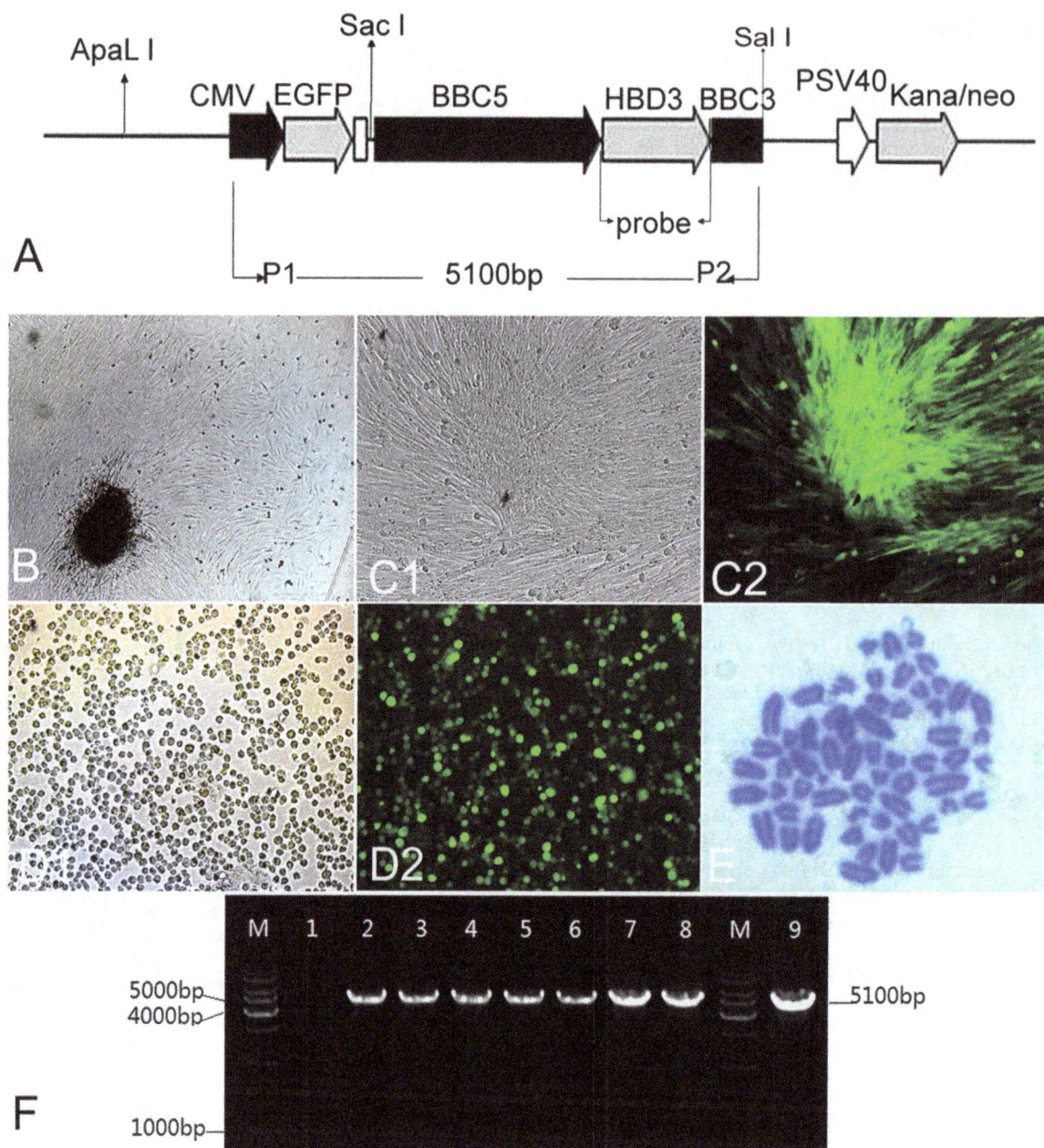

Figure 1. Generation of transgenic donor cells. (A) Schematic representation of the recombinant plasmid pEBB. Positions of primers (P1/P2) and the probe used for PCR and Southern blot are showed. (B) Primary culture of GFFs. (C) A G418-resistant colony expressing EGFP under bright field (C1) and fluorescence (C2). (D) Transgenic cells expressing EGFP under bright field (D1) and fluorescence (D2) digested by a 0.25% trypsin solution. (E) Chromosome number in transgenic cells (2n = 60). (F) PCR analysis of G418-resistant colonies. M, Marker; Lane 1, untransfected GFFs (negative control); Lanes 2–8, G418-resistant colonies; Lanes 9, pEBB vector (positive control).

Milk Collection and Composition Analysis

At about 16 months of age, transgenic goats were naturally mated, and pregnancy was detected by ultrasonography at around 60 days after fertilization. After delivery, milk samples were collected from transgenic and non-transgenic goats (negative control) of the same breed and age. The content of fat, protein, lactose, and dry matter was determined using a MilkoScan FT-120 (Foss, Hillerod, Denmark). The general profile of milk proteins was examined by Tricine-sodium dodecyl sulfate-polyacrylamide gel electrophoresis (Tricine-SDS-PAGE) and Coomassie Blue stain-

ing. Secretion of HBD3 in the milk of transgenic goats was analyzed according to the following procedures.

Tricine-SDS-PAGE and Western Blot Analysis of Transgenic Milk

Secretion of HBD3 in the milk of transgenic goats was analyzed by Tricine-SDS-PAGE and western blot analysis. Milk samples were centrifuged at 3,000 g for 15 min to remove the fat fractions. Liquids from the lower layer were then adjusted to pH 3.8–4.6 with 1 M HCl to eliminate the casein fraction. Then, 5 µl milk samples were mixed with the same volume of loading buffer and separated by 12% Tricine-SDS-PAGE. After electrophoresis, proteins were stained with Coomassie Brilliant Blue to examine the protein composition in the milk, or were transferred onto polyvinylidene difluoride membranes in a Bio-Rad trans-blot Cell (Bio-Rad, Hercules, CA, USA). After blocking, the membranes were reacted with a rabbit anti-HBD3 antibody (Sigma) at a 1:100 dilution, and then incubated with an alkaline phosphatase-conjugated goat anti-rabbit IgG (Beyotime Institute of Biotechnology, Shanghai, China) at a 1:1000 dilution. The membranes were washed extensively and exposed to Kodak XBT-1 film in a dark room. Milk from non-transgenic goats was used as a negative control, and 10 μg HBD3 standard (Premedical, Beijing, China) was used as a positive control. In addition, whole milk samples were analyzed by SDS-PAGE and western blotting using a rabbit anti-β-casein antibody (Ricky, Shanghai, China) to detect the expression of β-casein protein.

Concentration Measurement of HBD3 in Milk

Concentrations of HBD3 in transgenic milk were determined by a sandwich enzyme-linked immunosorbent assay (ELISA) kit (Adipo Bioscience, Santa Clara, CA, USA) according to the manufacturer's instructions. All samples were tested in duplicate, and the procedure was repeated three times.

Analysis of HBD3 Anti-bacterial Activity *in vitro*

The anti-bacterial activity of rHBD3 in the milk of transgenic goats, non-transgenic controls, and non-transgenic control milk containing the HBD3 standard was characterized by two methods as follows. Three classes of microorganisms, *Escherichia coli* (*E. coli*, ATCC25922), *Staphylococcus aureus* (*S. aureus*, ATCC25923) and *Streptococcus agalactiae* (*S. agalactiae*, ATCC12386), were obtained from the Chinese Institute of Veterinary Drug Control. The microorganisms were grown overnight at 37°C in 50 ml trypticase soy broth medium. Bacterial suspension was centrifuged at 3,000 g for 5 min, and the bacteria were washed and resuspended in phosphate buffered saline (PBS) to 1×10^7 colony forming units (CFU)/ml.

The anti-bacterial activity of transgenic milk was roughly estimated by inhibition zone assay. A bacterial suspension (100 μl) was spread onto a 100-mm agar dish and air dried for 10 min. Then, milk samples (20 μl) were spotted onto individual discs of quantitative filter paper (7 mm in diameter), which were placed on the agar dish containing *E. coli*, *S. aureus* or *S. agalactiae*. After incubation at 37°C for 24 h, inhibition zones around the filter paper discs indicated the anti-bacterial activity. Non-transgenic milk was spotted on the filter paper as negative controls, and non-transgenic control milk containing 100 or 200 μg/ml HBD3 standard was used as positive controls.

The minimal inhibitory concentration (MIC) of rHBD3 in milk from transgenic goats and the HBD3 standard was measured by a liquid growth inhibition assay as described previously [22] with slight modifications. Transgenic milk and the HBD3 standard were diluted in non-transgenic control milk at various final concentrations. Then, ten microliter bacterial suspension was added to 90 μl milk samples (1×10^5 CFU/ml), followed by incubation for 3 h at 37°C. The samples were then diluted serially in PBS, plated on agar dish, and incubated for 18 h at 37°C. The number of colonies was then counted. The MIC is the lowest concentration at which 99.9% of the viable cells are inhibited. All assays were conducted in duplicate and repeated four times.

In vivo Anti-bacterial Activity Assay

Mastitis resistance ability of HBD3 transgenic goats was examined by intramammary infusion of viable bacterial inoculums as described previously [7]. Six transgenic and 13 non-transgenic goats were used in this experiment. Before treatment (0 h), milk samples were collected from each gland, and the number of bacteria and somatic cell count (SCC) were determined to assess health of the animals. After milking, the right and left glands of each goats were infused with 1 ml *S. aureus* and *E. coli* inoculums (10^3 CUF/ml in PBS), respectively, via the streak canal. Six glands of three non-transgenic goats received 1 ml sterile PBS. At 12, 24 and 48 h after infusion, milk samples were collected and plated on agar dish to count the colony forming unit. The SSC of milk samples was determined on Fossomatic FC instrument (Foss, Hillerod, Denmark) according to the manufacturer's instructions. Glands containing fewer than 10 CFU/ml bacteria in milk were considered as not infected. After experiment, antibiotic therapy was administered to the infected goats.

Statistical Analysis

All data were analyzed using SPSS 16.0 statistical software (IBM Corporation, Somers, NY). The *in vivo* developmental rates of cloned embryos were tested by Chi-square analysis. Data of *in vitro* developmental rates and anti-bacterial activities were represented as the mean±SEM, and results were analyzed by one-way ANOVA and least-significant difference tests. For all analyses, $P<0.05$ was considered significant.

Results

Preparation of HBD3 Transgenic Donor Cells

Nine GFF cell lines were established from different fetuses. The primary GFF1 cell line from a female fetus was used in this study (Fig. 1B). About 3×10^6 cells were transfected with pEBB by electroporation. A total of 274 G418-resistant colonies were obtained, and 31 (14.4%) colonies expressed EGFP (Fig. 1C). We picked up 20 EGFP-positive colonies and transferred them to 48-well plates. Seven (35.0%) selected colonies were expanded to more than 2×10^6 cells, and all cell populations were positive for fluorescence (Fig. 1D). After expansion in culture, a portion of the clonal cells was frozen or used as donor cells for SCNT, and the remaining cells were passaged routinely to determine the chromosome number (Fig. 1E) and integration of the transgene. The percentage of cells with a normal chromosomal number (2n = 60) in G418-resistant colonies (F1HBDC1, F1HBDC2, F1HBDC3, F1HBDC4, F1HBDC5, F1HBDC6, and F1HBDC7) were 72.3±1.2(50/69), 71.1±2.0(44/62), 59.9±1.4(42/70), 49.3±1.6(32/65), 51.5±0.8(34/65), 57.9±1.8(40/69), and 51.6±1.9(35/68) %, respectively. PCR amplification indicated presence of the transgene in the seven transgenic cell lines (Fig. 1F).

Generation of Cloned Transgenic Goats by SCNT

Most of the cloned blastocysts from all the transgenic cell lines expressed EGFP (Fig. 2A), but the *in vitro* developmental rates of cloned embryos from F1HBDC1 and F1HBDC2 were significantly higher than those from F1HBDC4, F1HBDC5, F1HBDC6, and F1HBDC7 ($P<0.05$, Table 1). After comprehensive examination, transgenic cell lines F1HBDC1 and F1HBDC2 were used as donor cells to produce cloned transgenic goats. A total of 1817 early-stage embryos were transferred into the oviducts of 79 recipient goats (Table 2). No significant difference was observed in pregnancy rates between these two cell lines at 30 days of gestation. However, the delivery rate of cell line F1HBDC2 was higher than that of F1HBDC1 (23.5% vs. 9.1%, $P<0.05$). The cloning efficiencies (offspring produced per embryo transferred) of F1HBDC1 and F1HBDC2 were 0.4% and 1.6%, respectively.

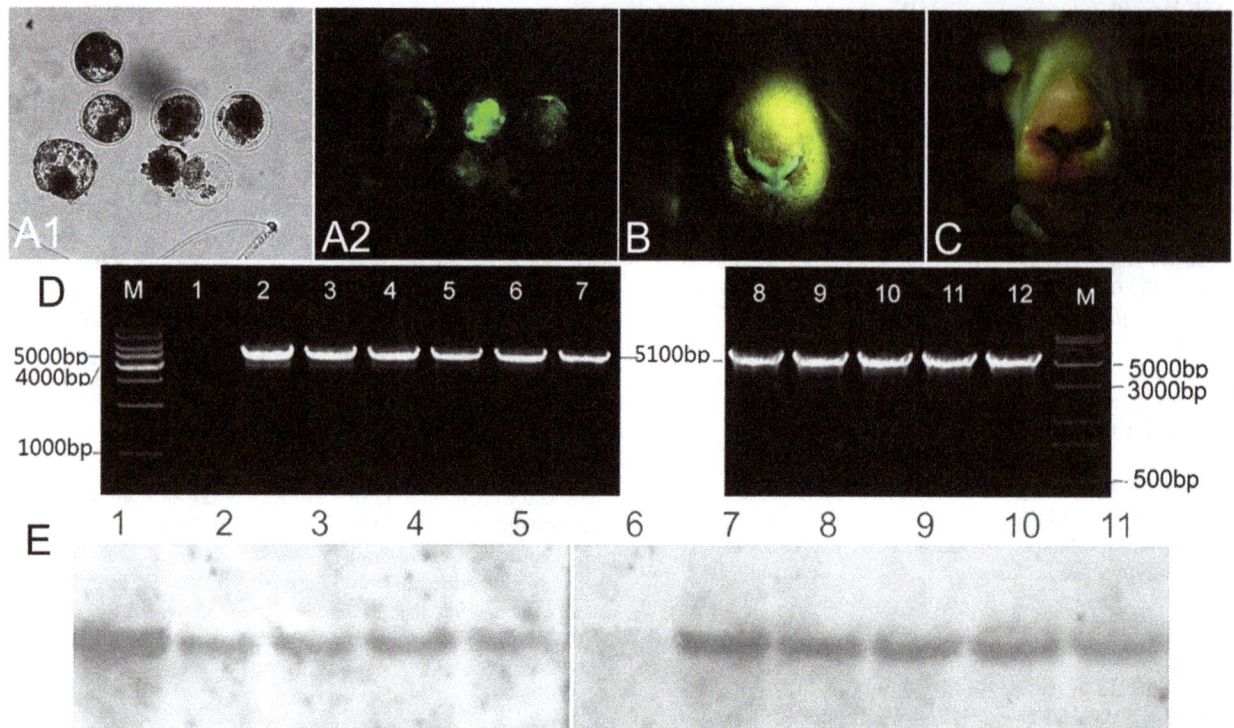

Figure 2. Identification of transgenic cloned goats. (A) Cloned transgenic blastocysts under bright field (A1) and fluorescence (A2). (B and C) EGFP expression in cloned goats was observed using a Dual Fluorescent Protein Flashlight (B), and no EGFP expression was observed in wild-type goats (C). (D) PCR analysis. M, Marker; Lane 1, wild-type goat (negative control); Lanes 2–3: transgenic cell lines F1HBDC1 and F1HBDC2 (positive control); Lanes 4–12: transgenic goats. (E) Southern blot analysis. Lane 1, pEBB vector (positive control); Lane 6, wild-type goat (negative control); Lanes 2–5 and 7–11, transgenic goats.

We obtained 26 cloned offspring, and the information on the cloned goats was summarized in Table 3. Gestation lengths ranged from 148 to 162 days, and birth weights ranged from 1.3 to 4.9 kg. Eighteen cloned goats were alive at birth, fourteen of which survived and no abnormality was observed. Mortality of the cloned goats was duo to placental defects, intrauterine infection,

abnormal joints, or anal atresia, while no abnormalities were observed in some of the dead goats. The expression of EGFP in cloned goats was observed using a Dual Fluorescent Protein Flashlight (NightSea DFP-1, MA, USA), which revealed that the skin of cloned goats expressed a high level of EGFP (Fig. 2B and C). PCR detection indicated that all the cloned goats have integration of transgene HBD3 (date not shown). Nine transgenic goats grew to adulthood and were naturally mated, and the integration of transgene was detected by PCR using primers P1/ P2 (Fig. 1A) and Southern blot using probe (Fig. 1A). Analysis of genomic DNA by PCR revealed integration of the transgene in cloned goats, whereas no transgene integration was detected in the surrogate female goat (Fig. 2D). Genomic Southern blot analysis further confirmed integration of the transgene in the genome of live offspring (Fig. 2E). After delivery, the nine transgenic goats were able to lactate normally. Six transgenic goats (TG3, TG7, TG10, TG14, TG15 and TG 18) at the same lactation stage (within 11 days) were used in the subsequent experiments.

Composition Analysis of Transgenic Milk

Composition analysis of whole milk samples was performed on a MilkoScan FT-120 (Foss, Hillerod, Denmark). The content of total fat, protein, lactose and dry matter in milk samples obtained from transgenic goats ($n = 6$) was similar to those in milk samples from control goats ($n = 6$) (Fig. 3A). Furthermore, the total protein profiles of milk samples (excluding the casein fraction) were similar between transgenic and non-transgenic goats (Fig. 3B).

Table 1. *In vitro* development of cloned embryos from different transgenic cell lines.

Donor cells	Fusion rate*	Cleavage rate**	Blastocyst rate***
F1HBDC1	(141/162)87.0±1.1[a]	(105)74.6±0.9[a]	(31)22.0±0.7[a]
F1HBDC2	(134/161)83.2±0.6[a]	(100)74.8±1.5[a]	(30)22.3±1.1[a]
F1HBDC3	(137/158)86.8±0.9[a]	(103)75.2±1.6[a]	(29)21.2±0.4[ab]
F1HBDC4	(144/167)86.3±1.3[a]	(94)65.3±1.5[b]	(23)16.0±0.6[c]
F1HBDC5	(136/162)83.9±1.5[a]	(90)67.1±1.8[b]	(22)16.2±0.1[c]
F1HBDC6	(145/166)87.5±2.3[a]	(106)73.0±2.0[a]	(28)19.3±0.7[b]
F1HBDC7	(138/161)85.8±0.7[a]	(100)72.8±2.8[a]	(27)19.6±1.0[b]
GFF1	(142/167)85.1±2.4[a]	(107)75.5±1.3[a]	(31)21.8±0.6[a]

*fusion rate = No. of fused embryos/No. of couplets.
**cleavage rate = No. of cleavage embryos/No. of fused embryos.
***blastocyst rate = No. of blastocyst/No. of fused embryos.
Four replicate experiments were performed per cells. Numbers in parentheses represent total embryo numbers of four replicates, while other numbers represent development rates (mean±SEM %).
[a–c]Within a column, values with different superscripts are significantly different from each other ($P < 0.05$).

Table 2. *In vivo* development of cloned embryos from different transgenic cell lines.

Donor cell	No. of embryos cultured	No. of embryos transferred	No. of recipients	No. of pregnant (%)*		No. of kids born (%)**
				Day 30	Term	
F1HBDC1	279	253	11	4(36.4)[a]	1(9.1)[a]	1(0.4)
F1HBDC2	1632	1564	68	28(41.2)[a]	16(23.5)[b]	25(1.6)

*Pregnancy rate = No. of pregnant recipients/No. of recipients used.
**Cloning efficiency = No. of kids born/No. of embryos transferred.
[a, b]Within a column, values with different superscripts are significantly different from each other ($P<0.05$).

Expression of HBD3 in the Milk of Transgenic Goats

Milk samples were diluted and analyzed by western blotting. Milk from non-transgenic goats was used as a negative control, and 10 μg HBD3 standard was used as a positive control. Western blot analysis confirmed that rHBD3 (5 kD) was only present in the milk of transgenic goats (Fig. 4A). Beta-casein (30 kD) was present in the milk of both transgenic and non-transgenic goats, which was used as an internal control (Fig. 4B). These results indicated that rHBD3 was expressed in the milk of transgenic goats.

The concentration of rHBD3 in the milk of transgenic goats was further quantified by ELISA (Table 4). To evaluate the stability of rHBD3 expression, milk samples were collected at different stages of lactation. The results showed that the expression level of rHBD3 in milk of the six transgenic goats was from 98 to 121 μg/ml at 15 days of lactation, and maintained 90 to 111 μg/ml during the following 2 months.

In vitro Anti-bacterial Activity of HBD3 Transgenic Milk

The anti-bacterial activity of rHBD3 in the milk of transgenic goats, non-transgenic goats, and non-transgenic control milk containing HBD3 standard was evaluated by inhibition zone and liquid growth inhibition assays.

Table 3. Summary of HBD3 transgenic goats.

Recipients	Cloned goats	Gestation length (day)	Birth weight (kg)	Postnatal status
55	TG1	151	2.7	stillborn
0340	TG2	152	2.0	normal
42	TG3	153	2.4	normal
42	TG4	153	1.7	normal
95	TG5	152	4.2	stillborn
54	TG6	162	3.1	stillborn
104	TG7	152	3.5	normal
03910	TG8	150	2.7	stillborn
03058	TG9	152	2.2	normal
87	TG10	151	3.5	normal
46	TG11	148	1.8	normal
46	TG12	148	2.1	normal
46	TG13	148	2.8	abnormal joints
0658	TG14	150	3.5	normal
0658	TG15	150	2.5	normal
52	TG16	154	1.9	spinal deformity
74	TG17	150	4.9	stillborn
100	TG18	151	3.3	abnormal joints
100	TG19	151	2.1	stillborn
018	TG20	151	2.8	stillborn
018	TG21	151	1.9	normal
0323	TG22	152	4.4	normal
124	TG23	151	2.0	normal
124	TG24	151	3.0	anal atresia
124	TG25	151	2.4	normal
124	TG26	151	1.3	stillborn

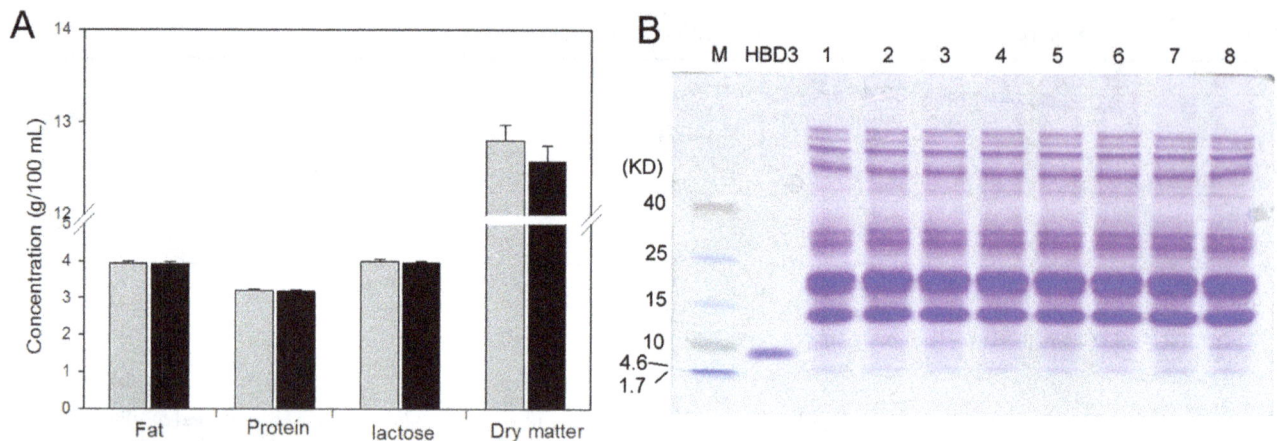

Figure 3. Composition analysis of transgenic milk. (A) Comparison of basic components between transgenic and non-transgenic milk. Gray bars, milk samples from transgenic goats (n = 6); black bars, milk samples from non-transgenic goats (n = 6). (B) Analysis of proteins in milk (excluding the casein fraction) by Tricine-SDS-PAGE and Coomassie Blue staining. Milk samples (5 µl) were loaded and analyzed. M, protein marker; HBD3, human β-defensin-3 standard; Lanes 1–2, milk samples from non-transgenic goats; Lanes 3–8, milk samples from transgenic goats.

The anti-bacterial activity of HBD3 could be roughly compared by measuring the diameter of the clear zone (Fig. 5). Inhibition zones around filters containing 20 µl milk samples from six transgenic goats were clearly visible after incubation for 24 h, which were similar to those of non-transgenic control milk containing 100 µg/ml HBD3 standard. No inhibition zone formed around filters containing milk from a non-transgenic goat. This result demonstrated that milk samples from HBD3 transgenic goats showed an inhibitory activity against *E. coli*, *S. aureus* and *S. agalactiae*, and the anti-bacterial activity against *E. coli* was higher than that against *S. aureus* and *S. agalactiae*.

The MIC of rHBD3 and the HBD3 standard in milk against *E. coli*, *S. aureus* and *S. agalactiae* was measured by a liquid growth inhibition assay (Table 5). The results indicated that the MICs of rHBD3 in transgenic milk against *E. coli*, *S. aureus* and *S. agalactiae* were 9.5–10.5, 21.8–23.0 and 17.3–18.5 µg/ml, respectively. The MIC of rHBD3 in milk was similar to that of the HBD3 standard ($P>0.05$). These results were consistent with those obtained from the inhibition zone assay. Thus, the concentration of rHBD3 in transgenic milk is sufficient for inhibiting the growth of *E. coli*, *S. aureus* and *S. agalactiae*.

In vivo Anti-bacterial Activity Assay

Mastitis resistance ability of HBD3 transgenic goats was investigated by intramammary infusion of viable bacterial inoculums (Table 6 and Fig. 6). Before treatment, the concentrations of rHBD3 in right and left glands of the six transgenic goats were 98.9±2.3 and 98.0±2.3 µg/ml, and no bacteria presented in the milk. The right and left glands of each transgenic (n = 6) and non-transgenic (n = 10) goats were infused with 1 ml *S. aureus* and *E. coli* inoculums (10^3 CUF/ml in PBS), respectively. Six glands of

three non-transgenic goats received 2 ml sterile PBS. Glands containing more than 10 CFU/ml were considered as infected. We observed that 9/10 and 8/10 glands of non-transgenic goats infused with *S. aureus* and *E. coli* were infected at 12 h after treatment, and the mean numbers of viable bacteria were 0.6×10^3 and 1.1×10^3 CFU/ml, and increased up to 2.9×10^3 and 95.4×10^3 CFU/ml at 48 h after infusion, respectively (Table 6). Furthermore, the mean SSC in infected glands significantly increased from 7.2×10^5 to 260.4×10^5 and 12.4×10^5 to 622.2×10^5 cells/ml in *S. aureus* and *E. coli* infected goats, respectively (Fig. 6). In contrast, no bacteria presented in glands of transgenic goats and PBS-infused controls, and the SSC did not significantly change throughout the period with a mean value from 3.1×10^5 to 5.8×10^5 cells/ml.

Discussion

The production of transgenic dairy animals that express anti-bacterial proteins in their milk has been proposed as an alternative approach to enhance mastitis resistance [5–8]. In addition, the transgenic milk can be used as a source to purify anti-bacterial proteins for pharmacological and therapeutic applications. In this study, we generated a transgenic goat expressing rHBD3 in the mammary gland. The results indicated that the expression of rHBD3 in the lactating mammary gland efficiently inhibited the growth of bacteria that could cause mastitis.

Several investigations have reported that β-defensins play an important role in the response to mastitis, and protect the body from bacterial invasion [11,12]. HBD3 has a broad-spectrum anti-bacterial activity against bacteria, fungi and enveloped viruses, and plays an important role in immunity [15]. Therefore, we hypothesized that exogenous expression of HBD3 in the mammary gland of dairy animals might enhance mastitis resistance. In this study, we examined the *in vitro* and *in vivo* anti-bacterial activity of rHBD3 in the milk of transgenic goats. The results indicated that rHBD3 in the milk of transgenic goats showed an efficient inhibitory activity against *E. coli*, *S. aureus* and *S. agalactiae in vitro*, and the anti-bacterial activity against *E. coli* was higher than those against *S. aureus* and *S. agalactiae*. The MICs of rHBD3 in transgenic milk against *E. coli*, *S. aureus* and *S. agalactiae* were similar to those of the HBD3 standard. In addition, the *in vivo* antibacterial activity assay has demonstrated that expression of

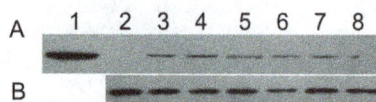

Figure 4. Expression of rHBD3 in transgenic goat milk. Protein expression of rHBD3 (A) and β-casein (B) were detected by western blotting. Lane 1, HBD3 standard; Lane 2, milk samples from non-transgenic goats; Lanes 3–8, milk samples from transgenic goats.

Table 4. Concentrations of HBD3 in milk from transgenic goats.

Transgenic goats	Concentration of HBD3 (μg/ml)		
	15 days of lactation	**45 days of lactation**	**65 days of lactation**
TG3	101.2±0.6	90.7±0.2	89.7±0.6
TG7	120.5±0.1	110.8±0.2	110.8±0.7
TG10	116.4±0.5	104.0±0.5	104.9±0.2
TG14	98.4±0.4	90.7±0.2	90.3±0.2
TG15	116.5±0.4	110.4±0.2	111.8±0.4
TG18	98.1±0.5	91.6±0.7	92.1±0.4

Based on three independent assays.

rHBD3 (mean 98 μg/ml) in mammary gland of goats could protect against infection of *E. coli* and *S. aureus*. Previous studies have demonstrated that the MIC of HBD3 against *E. coli* and *S. aureus* are 6 and 12 μg/ml, respectively [13,22], and the anti-bacterial and biochemical properties of recombinant and chemically synthesized HBD3 are indistinguishable from that of the isolated native peptide [13]. In contrast, the anti-bacterial activity of rHBD3 in this study was lower than those reported in previous reports. In this study, the anti-bacterial activity of rHBD3 and the HBD3 standards was measured in milk. Therefore, the reason for this discrepancy might be interference by the milk components.

Beta-casein is one of the most abundant proteins in milk. Therefore, the bovine β-casein promoter was used as the main control element to direct the expression of HBD3 in the mammary gland of transgenic goats. The mammary-specific expression vector has been assessed in mammary epithelial cells in a previous study [16]. In this study, transgenic fetal fibroblasts were used as donor cells for SCNT to produce transgenic goats. The expression level of rHBD3 in the mammary glands of transgenic goats cannot be determined before lactation. Furthermore, the transgenic technique is associated with random integration, and the positional effects may cause low expression of the transgene [23]. In the present study, two transgenic colonies with normal chromosomes and uniform small round cells were used as donor cells for SCNT, all live offspring were obtained from the transgenic cell line F1HBDC2. The concentrations of rHBD3 in milk ranged from 98 to 121 μg/ml at 15 days of lactation, and maintained at 90–

111 μg/ml during the following 2 months. Although the concentration of rHBD3 in the milk of transgenic goats was not high, it was sufficient for inhibiting the growth of *E. coli*, *S. aureus* and *S. agalactiae*.

Transgenic donor cell preparation is an important step for SCNT to produce transgenic animals. In current study, seven transgenic clonal cell lines were obtained after the selection of G418 and confirmation by EGFP expression and PCR screening. After SCNT, we observed that the developmental rate of cloned embryos derived from various transgenic clonal cells was different, and this result is consistent with the finding reported in a previous publication [24]. The cloning efficiencies of transgenic fetal fibroblasts F1HBDC1 and F1HBDC2 were 0.4 and 1.6%, respectively, similar to those obtained by other research groups [25–27]. The offspring mortality was 46.2% (12/26), close to the data reported in previous publications [17,27]. Although postnatal mortality was relatively high, seventy-eight percent (14/18) of the live born transgenic goats did not show abnormities. Composition analysis of the milk showed no significant difference in the content of total fat, protein, lactose, and dry matter between transgenic and non-transgenic goats. Moreover, the protein profile of the milk from transgenic and non-transgenic goats was similar. Expression of HBD3 in transgenic goats appeared to have no effect on the integrity of the mammary gland and milk production. This result is similar to the previous reports in bovine [6,7]. Unlike several cationic peptides, HBD3 does not exhibit a cytotoxic activity against eukaryotic cells. It exhibits a very low hemolytic

Figure 5. Anti-bacterial activity of rHBD3 in milk was measured by an inhibition zone assay. Bacterial suspensions *E. coli* (A), *S. aureus* (B) or *S. agalactiae* (C) were spread onto 100-mm agar dishes. Then, 20 μl milk samples were spotted on quantitative filter paper. The HBD3 standard was diluted in non-transgenic milk to 100 and 200 μg/ml. NC, milk samples from non-transgenic goats (negative control); PC, HBD3 standard (positive control); TG, milk samples from transgenic goats.

Table 5. The MIC of rHBD3 in the milk of transgenic goats.

Groups	The MIC of rHBD3 and HBD3 standards in milk (μg/ml)*		
	E. coli	S. aureus	S. agalactiae
HBD3 standard	9.5±0.3	21.5±0.6	18.0±0.4
TG3	10.3±0.3	21.8±0.5	18.3±1.0
TG7	10.0±0.4	22.8±0.5	17.3±0.9
TG10	9.5±0.5	22.8±0.8	18.0±0.7
TG14	10.0±0.4	23.0±0.4	18.3±0.6
TG15	10.5±0.3	22.8±0.5	17.8±0.9
TG18	9.5±0.3	23.0±0.4	18.5±0.6

*The minimal inhibitory concentration (MIC) is the lowest concentration of HBD3 at which 99.9% of the viable cells are inhibited. All assays were conducted in duplicate and repeated four times.
Within a column, No difference was found in MIC among the groups ($P>0.05$).

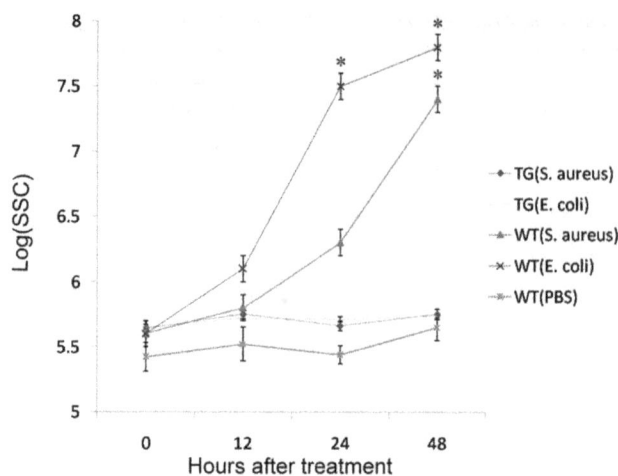

Figure 6. Somatic cell counts (SSC) in glands after infusion with E. coli or S. aureus. TG, transgenic goats (n = 6); WY, wild-type goats. Ten glands of wild-type goats were infused with E. coli or S. aureus, six glands of wild-type goats were infused with PBS. The data were represented as the mean±SEM, and analyzed by one-way ANOVA and least-significant difference tests. *$P<0.05$.

activity towards human erythrocytes when a high amount of the peptide (up to 500 μg/ml) is used [13,28]. We inferred that the expression of HBD3 in transgenic goats would be safe for mammary epithelial cells. However, apart from the anti-bacterial activity, HBD3 has been reported to induce the secretion of IL-18, a proinflammatory cytokine in human keratinocytes [29]. The safety of constitutively expressed HBD3 in the mammary glands of transgenic goats needs to be studied in the future.

It has been reported that the expression of a number of anti-bacterial proteins occurs in response to mastitis, and that these proteins play an important role in the first line of defense that inhibits or delays bacterial infection of mammary tissue [1]. Clearly, these mechanisms are insufficient to prevent mastitis. Several transgenic animal models have been produced, which express human lysozyme [5,6,30], human lactoferrin [31,32], bovine tracheal anti-bacterial peptide [33], and modified lysostaphin [7,34]. Milk from transgenic mice containing 0.38 mg/ml recombinant human lysozyme was found to be bacteriostatic against Pseudomonas fragi, Lactobacillus viscous and a mastitis-causing strain of S. aureus, but not against a pathogenic strain of E. coli. [30]. Milk from human lysozyme transgenic goats (270±84 μg/ml) is capable of slowing the growth of mastitis-causing strains of E. coli and S. aureus, but does not affect the growth of an organism involved in cheese-making, Lactococcus lacti [5]. However, high

concentrations of human lactoferrin (mean, 2.9 mg/ml) in the milk of transgenic cows does not protect them from infection by experimental E. coli [35]. Transgenic cows secreting lysostaphin at concentrations ranging from 0.9 to 14 μg/ml in their milk are resistant against S. aureus challenge in vivo [7], but lysostaphin is only active against S. aureus and it is thus not expected to be effective against non-staphylococcal infections. Our current study has demonstrated that rHBD3 in the milk of transgenic goats efficiently inhibits the growth of E. coli and S. aureus both in vitro and in vivo.

In conclusion, we successfully produced HBD3 transgenic goats by SCNT. The transgenic milk from these goats containing 90–121 μg/ml rHBD3 efficiently inhibited the growth of E. coli, S. aureus and S. agalactiae, which was similar to that of the HBD3 standards. The present study demonstrated that HBD3 could be an effective antibacterial protein to enhance the mastitis resistance of dairy animals.

Table 6. The colony forming unit in left and right glands post infusion with S. aureus and E. coli.

Groups	Concentration of rHBD3 before treatment	Glands treated	Glands infected*	The number of colony forming unit (×10³ CFU/ml)			
				0 h**	12 h	24 h	48 h
TG	98.9±2.3[a]	6(S. aureus)	0	0	0	0	0
TG	98.0±2.3[a]	6(E. coli)	0	0	0	0	0
WT	0	10(S. aureus)	9	0	0.6±0.1[a]	2.1±0.4[ab]	2.9±0.5[b]
WT	0	10(E. coli)	8	0	1.1±0.2[a]	7.6±0.9[a]	95.4±16.7[b]
WT	0	6(PBS)	0	0	0	0	0

*Glands containing fewer than 10 CFU/ml were considered as noninfected.
**Before treatment.
TG, transgenic goats; WT, wild-type goats.
[a, b]Within a row, values with different superscripts are significantly different from each other ($P<0.05$).

Acknowledgments

We are grateful to Mr. Hengde Zhang and Mr. Xueyao Bai for their excellent technical assistance during the embryo transfer experiments.

Author Contributions

Conceived and designed the experiments: JL YZ. Performed the experiments: JL YL HTG CQH YSW. Analyzed the data: JL YL JMS. Contributed reagents/materials/analysis tools: HZ FSQ MQG. Wrote the paper: JL YL.

References

1. Gray C, Strandberg Y, Donaldson L, Tellam R (2005) Bovine mammary epithelial cells, initiators of innate immune responses to mastitis. Animal Production Science 45: 757–761.
2. Seegers H, Fourichon C, Beaudeau F (2003) Production effects related to mastitis and mastitis economics in dairy cattle herds. Veterinary Research 34: 475–491.
3. Goni P, Vergara Y, Ruiz J, Albizu I, Vila J, et al. (2004) Antibiotic resistance and epidemiological typing of Staphylococcus aureus strains from ovine and rabbit mastitis. International Journal Antimicrob Agents 23: 268–272.
4. McDougall S, Parker KI, Heuer C, Compton CW (2009) A review of prevention and control of heifer mastitis via non-antibiotic strategies. Veterinary Microbiology 134: 177–185.
5. Maga EA, Cullor JS, Smith W, Anderson GB, Murray JD (2006) Human lysozyme expressed in the mammary gland of transgenic dairy goats can inhibit the growth of bacteria that cause mastitis and the cold-spoilage of milk. Foodborne Pathogens and Disease 3: 384–392.
6. Yang B, Wang J, Tang B, Liu Y, Guo C, et al. (2011) Characterization of bioactive recombinant human lysozyme expressed in milk of cloned transgenic cattle. PLoS One 6: e17593.
7. Wall RJ, Powell AM, Paape MJ, Kerr DE, Bannerman DD, et al. (2005) Genetically enhanced cows resist intramammary Staphylococcus aureus infection. Nature Biotechnology 23: 445–451.
8. Zhang JX, Zhang SF, Wang TD, Guo XJ, Hu RL (2007) Mammary gland expression of antibacterial peptide genes to inhibit bacterial pathogens causing mastitis. Journal of Dairy Science 90: 5218–5225.
9. Kerr D, Wellnitz O (2003) Mammary expression of new genes to combat mastitis. Journal of Animal Science 81: 38–47.
10. Rainard P, Riollet C (2006) Innate immunity of the bovine mammary gland. Veterinary Research 37: 369–400.
11. Roosen S, Exner K, Paul S, Schröder JM, Kalm E, et al. (2004) Bovine ß-defensins: Identification and characterization of novel bovine ß-defensin genes and their expression in mammarygland tissue. Mammalian Genome 15: 834–842.
12. Swanson K, Gorodetsky S, Good L, Davis S, Musgrave D, et al. (2004) Expression of a β-defensin mRNA, lingual antimicrobial peptide, in bovine mammary epithelial tissue is induced by mastitis. Infection and Immunity 72: 7311–7314.
13. Harder J, Bartels J, Christophers E, Schröder JM (2001) Isolation and characterization of human β-defensin-3, a novel human inducible peptide antibiotic. Journal of Biological Chemistry 276: 5707–5713.
14. Dunsche A, Acil Y, Dommisch H, Siebert R, Schröder JM, et al. (2002) The novel human beta-defensin-3 is widely expressed in oral tissues. European journal of oral sciences 110: 121–124.
15. Dhople V, Krukemeyer A, Ramamoorthy A (2006) The human beta-defensin-3, an antibacterial peptide with multiple biological functions. Biochimica et Biophysica Acta (BBA)-Biomembranes 1758: 1499–1512.
16. Liu J, Luo Y, Liu Q, Zheng L, Yang Z, et al. (2012) Production of cloned embryos from caprine mammary epithelial cells expressing recombinant human beta-defensin-3. Theriogenology. doi: 10.1016/j.theriogenology.2012.11.021.
17. Keefer CL, Baldassarre H, Keyston R, Wang B, Bhatia B, et al. (2001) Generation of dwarf coat (Capra hircus) clones following nuclear transfer with transfected and nontransfected fetal fibroblasts and in vitro-matured oocytes. Biology of Reproduction 64: 849–856.
18. Ross JW, Whyte JJ, Zhao J, Samuel M, Wells KD (2010) Optimization of square-wave electroporation for transfection of porcine fetal fibroblasts. Transgenic Research 19: 611–620.
19. Giraldo AM, Lynn JW, Godke RA, Bondioli KR (2006) Proliferative characteristics and chromosomal stability of bovine donor cells for nuclear transfer. Molecular Reproduction and Development 73: 1230–1238.
20. Zhang YL, Liu FJ, Sun DQ, Chen XQ, Zhang Y, et al. (2008) Phytohemag-glutinin improves efficiency of electrofusing mammary gland epithelial cells into oocytes in goats. Theriogenology 69: 1165–1171.
21. Liu J, Li LL, Du S, Bai XY, Zhang HD, et al. (2011) Effects of interval between fusion and activation, cytochalasin B treatment, and number of transferred embryos, on cloning efficiency in goats. Theriogenology 76: 1076–1083.
22. Hoover DM, Wu Z, Tucker K, Lu W, Lubkowski J (2003) Antimicrobial characterization of human β-defensin 3 derivatives. Antimicrobial Agents and Chemotherapy 47: 2804–2809.
23. Kong Q, Wu M, Huan Y, Zhang L, Liu H, et al. (2009) Transgene expression is associated with copy number and cytomegalovirus promoter methylation in transgenic pigs. PLoS One 4: e6679.
24. Kuhholzer B, Hawley RJ, Lai L, Kolber-Simonds D, Prather RS (2001) Clonal lines of transgenic fibroblast cells derived from the same fetus result in different development when used for nuclear transfer in pigs. Biology of Reproduction 64: 1695–1698.
25. Baguisi A, Behboodi E, Melican DT, Pollock JS, Destrempes MM, et al. (1999) Production of goats by somatic cell nuclear transfer. Nature Biotechnology 17: 456–461.
26. Amiri Yekta A, Dalman A, Eftekhari-Yazdi P, Sanati MH, Shahverdi AH, et al. (2013) Production of transgenic goats expressing human coagulation factor IX in the mammary glands after nuclear transfer using transfected fetal fibroblast cells. Transgenic Research 22: 131–142.
27. An LY, Yuan YG, Yu BL, Yang TJ, Cheng Y (2012) Generation of human lactoferrin transgenic cloned goats using donor cells with dual markers and a modified selection procedure. Theriogenology 78: 1303–1311.
28. Klüver E, Schulz-Maronde S, Scheid S, Meyer B, Forssmann WG, et al. (2005) Structure-activity relation of human β-defensin 3: influence of disulfide bonds and cysteine substitution on antimicrobial activity and cytotoxicity. Biochemistry 44: 9804–9816.
29. Niyonsaba F, Ushio H, Nagaoka I, Okumura K, Ogawa H (2005) The human β-defensins (-1,-2,-3,-4) and cathelicidin LL-37 induce IL-18 secretion through p38 and ERK MAPK activation in primary human keratinocytes. The Journal of Immunology 175: 1776–1784.
30. Maga EA, Anderson GB, Cullor JS, Smith W, Murray JD (1998) Antimicrobial properties of human lysozyme transgenic mouse milk. Journal of Food Protection® 61: 52–56.
31. Yang P, Wang J, Gong G, Sun X, Zhang R, et al. (2008) Cattle mammary bioreactor generated by a novel procedure of transgenic cloning for large-scale production of functional human lactoferrin. PLoS One 3: e3453.
32. van Berkel PHC, Welling MM, Geerts M, van Veen HA, Ravensbergen B, et al. (2002) Large scale production of recombinant human lactoferrin in the milk of transgenic cows. Nature Biotechnology 20: 484–487.
33. Yarus S, Rosen JM, Cole AM, Diamond G (1996) Production of active bovine tracheal antimicrobial peptide in milk of transgenic mice. Proc Natl Acad Sci U S A 93: 14118–14121.
34. Kerr DE, Plaut K, Bramley AJ, Williamson CM, Lax AJ, et al. (2001) Lysostaphin expression in mammary glands confers protection against staphylococcal infection in transgenic mice. Nature Biotechnology 19: 66–70.
35. Hyvonen P, Suojala L, Orro T, Haaranen J, Simola O, et al. (2006) Transgenic cows that produce recombinant human lactoferrin in milk are not protected from experimental Escherichia coli intramammary infection. Infection and Immunity 74: 6206–6212.

8

A New Transgenic Mouse Model for Studying the Neurotoxicity of Spermine Oxidase Dosage in the Response to Excitotoxic Injury

Manuela Cervelli[1]*, Gabriella Bellavia[1], Marcello D'Amelio[2], Virve Cavallucci[2], Sandra Moreno[1], Joachim Berger[3], Roberta Nardacci[4], Manuela Marcoli[5], Guido Maura[5], Mauro Piacentini[4], Roberto Amendola[6], Francesco Cecconi[2], Paolo Mariottini[1]

1 Dipartimento di Biologia, Università "Roma Tre," Rome, Italy, 2 Laboratory of Molecular Neuroembryology, Istituto di Ricovero e Cura a Carattere Scientifico (IRCCS) Fondazione Santa Lucia, Rome, Italy, 3 Faculty of Medicine, Nursing and Health Sciences, Monash University, Clayton, Australia, 4 Istituto Nazionale per le Malattie Infettive, IRCCS "L. Spallanzani," Rome, Italy, 5 Dipartimento di Farmacia, Sez. Farmacologia e Tossicologia, Centro di Eccellenza per la Ricerca Biomedica CEBR, Università di Genova, Genoa, Italy, 6 Agenzia nazionale per le nuove tecnologie, l'energia e lo sviluppo economico sostenibile (ENEA), Il Centro Ricerche Casaccia, Sezione Tossicologia e Scienze Biomediche (BAS-BIOTECMED), Rome, Italy

Abstract

Spermine oxidase is a FAD-containing enzyme involved in polyamines catabolism, selectively oxidizing spermine to produce H_2O_2, spermidine, and 3-aminopropanal. Spermine oxidase is highly expressed in the mouse brain and plays a key role in regulating the levels of spermine, which is involved in protein synthesis, cell division and cell growth. Spermine is normally released by neurons at synaptic sites where it exerts a neuromodulatory function, by specifically interacting with different types of ion channels, and with ionotropic glutamate receptors. In order to get an insight into the neurobiological roles of spermine oxidase and spermine, we have deregulated spermine oxidase gene expression producing and characterizing the transgenic mouse model JoSMOrec, conditionally overexpressing the enzyme in the neocortex. We have investigated the effects of spermine oxidase overexpression in the mouse neocortex by transcript accumulation, immunohistochemical analysis, enzymatic assays and polyamine content in young and aged animals. Transgenic JoSMOrec mice showed in the neocortex a higher H_2O_2 production in respect to Wild-Type controls, indicating an increase of oxidative stress due to SMO overexpression. Moreover, the response of transgenic mice to excitotoxic brain injury, induced by kainic acid injection, was evaluated by analysing the behavioural phenotype, the immunodistribution of neural cell populations, and the ultrastructural features of neocortical neurons. Spermine oxidase overexpression and the consequently altered polyamine levels in the neocortex affects the cytoarchitecture in the adult and aging brain, as well as after neurotoxic insult. It resulted that the transgenic JoSMOrec mouse line is more sensitive to KA than Wild-Type mice, indicating an important role of spermine oxidase during excitotoxicity. These results provide novel evidences of the complex and critical functions carried out by spermine oxidase and spermine in the mammalian brain.

Editor: Rudolf Kirchmair, Medical University Innsbruck, Austria

Funding: The financial support utilized for this work was mainly from the Università Roma Tre, the year budget was below 10,000 Euro. The funders had no role in study design, data collection and analysis, decision to publish, or preparation of the manuscript. No additional external funding was received for this study.

Competing Interests: The authors have declared that no competing interests exist.

* E-mail: cervelli@uniroma3.it

Introduction

Putrescine (Put), spermidine (Spd), and spermine (Spm) are endogenous polyamines (PAs) essential for cell growth, proliferation, regeneration, and differentiation [1–4]. The functional role of natural PAs in the normal and diseased brain is under active research [5–9]. Early reports on the effects of PAs on neuronal firing and transmitter release were followed by compelling evidences showing that PAs are potentially involved in the regulation of a number of metabolic and electrophysiological processes [10]. Alteration of PAs content and their synthetic enzyme ornithine decarboxylase (ODC) in response to injuries, such as ischemia, hypoglycaemia, epilepsy, or trauma have been reported [11–15]. Even though these results suggest that PAs play an important role in neurodegeneration, the mechanisms whereby they participate in neuronal death, as well as the role of endogenous PAs in normal brain functioning, are to be elucidated yet. Specific interactions of PAs, in particular Spm, with different types of ion channels, have been reported [16,17]. Intracellular PAs are able to block some types of K^+ and Na^+ channels and the glutamatergic AMPA (α-amino-3-hydroxy-5-methyl-4-isoxazole-propionic acid) and kainate receptors, while extracellular PAs modulate glutamatergic NMDA (N-methyl-D-aspartate) receptors [16,18–21]. The catabolism of polyamines is finely regulated by the concerted action of three enzymes: spermidine/spermine-N1-acetyltransferase (SSAT), which acetylates Spm and Spd; acetylpolyamine oxidase (APAO), which oxidizes these acetylated derivatives, regenerating Spd and Put, respectively; and the flavoprotein spermine oxidase (SMO), directly oxidizing Spm to produce Spd, 3-aminopropanal and hydrogen peroxide (H_2O_2).

While APAO is constitutively expressed, SSAT and SMO are inducible enzymes, and have therefore been more extensively investigated [9]. Interestingly, it was shown that transgenic activation of PAs catabolism not only profoundly disturbs PAs homeostasis in most tissues, but also creates a complex phenotype affecting skin, female fertility, fat depots, pancreatic integrity and regenerative growth [22]. The SSAT overexpression in the Central Nervous System (CNS) resulted in significantly elevated threshold to pentylenetetrazol-induced seizure activity and protection against kainate-induced toxicity of transgenic animals [23]. Since SSAT overexpression resulted in even greater expansion of Put pool in different regions of the brain, a neuroprotective role of Put has been suggested [24]. Consistent with these data, also ODC overexpression, leading to Put accumulation, is neuroprotective [25–28]. In this scenario, we investigated the effects of SMO overexpression, so far unexplored, in a mouse genetic model. Since, among PA, Spm is the strongest modulator of GluRs and some types of K^+ channels, we have generated a neocortex specific SMO overexpressing mouse model using a Cre/loxP-based recombination approach. This mouse model (named JoSMOrec) overexpresses SMO only in proneural populations of the hippocampus and the neocortex [29], allowing us to exclude any pleiotropic influence by other organs. In this work, we have studied the effect of SMO overexpression in the neocortex of young and aged mice, by analysing PA metabolism and glial stress markers expression. Aged SMO overexpressing mice show neuronal reduction, and light astrocyte and microglia activation in the cerebral cortex. To elucidate the possible role played by SMO in neuronal damage during excitotoxic insult, we have evaluated the effect of its overexpression after kainic acid (KA) systemic administration, known to induce epileptiform activity and excitotoxic mechanism activation in rodents [30]. Excitotoxicity refers to a process of neuronal death triggered by elevated levels of excitatory amino acid resulting in the opening of ionotropic glutamate receptors causing prolonged depolarization of neurons,

the subsequent influx of calcium, and the activation of enzymatic and nuclear mechanisms of cell death [30]. In the JoSMOrec mouse model, we have also examined in the neocortex neuronal degeneration and astrocyte and microglia activation, analyzing the immunohistochemical expression of different cell markers, as well as H_2O_2 production. Moreover, in the attempt to characterize the neurodegenerative process occurring in SMO overexpressing KA-treated animals, we have performed an ultrastructural analysis of the injured neocortex. It resulted that the transgenic SMO mouse line is more sensitive to KA than Wild-Type (WT) mice. The results presented in this work provide novel evidences of the complex and crucial functions carried out by SMO and Spm in mammalian brain in physiological and pathological conditions.

Materials and Methods

Ethics statement

The experiments were carried out in accordance with the ethical guidelines for the conduct of animal research of the European Community's Council Directive 86/609/EEC. Formal approval of these experiments was obtained from the Italian Ministry of Health (Official Italian Regulation D.L.vo 116/92, "Communication to Ministero della Salute no. 70-VI/1.1").

Construction of plasmids and generation of transgenic mice

A loxP-egfp-polyA cassette was cloned into the EcoRI site of Pcaggs [31] followed by a loxP-(XhoI)-IRES-lacZ-polyA cassette resulting in pJojo vector. The coding sequence of SMO gene (GenBankTM accession number AY033889) was amplified with the primers SMO-1F and SMO2-R (Table 1) to introduce the XhoI restriction site. Amplified PCR product was restricted by XhoI and ligated with the restricted XhoI pJojo vector resulting in pJoSMO vector. The plasmids were used to generate pJoSMO (GFP-SMO) mice by pronuclear microinjection. Transgenic mice were identified by GFP fluorescence. Genomic PCR was performed with the primers SMO-3F and SMO-4R (Table 1). The transgenic mice were produced by the standard pronuclear microinjection technique [32]. Fertilized oocytes were obtained from superovulated BALB/c x DBA/2 mice mated with males of the same strain.

Animals and kainate administration

All double transgenic animals JoSMOrec were obtained from a cross between JoSMO (BALB/cx DBA/2 as described above) and Dachshund-Cre (DBA/2) mice. Dachshund-Cre mouse line expresses Cre in proneural population of the nervous system [29] and leads to the recombination in the crossed progeny to produce JoSMOrec mice. Animals were housed under controlled temperature (20±1°C), humidity (55±10%), and on a 12-h light/dark schedule. Food and water were provided ad libitum. All experiments were performed on independent groups of mice. Kainic acid was dissolved in isotonic saline solution (50 mM NaPi pH 7.2, 100 mM NaCl) and administered subcutaneously at a dose of 25 mg/kg p.c. Following KA administration, mice were monitored continuously for 3–4 h for the onset and extent of seizure activity. Seizures were rated according to a previously defined scale by Schauwecker [30]: stage 1, immobility; stage 2, forelimb and/or tail extension, rigid posture; stage 3, repetitive movements, head bobbing; stage 4, rearing and falling; stage 5, continuous rearing and falling; stage 6 severe tonic-clonic seizures.

Table 1. Primers used in this study.

Target gene	Name	Primer sequence
SMO	SMO-1F	5'-TTTATACTCGAGCCTAGAAGGTGAG-CACGGAC-3'
SMO	SMO-2R	5'-AAATATCTCGAGGGAACACATTTGG-CAGTGAGG-3'
SMO	SMO-3F	5'-TCATCCCCTCGGGCTTCATG -3'
SMO	SMO-4R	5'-GGAACACATTTGGCAGTGAGG-3'
SMO	SMO-5F	5'-GTACCTGAAGGTGGAGAG-3'
SMO	SMO-6R	5'-TGCATGGGCGCTGTCTTGG-3'
APAO	APAO-1F	5'-GAGCCACCACTGCCTGCC-3'
APAO	APAO-2R	5'-CCATGTGTGGCTTCCCC-3'
ODC	ODC-1F	5'-TCCAGGTTCCCTGTAAGCAC-3'
ODC	ODC-2R	5'-CCAACTTTGCCTTTGGATGT-3'
SSAT	SSAT-1F	5'-CGTCCAGCCACTGCCTCTG-3'
SSAT	SSAT-2R	5'-GCAAGTACTCTTTGTCAATCTTG-3'
rpS7	rpS7-1F	5'-CGAAGTTGGTCGG -3'
rpS7	rpS7-2R	5'-GGGAATTCAAAATTAACATCC -3'
β-actin	β-actin-1F	5'-TGTTACCAACTGGGACGACA-3'
β-actin	β-actin-2R	5'-AAGGAAGGCTGGAAAAGAGC-3'

In situ detection of H_2O_2 in mouse neocortex and cerebellum

In situ detection of H_2O_2 in mouse neocortex and cerebellum was carried out by exploiting the fluorogenic peroxidase substrate AUR (Amplex UltraRed reagent, Invitrogen) that reacts in a 1:1 stoichiometry with H_2O_2 to produce a highly fluorescent reaction product (excitation/emission maxima approximately 568/581 nm). Neocortex and cerebellum samples from Tg and Sg mice were stained by incubation with 0.1 mM AUR, for 5 min under vacuum (-400 mbar). After washing cortex and cerebellum we observed under LSCM (HeNe laser emitting at wavelength of 543 nm). The selected emission bands ranged from 550 to 700 nm. The selected bands do not overlap with the excitation and emission wavelength from GFP (488 and 509 nm, respectively) protein expressed by Tg and Sg mice.

RT-PCR analysis

The relative levels of SMO, APAO, ODC, SSAT, β-actin and rpS7 transcripts were measured by RT-PCR with specific primers listed in Table 1. Total RNA was isolated from brain cortex and cerebellum as control by TRIZOL reagent (Gibco BRL), according to the manufacturer's instructions. Synthesis of the cDNAs from the RNAs of different mouse organs were performed by primer random examers in 20 μl reaction volume containing 1 μg of total RNA, according to the manufacturer's instructions (SuperScriptIII First-Strand Synthesis System for RT-PCR, Invitrogen). Aliquots of reverse-transcribed-RNA were amplified within with Taq DNA polymerase (M-Medical) in the linear range and in saturating experimental conditions by 20, 25, 30 or 35 PCR cycles: denaturation at 94°C for 1 min, annealing at 60°C for 30 sec and extension at 72°C for 1 min. The RT-PCRs were normalized by the comparison of the β-actin and rpS7 controls. Further control reaction mixtures, either without template (not shown) or RT enzyme (not shown), were uniformly negative.

Determination of SMO and APAO enzyme activity and PA content

Polyamine oxidase activity of SMO/APAO was assayed using a modification of the chemiluminesence analysis reported by Wang et al. [33]. Briefly, luminol-dependent chemiluminescence was determined using a Lumat LB 9507 G&G BERTHOLD luminometer. Luminol was prepared as a 100 mM stock solution in DMSO and diluted to 100 μM with H_2O, immediately prior to use. Tissue cortex and cerebellum extract was assayed in a 83 mM glycine buffer pH 8.3, 20 μg/ml horseradish peroxidase, 0.2 mM 2-bromoethylamine (catalase inhibitor), 15 μM deprenyl (copper-containing amine oxidase inhibitor), 0.15 μM clorgyline (mitochondrial oxidase inhibitor), and 500 μM Spm or 500 μM N^1-acetylSpm as substrate, to determine SMO or APAO activity respectively. All reagents, with the exception of substrate, were combined and incubated for 5 minutes at 37°C, then 5 nmol luminal was added and incubated again at 37°C for 2 minutes, transferred to the luminometer where spermine or $N1$-acetylspermine was added, and the resulting chemiluminescence was integrated over 40 seconds. Polyamine concentration was determined as described in Mates et al. [34].

Light and electron microscopy

Transgenic (Tg) and syngenic (Sg) mice were sacrificed 1 and 3 days after KA administration, or 1 day after vehicle injection. Animals were transcardially perfused at room temperature (RT) with 0.1 M phosphate buffer (PB), pH 7.3, followed by 4% freshly depolymerised paraformaldehyde in PB. Brains were removed 1 h after perfusion and sagittaly cut along the midline. The right halves of the brains were processed for immunohistochemical studies, while the left halves were collected for morphological analyses at the light and electron microscopic level.

Immunohistochemistry. Samples were dehydrated in graded ethanol, transferred to Bioclear (BioOptica, Milan, Italy), then to a 1:1 mixture of Bioclear and paraffin, and finally embedded in paraffin. Sagittal, 5 μm thick sections were then serially cut by a microtome and collected on Vectabond (Vector, Burlingame, CA, USA) pre-coated slides. Sagittal, 100 μm thick sections were obtained by a vibratome, and collected in phosphate buffer. Serial, sagittal brain sections from WT, Sg and Tg animals were deparaffinized by using xylene and graded ethanol and rehydrated. Slides were then immersed in 10 mM sodium citrate buffer, pH 6.1, and processed for the antigen-retrieval procedure, using a microwave oven operated at 720W for 10 min [35]. After cooling, slides were transferred to phosphate buffer saline (PBS) containing 5% (w/v) non-fat dry milk, for 1 h at RT, then incubated for 48 h at 4°C with either of the following antibodies, diluted in PBS containing 2.5% (w/v) non-fat dry milk: 1:100 AntiSMO (Proteintech Group), 1:500 NeuN (Chemicon), 1:500 GFAP (Dako Cytomation) and 1:500 Iba1 (Biocare Medical). In control sections, the primary antibody was omitted or substituted with normal rabbit serum. Slides were then incubated for 1 h at RT with biotinylated goat anti-rabbit IgG or goat anti-mouse IgG (Vector), diluted 1:200 in PBS containing 1% normal goat serum (Vector). Immuno-complexes were revealed by means of an avidin biotin system (Vectastain Elite ABC kit, Vector), using 3,3'-diamino-benzidine (DAB Substrate kit for Peroxidase, Vector), as the chromogen. Slides were finally dehydrated and mounted with Eukitt (Kindler GmbH & Co., Freiburg, Germany). Sections were observed under an Olympus BX 51 microscope, equipped with a Leica DFC 420 camera; electronic images were captured by a Leica Application Suite system, and composedin an Adobe Photoshop CS2 format.

Morphological analysis. After brain dissection, small pieces from the left neocortices of Sg and Tg mice were postfixed in 1% OsO_4 in PB, dehydrated, and embedded in epoxy resin. Specimens were cut on a Leica Reichert Supernova ultramicrotome. Semithin sections were stained with toluidine blue and observed in a LEITZ DMRB light microscope. Ultrathin sections were briefly contrasted with uranyl acetate and examined by a Zeiss CM 900 electron microscope. Images of semithin and ultrathin sections were electronically captured and composed in an Adobe Photoshop CS3 format.

Statistical analysis

Cell counting in each field (number of cells/0.24 mm^2) after immunohistochemical analyses were performed on mouse cortex slides (n = 6) from six independent individuals for each groups; data are presented as mean ± S.D. Data significance was assessed by one or two-way ANOVA tests. In particular, one-way analysis of variance and post-hoc test Bonferroni has been used for comparing several groups, while two-way analysis of variance and post-hoc test Bonferroni has been used for comparing different groups and treatment effects. Probabilities of $p < 0.05 = *$, $p < 0.01 = **$ and $p < 0.001 = ***$ were taken as levels of significance. Toluidine blue stained semithin sections from mouse neocortices of Sg (n = 6) and Tg mice (n = 6) were examined using a LEITZ DMRB light microscope, at low-to-high magnification. We evaluated the percentage of dying neurons *vs.* the total cell number, considering as altered those neurons showing high condensation and strong nuclear basophilia. Three different embedding blocks were analyzed for each condition, and a minimum of 200 cells per block were observed. Cell counting was performed on slides from three independent individuals; data are presented as mean ± S.D.

Results

Conditional activation of *SMO in vivo*

For conditional activation of *SMO* we generated a construct (*pJoSMO*) that contains a floxed *gfp*-stop cassette under control of the β-actin/CMV fusion promoter [31], driving ubiquitous expression of the *gfp* (Green Fluorescent Protein) reporter gene. Upon Cre recombination the *gfp*-stop cassette is excised, leading to simultaneous expression of *SMO* and of the second reporter gene, *lacZ* (β-galactosidase), via an IRES sequence (Figure 1A). The transgenic mouse line generated with this construct was named *JoSMO* and the characterization of the mice was carried out to select the founders possessing a single copy inserted transgene by Southern blot analysis (not shown) and overexpressing GFP in all tissues (not shown). After genotyping, the selected *JoSMO* mice exhibited widespread GFP fluorescence and Western blot analysis confirm the presence of GFP in different organs (Figure 1B–C). In order to test the recombination of the integrated construct, *JoSMO* were crossed with *Dachshund-Cre* mice expressing Cre [29] and directing recombination in proneural population in the nervous system. This cross produced a double transgenic mouse line hereafter named *JoSMOrec* (or Tg for the sake of simplicity), while single transgenic *JoSMO* mice, coming from the same offspring, were used as control and referred as syngenic animals (Sg). Before making the cross *JoSMO;Dachshund-Cre*, the Cre recombination was monitored with pJoSMO and pJoSMOrec plasmids transfected in HeLa cells. After transfection, GFP fluorescence and LacZ staining on HeLa cells were monitored (Figure 1D), and PCR was performed, using the primers JoP6F and JoP6R which bind 5' and 3' ends of the floxed *gfp*-stop cassette, the genetic constructs were confirmed (not shown). Expression of the second reporter LacZ was tested by staining of whole-mount and isolated brains (Figure 2A–B). The brain of *JoSMOrec* exhibited LacZ specifically in the cerebral cortex at E12.5 and E14.5 mouse developmental stages (Figure 2A–B). The SMO overexpression in the neocortex of these mice was assessed by semiquantitative RT-PCR, which revealed an approximate two-fold increase of SMO transcript level (Figure 2C–D).

Immunohistochemical analysis of *JoSMOrec* neocortex

To study the overall cytoarchitecture of Tg mouse neocortex, as compared to its Sg counterpart, we analysed the immunohistochemical distribution of SMO and of specific neural cell markers. Serial sagittal paraffin sections from Tg mice were immunostained using antibodies to SMO, to the neuronal marker NeuN (neuronal nuclei), to the astroglial marker GFAP (glial fibrillary acid protein), and to the microglial marker Iba1 (ionized calcium binding adapter molecule). SMO immunoreactivity was sensibly enhanced in both young and old Tg mice, compared to controls. Quantitative evaluation of immunohistochemical data, indicates increases of about 50% and ~15% in the number of positive cells in Tg young and old neocortex, with respect to Sg (Figure 3). As to the general cortical cytoarchitecture of Tg and Sg mice, we did not detect any difference in the number of NeuN-positive neurons, between Tg and Sg young mice. By contrast, a significant reduction of neurons was observed in old Tg mice (~30% lower than Sg) (Figure 3). Using an anti-GFAP antibody, we could observe in old Tg mouse neocortex a numerical increase of astrocytes, which also appear hypertrophic and highly ramified (Figure 3). We also observed a significant increase of Iba1-stained

cells in the aged neocortex, indicating a microglial activation. Interestingly, the neocortex of both young and old Tg mice showed more intense Iba1-immunoreactivity, when compared to Sg mice (Figure 3). As an additional control, the immunohistochemical analysis on neocortex with the same panel of antibodies used in Figure 3 was performed also on non-transgenic littermates (WT) (Figure S1).

SMO enzymatic activity and PA content in the neocortex of young and old *JoSMOrec* mice

SMO enzyme activity was measured in homogenates from the neocortex and cerebellum (chosen as a reference brain area) of young and old *JoSMOrec* mice. We observed a 3-fold-increase in the neocortex of Tg 3-month-old mice respect to aged-matched Sg, while no difference was detected between Tg and Sg mice at 20 month of age (Figure 4). In parallel, the activity of the most representative enzymes of PA metabolism, namely APAO, SSAT and ODC, were assayed. While no differences were observed in APAO and ODC enzymatic activities, SSAT showed significantly increased activity in the neocortex of both young and aged SMO transgenic mice, especially in 3-month-old animals (100% *vs.* a 20% increase, in young and old mice, respectively). As to PA content, only Put and Spd resulted to be slightly increased in young Tg mice (Figure 4).

In situ detection of H_2O_2 in the neocortex of *JoSMOrec* mice

The production of H_2O_2 of the neocortex of *JoSMOrec* mice have been evaluated by *in situ* AUR staining (Figure 5). Neocortex from Tg mice showed a three-fold increase of H_2O_2 production compared to Sg controls, no difference of staining was observed in the cerebellum from Tg and Sg mice (Figure 5A). Quantification of stained H_2O_2 molecules was plotted with a histogram shown in Figure 5B.

Excitotoxic conditions on *JoSMOrec* mice induced by kainate treatment and behavioural phenotype evaluation

Kainic acid (2-carboxy-4-isopropenylpyrrolidin-3-ylacetic acid, KA) is an acidic pyrolidine isolated from the seaweed *Digenea simplex*, and is the most potent of the common exogenous excitotoxins [36–38]. Its neurotoxic threshold is nearly two orders of magnitude lower than that of the other receptor-specific agonists, namely, N-methyl-D-aspartic acid and quisqualic acid [36]. To investigate the possible involvement of SMO in excitotoxicity, we administered KA at dose of 25 mg/Kg [39], to 3-month-old Tg and Sg mice whereas control animals were treated with saline solution (vehicle). After KA treatment, animals were monitored for 6 h to assess the onset time and the level of seizures activity, which was scored according to Schauwecker [30]. Behavioural response is represented in Table 2. Most of the Sg mice (~63%) displayed a milder phenotype (stage 1 to 3), than Tg mice. By contrast, within 30 min of injection 58% of Tg mice showed progressive seizures, ranging from stage 4 to 5, while over the next hour about 35% of animals even displayed stage 6. Overall, these data point out a higher sensitivity to KA of Tg mice, compared to their Sg counterpart. No seizure activity was observed in Tg and Sg animals treated with vehicle. After 6 h monitoring, all mice recovered a normal behaviour and were kept up to 1 or 3 days before being processed.

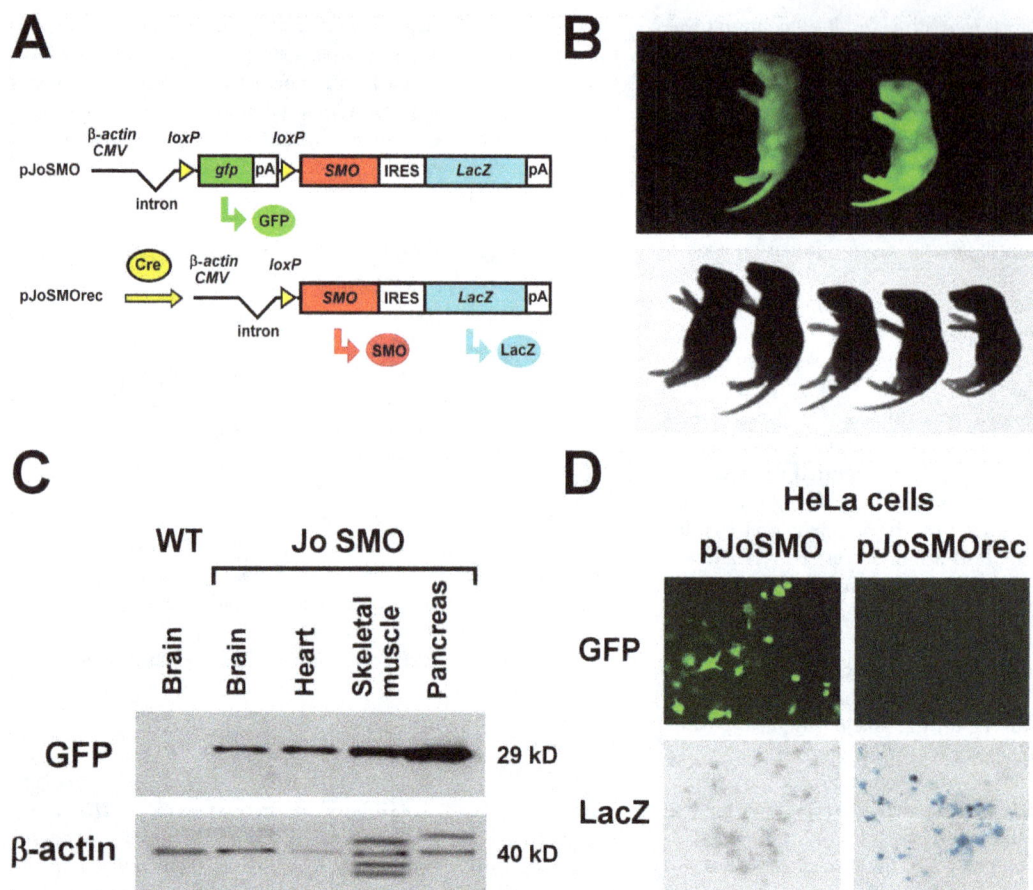

Figure 1. *JoSMO* **mouse line generation. A**. Scheme of pJoSMO and, upon Cre recombination, of pJoSMOrec plasmids. The β-actin/CMV fusion promoter drives the ubiquitous expression of the *gfp* (green arrow, Green Fluorescent Protein) reporter gene. Upon Cre recombination the *gfp*-stop cassette is excised, leading to simultaneous expression of *SMO* (red arrow) and of the second reporter gene, *lacZ* (blue arrow, β-galactosidase), via an IRES sequence. **B**. *JoSMO* mice exhibited widespread GFP fluorescence. **C**. Western blot analysis confirms the presence of GFP in different organs. **D**. The Cre recombination was analysed with pJoSMO and pJoSMOrec plasmids transfection in HeLa cells. After transfection, GFP fluorescence and LacZ staining on HeLa cells were monitored.

Morphological analysis of Sg and Tg neocortex after KA treatment

Neocortical samples from Tg and Sg mice injected with KA/vehicle were analyzed 1 and 3 days after treatment. Toluidine blue stained semithin sections were utilized to assess the extent of tissue damage caused by the toxic agent (Figure 6A, C, E, A′, C′, E′). The Tg brain resulted especially susceptible to KA induced damage, in that several neocortical neurons showed abnormal morphological features, including cytoplasmic condensation and strong nuclear basophilia (arrows in Figure 6E, A′, C′, E′). At the ultrastructural level, Tg neurons appeared highly electron dense in their cytoplasmic and nuclear compartments, the latter showing heterochromatin clumps and nuclear envelope invaginations (Figure 6B′, D′, F′). Quantification of dying neurons in semithin sections demonstrated a significantly higher number in KA treated Tg neocortices, compared to their Sg counterparts. Specifically, 1 day after KA injection, we detected a more than two-fold increase in the number of condensed neurons, while 3 days after injection an approximate three-fold increase was observed (Figure 6G).

Immunohistochemical analysis of KA-treated *JoSMOrec* neocortex

Immunohistochemical analysis using antibodies against SMO, NeuN, GFAP and Iba1 was performed on serial sagittal paraffin sections of the neocortex from Tg and Sg mice treated with vehicle or KA, and sacrificed 1 or 3 days after treatment. The Tg neocortex showed a higher SMO immunoreactivity in terms of both intensity and number of positive cells (∼30% increase), compared to Sg one (Figure 7). After KA injection SMO immunoreactive cells decreased in both Tg and Sg animals, compared to sham animals. Nevertheless, Tg injured mice displayed a higher number of positive cells than Sg treated animals (∼20% increase). Using NeuN antibodies, we could not observe any difference in the number of neurons between Tg and Sg mice treated with vehicle, while a reduction (∼20%) was detected in KA injected Tg mice (Figure 7), indicating a higher loss of neurons in Tg mice. Since KA mediated lesions induce astrogliosis, we examined GFAP immunodistribution in the neocortex from KA/vehicle injected mice. Sham animals showed little GFAP staining with no significant differences between Sg and Tg mice, whilst KA treatment resulted in intense GFAP reactivity, especially dramatic in Tg neocortex. In these samples, astrocytes are not only increased in number (around 200% compared to

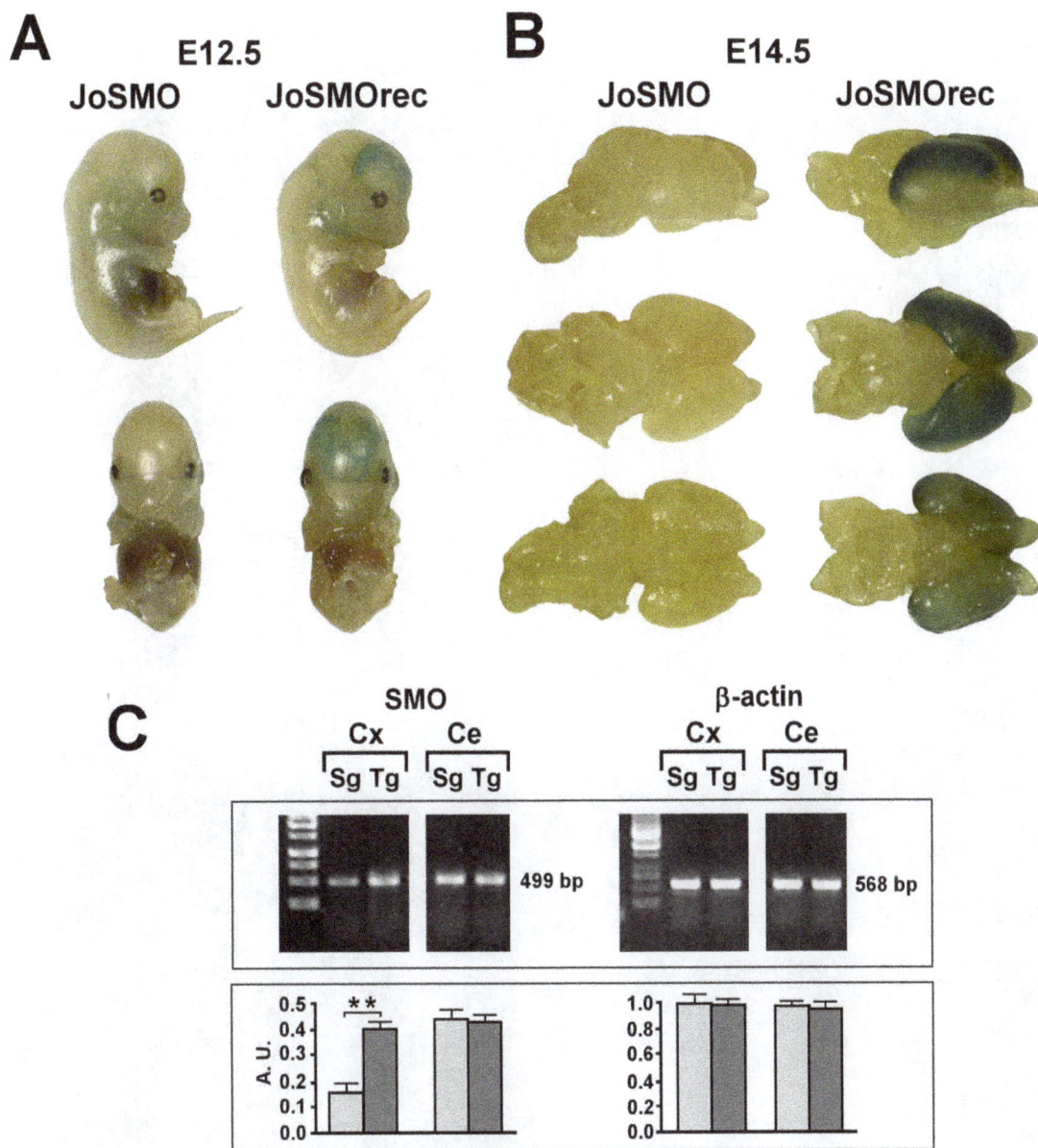

Figure 2. Functional analysis of the second reporter gene LacZ of whole-mount and isolated brains. A. LacZ staining of whole-mount at E12.5 mouse developmental stage. **B**. LacZ staining of brain from *JoSMOrec* mice at E14.5 developmental stage. **C**. SMO transcript level analyses by semiquantitative RT-PCR in the neocortex of *JoSMOrec* mice and densitometric. Representative RT-PCR experiments from three independent replicas are shown. Densitometric analyses of PCR gel bands, represent the measurements done on three separate experiments. The β-actin gene expression was used for normalization. An arbitrary densitometric unit bar graph (A.U.) is shown. The p values were measured with the one-way ANOVA test and post-hoc test Bonferroni (**, p<0.01). Sg, syngenic mice; Tg, transgenic mice; Cx, cortex; Ce, cerebellum.

sham), but show morphological features, including hypertrophy and wide ramification, typical of reactive astrogliosis. Consistently, the microglial marker Iba1, which is upregulated during activation of these cells, was highly expressed in the neocortex of Tg mice. Indeed, vehicle-injected Tg mice also showed a higher number (40% increase) of Iba1-positive cells, compared to Sg mice. After KA treatment this increase reached 60%, indicating a more dramatic microgliosis occurring in Tg than in Sg mice (Figure 7). Altogether, immunohistochemical results, demonstrating higher neuronal loss and a stronger astroglial and microglial activation in Tg mice compared to Sg mice, support the hypothesis that *JoSMOrec* mice are especially sensitive to KA excitotoxicity.

Activity assays of PA key metabolic enzymes

In order to investigate PA metabolism in KA-injected mice, SMO, APAO, SSAT and ODC enzymatic activities were analyzed in Tg and Sg neocortex. To this purpose, animals were sacrificed 1 or 3 days after KA/vehicle treatment. Both Tg and Sg mice treated with saline solution displayed SMO, APAO, SSAT and ODC activities comparable with the ones of untreated animals (Figure 8; compare with Figure 4), demonstrating that vehicle injection does not affect any of the above enzymatic activities (see below). KA treatment induced an increase of SMO enzymatic activity in both Tg and Sg neocortices at 1 and 3 days after injection. The sample taken 1 day after injury showed a 20% increase of SMO activity in Tg mice, while no difference was

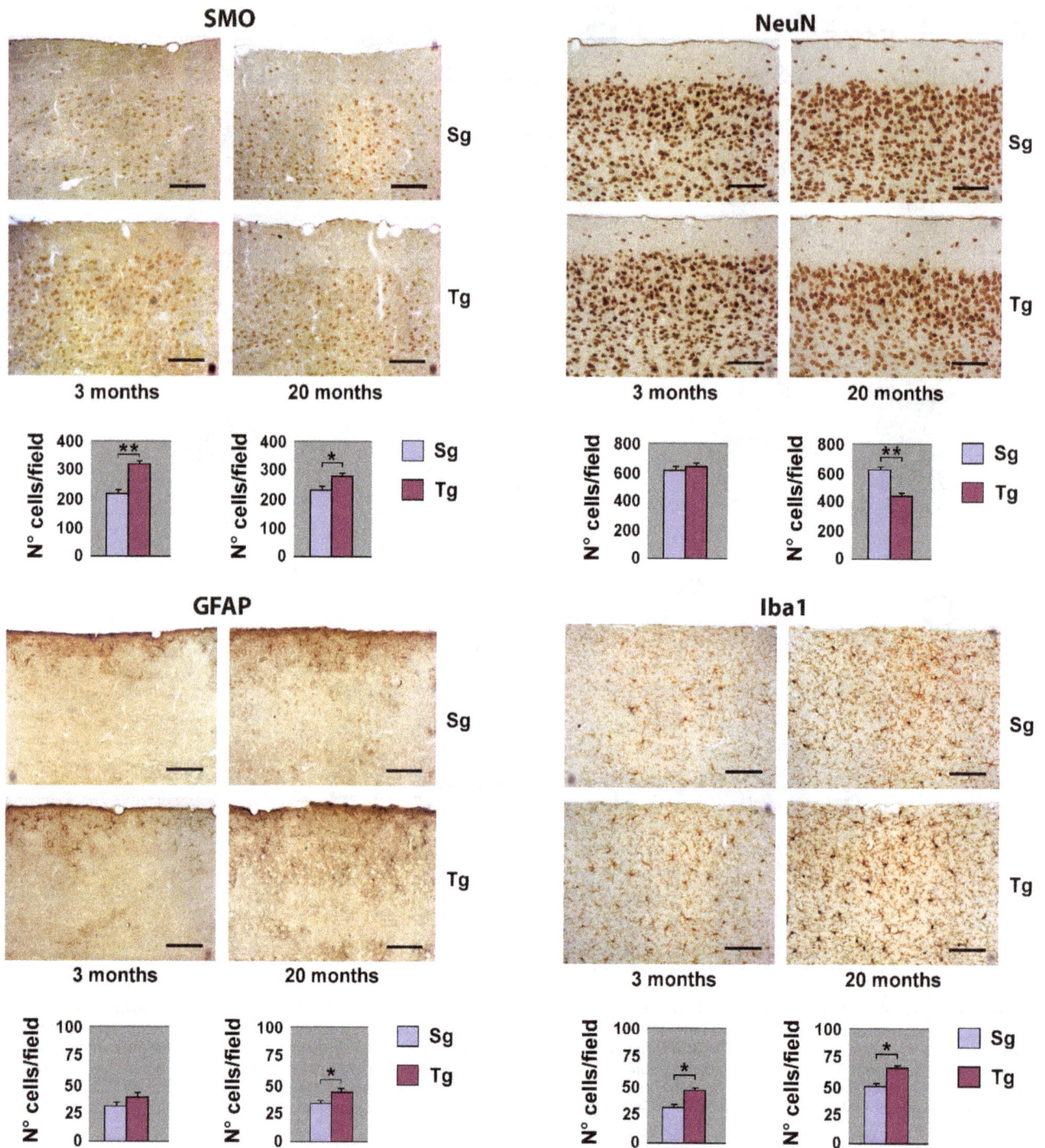

Figure 3. Immunohystochemical analysis of neocortex from Tg and Sg mice. Sagittal brain slices from *JoSMOrec* mice were stained with antibodies directed against SMO, NeuN, GFAP and Iba1. Slides of neocortex from 3 and 20 months old mice were analyzed. Cell counting is expressed as number of positive cells per 0.24 mm^2 area. The p values were measured with the one-way ANOVA test and post-hoc test Bonferroni (*, p<0.05; **, p<0.01; ***, p<0.001). Sg, syngenic mice; Tg, transgenic mice.

observed between Tg and Sg mice at 3 days (Figure 8A). While no difference was observed in APAO basal activity between Tg and Sg neocortices, 1 day after injury this enzyme showed a strong increase in both genotypes, especially dramatic in the Tg neocortex (300%). Furthermore, three days after KA injection, APAO activity decreased in both Tg and Sg samples, down to the basal level in Sg mice, but still higher by two folds than the basal

level in Tg mice. KA treatment resulted in a significant increase by 200% of the SSAT activity at 1 day and remained significantly higher at 3 days in the Tg mice respect to Sg mice (Figure 8A). ODC activity measured at 1 and 3 days after KA injury, was apparently unaffected in either Tg or Sg mice (Figure 8A).

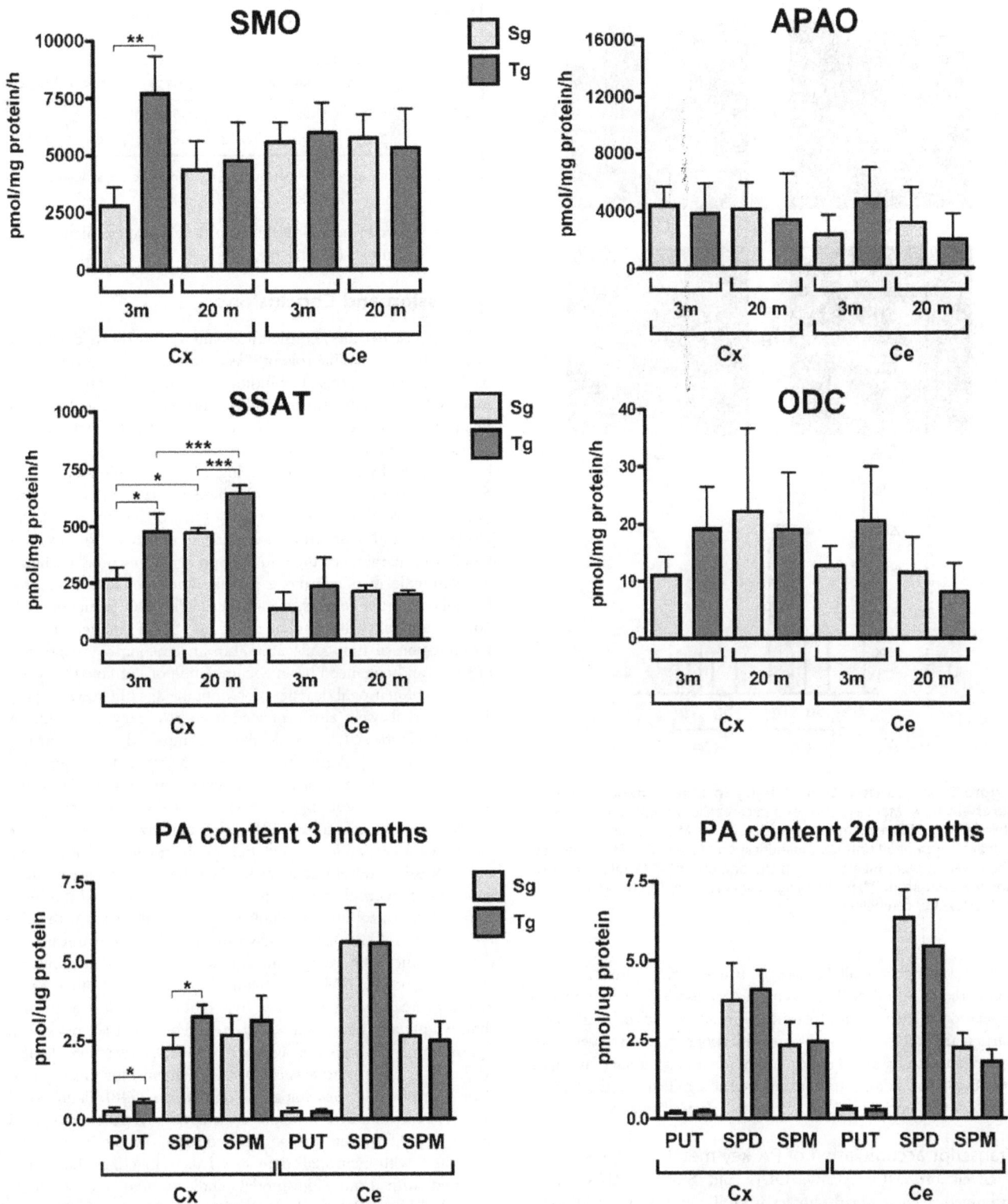

Figure 4. SMO, APAO, ODC and SSAT activities and PA content of Tg and Sg mice. Enzyme activities of SMO, APAO, ODC and SSAT and PA content from cortex and cerebellum of 3 (3 m) and 20 (20 m) months old mice were analyzed. The p values were measured with one-way ANOVA test and post-hoc test Bonferroni (*, p<0.05; **, p<0.01). Sg, syngenic mice; Tg, transgenic mice; Cx, cortex; Ce, cerebellum.

PA content analysis

To examine possible changes in PA content following KA treatment, Put, Spd and Spm levels in the neocortex were measured by HPLC (Figure 8B). Put level was higher in Tg mice compared to Sg mice in vehicle injected animals, while KA treatment leads to its decrease in Tg mice at 1 and 3 days. At difference Put levels in Sg mice were unchanged. As regards Spd

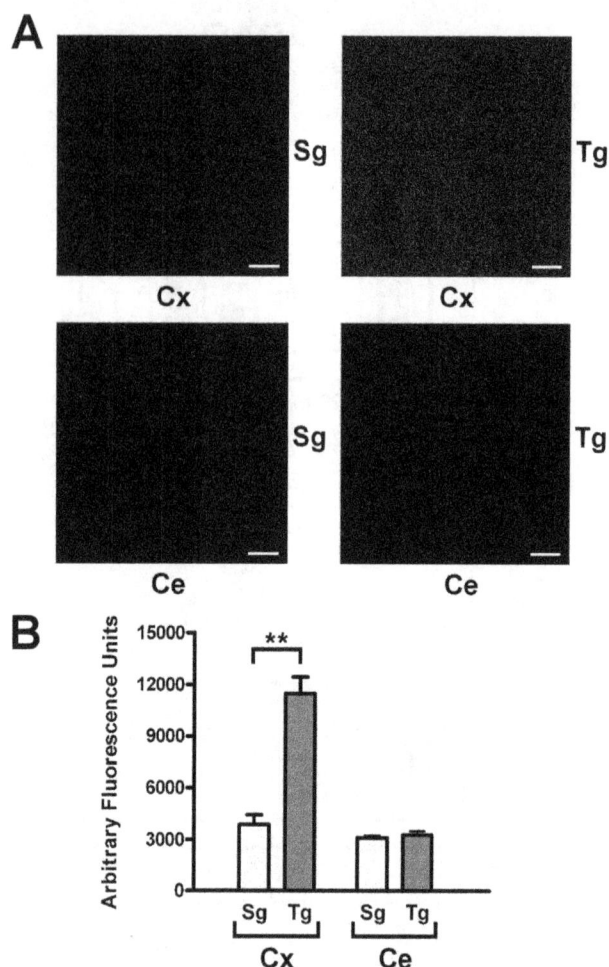

Figure 5. *In situ* detection of H$_2$O$_2$ in mouse neocortex and cerebellum. A. Mouse neocortex and cerebellum from Tg and Sg were stained with AUR reagent and analyzed by LSCM (HeNe laser emitting at wavelength of 543 nm). B. Quantification of stained H$_2$O$_2$ molecules. The p values were measured with the one-way ANOVA test and post-hoc test Bonferroni (**, p<0.01). Sg, syngenic mice; Tg, transgenic mice; Cx, cortex; Ce, cerebellum.

Table 2. Behavioural evaluation following KA treatment.

Mouse line	Stage 1	Stage 2	Stage 3	Stage 4	Stage 5	Stage 6
Tg %	5.38	3.85	23.08	15.38	7.69	34.62
Sg %	33.33	7.41	22.22	14.81	3.70	18.52

Scored mice are expressed as percentage. Tg, transgenic; Sg, syngenic.

neocortex, furthermore SSAT mRNA levels resulted higher in Tg than in Sg samples (Figure 9).

Discussion and Conclusions

In the last decades, considerable interest has been devoted to understand the possible role of PAs as modulators of several types of ion channels [16,41]. Studies on transgenic animals with modified PA metabolism have also contributed to elucidating the involvement of PAs in brain functioning and pathology. Several genes encoding enzymes of PA metabolic pathways have so far been targeted by genetic engineering, and transgenic models have been used to address the issue of whether altered PA metabolism in response to brain injury is a cause of neuronal damage or a sign of plasticity and neuroprotection. Indeed, enhanced Put accumulation with marginal changes in the Spd and Spm content is linked to a vast majority of neurotoxic insults, either chemical or physical, in different brain regions, usually through ODC induction [10], but occasionally also as a result of SSAT induction [42]. Examination of transgenic animals with life-long overexpression of ODC and enhanced brain accumulation of Put failed to reveal any signs of neuronal degeneration until the age of two years [43]. These animals were also protected from physically or chemically induced seizure activity, while showing impaired spatial learning and memory [44]. A number of studies performed on transgenic mice and rats have indicated that overexpression of ODC and grossly elevated Put brain levels, partially protects against ischemia-reperfusion injury [25–28]. Similarly, transgenic mice with greatly expanded brain Put pools resulting from SSAT overexpression were relatively less sensitive to kainate-induced general and neuronal toxicity [45] and showed elevated threshold to pentylentetrazol-induced convulsions, in comparison with wild-type animals [23]. Interestingly, the latter difference was not detected when the convulsant was administered concomitantly with ifenprodil, a NMDA antagonist [23]. On the other hand, SSAT overexpressing mice are hypomotoric and less aggressive than wild-type animals and show impaired spatial learning [46]. Spermidine and Spm are agonists of NMDA receptor [47,48], while Put is believed to acts as a weak antagonist for this receptor with questionable physiological significance [48]. Functional NMDA receptors are needed for synaptic plasticity and spatial learning [49,50], but prolonged activation of the receptor is associated with neuronal damage. Taken together the results obtained with transgenic models indicate that the strikingly expanded brain Put pools in the transgenic animals create a partial blockade of NMDA receptors [24]. In order to highlight the role played by the SMO enzyme in normal brain functioning and during neurodegenerative processes we have genetically engineered a mouse model overexpressing *SMO* gene in the cerebral cortex. This genetic model is conceptually new and makes it possible to express SMO conditionally, in particular, a transgenic mouse founder has been created (*JoSMO*), which ubiquitously expresses GFP at a high level, but does not overexpress SMO. When this model is bred with another

level, it was higher in all Tg mice (control and KA treated animals) compared to Sg mice. But, a general decrease of Spd content was observed following the treatment in either Sg or Tg mice, particularly after 3 days. No modifications in Spm level were found between Sg and Tg mice in both sham and treated animals. However, KA administration induced a significant reduction of Spm content in Sg and Tg mice after 3 days of treatment.

Transcript accumulation of PA key metabolic genes

Levels of SMO, APAO, ODC and SSAT mRNAs were examined in neocortical samples from Tg and Sg mice injected with KA/vehicle. The housekeeping control β-actin protein and ribosomal protein S7 (rpS7), were also probed to quantify the amplified samples [40]. Figure 9 shows the transcripts accumulation of *SMO* gene in Tg mice, which is significantly higher than Sg mice in basal conditions, and increases after KA treatment. By contrast, no significant differences were observed in either APAO or ODC transcript accumulation in Tg and Sg mice injected with KA or vehicle (Figure 9). Regarding SSAT transcript accumulation, we observed gene induction after injury in both Tg and Sg

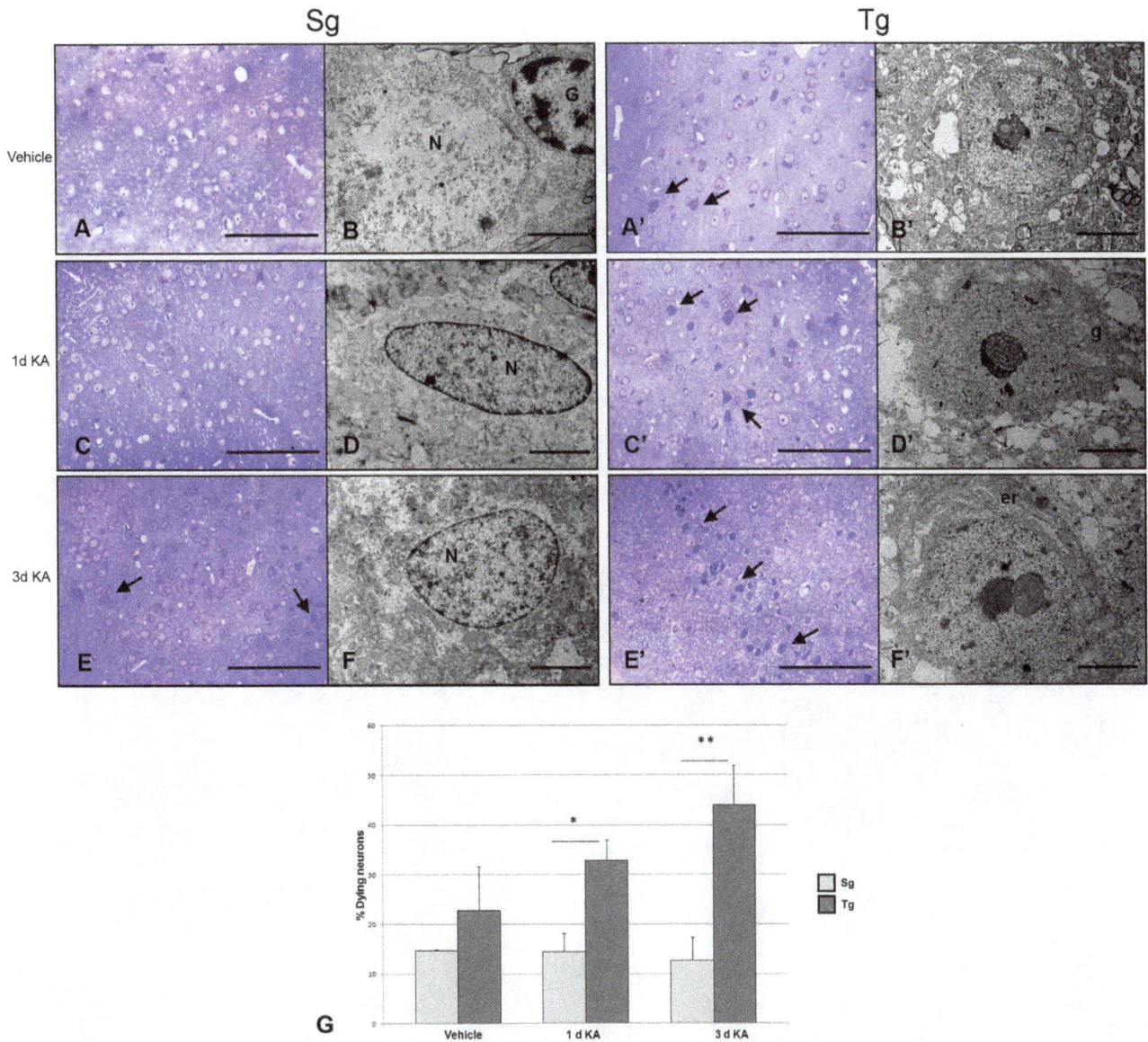

Figure 6. Light and electron microscopic analysis of Tg and Sg neocortex. A, C, E semithin and ultrathin **B, D, F** sections from Sg neocortex, showing cell morphology in all the examined conditions. Condensed neurons are marked by black arrows. At ultrastructural level, neurons show regular nuclei (N) and preserved cytoplasmic organelles. G, glial cells. Scale bars: 1 mm (**A, C, E**), and 5 μm (**B, D, F**). **A', C', E'** semithin and ultrathin **B', D', F'** sections from Tg neocortex. Neurons with abnormal morphological features, including cytoplasmic condensation and strong nuclear basophilia are marked by black arrows. The electron microscopic analysis of damaged neurons reveals irregular nuclear envelope, heterochromatin clumps, and cytoplasmic shrinkage (**D'** and **F'**). These alterations are accompanied by enlargement of Golgi apparatus (g) and endoplasmic reticulum cisternae (er). The graph shows the quantification of dying neurons in Tg and Sg neocortices at 1 day after vehicle injection, at 1 and 3 days after KA injection. The p values were measured with the two-way ANOVA test and post-hoc test Bonferroni (*, $p<0.05$; **, $p<0.01$; ***, $p<0.001$). Scale bars: 1 mm (**A', C', E'**), and 5 μm (**B', D', F'**). Sg, syngenic mice; Tg, transgenic mice.

transgenic line expressing Cre recombinase in a tissue specific way, SMO overexpression takes place in this specific tissue. This strategy has several advantages, among which the possibility to cross the founder with any Cre tissue specific expressing lines and creating double transgenic animals expressing specifically SMO potentially in any tissue. In our genetic model system *JoSMOrec*, SMO activity resulted higher in the neocortex of young mice compared to controls, leading to a higher neuronal death rate during ageing in this brain region. In fact, no difference of SMO activity was detected between *JoSMOrec* 20-months-old mice and aged-matched controls. In line with that, the immunohistochemical analyses using the neuronal marker NeuN showed a significant

neuronal loss in old *JoSMOrec* mice in respect to controls. Consistently, a relatively low number of SMO positive neurons was detected in old Tg mice. While no differences were observed in APAO and ODC enzymatic activities, SSAT showed increased activity in the neocortex of both young and old SMO Tg mice. This increase likely represents a cellular response to compensate SMO overexpression, possibly via Spd acetylation and export or/ and acetyl-Spd oxidation by APAO. The SSAT response, that should be neuroprotective, however could not overcome the SMO overexpression-induced neurotoxicity. We could not also rule out that increased SSAT activity could be responsible of Spm acetylation and consequently SMO substrate subtraction. This is

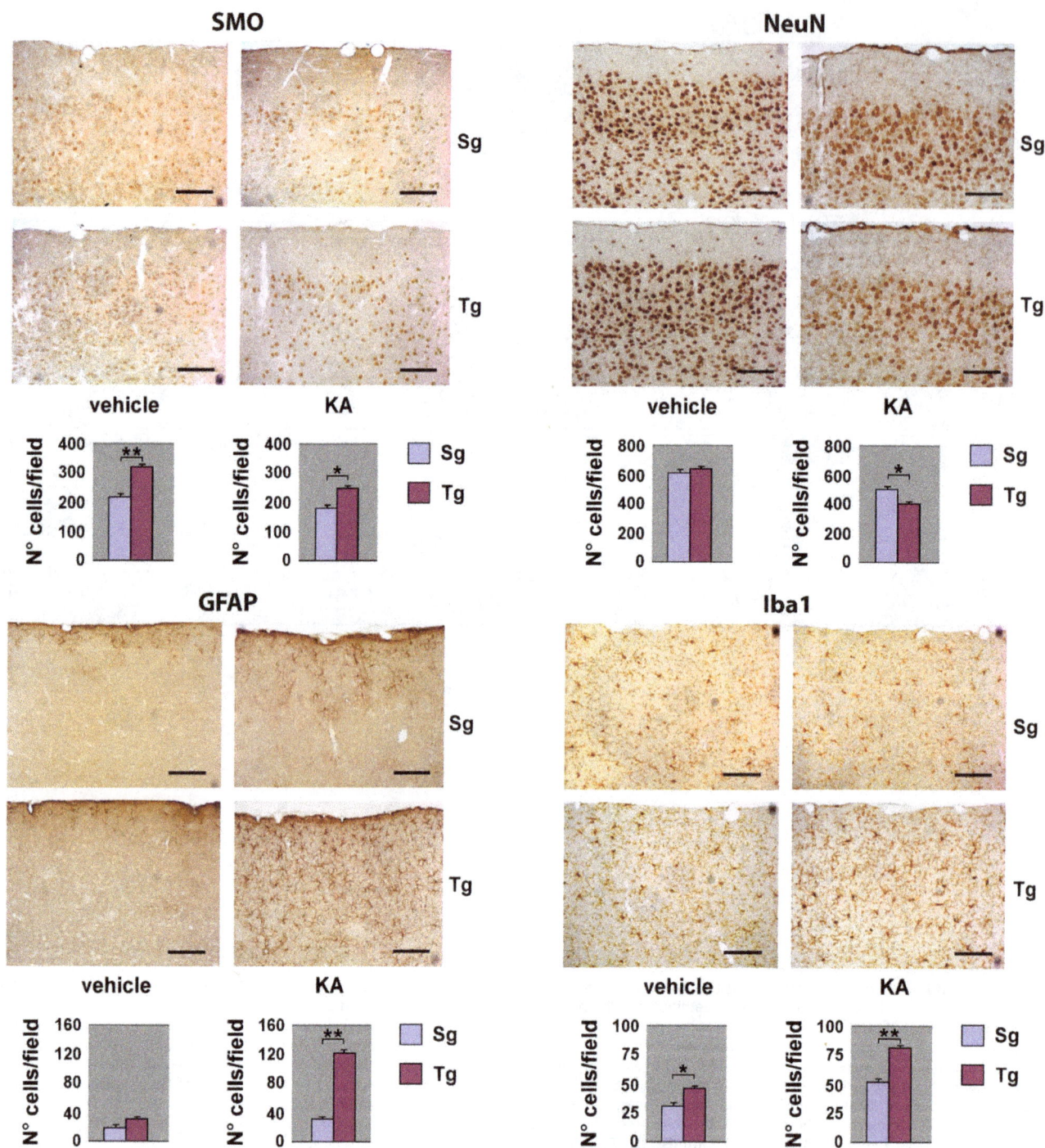

Figure 7. Immunohystochemical analysis of neocortex from KA treated Tg and Sg mice. Sagittal brain slices from *JoSMOrec* mice were stained with antibodies directed against SMO, NeuN, GFAP and Iba1. Slides of neocortex from KA treated 3 months old mice were analyzed. Cell counting is expressed as number of positive cells per 0.24 mm^2 area. The p values were measured with the two-way ANOVA test and post-hoc test Bonferroni (*, p<0.05; **, p<0.01). Sg, syngenic mice; Tg, transgenic mice.

in agreement with the higher Spd and Put content in the neocortex showed by young Tg mice. The immunohistochemical analysis of neocortex in *JoSMOrec* mice using GFAP and Iba1 markers showed a significant astroglial and microglial activation in old Tg mice. In order to rule out any hypothetical side-effects due to random transgene integration and consequently gene disruption, we performed immunohystochemical analyses of neocortex

from non-transgenic littermates (WT) using antibodies directed against SMO, NeuN, GFAP and Iba1. No different expression of the markers analyzed was observed between WT and Sg mice. This indicates that SMO overexpression is affecting Tg phenotype, leading to a more pronounced brain damage during ageing, possibly due to a SMO-derived overproduction of H$_2$O$_2$ when cellular antioxidant defences become less efficient in old mice. In

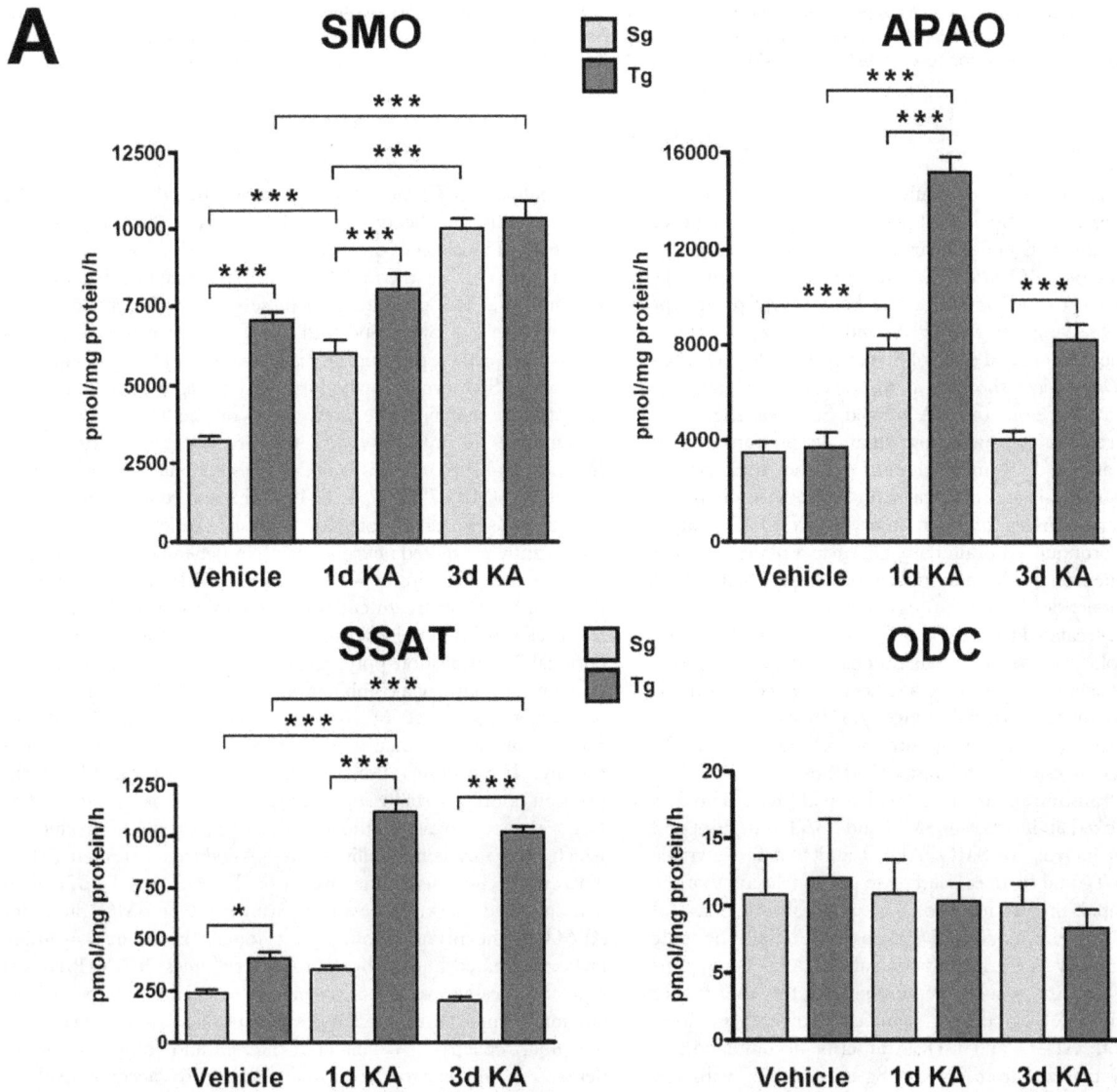

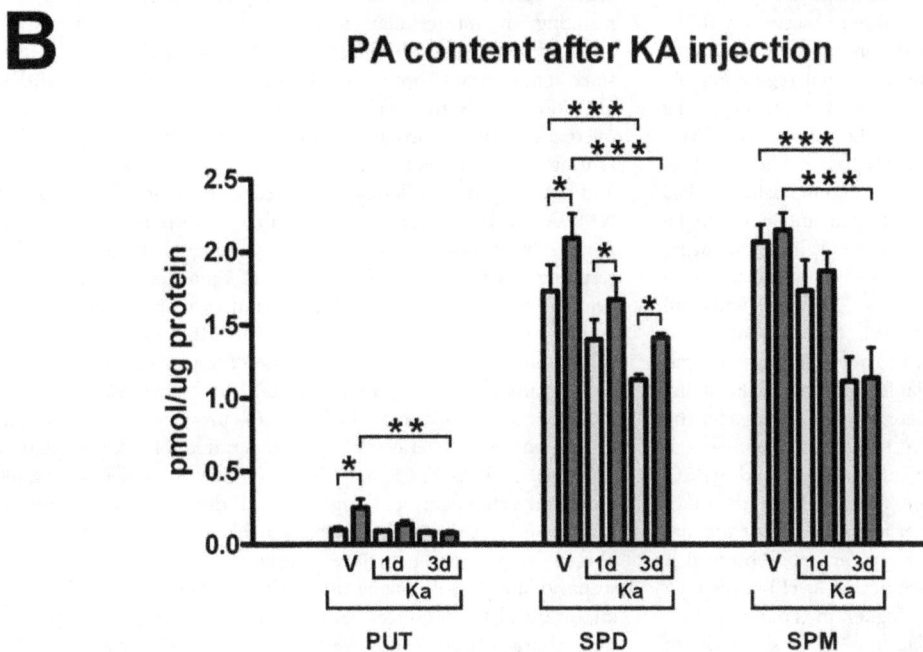

Figure 8. SMO, APAO, ODC and SSAT activities and PA content of KA treated Tg and Sg mice. Enzyme activities of SMO, APAO, ODC and SSAT and PA content from neocortex of 3 months old mice were analyzed after 1 and 3 days of KA treatment. The p values were measured with the two-way ANOVA test and post-hoc test Bonferroni (*, $p<0.05$; **, $p<0.01$; ***, $p<0.001$). Sg, syngenic mice; Tg, transgenic mice.

fact, H_2O_2 production resulted greatly enhanced in the neocortex of Tg mice compared to Sg controls, strongly suggesting that Tg mice were suffering a higher cellular oxidative stress. In order to evaluate the role of SMO in an excitotoxic condition, young Tg and Sg mice were treated with KA and behavioural phenotype was analyzed observing and scoring the induced seizures activity according to the well defined scale of Schauwecker [30]. The KA treated *JoSMOrec* mice showed a more severe behavioural phenotype with respect to the KA treated Sg ones, suggesting that SMO overexpression is affecting glutamatergic transmission. After KA treatment Tg mice showed in the neocortex a considerable astrogliosis and a stronger microgliosis compared to Sg mice, evident markers of brain injury. Also TEM analysis confirmed this pronounced brain damage, since a higher number of neocortical neurons with abnormal morphological features, as cytoplasmic condensation and strong nuclear basophilia, were observed in KA-treated Tg mice compared to KA-treated Sg ones. These morphological neuronal alterations describe a poorly understood degenerative state that has been detected in various pathological conditions [51-53]. Since *JoSMOrec* mice overexpressing SMO were clearly more sensitive to excitotoxic insult, PA metabolism was investigated to understand how it could affect glutamatergic transmission in mice treated with KA. The KA treatment provoked an increase of SMO and SSAT transcript and a more evident increase of SMO, APAO and SSAT enzymatic activity in both Tg and Sg mice, leading to the conclusion that the whole PA catabolism was induced. The noticeable increase of SMO and SSAT enzymatic activity observed could be only partially explained by transcript induction after KA injection, post-transcriptional control would be responsible for the higher enzymatic activity observed. No significant differences were observed in both APAO and ODC transcript accumulation in all the samples analyzed from Tg and Sg mice treated with KA and vehicle. While analyses of ODC transcript and enzymatic activity are completely matching, we could not observe any *APAO* gene induction and also in this case the increase of enzymatic activity measured is due to a post-transcriptional regulation. As expected, the increase in enzymatic activity is more evident in Tg than in Sg mice, this might be explained by the concomitant SMO induction and overproduction in *JoSMOrec* mice.The effect of SMO overexpression is clearly visible by immunohistochemical analysis of the Tg neocortices where astroglia and a microglia activation, as well as abnormal neurons, occur. When comparing PA content between Tg and Sg mice, it can be noticed that Spd and Put levels are higher in Tg mice than Sg ones, while no difference can be observed in the Spm level. This is pointing out that among PAs, Spm is well buffered in its cellular content and confirming that its homeostasis is crucial for physiological cell life. Nevertheless, we observed an increase of Spd that can be explained with Spm oxidation by SMO, while the increase of Put can be due to the concerted action of SSAT and APAO according to their enzymatic reaction. After KA treatment of *JoSMOrec* mice, the PAs content is further altered, as consequential of the whole PA catabolism activation. In fact, we observed a general decrease of PAs, but a difference in Spd level between Tg and Sg mice is still noticeable, being higher in Tg mice. This difference in Spm/Spdratio could explain the higher sensitivity to

KA treatment in Tg mice. It is well demonstrated that intracellular PAs can cause rectification of AMPA and KA receptors, acting as internal cell blockers of these receptor channels to prevent the flux of Na^+ and Ca^{2+} as the membrane is depolarized during synaptic activity [16]. The magnitude of these effects will depend on the relative levels of Spm, Spd and Put [16]. Since we observed a decrease of Spm/Spd ratio in *JoSMOrec* mice, most probably the increase of Spd level compete for Spm binding to AMPA and KA receptors, enhancing the KA effect on both receptors. This explanation is in line with the treatment of organotypic hippocampal slice cultures with N,N^4-bis(2,3-butadienyl)-1,4-butanediamine (MDL72,527), a SMO inhibitor, that resulted in a considerable neuronal protection against KA-induced toxicity and significantly prevented neuronal death in ischemia and mechanical injury models [54]. In the work of Liu et al [54] has been reported that while the pre-treatment with a combination of MDL72,527 and cyclosporin A (a blocker of the formation of the mitochondrial permeability transition pore) provided additive neuronal protection, on the contrary a combination of MDL72,527 and EUK-134 (a specific scavenger of hydrogen peroxide [55,56]) did not produce an additive neuronal protective effect on KA neuronal toxicity. These observations strongly suggest that SMO overexpression is only partially affecting the increase of apoptosis after KA treatment, while is the major source of H_2O_2 production which increases vulnerability to a KA-induced neuronal death. This is in line with the work of Pledgie et al [57] who demonstrated by knockdown experiments that SMO, and not APAO, is the primary source of cytotoxic H_2O_2 in polyamine analogue-treated human breast cancer cell lines. It is well known that an excessive release of excitatory neurotransmitters such as glutamate importantly contributes to neuronal damage in cerebral ischemia, epilepsy, Alzheimer's disease and other forms of dementia. This excitotoxic insult is frequently accompanied by excess calcium influx and followed by generation of ROS, resulting in intracellular membrane damage and triggering apoptotic pathways, ultimately leading to cell death [37]. In fact, since endogenous Spm can block and/or modulate GluRs, alteration of Spm levels in *JoSMOrec* mice due to SMO overexpression can produce changes in Ca^{2+} flux through GluRs. During KA treatment there is an excessive activation of AMPA and KA receptors producing excitotoxicity leading to a consequent NMDA receptors activation [58], that in turn may drive Spm release from synaptic vesicles to synaptic cleft [59,60]. The resulting neuronal and glial uptake of Spm could increase the cytoplasmic Spm level which is not compartmentalized in synaptic vesicles and as available as substrate for SMO enzyme activity. It can be noticed that there is a decrease of Spm content in both Sg and Tg mice after KA treatment, but the higher SMO level in *JoSMOrec* mice leads to an higher H_2O_2 production, contributing to the observed phenotype. The generation of Spm oxidation products, such as H_2O_2 and 3-aminopropanal, which spontaneously converts in acrolein, together with direct effects of Spm on AMPA and KA receptors, are likely involved in ROS increase and ultimately to neuronal degeneration and death. Within this scenario, and hypothesizing that SMO enzyme is one of the most important H_2O_2 producers in the brain, the transgenic *JoSMOrec* mice characterized in this work could represent a useful genetic

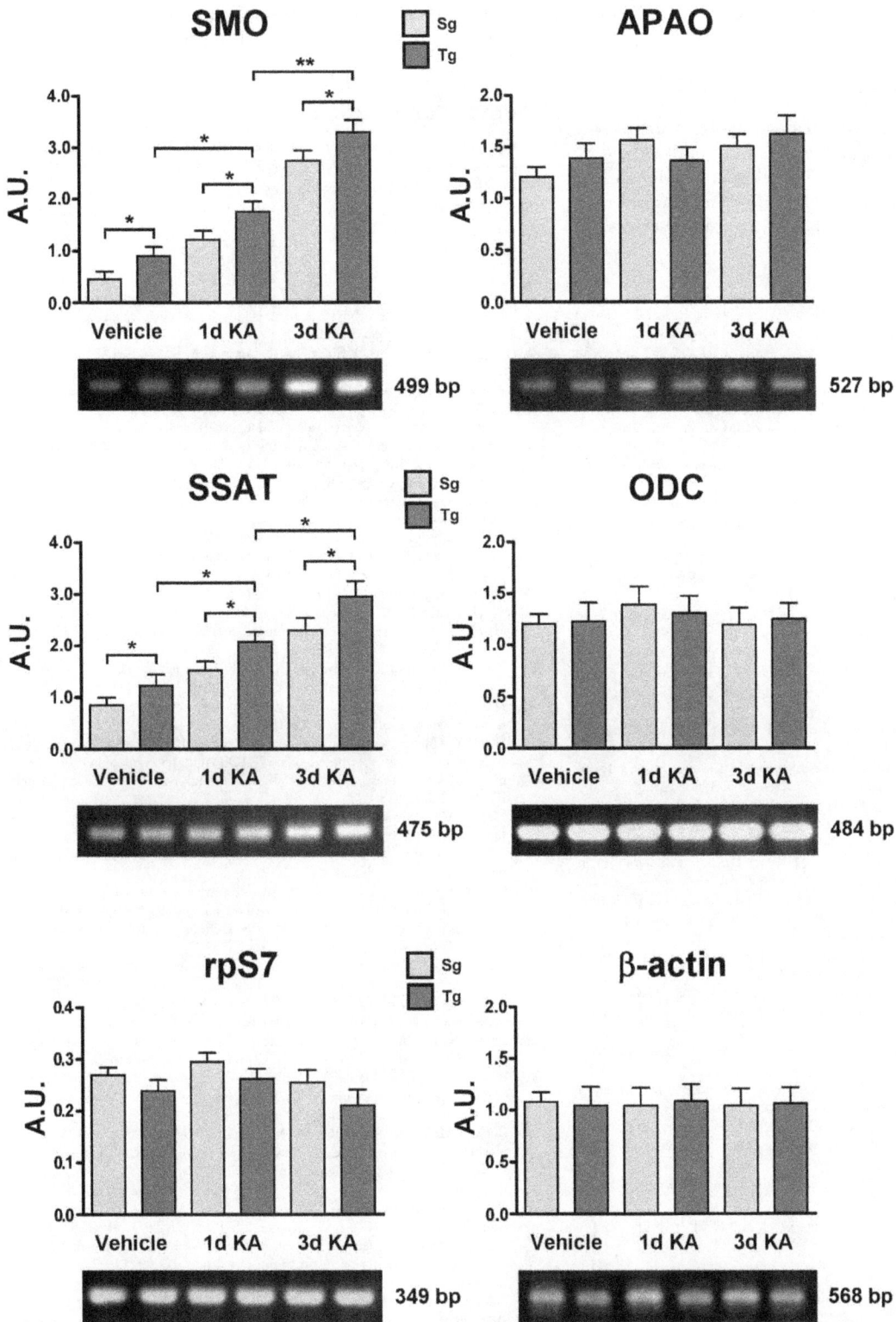

Figure 9. SMO, APAO, ODC and SSAT transcript level analyses of KA treated Tg and Sg mice. SMO, APAO, ODC and SSAT transcript accumulation of 3 months old mice were analyzed after 1 and 3 days of KA treatment. Representative RT-PCR experiments from three independent replicas are shown. Densitometric analyses of PCR gel bands, represent the measurements done on three separate experiments. The β-actin and/or rpS7 gene expression was used for normalization. An arbitrary densitometric unit bar graph (A.U.) is shown. The p values were measured with the two-way ANOVA test and post-hoc test Bonferroni (*, $p < 0.05$; **, $p < 0.01$). Sg, syngenic mice; Tg, transgenic mice.

model for studying brain pathologies such as epilepsy, Alzheimer's disease and other forms of dementia.

Supporting Information

Figure S1 Immunohystochemical analysis of neocortex from Sg and WT mice. Sagittal brain slices from WT and Sg mice were stained with antibodies directed against SMO, NeuN, GFAP and Iba1. Slides of neocortex from 12 months old mice were analyzed. Cell counting is expressed as number of positive cells per 0.24 mm^2 area. Statistical analyses were carried out with the one-way ANOVA test. WT, wild-type mice; Sg, syngenic mice.

Author Contributions

Conceived and designed the experiments: MC MD FC PM. Performed the experiments: MC VC GB MD SM RN. Analyzed the data: RA JB MC MM GM MP PM. Contributed reagents/materials/analysis tools: FC MC PM MP. Wrote the paper: MC MD PM. Obtained permission for animal house starvation: FC EJB.

References

1. Rea G, Bocedi A, Cervelli M (2004) Question: What is the biological function of the polyamines? IUBMB Life 56: 167–169.
2. Cervelli M, Fratini E, Amendola R, Bianchi M, Signori E, et al. (2008) Increased spermine oxidase (SMO) activity as a novel differentiation marker of myogenic C2C12 cells. Int J Biochem Cell Biol 41: 934–944.
3. Amendola R, Cervelli M, Fratini E, Polticelli F, Sallustio DE, et al. (2009) Spermine metabolism and anticancer therapy. Current Cancer Drug Targets 9: 118–130.
4. Casero RA, Pegg AE (2009) Polyamine catabolism and disease. Biochem J 421: 323–38.
5. Fiori LM, Wanner B, Jomphe V, Croteau J, Vitaro F, et al. (2010) Association of polyaminergic loci with anxiety, mood disorders, and attempted suicide. PLoS One 5: e15146.
6. Fiori LM, Bureau A, Labbe A, Croteau J, Noël S, et al. (2011) Global gene expression profiling of the polyamine system in suicide completers. Int J Neuropsychopharmacol 14: 595–605.
7. Yoshida M, Higashi K, Jin L, Machi Y, Suzuki T, et al. (2010) Identification of acrolein-conjugated protein in plasma of patients with brain infarction. Biochem Biophys Res Commun 391: 1234–1239.
8. Zahedi K, Huttinger F, Morrison R, Murray-Stewart T, Casero RA, et al. (2010) Polyamine catabolism is enhanced after traumatic brain injury. J Neurotrauma 27: 515–525.
9. Cervelli M, Amendola R, Polticelli F, Mariottini P (2012) Spermine oxidase: ten years after. Amino Acids 42: 441–450.
10. Kauppinen RA, Alhonen LI (1995) Transgenic animals as models in the study of the neurobiological role of polyamines. Progr Neurobiol 47: 545–563.
11. Paschen W, Hallmayer J, Mies G, Röhn G (1990) Ornithine decarboxylase activity and putrescine levels in reversible cerebral ischemia of Mongolian gerbils: effect of barbiturate. J Cereb Blood Flow Metab 10: 236–242.
12. Paschen W, Csiba L, Röhn G, Bereczki D (1991) Polyamine metabolism in transient focal ischemia of rat brain. Brain Res 566: 354–357.
13. de Vera N, Artigas F, Serratosa J, Martínez E (1991) Changes in polyamine levels in rat brain after systemic kainic acid administration: relationship to convulsant activity and brain damage. J Neurochem 57: 1–8.
14. Martínez E, de Vera N, Artigas F (1991) Differential response of rat brain polyamines to convulsant agents. Life Sci 48: 77–84.
15. Henley CM, Muszynski C, Cherian L, Robertson CS (1996) Activation of ornithine decarboxylase and accumulation of putrescine after traumatic brain injury. J Neurotrauma 13: 487–496.
16. Williams K (1997) Interactions of polyamines with ion channels. Biochem J 325: 289–297.
17. Igarashi K, Kashiwagi K (2000) Polyamines: mysterious modulators of cellular functions. Biochem Biophys Res Commun 271: 559–564.
18. Fleidervish IA, Libman L, Katz E, Gutnick MJ (2008) Endogenous polyamines regulate cortical neuronal excitability by blocking voltage-gated Na+ channels. Proc Natl Acad Sci USA 105: 18994–18999.
19. Igarashi K, Kashiwagi K (2010) Modulation of cellular function by polyamines. Int J Biochem Cell Biol 42: 39–51.
20. Mony L, Zhu S, Carvalho S, Paoletti P (2011) Molecular basis of positive allosteric modulation of GluN2B NMDA receptors by polyamines. EMBO J 30: 3134–3146.
21. Traynelis SF, Wollmuth LP, McBain CJ, Menniti FS, Vance KM, et al. (2010) Glutamate receptor ion channels: structure, regulation, and function. Pharmacol Rev 62:405–496.
22. Jänne J, Alhonen L, Pietilä M, Keinänen TA (2004) Genetic approaches to the cellular functions of polyamines in mammals. Eur J Biochem 271: 877–894.
23. Kaasinen SK, Gröhn OH, Keinänen TA, Alhonen L, Jänne J (2003) Over-expression of spermidine/spermine N1-acetyltransferase elevates the threshold to pentylenetetrazol-induced seizure activity in transgenic mice. Exp Neurol 183: 645–652.
24. Jänne J, Alhonen LI, Keinänen TA, Pietilä M, Uimari A, et al. (2005) Animal disease models generated by genetic engineering of polyamine metabolism. J Cell Mol Med 9: 865–882.
25. Lukkarinen J, Kauppinen RA, Koistinaho J, Halmekytö M, Alhonen L, et al. (1995) Cerebral energy metabolism and immediate early gene induction following severe incomplete ischaemia in transgenic mice overexpressing the human ornithine decarboxylase gene: evidence that putrescine is not neurotoxic in vivo. Eur J Neurosci 7: 1840–1849.
26. Lukkarinen J, Gröhn O, Sinervirta R, Järvinen A, Kauppinen RA, et al. (1997) Transgenic rats as models for studying the role of ornithine decarboxylase expression in permanent middle cerebral artery occlusion. Stroke 28: 639–645.
27. Lukkarinen JA, Kauppinen RA, Gröhn OHJ, Oja JME, Sinervirta R, et al. (1998) Neuroprotective role of ornithine decarboxylase activation in transient cerebral focal ischemia: A study using ornithine decarboxylase-overexpressing transgenic rats. Eur J Neurosci 10: 2046–2055.
28. Lukkarinen JA, Gröhn OH, Alhonen LI, Jänne J, Kauppinen RA (1999) Enhanced ornithine decarboxylase activity is associated with attenuated rate of damage evolution and reduction of infarct volume in transient middle cerebral artery occlusion in the rat. Brain Res 826: 325–329.
29. van den Bout CJ, Machon O, Røsok Ø, Backman M, Krauss S (2002) The mouse enhancer element D6 directs Cre recombinase activity in the neocortex and the hippocampus. Mech Dev 110: 179–182.
30. Schauwecker PE (2003) Differences in ionotropic glutamate receptor subunit expression are not responsible for strain-dependent susceptibility to exicitotoxin-induced injury. Mol Brain Res 112:70–81.
31. Niwa H, Yamamura K, Miyazaki J (1991) Efficient selection for high-expression transfectants with a novel eukaryotic vector. Gene 108: 193–199.
32. Hogan B (1983) Molecular biology. Enhancers, chromosome position effects, and transgenic mice. Nature 306: 313–314.
33. Wang Y, Hacker A, Murray-Stewart T, Fleischer JG, Woster PM, et al. (2005) Induction of human spermine oxidase SMO(PAOh1) is regulated at the levels of new mRNA synthesis, mRNA stabilization and newly synthesized protein. Biochem J 386: 543–547.
34. Mates JM, Marquez J, Garcia-Caballero M, Nunez de Castro I, Sanchez-Jimenez F (1992) Simultaneous fluorimetric determination of intracellular polyamines separated by reversed-phase high-performance liquid chromatography. Agents Actions 36: 17–21.
35. Shi SR, Key ME, Kalra KL (1991) Antigen retrieval in formalin-fixed, paraffin-embedded tissues: an enhancement method for immunohistochemical staining based on microwave oven heating of tissue sections. J Histochem Cytochem 39: 741–748.
36. Coyle JT (1987) Kainic acid: insights into excitatory mechanisms causing selective neuronal degeneration. Ciba Found Symp Review 126: 186–203.
37. Wang Q, Yu S, Simonyi A, Sun GY, Sun AY (2005) Kainic acid mediated excitotoxicity as a model for neurodegeneration. Mol Neurobiol 31: 3–16.
38. Zhang XM, Zhu J (2011) Kainic acid-induced neurotoxicity: targeting glial responses and glia-derived cytokines. Current Neuropharmacol 9: 388–398
39. Benkovic SA, O'Callaghan JP, Miller DB (2004) Sensitive indicators of injury reveal hippocampal damage in C57BL/6J mice treated with kainic acid in the absence of tonic-clonic seizures. Brain Res 1024: 59–76.
40. Cervelli M, Bellini A, Bianchi M, Marcocci L, Nocera S, et al. (2004.) Mouse spermine oxidase gene splice variants. Nuclear subcellular localization of a novel active isoform. Eur J Biochem 271: 760–770.
41. Dingledine R, Borges K, Bowie D, Traynelis SF (1999) The glutamate receptor ion channels. Pharmacol Rev 51: 7–61.
42. Najm I, El-Skaf G, Tocco G, Vanderklish P, Lynch G, et al. (1992) Seizure activity-induced changes in polyamine metabolism and neuronal pathology during thepostnatal period in rat brain. Dev Brain Res 69: 11–21.
43. Alhonen L, Halmekytö M, Kosma VM, Wahlfors J, Kauppinen R, et al. (1995) Life-long over-expression ofornithine decarboxylase (ODC) gene in transgenic mice does not lead to generally enhanced tumorigenesis or neuronal degeneration. Int J Cancer 63: 402–404.
44. Halonen T, Sivenius J, Miettinen R, Halmekytö M, Kauppinen R, et al. (1993) Elevatedseiure threshold and impaired spatial learning in transgenic mice with putrescine overproduction in the brain. Eur J Neurosci 5: 1233–1239.
45. Kaasinen K, Koistinaho J, Alhonen L, Jänne J (2000) Overexpression of spermidine/spermine N1-acetyltransferasein transgenic mice protects the animals from kainate-induced toxicity. Eur J Neurosci 12: 540–548.
46. Kaasinen SK, Oksman M, Alhonen L, Tanila H, Jänne J (2004) Spermidine/spermine N1-acetyltransferase overexpressionin mice induces hypoactivity and spatial learningimpairment. Pharmacol Biochem Behav 78: 3545.
47. Williams K, Romano C, Molinoff PB (1989) Effects of polyamines on the binding of [3H]MK-801 to the Nmethyl-D-aspartate receptor: pharmacological

evidence for the existence of a polyamine recognition site. Mol Pharmacol 36: 575–581.

48. Williams K, Dawson VL, Romano C, Dichter MA, Molinoff PB (1990) Characterization of polyamines having agonist, antagonist, and inverse effects at the polyamine recognition site of the NMDA receptor. Neuron 5: 199–208.

49. Morris RG, Anderson E, Lynch GS, Baudry M (1986) Selective impairment of learning and blockade of long-term potentiation by an N-methyl-D-aspartate receptor antagonist, AP5. Nature 319: 774–776.

50. Paschen W (1992) Polyamine metabolism in different pathologicalstates of the brain. Mol Chem Neuropathol 16: 241–271.

51. Iannicola C, Moreno S, Oliverio S, Nardacci R, Ciofi-Luzzatto A, et al. (2000) Early alterations in gene expression and cell morphology in a mouse model of Huntington's disease. J Neurochem 75: 830–839.

52. Mastroberardino PG, Iannicola C, Nardacci R, Bernassola F, De Laurenzi V, et al. (2002) Tissue transglutaminase ablation reduces neuronal death and prolongs survival in a mouse model of Huntington's disease. Cell Death Differ 9: 873–880.

53. Yang DS, Kumar A, Stavrides P, Peterson J, Peterhoff CM, et al. (2008) Neuronal apoptosis and autophagy cross talk in aging PS/APP mice, a model of Alzheimer's disease. Am J Pathol 173: 665–681.

54. Liu W, Liu R, Schreiber SS, Baudry M (2001) Role of polyamine metabolism in kainic acid excitotoxicity in organotypic hippocampal slice cultures. J Neurochem 79: 976–984.

55. Doctrow SR, Huffman K, Marcus CB, Musleh W, Bruce A, et al. (1997) Salen-manganese complexes: combined superoxide dismutase/catalase mimics with broad pharmacological efficacy. Adv Pharmacol 38: 247–269.

56. Melov S, Ravenscroft J, Malik S, Gill MS, Walker DW, et al. (2000) Extension of life-span with superoxide dismutase/catalase mimetics. Science 289: 1567–1569.

57. Pledgie A, Huang Y, Hacker A, Zhang Z, Woster PM, et al. (2005) Spermine oxidase SMO (PAOh1), not N1-acetylpolyamine oxidase PAO, is the primary source of cytotoxic H_2O_2 in polyamine analogue-treated human breast cancer cell lines. J Biol Chem 280: 39843–39851.

58. Zhu X, Jin S, Ng YK, Lee WL, Wong PT (2001) Positive and negative modulation by AMPA- and kainate-receptors of striatal kainate injection-induced neuronal loss in rat forebrain. Brain Res 922: 293–298.

59. Fage D, Voltz C, Scatton B, Carter C (1992) Selective release of spermine and spermidine from the rat striatum by N-methyl-D-aspartate receptor activation in vivo. J Neurochem 58: 2170–2175.

60. Masuko T, Kusama-Eguchi K, Sakata K, Kusama T, Chaki S, et al. (2003) Polyamine transport, accumulation, and release in brain. J Neurochem 84: 610–617.

9

Regulation of Rhodopsin-eGFP Distribution in Transgenic *Xenopus* Rod Outer Segments by Light

Mohammad Haeri[1¤]**, Peter D. Calvert**[1]**, Eduardo Solessio**[1]**, Edward N. Pugh, Jr.**[2]**, Barry E. Knox**[1]*

1 Departments of Neuroscience and Physiology, Biochemistry and Molecular Biology, and Ophthalmology, SUNY Upstate Medical University, Syracuse, New York, United States of America, **2** Center for Neuroscience, University of California, Davis, California, United States of America

Abstract

The rod outer segment (OS), comprised of tightly stacked disk membranes packed with rhodopsin, is in a dynamic equilibrium governed by a diurnal rhythm with newly synthesized membrane inserted at the OS base balancing membrane loss from the distal tip via disk shedding. Using transgenic *Xenopus* and live cell confocal imaging, we found OS axial variation of fluorescence intensity in cells expressing a fluorescently tagged rhodopsin transgene. There was a light synchronized fluctuation in intensity, with higher intensity in disks formed at night and lower intensity for those formed during the day. This fluctuation was absent in constant light or dark conditions. There was also a slow modulation of the overall expression level that was not synchronized with the lighting cycle or between cells in the same retina. The axial variations of other membrane-associated fluorescent proteins, eGFP-containing two geranylgeranyl acceptor sites and eGFP fused to the transmembrane domain of syntaxin, were greatly reduced or not detectable, respectively. In acutely light-adapted rods, an arrestin-eGFP fusion protein also exhibited axial variation. Both the light-sensitive Rho-eGFP and arrestin-eGFP banding were in phase with the previously characterized birefringence banding (Kaplan, Invest. Ophthalmol. Vis. Sci. 21, 395–402 1981). In contrast, endogenous rhodopsin did not exhibit such axial variation. Thus, there is an axial inhomogeneity in membrane composition or structure, detectable by the rhodopsin transgene density distribution and regulated by the light cycle, implying a light-regulated step for disk assembly in the OS. The impact of these results on the use of chimeric proteins with rhodopsin fused to fluorescent proteins at the carboxyl terminus is discussed.

Editor: Cheryl M. Craft, Doheny Eye Institute and Keck School of Medicine of the University of Southern California, United States of America

Funding: This work was supported in part by the National Institutes of Health grants EY-11256, EY-12975 (BEK), EY018421 (PDC) and EY02660 (ENP), Research to Prevent Blindness (Unrestricted Grant to SUNY UMU Department of Ophthalmology) and the Lions of CNY. The funders had no role in study design, data collection and analysis, decision to publish, or preparation of the manuscript.

Competing Interests: The authors have declared that no competing interests exist.

* E-mail: knoxb@upstate.edu

¤ Current address: Department of Molecular and Human Genetics Baylor College of Medicine, Houston, Texas, United States of America

Introduction

The vertebrate photoreceptor is a highly polarized neuron with a modified cilium specialized for light detection. The cilium contains an OS with a stack of hundreds of disks enclosed in the plasma membrane (Fig. 1) [1]. Rhodopsin is the major protein in the OS, comprising approximately 90% of the membrane protein complement [2]. New rhodopsin molecules are made in the ER, transported via a complex vesicular pathway to the base of the OS and inserted into new disk membranes [3–5]. Previously made disks then move apically and the oldest disks at the OS tip are shed and taken up via phagocytosis by retinal pigment epithelium. This disk renewal occurs every day [6,7]. Accordingly, the whole length of the OS is renewed in 10 days for mammals and ~4–6 weeks for frogs depending upon the temperature. Disk formation is stimulated by light [8,9], but rhodopsin synthesis does not appear to be diurnal, at least in *Xenopus*[10]. Ultrastructural studies reveal a homogeneous distribution of disk membranes throughout the length of the OS [1]. However, light microscopy demonstrates inhomogeneities along the OS length, as birefringence bands perpendicular to the rod axis that arise from anisotropy in refractive index along the OS axis [11–14]. First observed in amphibian rods, they are also found in mammals [15]. In *Xenopus*,

the birefringence banding pattern has a spatial period of about 1.0 –1.6 µm, which corresponds to ~35 – 60 disks [16]. They are more pronounced at the base of the OS. The spatial periodicity is modulated by the length of the light-dark cycle, and is abolished under constant illumination conditions [17]. While these observations suggest an underlying variation in the disks produced in light compared to dark [18–20], the origin of the light-cycle dependent banding remains unexplained.

Imaging of fluorescently tagged proteins offers a powerful method for quantifying protein expression in photoreceptors [21]. In combination with systems for expression of transgenes in *Xenopus* photoreceptors [22], detailed measurements have been made on the distribution of soluble proteins [21], light-dependent protein movement into the OS [23], targeting signals necessary for rhodopsin OS localization and trafficking of membrane proteins to the OS [24–28] and diffusion of both soluble [29] and membrane-bound [30] proteins. Previously, a an eGFP tagged rhodopsin rhodopsin eGFP fusion protein (Rho-eGFP) was shown to exhibit non-uniform fluorescence intensity along the OS axis both in fixed [24] and live [31] samples, suggesting a time-varying production of the transgene. We employed confocal imaging in live rods [21,31] to quantitate the Rho-eGFP distribution in the OS. We found that the variation in fluorescence intensity of this protein along the OS

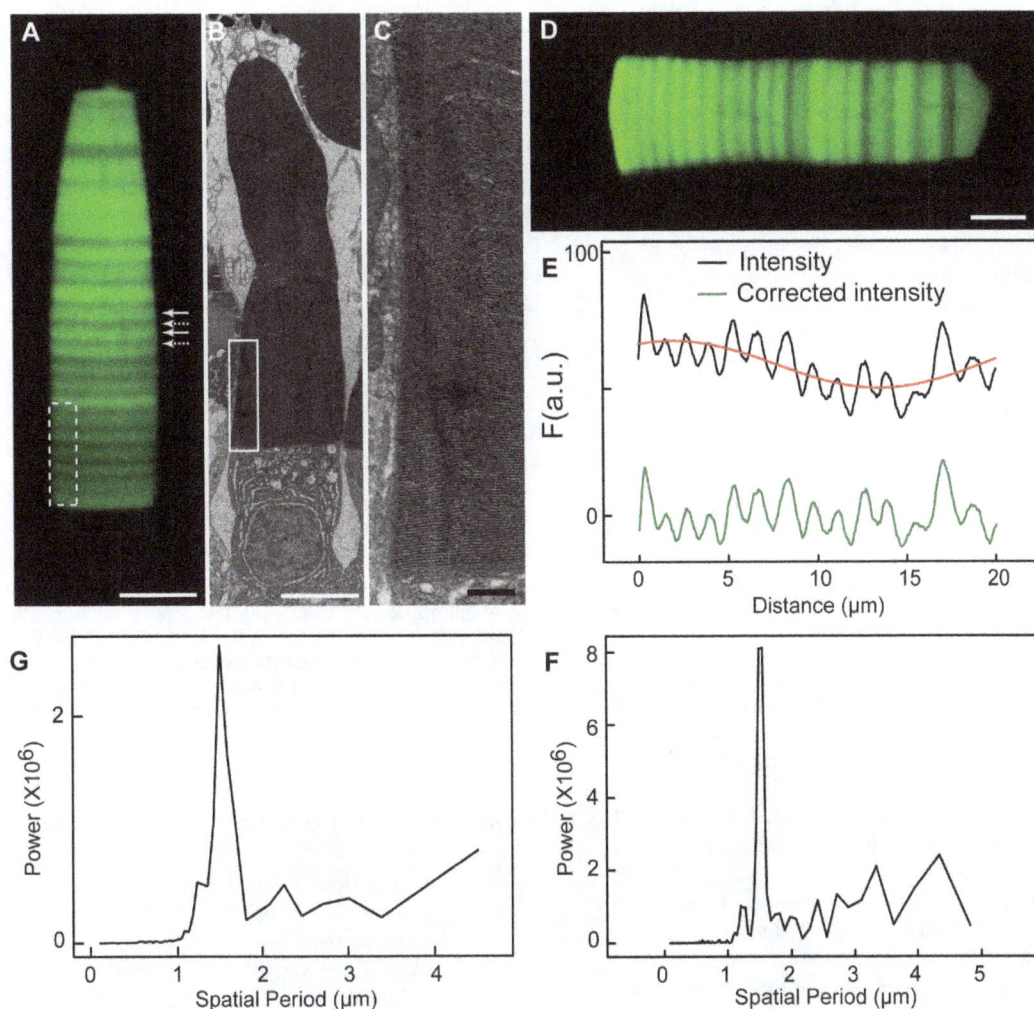

Figure 1. Periodic axial variation in fluorescence intensity in OS expressing a Rho-eGFP transgene. (A) The expression profile of OS (OS) demonstrates a varying level of Rho-eGFP in disk membranes along the OS axis in animals housed in a 24 h (12D:12L) cycle. Periodic axial variation is seen as alternating series of bright fluorescent regions (*solid arrows*) and dim regions (*dotted arrows*).(B, C) Electron micrographs of a rod photoreceptor, expressing Rho-eGFP, show equal spacing between disk membranes. Panel C is an enlargement of white box in panel B. The dotted white box in panel A illustrates a similar sized area for comparison. White bar is 5 μm and black bar is 200nm. (D, E) A rod expressing Rho-eGFP was analyzed to determine the intensity difference at peaks and troughs. The green trace represents the periodic variation in fluorescence intensity before (*black trace*) and after (*green trace*) subtraction of the aperiodic axial variation fit to a sinusoidal function (*red trace*). The amplitude of the variation between the maximum and minimum for each period was1.3±0.07 (mean ± S.D.) calculated from 90 periods taken from10 cells.(F) Fourier transform analysis of the proximal 20 μm of the OS shown in panel (B) demonstrates a peak at 1.5 μm representing the period of axial variation. The Fourier transform spectra are presented in terms of the spatial periods, and not as the more typical frequencies.(G) Fourier transform analysis of the proximal 20 μm of the OS averaged from five different cells also shows a period at 1.5 μm.

axis is coincident with the birefringence pattern and controlled by the light cycle. However, other integral membrane or membrane-associated fusion proteins, exhibit significantly reduced OS axial variation. Thus, these results suggest that there is a light-regulated pathway for trafficking membrane-associated proteins to the OS.

Results

Axial variation of Rho-eGFP distribution in *Xenopus* OS

The Rho-eGFP fusion protein binds 11-*cis* retinal, activates transducin and is transported predominantly to the OS [31–33]. We have previously reported that the expression levels of rhodopsin transgenes under control of the XOP promoter are substantially lower (<5%) than endogenous rhodopsin [31] and do not represent a significant overexpression of this membrane

protein. However, the distribution of fluorescence in the OS is not spatially uniform exhibiting two types of axial variation. First, there is a prominent periodic axial variation that appears as a regular pattern of alternating bright (Fig 1A, solid arrows) and dim fluorescent segments (Fig. 1A, dotted arrows) perpendicular to the rod axis. Second, there is a gradual axial variation in the general expression level along the OS axis. This can be seen in the OS shown in Fig. 1A as relatively brighter apical fluorescence compared to the dimmer basal fluorescence. The variation can sometimes be extreme (Fig. S1) and is less pronounced in F1 and subsequent generations of transgenic lines (*data not shown*). This variation, which we term asynchronous variation, does not align between rods from the same retina and is related to mosaic transgene expression.

The OS ultrastructure in transgenic animals expressing Rho-eGFP had a uniform distribution of disk membranes (Fig. 1B and C) and we did not detect any periodic axial variation in morphology. To determine the spatial period in the fluorescence variation, we computed the discrete Fourier transform on OS regions within 20 μm proximal to the base, where banding is most regular and not distorted by OS stretching or swelling. An example cell is shown (Fig. 1D) with its intensity profile(Fig. 1E) and power spectrum (Fig. 1G). Note, the data are presented in terms of spatial periods, and not as the more typical frequencies, of the FT components to facilitate comparisons with axial displacement values (see below). We found a prominent peak in the power spectrum at a spatial period of 1.5 μm (Fig. 1F), which is equivalent to ~54 disks [8]. The averaged power spectrum from 5 cells shows a similar peak at 1.5 μm (Fig. 1G). This spatial period is characteristic of periodic axial variation in all OS taken from animals housed in a 24 h (12D:12L) lighting cycle.

To determine the magnitude of the periodic axial variation, it was necessary to correct the OS fluorescence intensity for asynchronous variation in transgene expression. We found that in most but not all cells, the asynchronous variation could be fit by a sinusoidal function (Fig. 1E and Fig. S2). Unlike the periodic axial variation, there was a wide range of spatial periods, with a median of 27 μm, which is roughly half the length of the OS in animals housed in a 24 h (12D:12L) lighting cycle. Images were deconvolved to compensate for the PSF blurring. The ratio of corrected fluorescence intensity of peak to adjacent trough was 1.3±0.07 (N = 90 peaks from 10 cells). These results indicate ~30% higher Rho-eGFP density in regions of maximal fluorescence intensity compared to adjacent regions with minimal local intensity.

To determine whether Rho-eGFP found in the periodic bands are free to move laterally in the disk membrane, fluorescence recovery after photobleaching was used. Following the localized photobleaching across several bands of Rho-eGFP density, fluorescence intensity recovered substantially and stabilized after 130 s. Moreover, the recovered region remained in alignment with bands of similar intensity in neighboring non-bleached area (Fig. 2). Thus, consistent with the expected insertion of the Rho-eGFP transgene into discrete disc membranes, fluorescence redistribution after photobleaching was restricted to lateral mobility.

The periodic axial variation was also observed in animals expressing a Rho-mCherry transgene (Fig. 3A boxed region from the 24 h (12D:12L) lighting cycle period). The periodic axial variation of Rho-mCherry and Rho-eGFP was coincident in rods co-expressing both transgenes, although the asynchronous variation was not as tightly linked (Fig. S3A). We also compared periodic axial variation to the birefringence banding previously studied by Kaplan [17]. We obtained DIC images and fluorescence intensities from rods expressing Rho-mCherry (Fig. 3C–3F). In this example, the animal had been housed in a 24 h (12D:12L) lighting cycle for several weeks and then switched to constant light for a week (Fig. 3E, boxed region) prior to imaging. In the 24 h lighting cycle, Rho-mCherry fluorescence intensity was closely correlated with the alternating dark-light bands in the DIC image (Fig. 3F). However, when the animal was housed in constant light, which abolishes birefringence banding [17], the Rho-mCherry periodic axial variation was eliminated (Fig. 3E, boxed region). Thus, since birefringence banding is strongly regulated by the light cycle [17], we examined the periodic axial variation of rhodopsin transgenes in rods from animals housed in different lighting cycles.

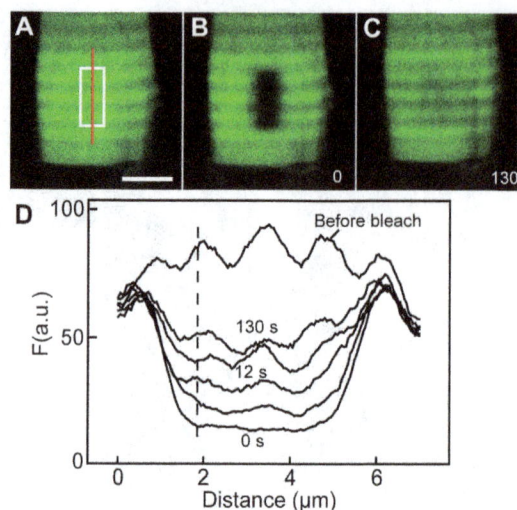

Figure 2. Fluorescence recovery after photobleaching (FRAP) in a live rod expressing Rho-eGFP. (A–C) Sequential fluorescent images from a cell are shown before (A), immediately after photobleaching (B) and 130 s later (C). The target area is highlighted in A with a white box.(D) The fluorescence intensity profile scanned through the photobleached area (red line in panel A) demonstrates the recovery of the banding pattern in register with the non-bleached neighboring area.

The light cycle rather than circadian factors, regulates axial variation of Rho-eGFP

When transgenic *Xenopus* were switched from a 24 h (12D:12L) lighting cycle to constant dark or light, Kaplan banding disappeared in regions synthesized under the constant conditions (Fig. 4A–C). These data indicate that the periodic axial banding is under control of the light cycle rather than a circadian clock. Moreover, extended light cycles produced axial variation with different periodicities of alternating fluorescence intensity. For example, animals housed in a 168 h (84D:84L) cycle exhibited a much larger spatial period than those housed in 24 h cycle (Fig. 4D). The spatial period of the axial variation from three sets of animals housed in different conditions was 1.3 μm for a 24 h (12D:12L) cycle, 2.2 μm for a 48 h (24D:24L) cycle and 4.1 μm for a 96 h (48D:48L) cycle (Fig. 4H). The average magnitude of the maximal variation in fluorescence intensity in the 168 h (84D:84L) lighting cycle was 3-fold (N = 30 peaks taken from 10 cells).The relative widths of the Rho-eGFP bands synthesized in the light were larger than those in the dark, most easily appreciated in the extended lighting cycles (e.g. Fig. 4D–G).This is consistent with previous studies that showed a higher rate of disk synthesis in the light than the dark [7]. For animals housed in extended light cycles and sacrificed at the end of the dark period (for example, Fig. 4D), we found bright fluorescence in the most recently synthesized disks, suggesting that the bright fluorescence bands are synthesized in the dark. Similarly, animals sacrificed at the end of the light period had regions of lower fluorescence intensity at the base of the OS (data not shown). To confirm the period during which the different intensity bands are synthesized, transgenic frogs expressing Rho-eGFP and Rho-mCherry simultaneously were housed in an asymmetric cycle of 168 h ((24D:24L)₄:48L) for 6 weeks (Fig. 4E–G). The fluorescent pattern had two cycles of axial variation whose widths were similar to animals housed in 48 h (24D: 24L) cycle followed by a low fluorescence region (Fig. 4E arrows) whose width was similar to

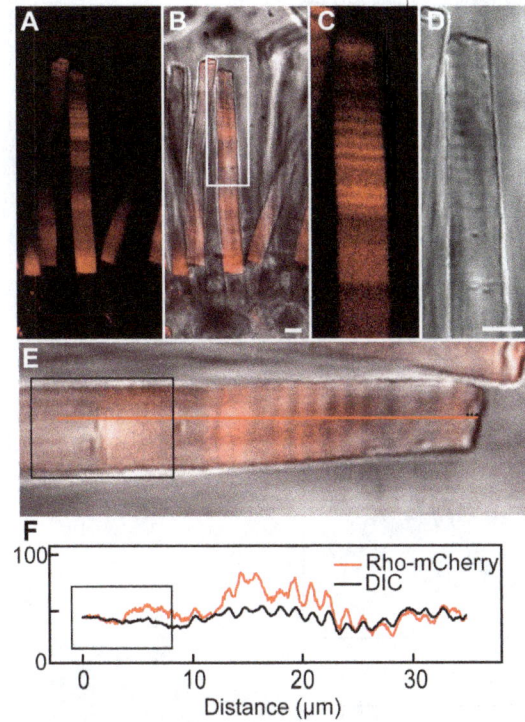

Figure 3. Comparison of periodic axial transgene fluorescence and refractive index variation measured by interferometry. (A–D) The fluorescent and DIC images of a live retinal chip (*white box*, enlargement of a region of an OS shown in C, D) expressing Rho-mCherry exhibit similar periodic banding in both preparations. (E, F) The intensity profile (F) of the fluorescent and DIC images for rod shown in E demonstrates a synchronized banding pattern. The animal from which the rod shown in E was taken was moved from a 24 h (12D:12L) cycle into constant light (approximate region is boxed in *black*) for over a week before sacrifice. White bar is 5 μm.

animals housed in a 96 h (48D:48L) cycle. Coincident fluorescent patterns were observed for both transgenes (Fig. 4E and F). These results show that the regions of lower fluorescence intensity (and hence Rho-eGFP and Rho-mCherry density) were synthesized in the light while the regions of higher fluorescence intensity were synthesized in the dark.

There was a transition between the OS zone with periodic axial banding and aperiodic fluorescence variation in animals that transitioned from constant conditions into 24 h (12D:12L) cycling conditions (Fig. 4A), or from cycling conditions into constant dark (Fig. 4B and 4D) or light (Fig. 4A and 4C) indicating that the light cycle is directly responsible for regulating the periodic axial variation. Moreover, animals maintained in constant light formed longer OS, as previously described [34]. The light cycle did not influence the asynchronous variation (Fig. 4A–C).

The periodic axial variation of Rho transgene density was very sensitive to light, and banding could be observed if animal housing was not completely light tight. In rods from animals housed in an extended cycle, axial banding from small light leaks can still be seen during the dark period (Fig. 5A, asterisks). In another experiment (Fig. 5B), transgenic frogs expressing Rho-eGFP were housed in a 24 h (12D:12L) light cycle, then moved to a darkened chamber for two weeks and finally into a totally light-sealed chamber for two weeks (Fig. 5B). In the first two periods, there was periodic axial banding with the expected spatial period of 1.0–1.5 μm. However, in the final period in complete darkness, no

banding was detected. The magnitude of the amplitude of the axial variation appeared to increase when brighter lights were used (e.g. compare the amplitude in normal light versus in the dim light), but the asynchronous variation precluded reliable quantification.

Axial density variation of endogenous rhodopsin

We investigated whether endogenous rhodopsin exhibits periodic axial variation in OS density similar to that of fluorescent rhodopsin transgenes using immunohistochemistry and rhodopsin densitometry. To improve spatial resolution via increased band width and axial variation amplitude as predicted by Rho-eGFP, we housed animals in an extended 168 h (84D:84L) cycle prior to sacrifice. We used antibodies that recognize the Rho-eGFP transgene alone via its C-terminal tag (1D4 and 3D6, [35]), antibodies that recognize endogenous Rho but not Rho-eGFP (K16-155C, [36]) and antibodies that recognize both endogenous and Rho-eGFP (4D2, [35]). Since the expression levels of Rho-eGFP are much lower than endogenous rhodopsin [30,31], the 4D2 antibody will primarily report the endogenous rhodopsin. Note that the C-terminal antibodies had better penetration into the OS than the N-terminal antibody (Fig. 6). Bands of high Rho-eGFP fluorescence were detected with C-terminal antibodies that recognize the transgene (Fig. 6A and B, for example asterisks). However, there was no correspondence (Fig. 6C and D, for example *arrows*) between bands with high Rho-eGFP fluorescence and bands of higher reactivity with N-terminal antibodies that recognize endogenous rhodopsin. These micrographs also show that the antibodies have some penetration into the OS and also show that there is little fluorescence bleed-through in the Cy3 channel. With an antibody that specifically recognizes Rho-eGFP, the expected axial banding pattern was detected although the magnitude of the fluorescence variation between regions synthesized in the dark and light are somewhat reduced (Fig. 6). We attribute this reduction to alterations in cellular morphology caused by fixation. Given the ~3-fold amplitude in Kaplan banding intensity of Rho-eGFP, we expect to be able to detect axial variation even at the margins of the OS (Fig. 6C and D). Thus, the endogenous and transgene rhodopsin axial distribution patterns were different.

We directly measured rhodopsin pigment concentrations using light-dark difference densitometry at 520 nm, which can reliably detect variations down to approximately 5% [30]. To amplify potential variations, we utilized rods derived from animals housed in an extended 168 h (84D:84L) cycle prior to sacrifice. Although the Rho-eGFP density exhibits ~3-fold variation between the dark and light periods, there is no apparent variation in endogenous rhodopsin density (Fig. 7). We conclude that endogenous rhodopsin does not exhibit axial variation with the same amplitude as Rho-eGFP transgenes although we cannot rule out small changes below our limits of detection. Thus, it appears that Rho-eGFP is reporting the light-sensitive regulation of membrane assembly that is apparently not tightly coupled to transport of rhodopsin to the OS.

We examined the expression level of selected transcripts in animals maintained in a 24 h (12D:12L) cycle using qRT-PCR to characterize changes in transcription throughout the light cycle. While both red cone opsin and nocturnin exhibited robust changes over 24 h (Fig. 8A and B), as previously reported [37,38], endogenous Rho did not show a significant change (Univariate ANOVA, F[6,7] = X, P<0.76, Fig. 8). We also did not find significant changes in the expression level of endogenous rhodopsin or Rho-transgenes in animals kept in 24 h light or dark (Fig. 8). However, since rhodopsin transgenes exhibit spatial

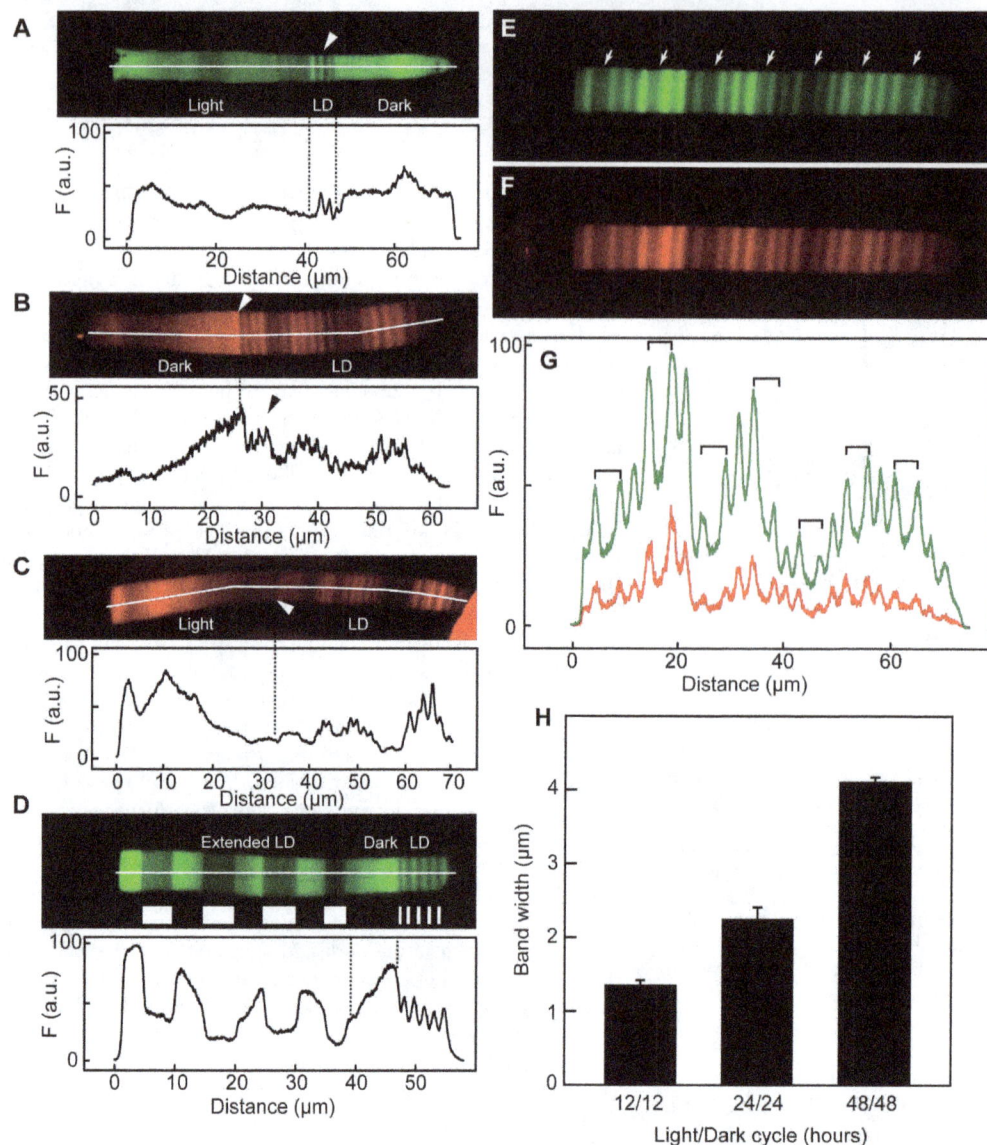

Figure 4. Effect of varying light-dark cycle on periodic axial variation. (A) Transgenic animals expressing Rho-eGFP were housed for 2 weeks in constant darkness, switched to 24 h (12D:12L) for 3 days (LD) and then in constant light for 2 weeks prior to imaging. The intensity profile of the cell is shown below. While axial banding with spatial period 1-1.4 µm are seen in the 24 h light cycle (*arrow*), little variation with this spatial period was observed in constant dark or light. Note that the rate of membrane addition to the OS is higher in the light than in either cycling or constant dark conditions. (B) Transgenic animals expressing Rho-mCherry were housed in 24 h (12D:12L) light cycle for over 4 weeks (LD) and switched (approximate position indicated by the *arrow*) to constant darkness for over 2 weeks before imaging. Axial banding with spatial period 1–1.4 µm is seen in the 24 h light cycle (*arrow*), no variation with this period was observed in constant dark. The intensity profile of the cell is shown below. (C) Transgenic animals expressing Rho-mCherry were housed in 24 h (12D:12L) cycle for over 4 weeks (LD) and switched (approximate position indicated by the arrow) to constant light for over 2 weeks before imaging. Axial banding with spatial period 1–1.4 µm is seen in the 24 h light cycle (LD), no variation with this period was observed in constant dark. The intensity profile of the cell is shown below. (D) Transgenic animals were first housed in a 24 h (12D:12L) cycle for over 4 weeks, switched to constant dark for 7 days and then maintained in a 168 h (84D:84L) cycle until imaging. The animals were sacrificed during the dark period. The approximate regions synthesized in different lighting periods (brighter regions synthesized in the dark) are indicated. A spatial period of 4–6 µm was observed in the extended lighting cycle. The intensity profile of the cell is shown below. (E–G) Transgenic animals expressing Rho-mCherry and Rho-eGFP simultaneously were housed in an asymmetric cycle of 144 h ((24D:24L)$_4$:48L) for 6 weeks. There were two different widths of bands assembled in the light, one with ~2 µm and a wider one (arrows) with ~4 µm. These are associated with the 24 h and 48 h light periods, respectively. (H) Frogs were kept in 24 h (12D:12L) cycle for 4 weeks and then moved to an asymmetric cycle of 96 h (24L:24D:24L:24D:48L). The widths of dark bands (error bars are SD) in cells (N = 3) from these animals were determined. Each light cycle width is statistically different from the others (p<0.5).

variegation across the retina and also asynchronous (temporal) variation, whole retinal analysis of transgene transcripts is not definitive.

We next examined the OS distribution of Rho transgenes controlled by several other rod-specific promoters, including rod transducin alpha subunit and rod arrestin, and found identical periodic axial variation as that in animals utilizing the rhodopsin

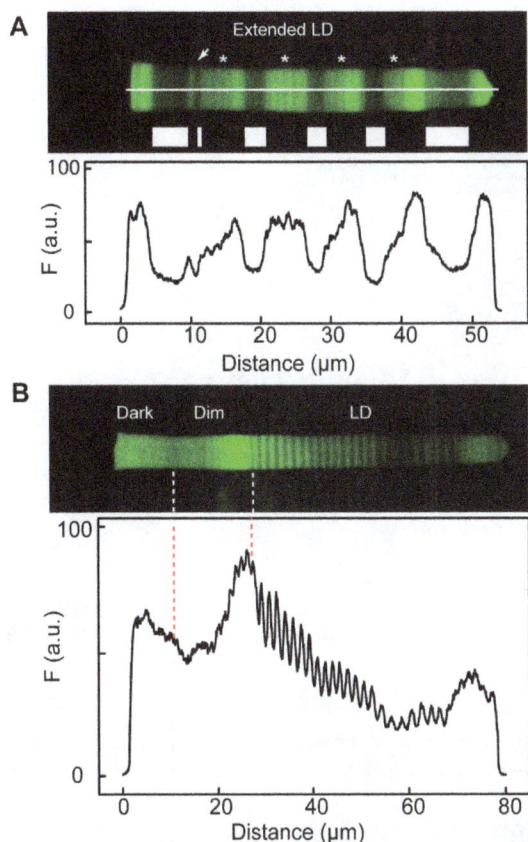

Figure 5. Periodic axial variation in dim light cycles. (A) Transgenic frogs expressing Rho-eGFP were kept for six weeks in 168 h (120D:48L) cycle. During the dark periods, there was dim light exposure since the incubator was not completely light-sealed. In these dark periods (asterisks), there was an increase in fluorescent intensity and an additional axial variation superimposed, with a spatial period of 4–6 μm. At the arrow, the animals experienced one 24 h (12D:12L) cycle and followed by 4 days light and finally a 3-day dark period. Animals were sacrificed during the dark period. The intensity profile of the cell is shown below. (B) Transgenic frogs expressing Rho-eGFP were kept first in a 24 h (12D:12L) cycle (LD) and then moved to an incubator was not completely light-sealed(Dim) for two weeks and were moved to a totally light-sealed chamber (Dark). Cells (N = 23) were imaged and a representative cell is shown. The light-sealed chamber did exhibit periodic axial banding as found in the other two regions. The approximate locations of the transitions in lighting are indicated. The intensity profile of the cell is shown below.

promoter (Fig. S3). Thus, while we cannot rule out regulation of Rho transgene transcription by light, the reproducibility of the spatial period and magnitude support an argument for a post-transcriptional mechanism to explain periodic axial variation.

Banding pattern of membrane-associated fluorescent proteins

To further explore the light-sensitive regulation of membrane protein distribution in the OS we expressed two other membrane or membrane-associated proteins under control of the XOP promoter: an eGFP with two protein prenylation signals at the C-terminus, resulting in the attachment of two geranylgeranyl (C_{20}) moietie sand an eGFP fused to the syntaxin-3 transmembrane[26]. We compared the OS fluorescence intensity distribution to Rho-eGFP and RhoΔPalm-eGFP, in which two highly conserved vicinal cysteines (C322–C323) in the C-terminus were changed to

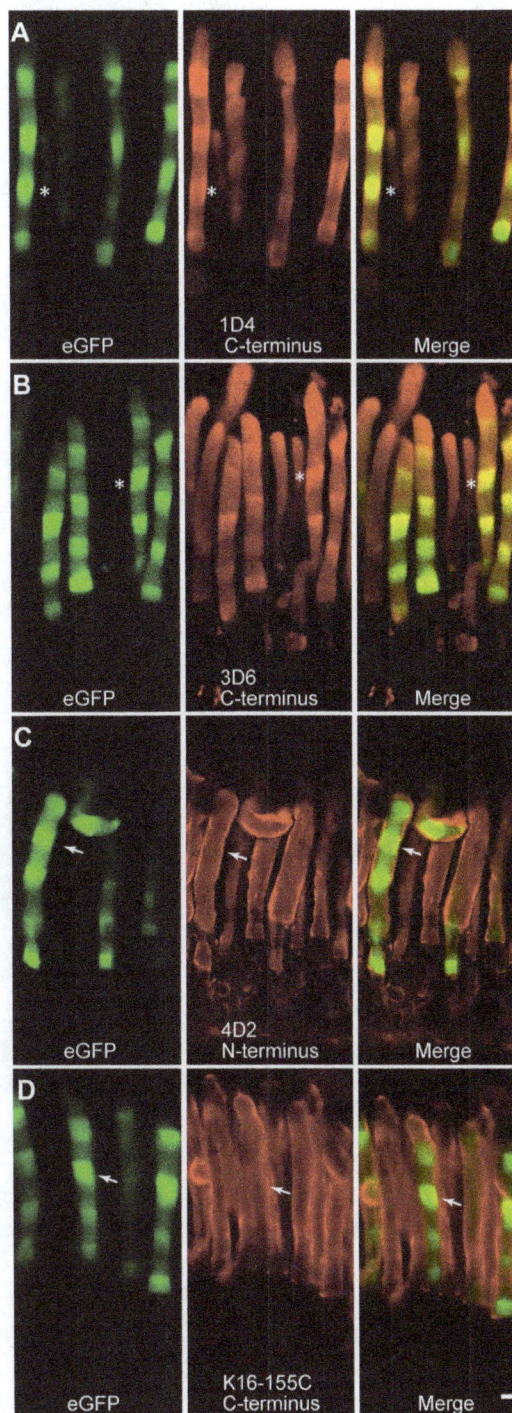

Figure 6. Immunostaining of transgenic retina expressing Rho-eGFP distributed in axial bands with anti-rhodopsin antibodies. Adult transgenic frogs expressing Rho-eGFP, which has been modified to contain an epitope for the monoclonal antibody 1D4, were kept for 8 weeks in a 168 h (84L–84D) light cycle and then eyes were fixed and immunostained with the indicated antibodies. Rho-eGFP signal (green) is the intrinsic from eGFP fluorescence. The antibodies were detected using a secondary antibody labeled with Cy3 (red). Confocal microscope images of individual and merged channels are presented. Examples of corresponding regions with high Rho-eGFP fluorescence are indicated in A and B by asterisks and in C and D by arrows. White bar is 5 μm.

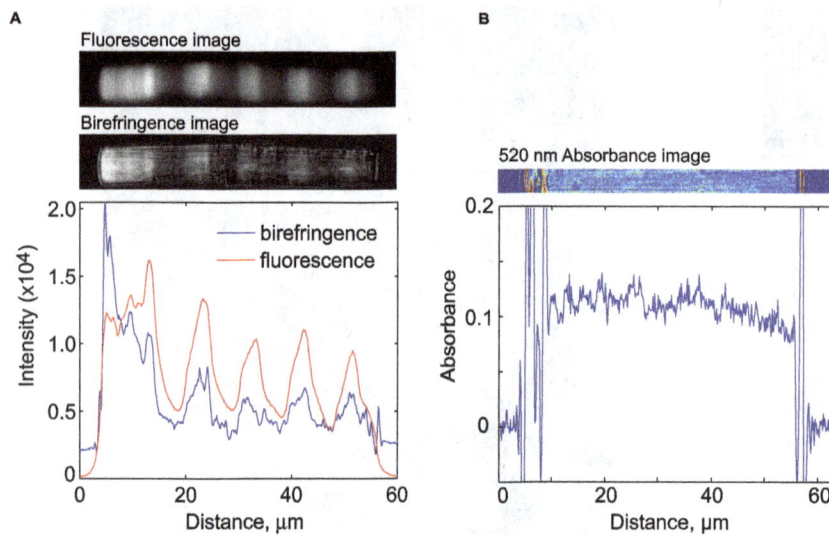

Figure 7. Axial rhodopsin absorbance is invariant despite large variation in Rho-EGFP levels. (A) A fluorescence image (top) and birefringence image (middle) of an outer segment of a Rho-EGFP–expressing rod from an animal housed on an extended 168 h (84D:84L) cycle. The intensity of the fluorescence and birefringence are plotted (bottom). (B) A 520-nm absorbance image (top) and the axial fluorescence profile (bottom).

serine and threonine, respectively. These mutations removed the two palmitoylation sites that anchor a portion of the C-terminus to the disk membrane surface. The RhoΔPalm-eGFP distribution had an axial variation similar in magnitude (compare Fig. 9A and B) and spatial period (compare 9E and F) as found in Rho-eGFP. Both Rho-dGryGry and eGFP-Syntaxin(TMD) were found predominantly in the OS (Fig. 9C and D), but there was significantly reduced periodic axial variation compared to Rho transgenes (Fig. 9A and B).For eGFP-dGryGry, the spatial period was significantly higher (2.3 μm) than for Rho transgenes. There is much less power (~20%) in the normalized spectra at this spatial period than in Rho-eGFP. eGFP-Syntaxin(TMD) had very weak banding, with the major peak occurring at 4 μm, substantially higher than for Rho-eGFP. There was very little power (<5%) in spatial periods associated with the light-sensitive variation. Even in lighting cycles with extended dark and light durations, the magnitudes of periodic axial variation were significantly reduced for both dGryGry and eGFP-Syntaxin(TMD) (Fig. 9).

Axial variation of Arrestin-eGFP distribution in light-adapted OS

The birefringence and Rho transgene periodic axial variation suggest that there may be associated axial variation in other disk membrane properties. Arrestin-eGFP [21] binds to rhodopsin in a light-dependent fashion and undergoes light-dependent transloca-tion to the OS[39]. Transgenic rods expressing arrestin-eGFP (eGFP at the C-terminus of arrestin) were light-adapted and OS fluorescence intensity was determined in live rods at various times after light exposure (Fig. 10). The OS distribution of arrestin-eGFP exhibited a similar periodic axial variation pattern as observed with Rho transgenes and was in phase with the birefringence banding pattern. Following photobleaching, the fluorescence recovered after 65 s to reproduce the pre-bleach periodic axial variation in alignment with bands in neighboring non-bleached area. We measured the rate of recovery of arrestin-eGFP in the IS and OS and found that the half-time of recovery for arrestin-eGFP in the OS is longer than that of the inner segment. Thus, arrestin

mobility is primarily lateral, revealing it to be bound to rhodopsin or to another component of the disc membranes.

Discussion

Two types of variation in rhodopsin transgene expression

Using transgenic *Xenopus* and live cell confocal imaging, we have exploited the rod OS to examine membrane synthesis over many weeks. We have observed two types of variation in rhodopsin transgene density along the OS axis. One type is a slow modulation of the overall expression level that was not synchro-nized with the light cycle or between cells in the same retina. This variation is likely related to transgenic effects such as position effects (e.g. [40]) and/or epigenetic modifications of the transgene locus (e.g. [41]). However, this is this first time to the best of our knowledge that such slow temporal variation in transgene expression has been reported. The second type is a light-sensitive fluctuation between higher densities of rhodopsin fusion protein in membranes synthesized at night and lower densities in membranes synthesized during the day. The periodic axial variation disap-pears in constant light or dark conditions. The axial density distributions of rhodopsin fusion protein exhibited a spatial period of 1–1.5 μm in 24 h (12L:12D) lighting cycles at 18–20°C, which is very similar to the spatial period for birefringence banding [16] and to the daily increase in OS length [8].Because of the common properties shared between the periodic axial variation of Rho-eGFP/Rho-mCherry and the birefringence axial, we will refer to them both as Kaplan banding. Another transmembrane protein, peripherin-eGFP, appears to also exhibit Kaplan banding[28]. By contrast, the magnitude of the Kaplan banding was significantly reduced with other membrane-associated fusion proteins such as eGFP-Syntaxin(TMD) or eGFP-dGryGry. These differences cannot be accounted for by differences in transgene expression levels. Rather, our data suggests that the magnitude of the variation may reflect differences in the rate of protein biosynthesis and/or membrane assembly for each transgene.

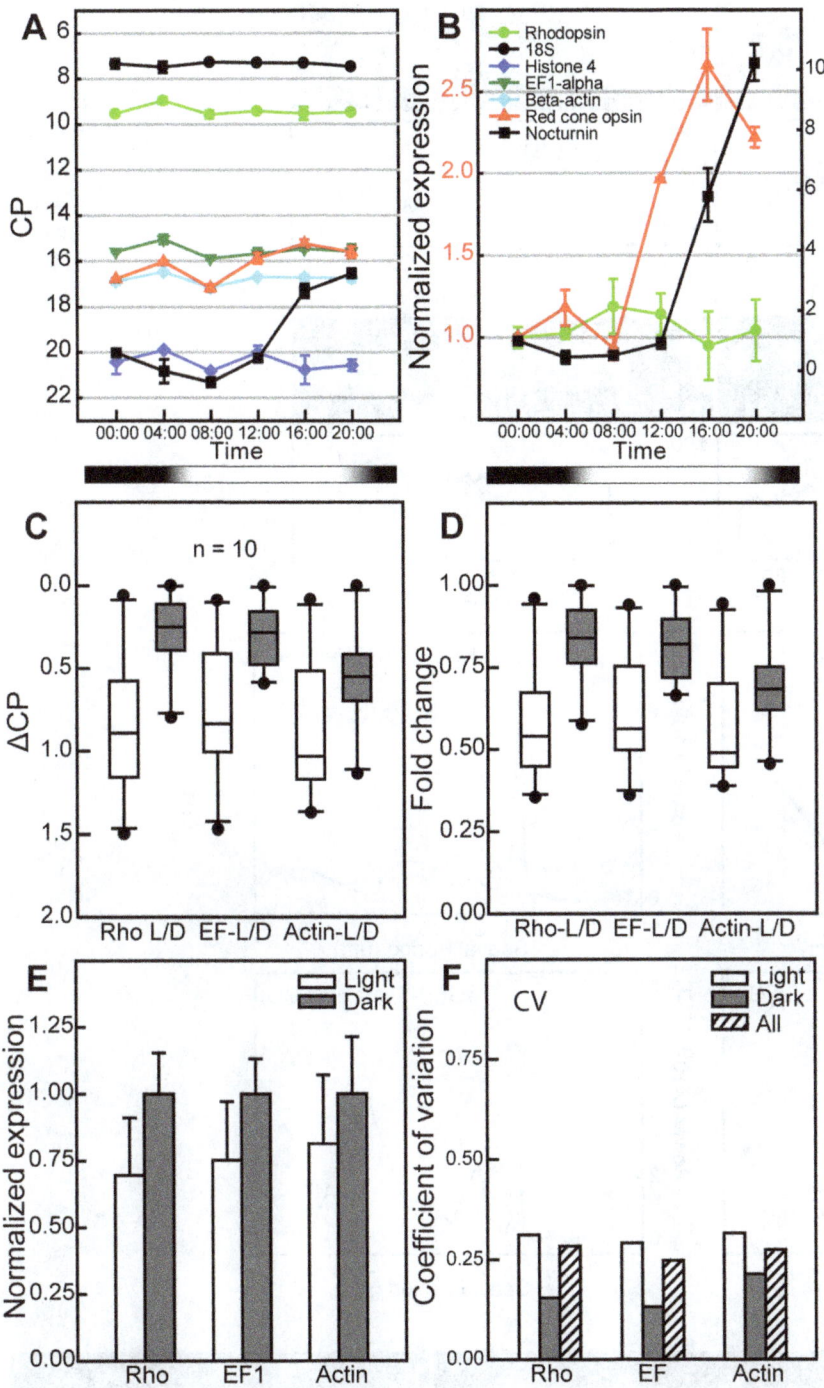

Figure 8. Expression level of rhodopsin transcripts at points in the light cycle. (A) Non-transgenic frogs were kept in 24 h (10D:14L) cycle and sacrificed at 4:00 hr time intervals and RNA extracted from eyes. Real-time PCR was used to quantify transcript levels using (A) PCR crossing point (CP) values. There is a slight increase in the transcript levels of rhodopsin, red cone opsin, and control genes, such as EF1-α and β-actin during the end of dark cycle. Black/white bar shows the dark/light periods (B) CP values normalized to housekeeping genes (*see methods for details*). The expression levels of rhodopsin and red cone opsin use the left axis scale while nocturnin uses the right Y-axis. The expression levels of rhodopsin, red cone opsin and nocturnin are shown. The expression of red cone opsin increased shortly after light onset, peaked a few hours afterwards and dropped in the dark. The expression of nocturnin increased before dark onset and peaked during the dark period and then returned to baseline. The expression level of rhodopsin, however, did not change significantly. (C, D) The expression levels of rhodopsin transcript and two control genes, EF1-α and β-actin, were determined during both a 24 hr light and 24 h dark cycle. The ΔCP (C) and fold change (D) in the light (L) and dark (D) values are shown as box plots. There is a higher expression level of all three genes during the dark cycle. The variation of the gene expression is much less during the dark cycle. (E) Expression levels normalized to 18S show ∼30% increase in all three genes during the dark cycle. (F) The coefficient of variation in the expression level of genes is shown during the dark or light cycle and combined (All).

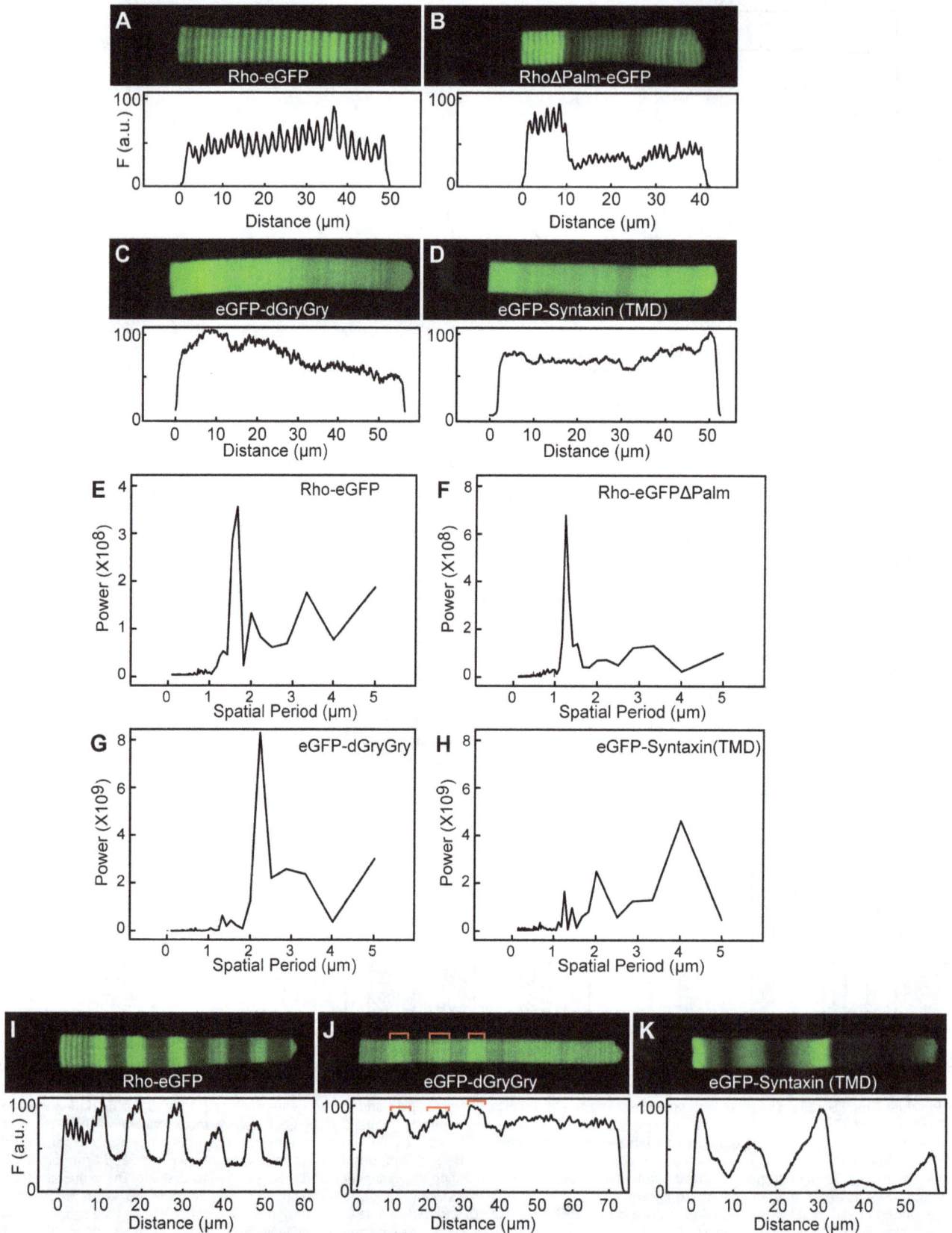

Figure 9. Axial variation of rhodopsin C-terminal mutants and membrane associated eGFP transgenes. Transgenic animals expressing the indicated transgene were housed in either a 24 h (12D:12L, A–D) or 168 h (84L:84D, I–K) light cycle. Representative cells and intensity profilesfrom animals housed in a 24 h (12D:12L) are shown. (E–H) Power spectra of the axial fluorescence intensity distribution from cells expressing the indicated transgenes. To enable comparisons of the power between different cells, the total OS fluorescence intensity for each cell was set to 1 and then the

fluorescence intensity distribution was normalized to that value. Note the scales in E and F are different than G and H. RhoΔPalm-eGFP has the palmitoylation sites deleted (C322S and C323T), eGFP-dGyrGy contains two geranylgeranyl acceptor sites and eGFP-Syntaxin(TMD) has eGFP fused with the transmembrane domain from syntaxin.

The light-sensitive variation is tightly related to axial variation in birefringence

Previously, we have shown that Rho-eGFP has a uniform lateral distribution in disk membranes, while excluded from incisures ([30,31]). Depending upon the light cycle, here we have observed between 30–300% variation in peak Rho-eGFP fluorescence intensity levels in dark and light. By contrast, the maximal variation in endogenous rhodopsin was <5% even in an extended light cycle. Thus, the Rho-eGFP subcellular distribution is like the endogenous rhodopsin distribution in being restricted to the OS discs but does not match the Kaplan banding in the OS. One possible explanation for this difference between native and transgenic rhodopsin is that the axial density of disk membranes along the OS may not be uniform — disks assembled in the dark may be packed more densely than those made in the light. In acutely calcium-depleted *Rana pipiens* rods, birefringence band

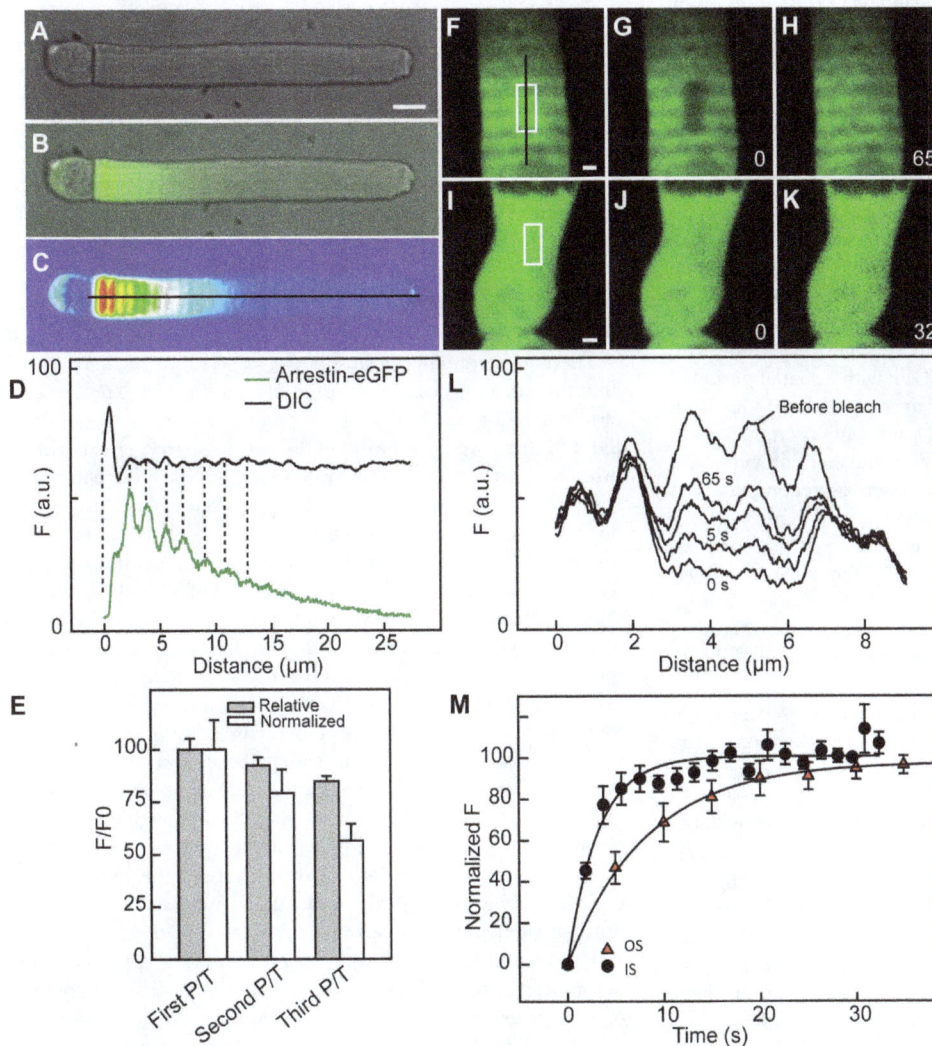

Figure 10. Axial variation of arrestin-eGFP transgenes in the OS. (A–E) A retinal explant from transgenic frogs expressing soluble arrestin-eGFP in rods was dark adapted and then moved to light for an hour before the live imaging. Arrestin-eGFP demonstrated axial variation in fluorescence at the base of the OS with a period of 1–1.4 μm that was synchronized with the DIC variation. The intensity profiles of the cell for both variations are shown (D). The intensity of the fluorescence decreased along the OS and the amplitude of the variation between peak (P) and neighboring trough (T) also decreased. The first peak was ~35% brighter than the next trough (N = 5 cells, p<0.03). Peak/trough ratios (P/T) were calculated for the first three light cycles closest to the base and then normalized to the first peak which was set 100% (Relative) and also after normalization to the mean (Normalized). Error bars represent standard deviation and each of the three bands were statistically different (N = 5 cells, ANOVA p<0.0003). (F–M) Fluorescence recovery after photobleaching. The light adapted rod with arrestin-eGFP in the OS (F–H, L) and IS (I–K) was photobleached within an area near the base of the OS (*white box*) in panels (F, I) and the fluorescence recovery of the photobleached area was recorded continuously. The intensity profile of this recovery along the black line in panel (F) is shown in panel (L) as a function of time. There is recovery after photobleaching within 65 seconds in the OS, regenerating the axial banding pattern with the same spatial period as neighboring regions. The IS arrestin-eGFP recovered faster that that in the OS. White bars are 1 μm.

contrast (i.e. the magnitude of the variation) increased permanently, suggesting that the lighting cycle controls a calcium-sensitive component of OS synthesis [17]. This component has not yet been identified. It could be related to the intradiscal space, which sequesters calcium [42], possibly by regulating bilayer interactions across the intradiscal space or altering molecular conformations that influence disk density. In electron micrographs of transgenic rods (Fig.1), no axial periodicity in disk membrane spacing was observed. However, the cells studied by EM were fixed in glutaraldehyde, which could have disrupted disk spacing and caused shrinkage of OS disks [17]. On the other hand, Rho-eGFP banding in fixed samples has been observed (e.g. [24] and Fig. 6), albeit with some reduction in the magnitude of the variation. Therefore, it seems unlikely that axial packing density variation of disks is responsible for the banding of rhodopsin fusion proteins.

The light-dependent axial variation may be a consequence of light-dependent rates of disc membrane synthesis

Alternatively, the Kaplan banding of Rho-eGFP could arise from periodic variation in Rho-eGFP density in disks made during different phases of the light cycle. If so, this represents a significant difference between rhodopsin fusion proteins expressed from transgenes and endogenous rhodopsin, the basis for which is not clear. We were not able to rule out transcriptional variation of Rho-eGFP transgenes over the course of the light cycle. However, since periodic axial variation of Rho-eGFP with a spatial period of 1–1.5 μm and similar amplitudes was observed in OS from animals harboring transgenes using different promoters and reliably in many different individual F0 transgenic individuals, it seems unlikely the cause of periodic variation is predominantly at the level of transcription. Moreover, there was extremely low or undetectable periodic axial variation in OS from animals expressing eGFP-Syntaxin(TMD) driven by the same promoter, XOP, as Rho-eGFP. Thus, if Rho-eGFP distribution reflects variation in Rho-eGFP content, then we conclude that there is a light-regulated variation in the assembly rate of Rho-eGFP into disk membranes.

Endogenous rhodopsin and Rho-eGFP are expressed at vastly different levels [30,31]. Furthermore, endogenous rhodopsin gene does not exhibit significant light-sensitive variation in transcription (Fig. 8) and others have shown that rhodopsin translation rates do not change in response to changes in light [9]. One possible explanation for this difference could be that the rate of Rho-eGFP production is some how limited and cannot increase in the light as endogenous rhodopsin (Fig. 11A). Thus, as disk synthesis increase, Rho-eGFP density would decrease compared to those made in the dark leading to two regions of different density (Fig. 11B). Rhodopsin, on the other hand, can increase production in the light and this maintains a constant density throughout the cycle.

In amphibians, the lighting cycle has a profound effect on the rate of disk assembly and OS elongation [7–9,34,43]. Light onset triggers the assembly of disks at the base of the OS [34,43]. In 24 h (12L:12DL) light cycles, Xenopus rods produce 29–91 disks per day, depending on the temperature [10]. The rate of disk displacement is 1.2–1.8 times greater in the light than in the dark, again depending upon the temperature [7]. The relative widths of the Rho-eGFP bands synthesized in the light compared to those in the dark fit within this range. Thus, it appears that rhodopsin fusion proteins are responding to the same light signals as those that regulate disk assembly and displacement.

It is not obvious why the Rho-eGFP or Rho-mCherry fusion proteins should behave differently than endogenous rhodopsin.

Although eGFP and mCherry are fused to the carboxyl terminus in a flexible region of rhodopsin [44] and are synthesized after rhodopsin translation and membrane insertion is complete, it is possible that they somehow alter folding, processing or transport to the OS, thus potentially reducing the rate of incorporation of Rho-eGFP into the OS compared to endogenous rhodopsin. Since we do not observe much Rho-eGFP in the IS (also see [31]), there would have to be significant degradation of the fusion protein within the first hour post synthesis, since the eGFP maturation time is approximately one hour at these temperatures [45]. Moreover, since endogenous rhodopsin is found at uniform densities throughout the OS even in extended lighting cycles (Fig. 6 and 7), the rate of endogenous rhodopsin synthesis, trafficking (through the secretory pathway) and assembly into disks is closely matched to the light-dependent rate of disc membrane synthesis. Thus, when the latter is sped up, the incorporation of rhodopsin matches. One possibility is that the rhodopsin synthesis rate varies, and that this leads to more membrane transport and higher rate of disc synthesis in darkness. However, rhodopsin synthesis measured in retinal explants for 4 hours in the light and dark were very similar [9]. This suggests other steps may be regulated by the light cycle. Apparently rhodopsin fusion proteins are unable to adjust the rate of assembly in response to light.

Possible explanations of the rhodopsin fusion protein light-dependent variation

There are a number of possible ways eGFP or mCherry could limit the rate of processing or assembly into the OS. The carboxyl terminus is also involved in sorting and transport of rhodopsin to the OS (reviewed in [46]). eGFP or mCherry could disturb processing through the ER and Golgi or its interaction with components of the vesicular transport machinery, leading to degradation. Recently, rhodopsin's carboxyl terminus has been implicated in rhodopsin dimer formation in membranes [47]. Thus, it is possible that the oligomeric state of Rho-eGFP may be different than endogenous rhodopsin, again potentially leading to an alteration of the rate of rhodopsin fusion proteins into disk membranes. Interestingly, an eGFP fusion protein containing the last 44 amino acids of rhodopsin at its carboxyl terminus was found in the OS, exhibiting a banding pattern similar to the periodic axial variation seen with rhodopsin transgenes [25]. However, it is not possible to quantitatively compare those images of fixed tissue with the live cell imaging. So, although the carboxyl terminus has an important role in banding, we cannot determine how important other regions of rhodopsin are for periodic axial variation. This highlights a potential complication for studies utilizing rhodopsin with carboxyl terminal alterations.

The banding observed with an arrestin-eGFP transgene does not appear to be explained by OS regions with different contents of endogenous rhodopsin. The correspondence of the arrestin-eGFP banding in light-adapted OS and the previously characterized birefringence banding suggests that there is an inhomogeneity in membrane composition or properties along the rod OS that may have physiological implications. For example, it is clear that cholesterol content changes with axial position [48]. So, it is possible that the disk assembly pathways operating in the dark or light could produce membranes with different lipid compositions potentially leading to alterations in how arrestin-eGFP interacts with rhodopsin or the membrane itself.

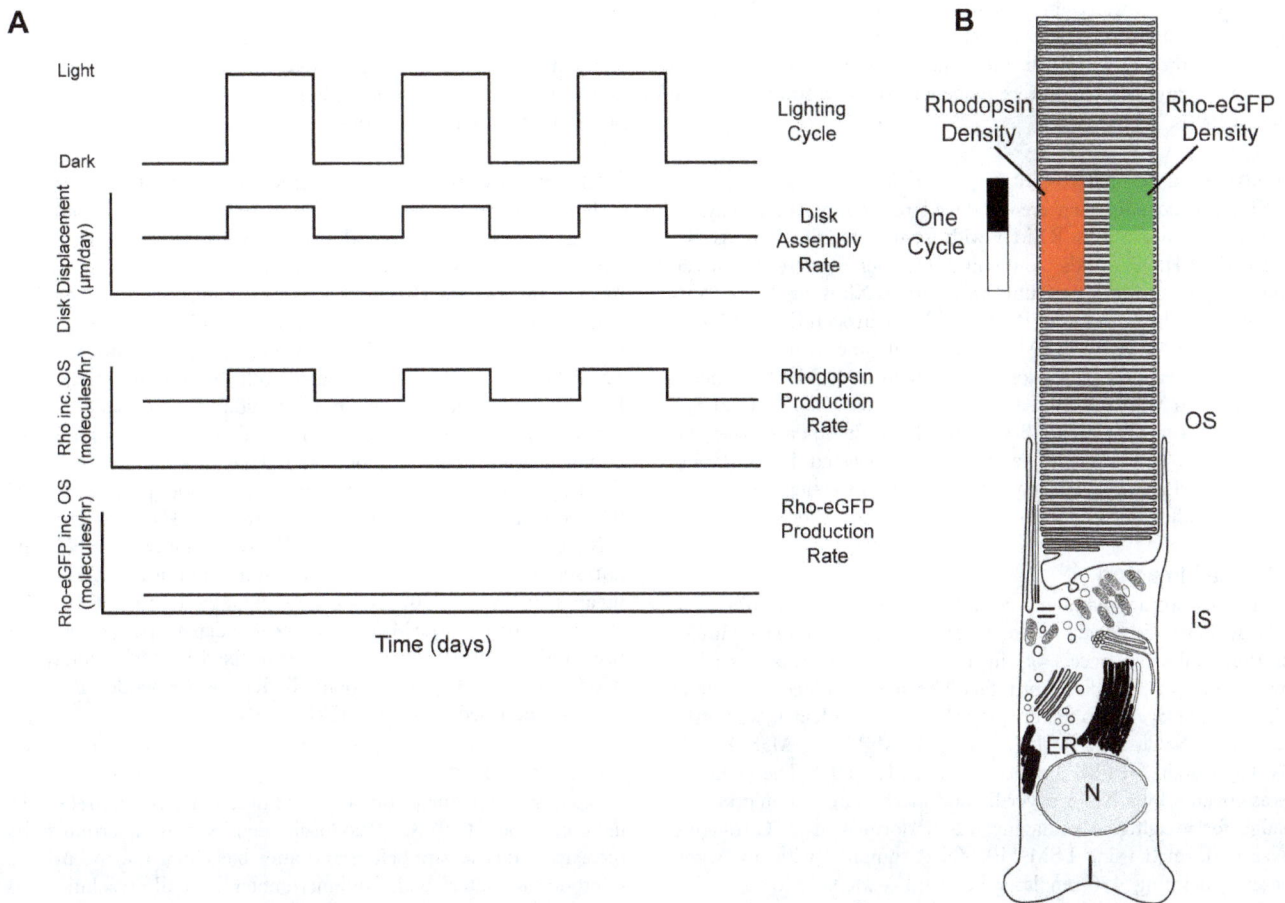

Figure 11. Schematic diagram of light-regulated disk assembly. A. Rods housed in an alternating light-dark cycle (*top panel*) will vary the rate of disk assembly and displacement (*second panel*) in response to the phase of the cycle [6,7]. Since endogenous rhodopsin density in OS membranes is constant, the rate of rhodopsin incorporation (*third panel*) must change in phase with the lighting cycle and disk assembly. By contrast, Rho-eGFP appears to have a relatively constant rate of incorporation into the OS (*bottom panel*). This would then lead to a periodic variation in Rho-eGFP content, with higher densities assembled at night. B. Schematic diagram illustrating the density variation of Rho-eGFP (*green*) throughout a light cycle and the absence of variation in endogenous rhodopsin (*red*). The length of the lighting cycle and the size of the cell are not to scale.

Light/dark variation in disc membrane synthesis may contribute to stable OS

The stability of the OS is an important factor in photoreceptor health and survival. We recently presented a theoretical model for the mechanical properties of the OS and derived parameters regulating their flexural rigidity [49]. Axial density variations similar to the banding considered here were an integral part of the model. Calculations showed that such density variations, which contribute to OS stiffness, increase the flexibility of the OS to bending and makes it less fragile. Thus, the axial banding may have evolved to improve the mechanical properties of OS, permitting larger and longer OS containing higher amounts of rhodopsin. Furthermore, OS were predicted to have a tendency to break in higher density regions. This could have implications for disk shedding at the apical surface. In many species, there is a strong light-dependence to shedding, with most of the disks being shed soon after light onset in large packets [7] reminiscent of the bands examined here. In addition, both disk shedding and axial variation are reduced in constant light. Thus, if OS disks are shed in packets that coincide with periodic Rho-eGFP bands, then Kaplan banding may provide a means to determine how many disks will be shed or to organize and facilitate membrane breakdown. Future experiments on the structural basis of Kaplan banding will be required to determine its potential role in OS renewal.

Methods

Animal Husbandry

Care and feeding (three times per week) of *Xenopus* have been described in detail (Solessio et al., 2009). Animals were housed at 18–20°C in a 24 h (12L:12D) lighting cycle unless otherwise stated. All animal handling and experiments were in agreement with the animal care and use guidelines at Association for Research in Vision and Ophthalmology (ARVO). This study was done under the approval of the SUNY Upstate Medical University Committee on the Human Use of Animals (CHUA no. 209).

Transgenic Constructs

Plasmids for transgene expression in *Xenopus* were based upon the pEGFP-N1 (Stratagene) backbone as previously described [22]. DNA fragments from *Xenopus* opsin (545 or 5417 bp) [22], *Xenopus* arrestin (XAR7, 285 bp) [50] or *Xenopus* α-transducin upstream (4996 bp) sequences (unpublished data) were generated by PCR and subcloned into the pEGFP(-) vector [22] at the XhoI-BamHI site. Dual transgene expression constructs were made by assembling each transgene protein cassette separately by PCR and then sub-cloning both into a vector containing duplicate XOP(-504/+41) promoters [50] in the same transcription direction. All protein coding sequences were sequenced on both strands prior to transgenesis.

Transgenic *Xenopus*

Transgenic animals were produced using restriction enzyme-mediated integration (REMI) with some modifications as described [51]. Plasmids containing transgenes were linearized outside the transcription unit with either XhoI or NheI (New England Biolabs), purified (PureLink™, Invitrogen,Carlsbad, CA) and then used in the REMI reaction with the same enzyme at 0.5 U per reaction. Embryos were kept in 0.1X MMR (concentrations in mM, NaCl, 10; KCl, 2; MgSO$_4$, 1; CaCl$_2$, 2; HEPES, 5 at pH = 7.4) for 6 days at 16°C with a 12/12 light/dark cycle and then at 20°C.Transgenic tadpoles were selected by observing whole eye fluorescence using a fluorescence dissecting microscope on day 6 after nuclei injection and again on day 20.

Live Cell Imaging

The recording chamber was fabricated in the center of a 5 cm plastic petri dish with a No. 1 coverslip forming the chamber bottom, allowing access of the microscope lenses as described previously [21,31]. Pieces of retina (chips) were minced into small pieces and placed into the imaging chamber with Ringer's solution (in mM: NaCl 111, KCl 2, CaCl$_2$ 1, MgCl$_2$ 1, MgSO$_4$ 0.5, NaH$_2$PO$_4$0.5, HEPES 3, glucose 10, EDTA 0.01). The chamber was covered by a No. 1 coverslip and placed onto the microscope stage for imaging. All imaging was performed at 20°C.Imaging was performed using LSM-510 (Zeiss) equipped with an Argon laser generating 488 nm laser line and a HeNe 543 laser. To reduce contamination signals across the two fluorescent channels, 500–535 nm and 655–710 nm bandpass filters were used to filter fluorescence excited by Ar and HeNe lasers respectively. The scanning objective for both channels was a Plan-Neofluor 63x/1.4 N.A. oil lens (Zeiss). The scan settings were: resolution, 0.04×0.04 µm in the xy plane, pixel time 3 µs, pinhole diameter 1.4 Airy units, amplifier offset 0.1 and amplifier gain 1. At least 5 z scans from the central area of each photoreceptor using a 0.5 µm interval were obtained. The LSM-510 software was set to correct for the z-plane while scanning dual channels. Intensities of images were measured using AxioVision™ software version 4.7 (Zeiss), corrected for the background intensity, and averaged values of at least 3 z sections were used for fluorescence intensity measurements. Intensities were collected in arbitrary units (0–255) and then were constrained normalized to 0-100.The magnitude of the density variation in OS bands was estimated as follows. Images were selected, and four to five central z planes were deconvolved using theoretical mode in Axiovision 4.7 deconvolution module (Zeiss). For a given cell, three consecutive bands with similar amplitudes were selected and used to determine the average ratio of maximum to minimum density for those bands. Ratios were determined for more than 90 bands from 10 cells, and then averaged to calculate the fluorescence intensity differences between the maximum and minimum levels in a band.

FRAP analysis by confocal microscopy

The recording chamber was initially searched for retinal chips with the rod axis parallel to the coverslip. A single rod photoreceptor was then centered in the imaging window and 8 bit images in gray scale were scanned from a 13.2×13.2 µm in xy plane (256×256 pixel). Photobleaching was performed in a 1.5 µm-wide rectangular in the central region of the rod OS for

40 ms with the 488 nm laser line. Images were acquired before and after photobleaching. All laser scanning and bleaching was at the same z axis. The images in the recovery phase were taken immediately after photobleaching and then every 3–5 s in time series of at least 25 recovery scans.

Immunohistochemistry and electron microscopy

For immunohistochemistry, eyes were fixed in 4% paraformaldehyde in PBS at 4°C and frozen in OCT (TissueTek Inc.). Frozen blocks were sectioned into 12–20 µm slices with a Microm560 cryostat (Richard Allen Scientific). The tissue on slide was permeabilized with 0.5% Triton X-100 in 1X PBS at room temperature, blocked with5% goat serum, 0.1% Triton X-100 in 1X PBS at 4°Cand immunostained with the primary antibody in blocking buffer for 96 hrs in a humidified chamber at room temperature, washed and then incubated with appropriate conjugated secondary antibody, and finally, washed and mounted. Primary antibodies used were mouse anti-rhodopsin K16-155C (C-terminus, 1:100) mouse anti-rhodopsin 4D2 (N-terminus, 1:20,000), mouse anti-rhodopsin 1D4 (C-terminus, 1:1000), and anti-mouse secondary antibody (Jackson ImmunoResearch) conjugated to Cy3 (1:750). Slides were examined with either a C2 series (Nikon) or LSM-510 confocal microscope (Zeiss). For preparation of EM sections, eyes were fixed in 2.5% glutaraldehyde and 1% OsO$_4$ in phosphate buffer and embedded in epoxy resin as described previously [52].

Microspectroscopy

Rods were obtained from animals housed in a 168 h (84L–84D) light cycle at 20–22°C. Rhodopsin density was determined by measuring the density before and after bleaching the rod using a microdensitometer with submicrometer spatial resolution as previously described [30].

Quantitative real-time PCR

To study the variation of light-sensitive gene expression, male *Xenopus* (8–10 cm long) were kept at 25°C with a 24 h (10D/14L) light cycle for at least 2 weeks prior to any experiment or in constant conditions for 24 h. Animals were sacrificed at 4 h intervals, three pairs of retina were separated in the dark from the RP, pooled and stored at −80°C. Total RNA was isolated from one (examination of endogenous rhodopsin in extended 24 h light or dark period) or three pairs (examination of genes every 4 h) of retinas using RNeasy kit (Qiagen) at each time point. RNA integrity was verified using a Bioanalyzer (Agilent 2100). cDNA was synthesized using QuantiTect Reverse Transcription Kit (Qiagen). Primers were designed using Primer3 (http://frodo.wi.mit.edu/): β-actin, 5′GCACCCCTGAATCCTAAAGC3′ and 5′TTGGCACAGTGTGGGTTACA3′; EF1-α, 5′GATTGATCGCCGTTCTGGTA3′ and 5′GCTTTCCTGGGATCATGTCA3′; rhodopsin, 5′ATGACCGTCCCAGCTTTCTT3′ and 5′CACCTGGCTGGAAGAGACAG3′; red cone opsin, 5′TCTTTGCCTGTTTTGCTGCT3′ and 5′TCCATCATCGACCTTTTTGC3′; nocturnin, 5′GCTGTGCCTTGTTCTTCCTG3′ and 5′TTAGATGGGTGACCGCAAAG3′). The accession numbers for the *Xenopus* genes used in this study are: L07770 (rhodopsin), BC081156 (red cone opsin), U74761 (nocturnin), AF079161 (β-actin), X02995 (18S ribosomal RNA), NM_001087442 (EF1α), X03017 (histone 4, [53]). Real-time PCR (cDNA from 5 ng reverse transcribed total RNA and 2.5 µM of each primer mix containing SYBR Green I) was performed using Roche LightCycler 480 (Roche Diagnostic). PCR amplification was performed following denaturation (95°C for 10 min) for 45 cycles of 95°C for 15 s, 60°C for 15 s, 72°C for

30 s, 80°C for 5 s. Finally, a product melting curve was obtained by heating the samples from 60–95°C (heating rate of 0.1°C per second) with continuous fluorescence measurement, followed by a cooling step to 25°C. RT PCR data analysis was performed as described [54]. The crossing point (CP) was determined using LightCycler software 3.3 (Roche Diagnostics) for each primer-pair PCR reaction. To determine the relative expression of a gene at different times of day, the difference between the CP value for a particular time (T) was subtracted from a standard time point, chosen arbitrarily as 8 AM. This difference was termed $\Delta CP = CP^T - CP^{8\ AM}$. To normalize for potential variation in tissue harvesting or RNA processing, comparisons of ΔCP from target genes were referenced to β-actin. Similar results were obtained when target genes were referenced to EF1α (not shown). The PCR amplification efficiencies (E_{target} and $E_{\beta actin}$), were determined from the slope of the real time PCR curves based on the relationship, $E = 10^{[-1/slope]}$ [55]. Thus, the relative RNA abundance at time T (R^T) for a target RNA compared to β-actin RNA was estimated using the following formula: Relative Expression = $(E\Delta^{CP}_{target})^{(t}8AM^{-t)})/(^{E\Delta CP}_{actin})^{(t}8AM^{-t)})$. For each time point, retina from three animals were collected, pooled and analyzed by real-time PCR. In addition, each time point was repeated with a separate group of animals sacrificed on different days. R^T from the two groups at each time was averaged (with standard deviation). One-way analysis of variance (ANOVA) between the different time points was used to determine significance of differences across the light cycle. Statistics on relative expression of endogenous rhodopsin, nocturnin, and red cone opsin after normalization to β-actin are: Rho F(6, 7) = 0.54, p = 0.7630; nocturnin F(6, 7) = 12.41, p = 0.0020; red cone opsin F(6, 7) = 6.34, p = 0.0141.

Supporting Information

Figure S1 Axial variation in cells expressing either soluble eGFP (A) or Rho-eGFP (B) transgenes. (A) eGFP fluorescence is found in both IS and OS and does not exhibit periodic axial variation. The intensity profile of the fluorescence along the OS axis (white line) is shown below. (B) Some rods expressing the Rho-eGFP transgene exhibit wide variation in fluorescence intensity along with a superimposed periodic axial variation with a spatial period of ~1.5 μm. The intensity profile of the fluorescence along the OS axis (white line) is shown below. Animals were housed in a 24 h (12D:12L) cycle. Compare the average fluorescence intensity in the region between 0–22 μm to that in the region between 25–45 μm. This arises from mosaic transgene expression.

Figure S2 Axial variation in cells expressing Rho-eGFP. A. Selected cells expressing Rho-eGFP under control of a Xenopus opsin promoter (~0.6 kb) from four different transgenic lines (F₁) housed in the same 24 h (12D:12L) cycle exhibit in-phase axial banding but asynchronous slower variation. Scale bar, 10 μm. (B–E) To characterize the asynchronous temporal variation, the fluorescence intensity profile (red circles) from the 20–30 μm proximal to the OS base was fit to a sinusoidal function (black lines) using SigmaPlot12 (Jandel Scientific). (F–G) A box plot (F) and histogram show the spatial period (μm) of the best fit sinusoidal function for each of the cells in (A). There was a range of frequencies across the cells, with a median of 27 μm corresponding to temporal frequency of ~18 days.

Figure S3 Axial variation in OS fluorescence in rods expressing of Rho-eGFP and Rho-mCherry under the control of rod-specific promoters. Top Right, a schematic diagram of plasmid constructs (#1–#8, see Methods for details) used for transgenesis in this study is shown. Representative images of OS fluorescence from rods harboring transgenes that expresses Rho-eGFP and/or Rho-mCherry under control of a various promoters are shown. (A) A dual transgene that expresses Rho-eGFP and Rho-mCherry under control of a Xenopus opsin promoter (~0.6 kb) from the same locus. The axial variation is precisely in phase while the asynchronous variation is less tightly coupled. (B) A dual transgene that expresses Rho-eGFP under control of a Xenopus opsin short promoter (~0.6 kb) and Rho-mCherry under the control of a Xenopus opsin long promoter (~5.5 kb). (C) A transgene that expresses Rho-eGFP under control of a Xenopus arrestin promoter. (D) A transgene that expresses Rho-eGFP under control of a Xenopus rod transducin promoter. (E) A dual transgene that expresses Rho-eGFP under the control of a Xenopus arrestin promoter and Rho-mCherry under control of a short Xenopus opsin promoter. Scale bar, 5 μm.

Acknowledgments

We would like to thank Drs. A. Ahmadi and M. Haeri for their help in the preparation of Fourier transform data and Dr. S. Reks for her assistance in immunohistochemistry.

Author Contributions

Conceived and designed the experiments: MH ENP PDC BEK. Performed the experiments: MH PDC. Analyzed the data: MH PDC ES BEK. Contributed reagents/materials/analysis tools: MH PDC BEK. Wrote the paper: MH BEK.

References

1. Steinberg RH, Fisher SK, Anderson DH (1980) Disc morphogenesis in vertebrate photoreceptors. J Comp Neurol 190: 501–508.

2. Papermaster DS, Dreyer WJ (1974) Rhodopsin content in the outer segment membranes of bovine and frog retinal rods. Biochemistry 13: 2438–2444.

3. Insinna C, Besharse JC (2008) Intraflagellar transport and the sensory outer segment of vertebrate photoreceptors. Dev Dyn 237: 1982–1992.

4. Pearring JN, Salinas RY, Baker SA, Arshavsky VY (2013) Protein sorting, targeting and trafficking in photoreceptor cells. Prog Retin Eye Res.

5. Roepman R, Wolfrum U (2007) Protein networks and complexes in photoreceptor cilia. Subcell Biochem 43: 209–235.

6. Young RW (1976) Visual cells and the concept of renewal. Invest Ophthalmol Vis Sci 15: 700–725.

7. Besharse JC (1980) The effects of constant light on visual processes. In: Williams TP, editor. Light and Membrane Biogenesis. New York: Plenum. pp. 409–431.

8. Hollyfield JG, Rayborn ME (1982) Membrane assembly in photoreceptor outer segments: progressive increase in 'open' basal discs with increased temperature. Exp Eye Res 34: 115–119.

9. Hollyfield JG, Rayborn ME, Verner GE, Maude MB, Anderson RE (1982) Membrane addition to rod photoreceptor outer segments: light stimulates membrane assembly in the absence of increased membrane biosynthesis. Invest Ophthalmol Vis Sci 22: 417–427.

10. Hollyfield JG, Anderson RE (1982) Retinal protein synthesis in relationship to environmental lighting. Invest Ophthalmol Vis Sci 23: 631–639.

11. Kaplan MW (1981) Concurrent birefringence and forward light-scattering measurements of flash-bleached rod outer segments. J Opt Soc Am 71: 1467–1471.

12. Kaplan MW (1982) Birefringence and birefringence gradients in rod outer segments. Methods Enzymol 81: 655–660.

13. Kaplan MW, Deffebach ME (1978) Birefringence measurements of structural inhomogeneities in Rana pipiens rod outer segments. Biophys J 23: 59–70.

14. Kaplan MW, Robinson DA, Larsen LD (1982) Rod outer segment birefringence bands record daily disc membrane synthesis. Vision Res 22: 1119–1121.

15. Andrews LD (1985) Structural periodicities observed in mammalian rod outer segments with Nomarski optics. Invest Ophthalmol Vis Sci 26: 778–782.

16. Kaplan MW, Iwata RT (1989) Temperature-dependent birefringence patterns in Xenopus rod outer segments. Exp Eye Res 49: 1045–1051.

17. Kaplan MW (1981) Light cycle—dependent axial variations in frog rod outer segment structure. Invest Ophthalmol Vis Sci 21: 395–402.

18. Corless JM, Kaplan MW (1979) Structural interpretation of the birefringence gradient in retinal rod outer segments. Biophys J 26: 543–556.

19. Roberts NW (2006) The optics of vertebrate photoreceptors: anisotropy and form birefringence. Vision Res 46: 3259–3266.

20. Kretzer F, Cohen H (1982) Imaging of outer segment periodicities in unstained cryoultramicrotomy sections of the frog retina. J Microsc 128: 287–300.

21. Peet JA, Bragin A, Calvert PD, Nikonov SS, Mani S, et al. (2004) Quantification of the cytoplasmic spaces of living cells with EGFP reveals arrestin-EGFP to be in disequilibrium in dark adapted rod photoreceptors. J Cell Sci 117: 3049–3059.

22. Knox BE, Schlueter C, Sanger BM, Green CB, Besharse JC (1998) Transgene expression in Xenopus rods. FEBS Lett 423: 117–121.

23. Peterson JJ, Tam BM, Moritz OL, Shelamer CL, Dugger DR, et al. (2003) Arrestin migrates in photoreceptors in response to light: a study of arrestin localization using an arrestin-GFP fusion protein in transgenic frogs. Exp Eye Res 76: 553–563.

24. Moritz OL, Tam BM, Papermaster DS, Nakayama T (2001) A functional rhodopsin-green fluorescent protein fusion protein localizes correctly in transgenic Xenopus laevis retinal rods and is expressed in a time-dependent pattern. J Biol Chem 276: 28242–28251.

25. Tam BM, Moritz OL, Hurd LB, Papermaster DS (2000) Identification of an outer segment targeting signal in the COOH terminus of rhodopsin using transgenic Xenopus laevis. J Cell Biol 151: 1369–1380.

26. Baker SA, Haeri M, Yoo P, Gospe SM, 3rd, Skiba NP, et al. (2008) The outer segment serves as a default destination for the trafficking of membrane proteins in photoreceptors. J Cell Biol 183: 485–498.

27. Luo W, Marsh-Armstrong N, Rattner A, Nathans J (2004) An outer segment localization signal at the C terminus of the photoreceptor-specific retinol dehydrogenase. J Neurosci 24: 2623–2632.

28. Tam BM, Moritz OL, Papermaster DS (2004) The C terminus of peripherin/rds participates in rod outer segment targeting and alignment of disk incisures. Mol Biol Cell 15: 2027–2037.

29. Calvert PD, Schiesser WE, Pugh EN, Jr. (2010) Diffusion of a soluble protein, photoactivatable GFP, through a sensory cilium. J Gen Physiol 135: 173–196.

30. Najafi M, Haeri M, Knox BE, Schiesser WE, Calvert PD (2012) Impact of signaling microcompartment geometry on GPCR dynamics in live retinal photoreceptors. J Gen Physiol 140: 249–266.

31. Haeri M, Knox BE (2012) Rhodopsin mutant P23H destabilizes rod photoreceptor disk membranes. PLoS One 7: e30101.

32. Moritz OL, Tam BM, Hurd LL, Peranen J, Deretic D, et al. (2001) Mutant rab8 Impairs docking and fusion of rhodopsin-bearing post-Golgi membranes and causes cell death of transgenic Xenopus rods. Mol Biol Cell 12: 2341–2351.

33. Jin S, McKee TD, Oprian DD (2003) An improved rhodopsin/EGFP fusion protein for use in the generation of transgenic Xenopus laevis. FEBS Lett 542: 142–146.

34. Besharse JC, Hollyfield JG, Rayborn ME (1977) Photoreceptor outer segments: accelerated membrane renewal in rods after exposure to light. Science 196: 536–538.

35. MacKenzie D, Arendt A, Hargrave P, McDowell JH, Molday RS (1984) Localization of binding sites for carboxyl terminal specific anti-rhodopsin monoclonal antibodies using synthetic peptides. Biochemistry 23: 6544–6549.

36. Adamus G, Zam ZS, Arendt A, Palczewski K, McDowell JH, et al. (1991) Anti-rhodopsin monoclonal antibodies of defined specificity: characterization and application. Vision Res 31: 17–31.

37. Pierce ME, Sheshberadaran H, Zhang Z, Fox LE, Applebury ML, et al. (1993) Circadian regulation of iodopsin gene expression in embryonic photoreceptors in retinal cell culture. Neuron 10: 579–584.

38. Green CB, Besharse JC (1996) Use of a high stringency differential display screen for identification of retinal mRNAs that are regulated by a circadian clock. Brain Res Mol Brain Res 37: 157–165.

39. Gurevich VV, Hanson SM, Song X, Vishnivetskiy SA, Gurevich EV (2011) The functional cycle of visual arrestins in photoreceptor cells. Prog Retin Eye Res 30: 405–430.

40. Yan C, Boyd DD (2006) Histone H3 acetylation and H3 K4 methylation define distinct chromatin regions permissive for transgene expression. Mol Cell Biol 26: 6357–6371.

41. Mehta AK, Majumdar SS, Alam P, Gulati N, Brahmachari V (2009) Epigenetic regulation of cytomegalovirus major immediate-early promoter activity in transgenic mice. Gene 428: 20–24.

42. Schnetkamp PP, Szerencsei RT (1993) Intracellular Ca2+ sequestration and release in intact bovine retinal rod outer segments. Role in inactivation of Na-Ca+K exchange. J Biol Chem 268: 12449–12457.

43. Besharse JC, Hollyfield JG, Rayborn ME (1977) Turnover of rod photoreceptor outer segments. II. Membrane addition and loss in relationship to light. J Cell Biol 75: 507–527.

44. Palczewski K, Kumasaka T, Hori T, Behnke CA, Motoshima H, et al. (2000) Crystal structure of rhodopsin: A G protein-coupled receptor. Science 289: 739–745.

45. Tsien RY (1998) The green fluorescent protein. Annu Rev Biochem 67: 509–544.

46. Sung CH, Chuang JZ (2010) The cell biology of vision. J Cell Biol 190: 953–963.

47. Knepp AM, Periole X, Marrink SJ, Sakmar TP, Huber T (2012) Rhodopsin forms a dimer with cytoplasmic helix 8 contacts in native membranes. Biochemistry 51: 1819–1821.

48. Boesze-Battaglia K, Fliesler SJ, Albert AD (1990) Relationship of cholesterol content to spatial distribution and age of disc membranes in retinal rod outer segments. J Biol Chem 265: 18867–18870.

49. Haeri M, Knox BE, Ahmadi A (2013) Modeling the flexural rigidity of rod photoreceptors. Biophys J 104: 300–312.

50. Mani SS, Besharse JC, Knox BE (1999) Immediate upstream sequence of arrestin directs rod-specific expression in Xenopus. J Biol Chem 274: 15590–15597.

51. Haeri M, Knox BE (2012) Generation of transgenic Xenopus using restriction enzyme-mediated integration. Methods Mol Biol 884: 17–39.

52. Pazour GJ, Baker SA, Deane JA, Cole DG, Dickert BL, et al. (2002) The intraflagellar transport protein, IFT88, is essential for vertebrate photoreceptor assembly and maintenance. J Cell Biol 157: 103–113.

53. Hollemann T, Bellefroid E, Pieler T (1998) The Xenopus homologue of the Drosophila gene tailless has a function in early eye development. Development 125: 2425–2432.

54. Pfaffl MW (2001) A new mathematical model for relative quantification in real-time RT-PCR. Nucleic Acids Res 29: e45.

55. Bustin SA (2000) Absolute quantification of mRNA using real-time reverse transcription polymerase chain reaction assays. J Mol Endocrinol 25: 169–193.

10

The POZ-ZF Transcription Factor Kaiso (ZBTB33) Induces Inflammation and Progenitor Cell Differentiation in the Murine Intestine

Roopali Chaudhary[1]ↄ, Christina C. Pierre[1]ↄ, Kyster Nanan[2], Daria Wojtal[1], Simona Morone[3], Christopher Pinelli[4], Geoffrey A. Wood[4], Sylvie Robine[5], Juliet M. Daniel[1]*

1 Department of Biology, McMaster University, Hamilton, Ontario, Canada, 2 Department of Pathology & Molecular Medicine, Queen's University, Kingston, Ontario, Canada, 3 Department of Medical Sciences, University of Torino, Torino, Italy, 4 Department of Pathobiology, University of Guelph, Guelph, Ontario, Canada, 5 Department of Morphogenesis and Intracellular Signalling, Institut Curie-CNRS, Paris, France

Abstract

Since its discovery, several studies have implicated the POZ-ZF protein Kaiso in both developmental and tumorigenic processes. However, most of the information regarding Kaiso's function to date has been gleaned from studies in *Xenopus laevis* embryos and mammalian cultured cells. To examine Kaiso's role in a relevant, mammalian organ-specific context, we generated and characterized a Kaiso transgenic mouse expressing a murine Kaiso transgene under the control of the intestine-specific *villin* promoter. Kaiso transgenic mice were viable and fertile but pathological examination of the small intestine revealed distinct morphological changes. Kaiso transgenics ($Kaiso^{Tg/+}$) exhibited a crypt expansion phenotype that was accompanied by increased differentiation of epithelial progenitor cells into secretory cell lineages; this was evidenced by increased cell populations expressing Goblet, Paneth and enteroendocrine markers. Paradoxically however, enhanced differentiation in $Kaiso^{Tg/+}$ was accompanied by reduced proliferation, a phenotype reminiscent of Notch inhibition. Indeed, expression of the Notch signalling target HES-1 was decreased in $Kaiso^{Tg/+}$ animals. Finally, our Kaiso transgenics exhibited several hallmarks of inflammation, including increased neutrophil infiltration and activation, villi fusion and crypt hyperplasia. Interestingly, the Kaiso binding partner and emerging anti-inflammatory mediator p120^{ctn} is recruited to the nucleus in $Kaiso^{Tg/+}$ mice intestinal cells suggesting that Kaiso may elicit inflammation by antagonizing p120^{ctn} function.

Editor: Pierre-Antoine Defossez, Université Paris-Diderot, France

Funding: This work was supported by CIHR grant (MOP-84320) to Juliet Daniel. R. Chaudhary was the recipient of an Ontario Graduate Scholarship and C. Pierre was the recipient of a McMaster Prestige Award. The funders had no role in study design, data collection and analysis, decision to publish, or preparation of the manuscript.

Competing Interests: The authors have declared that no competing interests exist.

* E-mail: danielj@mcmaster.ca

ↄ These authors contributed equally to this work.

Introduction

Since its discovery as a binding partner for the Src kinase substrate and cell adhesion protein p120^{ctn}, mounting evidence suggests that the POZ-ZF transcription factor Kaiso functions in vertebrate development and tumorigenesis [1,2,3,4,5,6,7,8]. To date however, Kaiso's role in these processes in mammalian systems remains unclear, and much controversy surrounds several aspects of Kaiso's function; this includes the mechanism by which it binds DNA [9,10,11,12,13,14,15,16,17] and its function in regulating the canonical Wnt signalling pathway that plays a key role in vertebrate development and tumorigenesis [8,11,14,18,19].

One study investigated the effect of Kaiso depletion on murine development and found that Kaiso null mice exhibited no overt developmental phenotypes [8]. This unexpected lack of a developmental phenotype may be attributed to the existence of two Kaiso-like proteins in mammals, ZBTB4 and ZBTB38, that may function redundantly with Kaiso [16,20], and highlights what may be an important consideration in deciphering Kaiso's role in mammalian systems. Surprisingly however, Kaiso depletion

extended the lifespan, and delayed tumour onset in the $Apc^{Min/+}$ model of intestinal tumorigenesis [8]. This observation implicated Kaiso as an oncogene and is consistent with the report that Kaiso binds and represses methylated tumour suppressor and DNA repair genes in colon cancer cells [7]. Given that constitutive Wnt signalling resulting from mutation of *APC* functions as the first "hit" in $Apc^{Min/+}$-mediated tumorigenesis, the $Kaiso$-null/$Apc^{Min/+}$ phenotype suggests that Kaiso is a positive regulator of Wnt signalling. This result is surprising, since Kaiso has been implicated as a negative regulator of canonical Wnt signalling in *Xenopus laevis* embryos and in mammalian cultured cells [19,21,22,23]. However it remains possible that Kaiso may potentiate intestinal tumorigenesis in the $Apc^{Min/+}$ model via a non-Wnt related mechanism.

Consistent with this possibility, studies to elucidate the role of the Kaiso binding partner p120^{ctn} in the intestine hinted at a non-cell autonomous mechanism for p120^{ctn}-mediated tumorigenesis [24,25]. Smalley Freed *et al.* found that mice with limited ablation of p120^{ctn} developed adenomas in addition to an intestinal barrier defect and chronic inflammation [25]. Surprisingly, conditional

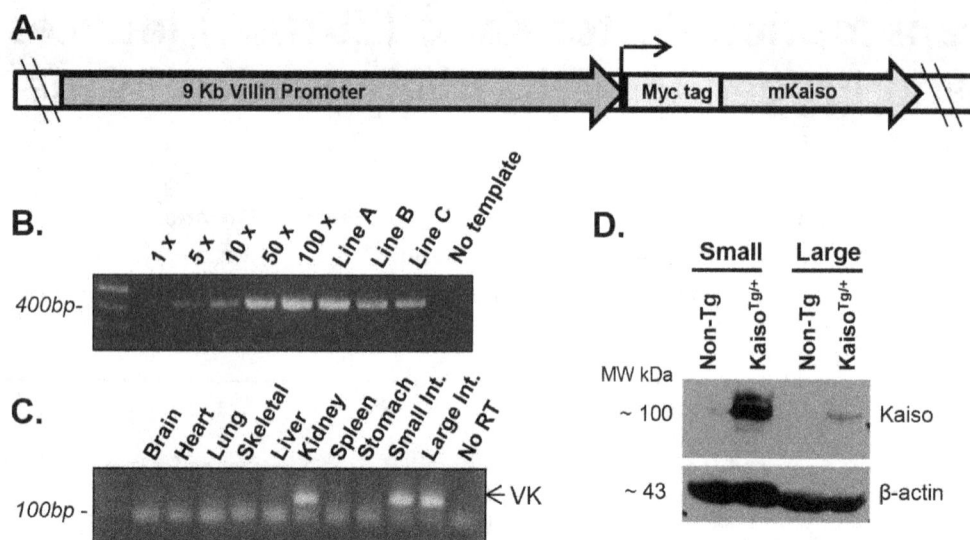

Figure 1. Generation of transgenic mouse lines ectopically expressing *villin*-Kaiso. (**A**) Myc-tagged murine *Kaiso* cDNA was cloned downstream of the 9 kb *villin* promoter sequence. (**B**) The transgene copy number in each transgenic line was evaluated via PCR. Line A transgenic animals have the greatest copy number. (**C**) RT-PCR confirmed expression of the Kaiso transgene in *villin*-expressing tissues of transgenic mice, *i.e.* the small intestine, large intestine, and kidneys. (**D**) Immunoblot analysis shows increased Kaiso expression in both small and large intestines in Kaiso transgenic (*Kaiso*[Tg/+]) Line A mice compared to non-transgenic (Non-Tg) siblings.

depletion of p120[ctn] in the murine intestine resulted in severe **i**nflammatory **b**owel **d**isease (IBD) and lethality [24,25]. Thus it was postulated that the adenomas arising in mice with limited p120[ctn] ablation was a result of chronic inflammation, which is considered a risk factor for colorectal cancer [26].

Since studies have implicated Kaiso in intestinal cancer development and progression [7,8], we generated an intestinal-specific Kaiso overexpression mouse model to clarify Kaiso's role in the context of murine intestinal epithelium development. We generated multiple Kaiso transgenic (*Kaiso*[Tg/+]) founder lines, each with varying copy numbers of the transgene. *Kaiso*[Tg/+] mice were viable and fertile with no deleterious developmental phenotypes. However we noticed several phenotypes in the intestines of *Kaiso*[Tg/+] mice that were reminiscent of Notch inhibition. *Kaiso*[Tg/+] mice exhibited increased differentiation of intestinal epithelial progenitor cells into secretory cell lineages (Paneth, Goblet, enteroendocrine) accompanied by reduced proliferation, a phenotype consistent with Notch inhibition [27,28,29]. Indeed, expression of the Notch signalling target HES-1 was also reduced in *Kaiso*[Tg/+] mice. Interestingly, p120[ctn] localized mainly to the nucleus in the small intestine in *Kaiso*[Tg/+] mice, and this was accompanied by increased infiltration of inflammatory cells and myeloperoxidase activity (a surrogate marker for inflammation) suggesting that *Kaiso*[Tg/+] mice are more susceptible to inflammation. Together these data suggest that Kaiso functions in a pro-inflammatory role in the murine intestine by antagonizing the anti-inflammatory functions of p120[ctn].

Materials and Methods

Ethics Statement

All mouse work was conducted according to the guidelines of the McMaster University Animal Research Ethics Board (AREB). Protocols for mouse husbandry, breeding, genotyping and euthanasia were approved by AREB under Animal Utilization Protocol (AUP) 10-05-32. Euthanasia was achieved via CO_2 asphyxiation followed by cervical dislocation.

Generation of Villin-Kaiso Transgenic Mice

Kaiso transgenic mice were created at the London Regional Transgenic Facility, University of Western Ontario. Myc-tagged murine *Kaiso* (*mKaiso-MT*) was cloned downstream of the murine 9 Kb intestinal-specific *villin* promoter fragment in the pBluescript II vector provided by Dr. Sylvie Robine (Institut Curie, Paris, France) [30]. The *villin-mKaiso-MT* fragment was excised from the plasmid by restriction enzyme digest with *SalI*. The isolated fragment was microinjected into 1-cell C57BL6/CBA hybrid mouse embryos *in vitro*, which were then implanted into pseudo-pregnant foster mothers to produce transgenic founders. Transgenic pups were identified by **p**olymerase **c**hain **r**eaction (PCR) analysis of DNA from tail biopsies using primer pairs corresponding to sequences in the Myc tag and murine Kaiso (forward 5'-*ATC ATC AAA GCC GGG TGG GCA-3'* and reverse 5'-*TTT TCT ACT CTC CAT TTC ATT CAA GTC CTC-3'*). The transgenic lines were backcrossed with C57BL/6N mice (Taconic) for a minimum of 8 generations to obtain stable transgenic offspring, which initially produced three transgenic founder lines, followed by an additional four transgenic founder lines. All transgenic offspring were genotyped by PCR using DNA obtained from ear snips upon weaning. Mice were fed a standard mouse chow diet and breeders were housed in the disease-free barrier facility, while post-genotyping pups were housed in a specific pathogen free (SPF) room with 12 h/12 h light/dark cycle in accordance with McMaster Central Animal Facility's (CAF) Standard Operating Procedures (SOPs).

Transgene Copy Number

Copy number standards were prepared by spiking wild-type tail DNA with specified amounts of purified transgenic DNA. PCR was performed using standard DNA and transgenic DNA from each founder line using the primers described above. The intensity of the band amplified in each of the transgenic animals was compared to that of the standards to estimate transgene copy number.

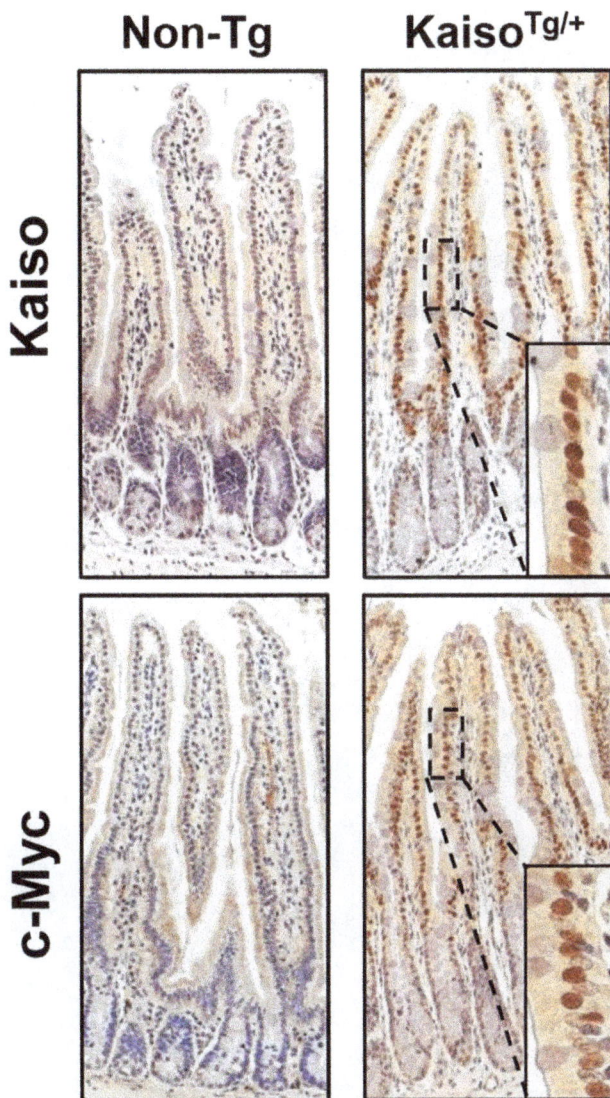

Non-Tg   Kaiso^(Tg/+)

Kaiso

c-Myc

Figure 2. Subcellular localization and expression of ectopic Kaiso in Line A *Kaiso^(Tg/+)* **small intestines.** *Kaiso^(Tg/+)* mice display strong nuclear Kaiso in the villi and crypt cells, compared to non-transgenic mice (Non-Tg), which mainly display weak Kaiso staining in the cytoplasm. Additionally, *Kaiso^(Tg/+)* mice display strong nuclear c-Myc staining corresponding to ectopic myc-tagged Kaiso expression, while Non-Tg mice display cytoplasmic c-Myc expression.

Mouse Tissue Harvest

Mice were sacrificed via CO_2 asphyxiation according to the McMaster CAF SOPs. Small and large intestines were immediately removed from the sacrificed animals and flushed with cold phosphate-buffered saline (PBS) on ice. Tissues were either flash frozen in liquid nitrogen for long term storage or rolled into "Swiss rolls" for fixation in 10% neutral-buffered formalin for 48 hours, followed by 70% ethanol dehydration at room temperature. The small intestine was divided into four equal sections for formalin fixation. Fixed tissues were sent to McMaster Core Histology Research Services for paraffin-embedding and sectioning at 5 μm within one week of tissue harvest, and placed onto glass slides for immunohistochemical (IHC) analysis as outlined below.

Morphological Analysis

Crypt depth and villi length were evaluated using haematoxylin and eosin (H&E) stained slides from both transgenic lines (n = 3 mice per genotype/founder line). Paneth cells were counted as eosin-filled cells at the base of the crypts. Periodic Acid-Schiff (PAS) stain for Goblet cells was performed by the McMaster Core Histology Research Services according to standard protocols. All images were collected using the Aperio ScanScope system, and ImageScope software was used for all measurements. For each small intestine, 800 open crypts and 80 complete villi were assessed per mouse by two independent blind observers. Student's T-test was used to compare any observed differences for statistical significance using GraphPad Prism.

Immunohistochemistry

Tissue slides were incubated in xylenes at room temperature for 10 min (2 washes) to remove paraffin, followed by rehydration in an ethanol gradient. Tissue was permeabilized with Tris-buffered saline with 0.05% Tween-20 (TBS-T), and antigen retrieval was accomplished by boiling samples in 10 mM sodium citrate buffer (pH 6.0). Endogenous peroxidase activity was quenched with 3% hydrogen peroxide in TBS. Slides were incubated in 5% normal goat serum (NGS), 10% bovine serum albumin (BSA) in TBS-T with avidin blocking solution (Vector Laboratories) for 1 hour at room temperature. For Lysozyme staining (Pierce), antigen retrieval was performed by treating tissues with 200 μg/mL of Proteinase K (Roche) solution in 50 mM Tris, pH 7.4 for 5 minutes, and blocked in 10% NDS in PBS with avidin blocking solution for 1 hour at room temperature. The slides were then incubated with biotin blocking solution (Vector Laboratories) and primary antibodies: rabbit anti-Kaiso polyclonal (gift from Dr. Albert Reynolds) at 1:1000 dilution, and mouse anti-c-Myc (Santa Cruz) at 1:60, rabbit anti-Lysozyme (Peirce) 1:50 at 4°C overnight. For rat anti-Ki67 (DAKO at 1:20 dilution), mouse anti-Synaptophysin (DAKO at 1:20 dilution), rabbit anti-HES-1 (Santa Cruz at 1:75 dilution) and rabbit anti-Cyclin D1 (US Biological at a 1:100 dilution) staining, antigen retrieval was accomplished by boiling samples at 95°C in Target Retrieval Solution Citrate pH 6.0 (DAKO). Slides were blocked in 5% normal donkey serum (NDS) in TBS-T for Ki67 and Cyclin D1, in 5% NDS, 10% BSA in PBS for HES-1, and in 10% NGS, 10% BSA in PBS for Synaptophysin. Primary antibody incubation was performed for 2 hours at room temperature. After three 2-min washes in TBS-T, and one in TBS, slides were incubated in secondary antibodies (biotinylated donkey anti-rabbit [Vector Laboratories] at a 1:1000 dilution, biotinylated goat anti-mouse [Vector Laboratories] at a 1:1000 dilution, or biotinylated rabbit anti-rat [DAKO] at a 1:200 dilution) for 2 hours at room temperature. Slides were washed as before, and incubated for 30 min in an avidin-biotin horseradish peroxidase complex, Elite ABC (Vector Laboratories). After a brief wash in TBS, Vectastain DAB substrate (Vector Laboratories) was applied for 3 minutes for satisfactory colour development. Ki67 and Cyclin D1 staining required a DAB time of 7 minutes. Tissues were counterstained with Harris hematoxylin (Sigma), differentiated in acid ethanol (0.3% HCl in 70% ethanol), blued in Scott's tap water substitute, and dehydrated in a gradient of ethanol. Slides were then dried in xylenes and mounted using PolyMount (Polysciences Inc). Images were acquired using the Aperio ScanScope, and processed using ImageScope.

Immunofluorescence

Tissue slides were incubated in xylenes at room temperature for 10 min (2 washes) to remove paraffin, followed by rehydration in

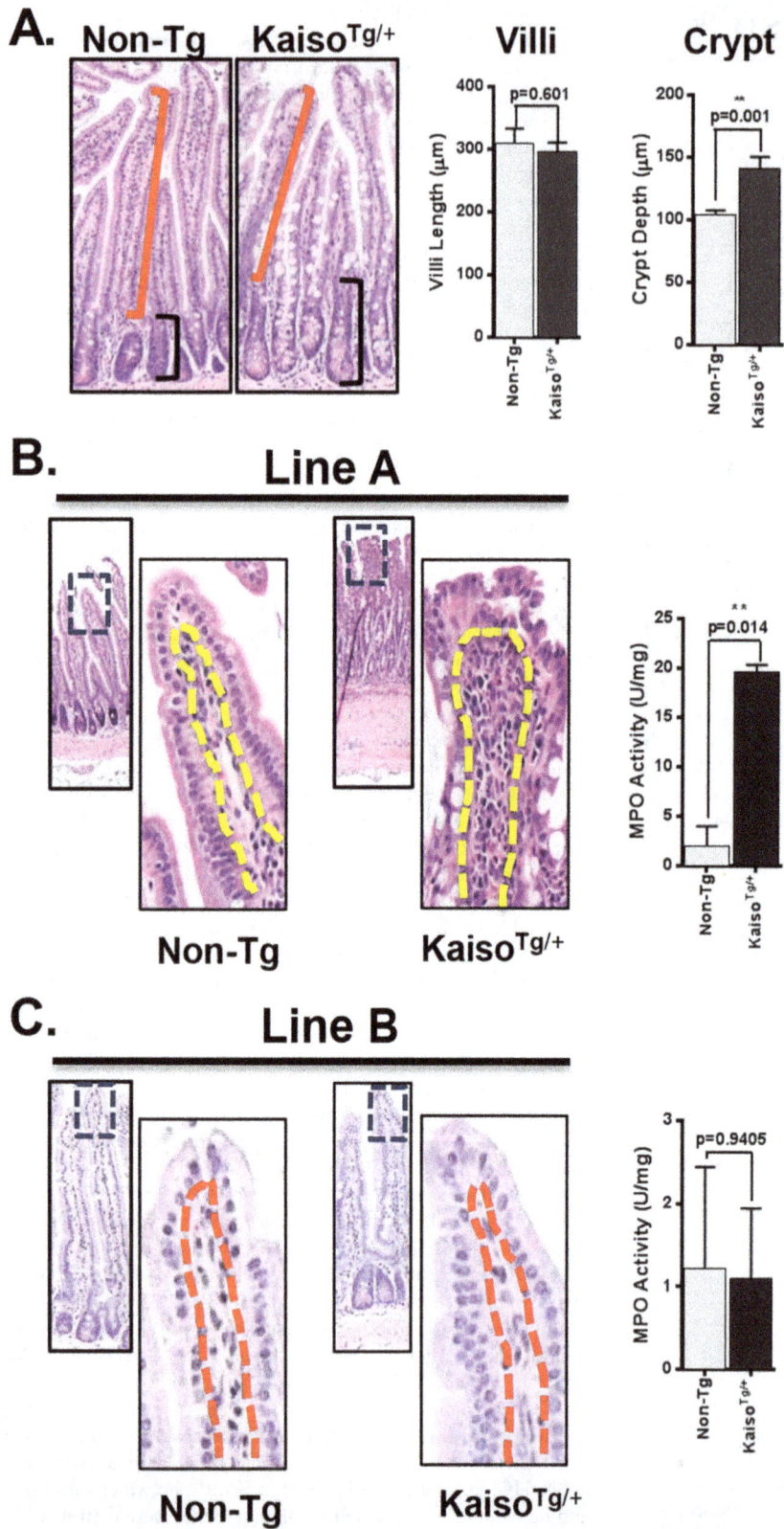

Figure 3. Kaiso transgenic mice exhibit inflammation of the intestinal mucosa. (A) Hematoxylin and eosin (H&E) stained sections were used to measure villi length (red bracket; ~80 villi/mouse) and crypt depth (black bracket; ~800 open crypts/mouse). $Kaiso^{Tg/+}$ display increased crypt depth compared to their Non-Tg siblings, p = 0.001. (B) $Kaiso^{Tg/+}$ mice exhibit increased immune cell infiltration of the lamina propria (yellow demarcated area) accompanied by increased MPO activity compared to their Non-Tg siblings, p = 0.014. (C) Line B mice do not exhibit immune cell infiltration or enhanced MPO activity compared to Non-Tg siblings. ** represents significance.

IF: p120

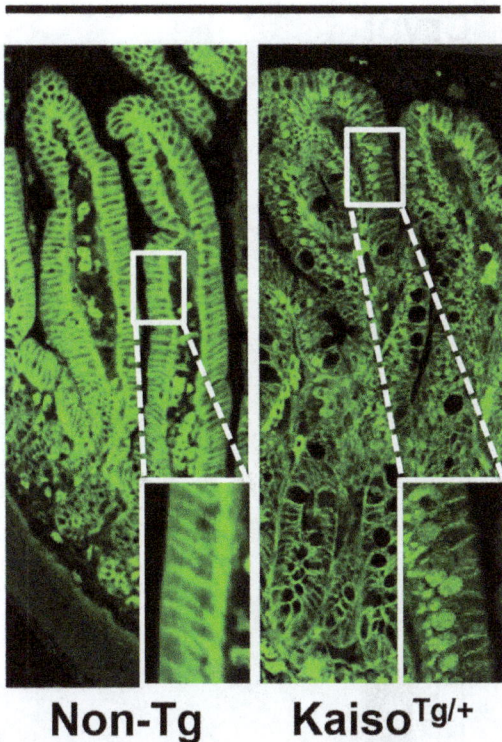

Non-Tg   Kaiso^{Tg/+}

Figure 4. *Kaiso^{Tg/+}* **mice display nuclear p120^ctn in villi of the small intestine.** Immunofluorescence staining for p120^ctn showed nuclear localization of p120^ctn in epithelial cells of villi overexpressing Kaiso (*Kaiso^{Tg/+}*), while Non-Tg mice displayed membrane localized p120^ctn.

an ethanol gradient as described above. Tissue was permeabilized with 0.05% TBS-T, and antigen retrieval was accomplished by boiling samples in 10 mM sodium citrate buffer (pH 6.0). Tissues were incubated in 5% normal goat serum, 10% bovine serum albumin in TBS-T for 1 hour at room temperature. The slides were then incubated with mouse monoclonal anti-p120 (BD Biosciences) at a dilution of 1:500 at 4°C overnight. After three 10 min washes in TBS-T, and one in TBS, slides were incubated in secondary antibodies (Alexa-488 goat anti-mouse [Invitrogen], at a dilution of 1:500) for 2 hours in the dark at room temperature. Slides were washed as before, and incubated for 30 min in the dark with TOTO-3 dye (Invitrogen; 1:1000) to stain the nuclei. Slides were mounted in ProLong Gold (Invitrogen) overnight in the dark and stored at −20°C until imaging. Images were captured and processed using a Leica Confocal Microscope.

Protein Isolation and Immunoblot

50 mg of flash frozen mouse tissue was minced with a sterile blade and homogenized in 1 mL cold RIPA buffer (1% NP-40, 50 mM Tris, 150 mM NaCl, 0.5% sodium deoxycholate, 1% SDS, 0.5% Na_3VO_4 and cOmplete ULTRA Tablet (1 tablet/5 mL buffer) [Roche]) in a chilled tissue grinder (Kontes). Harvested lysates were poured into chilled microfuge tubes followed by further homogenization using a 21 Gauge syringe. Lysates were incubated on ice for 30 minutes, followed by centrifugation at 13,000 RPM for 10 min at 4°C. The supernatants were transferred to new pre-chilled microfuge tubes. Total

protein content was quantified by Bradford assay, and 25 μg of protein was resuspended in Laemmli sample buffer, boiled for 5 minutes and subjected to electrophoresis in an SDS polyacrylamide gel. Proteins were transferred to a nitrocellulose membrane using a Hoeffer semi-dry transfer apparatus (Amersham Biosciences). To prevent non-specific antibody binding, the membranes were blocked with 3% skimmed milk/TBS (pH 7.4) and incubated at 4°C overnight with antibody diluted in 3% milk/TBS. Antibodies used were as follows: anti-Kaiso rabbit polyclonal antibody at a 1:30,000 dilution, anti-Cyclin D1 rabbit polyclonal antibody (US Biological) at a 1:5,000 dilution, anti-β-actin mouse monoclonal antibody (Sigma Aldrich) at a 1:30,000 dilution. The membranes were washed 5×5 minutes each with TBS and incubated at room temperature with HRP-conjugated donkey anti-mouse or goat anti-rabbit secondary antibody both at a dilution of 1:40,000 in 3% milk/TBS. Membranes were washed as previously described and processed with Enhanced Chemiluminescence (Amersham Biosciences) according to the manufacturer's protocol.

RNA Isolation

Mouse tissue was homogenized and total RNA purified using the RNeasy Kit (Qiagen). Briefly, ~20 mg frozen tissues were chopped finely with a clean blade, resuspended in 600 μl Qiagen Buffer RLT, and homogenized on ice in a glass tissue grinder. Lysates were further homogenized using a 21 Gauge needle and syringe on ice. Total RNA was then purified from the homogenized lysate using the RNeasy kit according to manufacturer's instructions.

RT-PCR

Reverse transcriptase PCR (RT-PCR) analysis was performed using SuperScriptII One-Step RT-PCR with Platinum Taq (Invitrogen). Briefly, 1 μg of RNA was DNaseI treated (Invitrogen) to remove any genomic DNA contamination. 100 ng total RNA was used for each reaction with primers specific to the *villin-mKaiso* transcript and transcription factor II D (TFIID) as a loading control. The primer pairs used were as follows: *villin-mKaiso*: forward 5'-CAA CTT CCT AAG ATC TCC CAG GT-3' and reverse 5'-CAA GGA GTT CAG CAG ACT GG -3'; TFIID: forward 5'-CCA CGG ACA ACT GCG TTG AT-3' and reverse 5'-GGC TCA TAG CTA CTG AAC TG-3'. The RT-PCR program included one round of cDNA synthesis at 50°C for 30 minutes, followed by denaturation at 95°C for 2 minutes. Twenty five cycles of DNA amplification was performed as follows: denaturation at 95° for 30 sec, annealing at 56°C for 30 sec, and extension at 72°C. Final extension occurred at 72°C for 10 mins.

Quantitative RT-PCR

Total RNA was purified from ~ 20 mg of small intestinal tissue as described above. 1 μg of RNA was DNaseI treated (Invitrogen) to remove any genomic DNA contamination, and cDNA synthesis was accomplished using the SuperScript III First-Strand Synthesis System (Invitrogen). RNA abundance was compared using PerfeCTa SYBR Green SuperMix Reaction Mixes (Quanta Biosciences). The standard curve method was used to calculate relative expression of HES1 and Kaiso following normalization to the housekeeping gene, GAPDH, and then normalizing to the non-Tg tissue level. Primer sequences used are as follows: *villin-mKaiso* as stated above; *mHES1*: forward 5'-AAA ATT CCT CCT CCC CGG TG-3' and reverse 5'-TTT GGT TTG TCC GGT GTC G-3'; and *mGAPDH*: *forward* 5'-ATG ACC ACA GTC CAT GCC ATC-3' and reverse 5'-CCT GCT TCA CCA CCT TCT TG-3'.

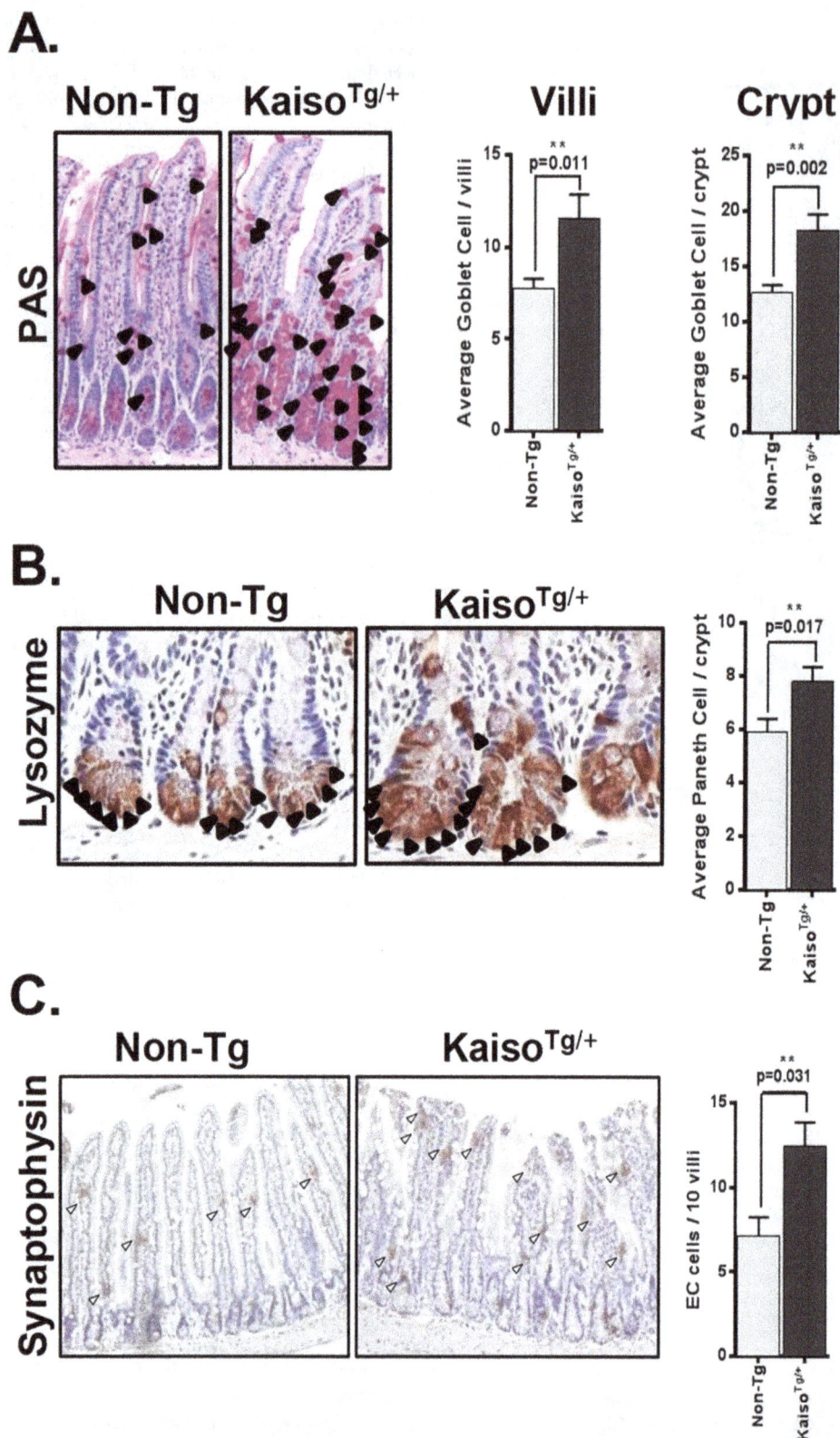

Figure 5. Secretory cell lineages are expanded in the intestines of *Kaiso^Tg/+* mice. (**A**) PAS stain for Goblet cells (black arrowheads) revealed increased numbers of Goblet cells in both the villi and crypts of *Kaiso^Tg/+* intestines, p = 0.011 & 0.002. (**B**) Lysozyme staining revealed increased Paneth cell numbers in *Kaiso^Tg/+* mice, p = 0.017. (**C**) Synaptophysin positive enteroendocrine cells (arrowheads) are increased in *Kaiso^Tg/+* mice, p = 0.031. n = 3 mice/genotype; measurements performed by two independent blind observers; T-test used for p-value. ** represents significance.

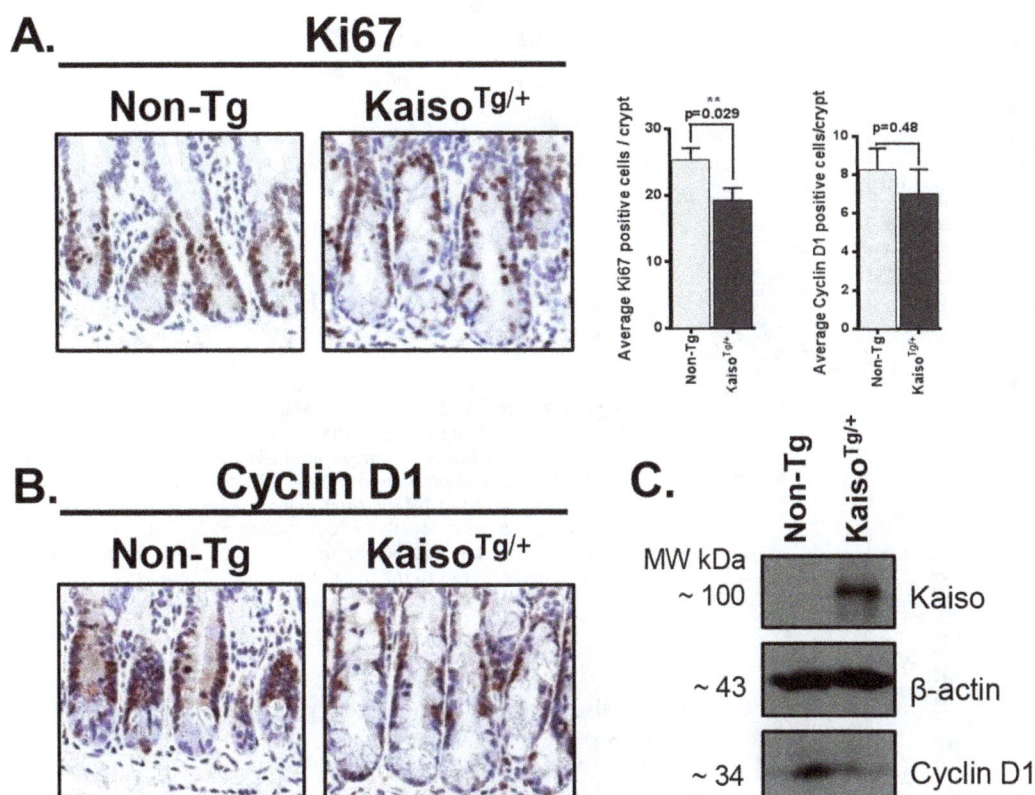

Figure 6. Cell proliferation is decreased in _Kaiso^Tg/+_ mice. Cell proliferation was evaluated by Ki67 (**A**) and Cyclin D1 (**B**) staining. Both markers exhibited reduced staining in _Kaiso^Tg/+_ mice compared to their Non-Tg siblings. Reduced CyclinD1 expression was also confirmed by immunoblot analysis of 3 different mice intestines (**C**). ** represents significance.

Student's T-test was used to determine significance using GraphPad Prism.

Myeloperoxidase (MPO) Assay

Approximately 50 mg of flash frozen ileum and colon were homogenized in 50 mg/mL of 0.5% HTAB buffer (0.5% hexadecyltrimethylammonium bromide in 50 mM phosphate buffer, pH 6.0) via sonication at 30 Hz for 4 minutes. Homogenates were cleared by centrifugation at 12,000 rpm for 15 minutes at 4°C. MPO Assay was carried by adding 200 μL of o-dianisidine dihydrochloride solution (16.8 mg/mL o-dianisidine dihydrochloride in 5 mM phosphate buffer, pH 6.0 with 50 μL of 1.2% H_2O_2) to 96-well plates. Samples (7 μL) were added to each well of the 96-well plate in triplicate, and absorbance measured at 450 nm every 30 sec (3 readings). The MPO activity was measured in units (U), where 1 U represents the amount of MPO needed to degrade 1 μmoL of H_2O_2/minute at 25°C, which gives an absorbance of 1.13×10^{-2} nm/min. MPO activity in each sample was determined as the change in absorbance $[\Delta A(t_2-t_1)]/\Delta min]/(1.13\times10^{-2})$. MPO activity/mg of tissue was calculated by dividing MPO U by 0.35 mg of tissue (7 μL homogenate ×50 mg/mL buffer). Student's T-test was used to compare any observed differences for statistical significance using GraphPad Prism.

Results

Generation of _villin-Kaiso_ Transgenic Mice

Kaiso transgenic (_Kaiso^Tg/+_) mice were generated by cloning the sequence encoding N-terminal myc-tagged murine _Kaiso_ downstream of a 9 Kb regulatory promoter region of the mouse _villin_

gene (Figure 1A). The _villin-Kaiso_ construct was injected into fertilized C57BL6/CBA embryos that were subsequently transferred to pseudopregnant foster mothers and resulted in four transgenic founder mice (Line A, B, C, D). Upon backcrossing with C57BL/6N mice, only lines A, B and C transmitted the transgene to their progeny at rates of 15%, 32% and 57%, respectively. Since pronuclear injections result in random genome integration, transgene copy number was estimated by PCR (Figure 1B). The three founders possessed varying copy numbers of the Kaiso transgene, with line A having the highest copy number and line C having the lowest copy number. Unfortunately, Line C died prior to being established and thus Lines A and B were used for further analysis. Upon founder line establishment (8 generations of backcrossing), Lines A and B transmitted the transgene at rates of 33.8% and 35.9% respectively, which is lower than the expected Mendelian rate of 50%.

To confirm tissue-specific expression of the _Kaiso_ transgene, RT-PCR was performed with transgene-specific primers. As expected, the transgene was detected in all 3 villin-positive tissues: kidneys, small intestine and large intestine (Figure 1C). Kaiso protein expression was confirmed by Western blot analysis of protein harvested from small and large intestine (Figure 1D). Consistent with the transgene copy number observed via PCR, higher Kaiso protein expression was detected in Line A transgenics compared to Line B, with the lowest protein expression in Line C (data not shown).

To further evaluate and confirm Kaiso expression and localization in _Kaiso^Tg/+_ and Non-Tg tissues, IHC was performed on tissues harvested from small and large intestines of Line A and Line B mice using a Kaiso-specific antibody. Line A _Kaiso^Tg/+_ mice

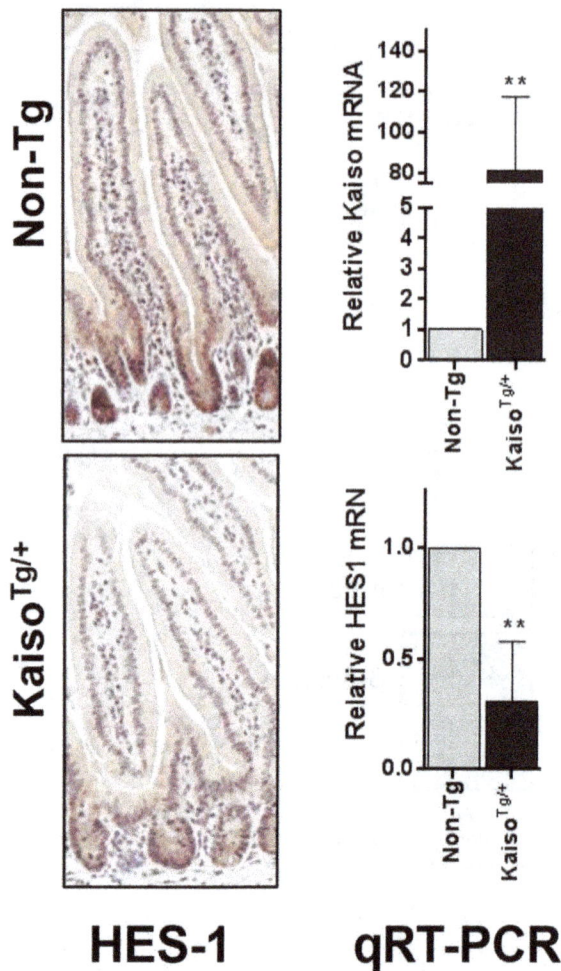

Figure 7. *Kaiso^{Tg/+}* **mice display decreased HES-1 expression in the small intestine.** Both Non-Tg and *Kaiso^{Tg/+}* tissues displayed nuclear HES-1 expression in the crypts of the small intestine, however *Kaiso^{Tg/+}* tissue displays significantly decreased HES-1 expression in the villi. Quantitative RT-PCR showed a significant decrease in HES-1 expression in *Kaiso^{Tg/+}* mice. Values were first normalized to the GAPDH housekeeping gene, followed by normalizing to non-Tg HES-1 expression (** represents $p < 0.05$).

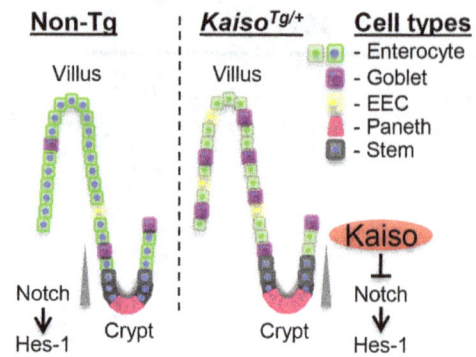

Figure 8. Schematic model of Kaiso's postulated effects in the intestine. Notch signalling in the crypts modulates differentiation of progenitor cells into the various epithelial cell lineages: enterocytes, Goblet, Paneth and enteroendocrine (EEC) cells. The gradient of Notch signaling is indicated by the grey triangle. HES-1 is necessary for the proper specification of these cell types. p120^{ctn} localizes to the membrane in the enterocytes of Non-Tg mice (green-membraned cells), but is recruited to the nucleus in *Kaiso^{Tg/+}* mice (green nucleated cells), which inhibits Notch signaling and Hes-1 expression, thus inducing inflammation.

Kaiso Transgenic Mice Exhibit Symptoms of Inflammation in the Intestinal Mucosa

After establishing that Kaiso was robustly expressed in the intestine via our transgene we next sought to determine the effect of ectopic Kaiso on intestinal morphology and function. Examination of H&E stained sections from small and large intestinal tissues of 1-year old Line A mice revealed longer crypts with no difference in villi length in the small intestine (Figure 3A), although this phenotype was not observed in Line B mice. We also noticed that several villi were fused and blunted in our Line A *Kaiso^{Tg/+}* mice in comparison to the characteristic elongated, finger-like appearance of villi in Non-Tg mice (Figure 3). To rule out the possibility that this phenotype was an artefact resulting from the transgene insertion site, we examined H&E sections from additional *Kaiso^{Tg/+}* lines that had been backcrossed for only 3 generations (Lines D, E, F & G). Two of these lines, Lines E and F, exhibited even more robust Kaiso expression than Line A mice, concomitant with extensive villi fusion and blunting (Figure S2).

Crypt hyperplasia accompanied by fused, blunted villi has been previously reported in both humans and mice exhibiting chronic inflammation of the intestinal mucosa [31,32,33,34], suggesting that ectopic Kaiso expression may cause intestinal inflammation. Indeed, closer examination of *Kaiso^{Tg/+}* intestines (Line A, E and F) revealed increased immune cell infiltration of the lamina propria compared to their Non-Tg siblings (Figure 3B and Figure S2); however no such phenotype was observed in Line B mice with low ectopic Kaiso expression (Figure 3C). We also measured the levels of **m**yelo**pero**xidase (MPO), which is a surrogate marker for inflammation, in *Kaiso^{Tg/+}* and Non-Tg intestinal tissues. MPO activity was increased in the distal small intestine (ileum) of Lines A, E and F *Kaiso^{Tg/+}* mice compared to their age-matched Non-Tg siblings (Figure 3B and Figure S2), while no change in MPO activity was detected in Line B mice (Figure 3C). Interestingly, the proximal colon of Lines E and F also exhibited increased MPO activity while mice from Lines A and B exhibited no such change (data not shown). These data suggest that ectopic Kaiso expression may predispose the murine intestine to inflammation, but this effect may be dose-dependent.

exhibited stronger nuclear Kaiso expression in the villi and increased nuclear expression in the crypts of the small intestine compared to their Non-Tg siblings (Figure 2). However, Line B *Kaiso^{Tg/+}*, which overexpressed less Kaiso than Line A, exhibited predominantly cytoplasmic localized Kaiso (Figure S1A). In the large intestine, both transgenic lines exhibited stronger Kaiso nuclear staining than their Non-Tg siblings (Figure S1B). Furthermore, strong nuclear Kaiso expression was observed in the epithelial cells near the top of the crypts, with lower expression at the bottom of the crypts (Figure 2). To confirm that increased Kaiso expression in *Kaiso^{Tg/+}* mice was due to the transgene rather than an enhancement of endogenous *Kaiso* gene expression, we evaluated c-Myc expression in Line A small intestines. Indeed, *Kaiso^{Tg/+}* mice exhibited stronger staining in comparison to Non-Tg mice, consistent with the expression of myc-tagged Kaiso (Figure 2). All subsequent analyses were performed on Line A *Kaiso^{Tg/+}* (unless noted otherwise).

Ectopic Kaiso Overexpression Results in Nuclear Accumulation of p120ctn

Given that Kaiso$^{Tg/+}$ mice exhibited an inflammatory response similar to that elicited by limited p120ctn depletion [24], albeit less severe, we examined p120ctn expression in the small intestines of our Kaiso$^{Tg/+}$ mice. Interestingly, in Kaiso$^{Tg/+}$ mice we observed nuclear localization of p120ctn and reduced p120ctn staining at the membrane in the distal small intestine (Figure 4). However in Non-Tg siblings, p120ctn was largely membrane bound (Figure 4). Taken together this data suggests that Kaiso overexpression results in nuclear accumulation of p120ctn, and decreased membrane-bound p120ctn, which phenocopies the consequences of p120ctn depletion [25].

Kaiso$^{Tg/+}$ Mice Exhibit Enhanced Differentiation of Progenitor Cells into Secretory Cell Fates

While characterizing the effect of ectopic Kaiso expression on intestinal morphology, we noted a significant expansion of Goblet cells in both the small and large intestine of Line A Kaiso$^{Tg/+}$ mice. Thus, we performed PAS staining for the Goblet cell-specific marker, Mucin, and quantification of Mucin positive (+) cells confirmed a significant increase in the Goblet cell population in both the small and large intestines of Line A mice compared to Line B and Non-Tg mice (Figure 5A & Figure S3). Interestingly, staining for the Paneth and enteroendocrine markers, lysozyme and synaptophysin respectively, revealed that these cell populations were also expanded in the small and large intestine of Line A Kaiso$^{Tg/+}$ mice but not in Line B or Non-Tg mice (Figure 5B, C & Figure S3).

The expansion of secretory cell lineages in our Kaiso$^{Tg/+}$ mice led us to hypothesize that Kaiso may be driving progenitor cell differentiation. However, since we also observed crypt expansion in Kaiso$^{Tg/+}$ mice, we questioned whether the increase in secretory cells was indicative of increased progenitor cell proliferation. Hence we examined the expression of the cell proliferation marker Ki67. Surprisingly, Ki67 expression was decreased in Line A mice and Ki67 positive cells were localized more apical to the normal crypt/villus boundary (Figure 6A). We next evaluated the expression of the Kaiso target gene cyclin D1 [4,21] that has been shown to drive proliferation in the intestinal epithelium and is frequently overexpressed in colon cancer [35]. Similar to Ki67, Cyclin D1 expression was also decreased in Line A Kaiso$^{Tg/+}$ mice but surprisingly the apparent decreased numbers of Cyclin D1 positive (+) cells in Kaiso$^{Tg/+}$ intestines was not statistically significant (Figure 6B, C).

Previous studies have reported an expansion of secretory cell lineages and a reduction in the number of proliferating columnar base cells upon inhibition of the Notch signalling pathway in the intestine [28,29,36,37]. Specifically, depletion of the Notch target gene Hes-1 resulted in increased expression of secretory cell markers in the intestine of Hes-1 null mice, suggesting that Hes-1 is necessary for specification of secretory cells in the intestine [37]. This prompted us to examine the expression of Hes-1 in our Kaiso$^{Tg/+}$ mice. Line A Kaiso$^{Tg/+}$ mice exhibited decreased Hes-1 staining and reduced expression of Hes-1 mRNA compared to Non-Tg littermates (Figure 7). Together our data demonstrate that ectopic Kaiso elicits enhanced differentiation of progenitor cells into secretory lineages, perhaps through the down-regulation of the Notch target Hes-1.

Discussion

Since Kaiso's discovery over a decade ago, several studies have utilized Xenopus laevis and cultured cells as models to elucidate Kaiso's biological roles [1,2,8,15,38,39,40,41]. Here we describe the first study to examine the role of Kaiso in a relevant organ-specific context, the murine intestine. Using the murine villin promoter we were able to successfully drive intestinal-specific expression of the Kaiso transgene. In all founder lines, Kaiso was expressed along the entire crypt-villus axis with the most robust expression in the villi, which is consistent with the normal expression pattern of villin [30].

A previous report examining the effect of Kaiso depletion on Apc$^{Min/+}$-mediated tumorigenesis found that Kaiso depletion resulted in fewer tumours [8], suggesting that Kaiso functions as an oncogene. However ectopic Kaiso expression was not sufficient to drive spontaneous tumour formation in our mouse model. Nonetheless, our Kaiso$^{Tg/+}$ Line A, mice exhibited enlarged crypts accompanied by fused, blunted villi, increased immune cell infiltration and increased MPO activity (indicative of neutrophil accumulation and inflammation) suggesting that Kaiso$^{Tg/+}$ mice have greater susceptibility to inflammation. Indeed, preliminary cytokine analysis of Kaiso$^{Tg/+}$ intestinal tissue revealed increased activity of the pro-inflammatory cytokine TNF-α compared to Non-Tg intestines (data not shown). Analysis of additional Kaiso transgenic lines (Lines E and F) revealed a similar intestinal phenotype to Line A, with concomitant increased neutrophil activation as measured by MPO activity. Increased MPO activity is often correlated with **u**lcerative **c**olitis (UC), a form of IBD and patients with IBD are at a higher risk of colon cancer [26,42,43,44]. Thus in accordance with Knudson's multiple hit theory of tumorigenesis, it is possible that Kaiso's full oncogenic potential may only be unmasked in the presence of a second oncogenic insult such as Apc mutation or p53 loss of function. Intriguingly, preliminary analysis of intestinal tissues from a **d**extran **s**odium **s**ulfate (DSS)-induced model of colitis (kind gift of Dr. Elena Verdú), revealed increased expression of Kaiso compared to non-DSS treated mice (Figure S4), further supporting the notion that Kaiso overexpression plays a role in intestinal inflammation.

The enhanced inflammation observed in Kaiso$^{Tg/+}$ mice may be linked to altered p120ctn function. Kaiso overexpression resulted in the nuclear localization of p120ctn, suggesting that Kaiso may somehow recruit or sequester p120ctn to the nucleus. Given that p120ctn was mainly localized to the cytoplasm and the cell membrane in non-transgenic mice, this change in localization may be indicative of altered or reduced p120ctn function that may phenocopy p120ctn loss observed by Smally Freed et al. [25]. Future studies are needed to determine whether p120ctn directly contributes to the Kaiso overexpression phenotype.

Interestingly, the phenotypes observed in Lines A, E and F mice were not observed in Line B mice which express significantly lower levels of ectopic Kaiso; this suggests that a threshold level of Kaiso expression is necessary for the observed inflammatory phenotype. Additionally, no change in MPO activity was seen in Line B mice, further supporting our hypothesis of threshold effects of Kaiso expression. This is not surprising since varying amounts of Kaiso were shown to have completely opposite effects in Xenopus laevis embryos [18]. Hence in Line B mice, it is likely that Kaiso expression is below the threshold at which it elicits inflammation and leads to expanded crypts.

Finally, Kaiso$^{Tg/+}$ mice exhibited increased populations of Goblet, Paneth and enteroendocrine cells. This expansion of secretory cell populations accompanied by decreased cell proliferation is consistent with the phenotype observed upon pharmacological inhibition of Notch signalling [28] and in Hes-1^{null} mice [37]. One study found that Notch signalling is activated in intestinal epithelium in response to inflammation and is required for proper regeneration of the intestinal epithelium following colitis induced damage [45]. It should be noted that 90-day old $Kaiso^{Tg/+}$ mice exhibit increased Goblet cells but do not exhibit any overt signs of inflammation or myeloperoxidase activity (data not shown). This suggests that inflammation in these mice develops over time although Notch inhibition is present at a very early age. Thus it is possible that the intestinal epithelium in our $Kaiso^{Tg/+}$ mice is incapable of regeneration following bacterial or physical insult and consequently develops chronic inflammation over time.

In summary, Kaiso overexpression promotes inflammation and inhibits Notch signalling in the murine intestine. These findings support a model in which $Kaiso^{Tg/+}$ mice develop inflammation, possibly by altering p120^{ctn} localization and consequently function (Figure 8). Kaiso's inhibition of the Notch pathway may hinder the ability of these mice to repair and regenerate the epithelium in response to inflammation, resulting in chronic inflammation that increases in severity over time, thus making the mice more susceptible to inflammation-induced tumorigenesis.

Supporting Information

Figure S1   Ectopic Kaiso expression in the intestine of $Kaiso^{Tg/+}$ mice. (A) Line B $Kaiso^{Tg/+}$ mice display sporadic nuclear expression and strong cytoplasmic Kaiso expression in the epithelial cells of the villi but lack Kaiso expression in the crypts, compared to Non-Tg mice. **(B)** In the colon, Non-Tg mice display low nuclear Kaiso expression in the crypts, while Line A and B $Kaiso^{Tg/+}$ show strong nuclear Kaiso expression, with the apical epithelial cells displaying the most Kaiso expression. Line A colons show greater Kaiso expression than Line B colons.

Figure S2   Ectopic Kaiso expression in the small intestine of multiple Kaiso transgenic lines induces inflammatory cell infiltration. $Kaiso^{Tg/+}$ mice display strong nuclear Kaiso expression in the villi and crypt cells, however Non-Tg mice display weak Kaiso expression with most Kaiso localizing to the cytoplasm. Line E and F (generation 3) show strong Kaiso

expression from the base of the crypts to the top of the villi. Interestingly, in all three $Kaiso^{Tg/+}$ lines analysed, ectopic Kaiso expression also appears to induce villi fusion (black arrows). Histological analysis showed increased neutrophil infiltration into the villi of Lines A, E and F $Kaiso^{Tg/+}$ mice (yellow demarcated area). An MPO assay of Line A, E and F ileums show increased MPO activity when compared to age-matched Non-Tg. Immunofluorescence revealed nuclear p120^{ctn} in both Line E and F in the villi.

Figure S3   Line A $Kaiso^{Tg/+}$ mice display increased numbers of differentiated cells in the colon. $Kaiso^{Tg/+}$ mice display a significant increase in Goblet (PAS stain), and enteroendocrine cells (synaptophysin) in the large intestine (colon) compared to their Non-Tg littermates.

Figure S4   Kaiso expression is increased in DSS-treated murine colon tissues. Preliminary analysis of DSS-induced murine colitis model intestinal tissues revealed increased Kaiso nuclear expression in DSS-treated colon tissues whereas non-treated mice show low cytoplasmic Kaiso expression.

Acknowledgments

We would like to thank Dr. Albert Reynolds (Vanderbilt University) for the anti-Kaiso rabbit polyclonal antibody and Dr. Sylvie Robine (Institute Curie, Paris, France) for the villin construct. Additionally, we would like to thank Christina Hayes and Dr. Elena Verdú (McMaster University) for their assistance with the MPO Assay.

Author Contributions

Conceived and designed the experiments: RC CCP KN JMD. Performed the experiments: RC CCP KN DW SM. Analyzed the data: RC CCP GW CP JMD. Contributed reagents/materials/analysis tools: SR. Wrote the paper: RC CCP JMD.

References

1. Cofre J, Menezes JR, Pizzatti L, Abdelhay E (2012) Knock-down of Kaiso induces proliferation and blocks granulocytic differentiation in blast crisis of chronic myeloid leukemia. Cancer Cell Int 12: 28.
2. Vermeulen JF, van de Ven RA, Ercan C, van der Groep P, van der Wall E, et al. (2012) Nuclear Kaiso expression is associated with high grade and triple-negative invasive breast cancer. PLoS One 7: e37864.
3. Wang Y, Li L, Li Q, Xie C, Wang E (2012) Expression of P120 catenin, Kaiso, and metastasis tumor antigen-2 in thymomas. Tumour Biol.
4. Jiang G, Wang Y, Dai S, Liu Y, Stoecker M, et al. (2012) P120-catenin isoforms 1 and 3 regulate proliferation and cell cycle of lung cancer cells via beta-catenin and Kaiso respectively. PLoS One 7: e30303.
5. Dai SD, Wang Y, Jiang GY, Zhang PX, Dong XJ, et al. (2010) Kaiso is expressed in lung cancer: its expression and localization is affected by p120ctn. Lung Cancer 67: 205–215.
6. Dai SD, Wang Y, Miao Y, Zhao Y, Zhang Y, et al. (2009) Cytoplasmic Kaiso is associated with poor prognosis in non-small cell lung cancer. BMC Cancer 9: 178.
7. Lopes EC, Valls E, Figueroa ME, Mazur A, Meng FG, et al. (2008) Kaiso contributes to DNA methylation-dependent silencing of tumor suppressor genes in colon cancer cell lines. Cancer Res 68: 7258–7263.
8. Prokhortchouk A, Sansom O, Selfridge J, Caballero IM, Salozhin S, et al. (2006) Kaiso-deficient mice show resistance to intestinal cancer. Mol Cell Biol 26: 199–208.
9. Blattler A, Yao L, Wang Y, Ye Z, Jin VX, et al. (2013) ZBTB33 binds unmethylated regions of the genome associated with actively expressed genes. Epigenetics Chromatin 6: 13.
10. Daniel JM, Spring CM, Crawford HC, Reynolds AB, Baig A (2002) The p120(ctn)-binding partner Kaiso is a bi-modal DNA-binding protein that recognizes both a sequence-specific consensus and methylated CpG dinucleotides. Nucleic Acids Res 30: 2911–2919.

11. Donaldson NS, Pierre CC, Anstey MI, Robinson SC, Weerawardane SM, et al. (2012) Kaiso Represses the Cell Cycle Gene cyclin D1 via Sequence-Specific and Methyl-CpG-Dependent Mechanisms. PLoS One 7: e50398.
12. Prokhortchouk A, Hendrich B, Jorgensen H, Ruzov A, Wilm M, et al. (2001) The p120 catenin partner Kaiso is a DNA methylation-dependent transcriptional repressor. Genes Dev 15: 1613–1618.
13. Ruzov A, Dunican DS, Prokhortchouk A, Pennings S, Stancheva I, et al. (2004) Kaiso is a genome-wide repressor of transcription that is essential for amphibian development. Development 131: 6185–6194.
14. Ruzov A, Hackett JA, Prokhortchouk A, Reddington JP, Madej MJ, et al. (2009) The interaction of xKaiso with xTcf3: a revised model for integration of epigenetic and Wnt signalling pathways. Development 136: 723–727.
15. Ruzov A, Savitskaya E, Hackett JA, Reddington JP, Prokhortchouk A, et al. (2009) The non-methylated DNA-binding function of Kaiso is not required in early Xenopus laevis development. Development 136: 729–738.
16. Sasai N, Nakao M, Defossez PA (2010) Sequence-specific recognition of methylated DNA by human zinc-finger proteins. Nucleic Acids Res 38: 5015–5022.
17. Yoon HG, Chan DW, Reynolds AB, Qin J, Wong J (2003) N-CoR mediates DNA methylation-dependent repression through a methyl CpG binding protein Kaiso. Mol Cell 12: 723–734.
18. Iioka H, Doerner SK, Tamai K (2009) Kaiso is a bimodal modulator for Wnt/beta-catenin signaling. FEBS Lett 583: 627–632.
19. Park JI, Kim SW, Lyons JP, Ji H, Nguyen TT, et al. (2005) Kaiso/p120-catenin and TCF/beta-catenin complexes coordinately regulate canonical Wnt gene targets. Dev Cell 8: 843–854.
20. Filion GJ, Zhenilo S, Salozhin S, Yamada D, Prokhortchouk E, et al. (2006) A family of human zinc finger proteins that bind methylated DNA and repress transcription. Mol Cell Biol 26: 169–181.
21. Donaldson NS, Pierre CC, Anstey MI, Robinson SC, Weerawardane SM, et al. (2012) Kaiso represses the cell cycle gene cyclin D1 via sequence-specific and methyl-CpG-dependent mechanisms. PLoS One In Press.

22. Park JI, Ji H, Jun S, Gu D, Hikasa H, et al. (2006) Frodo links Dishevelled to the p120-catenin/Kaiso pathway: distinct catenin subfamilies promote Wnt signals. Dev Cell 11: 683–695.

23. Spring CM, Kelly KF, O'Kelly I, Graham M, Crawford HC, et al. (2005) The catenin p120ctn inhibits Kaiso-mediated transcriptional repression of the beta-catenin/TCF target gene matrilysin. Exp Cell Res 305: 253–265.

24. Smalley-Freed WG, Efimov A, Burnett PE, Short SP, Davis MA, et al. (2010) p120-catenin is essential for maintenance of barrier function and intestinal homeostasis in mice. J Clin Invest 120: 1824–1835.

25. Smalley-Freed WG, Efimov A, Short SP, Jia P, Zhao Z, et al. (2011) Adenoma formation following limited ablation of p120-catenin in the mouse intestine. PLoS One 6: e19880.

26. Terzic J, Grivennikov S, Karin E, Karin M (2010) Inflammation and colon cancer. Gastroenterology 138: 2101–2114 e2105.

27. Ogaki S, Shiraki N, Kume K, Kume S (2013) Wnt and notch signals guide embryonic stem cell differentiation into the intestinal lineages. Stem Cells 31: 1086–1096.

28. VanDussen KL, Carulli AJ, Keeley TM, Patel SR, Puthoff BJ, et al. (2012) Notch signaling modulates proliferation and differentiation of intestinal crypt base columnar stem cells. Development 139: 488–497.

29. Zecchini V, Domaschenz R, Winton D, Jones P (2005) Notch signaling regulates the differentiation of post-mitotic intestinal epithelial cells. Genes Dev 19: 1686–1691.

30. Pinto D, Robine S, Jaisser F, El Marjou FE, Louvard D (1999) Regulatory sequences of the mouse villin gene that efficiently drive transgenic expression in immature and differentiated epithelial cells of small and large intestines. J Biol Chem 274: 6476–6482.

31. Ostanin DV, Pavlick KP, Bharwani S, D'Souza D, Furr KL, et al. (2006) T cell-induced inflammation of the small and large intestine in immunodeficient mice. Am J Physiol Gastrointest Liver Physiol 290: G109–119.

32. Kuhnert F, Davis CR, Wang HT, Chu P, Lee M, et al. (2004) Essential requirement for Wnt signaling in proliferation of adult small intestine and colon revealed by adenoviral expression of Dickkopf-1. Proc Natl Acad Sci U S A 101: 266–271.

33. Jacques P, Elewaut D (2008) Joint expedition: linking gut inflammation to arthritis. Mucosal Immunol 1: 364–371.

34. Goldstein NS (2006) Isolated ileal erosions in patients with mildly altered bowel habits. A follow-up study of 28 patients. Am J Clin Pathol 125: 838–846.

35. Yang R, Bie W, Haegebarth A, Tyner AL (2006) Differential regulation of D-type cyclins in the mouse intestine. Cell Cycle 5: 180–183.

36. Milano J, McKay J, Dagenais C, Foster-Brown L, Pognan F, et al. (2004) Modulation of notch processing by gamma-secretase inhibitors causes intestinal goblet cell metaplasia and induction of genes known to specify gut secretory lineage differentiation. Toxicol Sci 82: 341–358.

37. Jensen J, Pedersen EE, Galante P, Hald J, Heller RS, et al. (2000) Control of endodermal endocrine development by Hes-1. Nat Genet 24: 36–44.

38. van Roy FM, McCrea PD (2005) A role for Kaiso-p120ctn complexes in cancer? Nat Rev Cancer 5: 956–964.

39. Daniel JM (2007) Dancing in and out of the nucleus: p120(ctn) and the transcription factor Kaiso. Biochim Biophys Acta 1773: 59–68.

40. Martin Caballero I, Hansen J, Leaford D, Pollard S, Hendrich BD (2009) The methyl-CpG binding proteins Mecp2, Mbd2 and Kaiso are dispensable for mouse embryogenesis, but play a redundant function in neural differentiation. PLoS One 4: e4315.

41. Jones J, Wang H, Zhou J, Hardy S, Turner T, et al. (2012) Nuclear Kaiso indicates aggressive prostate cancers and promotes migration and invasiveness of prostate cancer cells. Am J Pathol 181: 1836–1846.

42. Rubin DC, Shaker A, Levin MS (2012) Chronic intestinal inflammation: inflammatory bowel disease and colitis-associated colon cancer. Front Immunol 3: 107.

43. Saleh M, Trinchieri G (2011) Innate immune mechanisms of colitis and colitis-associated colorectal cancer. Nat Rev Immunol 11: 9–20.

44. Xavier RJ, Podolsky DK (2007) Unravelling the pathogenesis of inflammatory bowel disease. Nature 448: 427–434.

45. Okamoto R, Tsuchiya K, Nemoto Y, Akiyama J, Nakamura T, et al. (2009) Requirement of Notch activation during regeneration of the intestinal epithelia. Am J Physiol Gastrointest Liver Physiol 296: G23–35.

11

Rasagiline Ameliorates Olfactory Deficits in an Alpha-Synuclein Mouse Model of Parkinson's Disease

Géraldine H. Petit[1]*, **Elijahu Berkovich[2]**, **Mark Hickery[3]**, **Pekka Kallunki[4]**, **Karina Fog[4]**, **Cheryl Fitzer-Attas[2]**, **Patrik Brundin[1,5]**

1 Neuronal Survival Unit, Wallenberg Neuroscience Center, Department of Experimental Medical Science, BMC B11, Lund University, Lund, Sweden, 2 Teva Pharmaceutical Industries Ltd., Global Innovative Products, Petach-Tikva, Israel, 3 H. Lundbeck A/S, Neurology, Copenhagen, Denmark, 4 H. Lundbeck A/S, Neurodegeneration-1, Copenhagen, Denmark, 5 Van Andel Research Institute, Center for Neurodegenerative Science, Grand Rapids, Michigan, United States of America

Abstract

Impaired olfaction is an early pre-motor symptom of Parkinson's disease. The neuropathology underlying olfactory dysfunction in Parkinson's disease is unknown, however α-synuclein accumulation/aggregation and altered neurogenesis might play a role. We characterized olfactory deficits in a transgenic mouse model of Parkinson's disease expressing human wild-type α-synuclein under the control of the mouse α-synuclein promoter. Preliminary clinical observations suggest that rasagiline, a monoamine oxidase-B inhibitor, improves olfaction in Parkinson's disease. We therefore examined whether rasagiline ameliorates olfactory deficits in this Parkinson's disease model and investigated the role of olfactory bulb neurogenesis. α-Synuclein mice were progressively impaired in their ability to detect odors, to discriminate between odors, and exhibited alterations in short-term olfactory memory. Rasagiline treatment rescued odor detection and odor discrimination abilities. However, rasagiline did not affect short-term olfactory memory. Finally, olfactory changes were not coupled to alterations in olfactory bulb neurogenesis. We conclude that rasagiline reverses select olfactory deficits in a transgenic mouse model of Parkinson's disease. The findings correlate with preliminary clinical observations suggesting that rasagiline ameliorates olfactory deficits in Parkinson's disease.

Editor: David R. Borchelt, University of Florida, United States of America

Funding: This work was supported by the Swedish Research Council [K2010-61X-11286-16-3, 315-2004-6629] (www.vr.se), the Swedish Parkinson Foundation [253/08] (www.parkinsonfonden.se), Swedish Brain Power (www.swedishbrainpower.se), H. Lundbeck A/S (www.lundbeck.com) and TEVA Pharmaceutical Industries Ltd (www.tevapharm.com). The authors EB, CFA (TEVA Pharmaceutical Industries Ltd) and MH, PK and KF (H. Lundbeck A/S) participated in the study design, the manuscript preparation and they approved the manuscript to be published. The other funders (the Swedish Research Council, the Swedish Parkinson Foundation and Swedish Brain Power) had no role in study design, data collection and analysis, decision to publish, or preparation of the manuscript.

Competing Interests: Teva Pharmaceutical Industries Ltd and H.Lundbeck A/S, which market rasagiline, contributed to the funding of this study. The authors EB and CFA work for Teva Pharmaceutical Industries Ltd. The authors MH, PK and KF work for H.Lundbeck A/S. PB has received speaking honoraria from and provided consulting services to Teva Pharmaceutical Industries Ltd and H.Lundbeck A/S. Teva Pharmaceutical Industries Ltd. covered the costs of GHP to attend the MDS congress (Toronto 2011). The use of rasagiline for the treatment of olfactory dysfunction is covered by patent number US 2012/0029087 A1.

* E-mail: geraldine.petit@med.lu.se

Introduction

Parkinson's disease (PD) patients not only exhibit motor dysfunction, but also multiple non-motor symptoms [1]. Hyposmia, i.e. impaired detection, discrimination and/or identification of odors affects 70–6% of PD patients [2,3], typically several years before onset of motor symptoms [4]. Hyposmia might be an useful sign when predicting who will develop PD later [5].

The causes of olfactory impairments in PD are not understood. In PD, Lewy bodies and Lewy neurites are present in mitral cells and in the inner plexiform layer of the olfactory bulb (OB), and in cells along the olfactory neural pathways [6]. Braak et al. (2003) have suggested that these α-synuclein aggregates appear before the onset of motor symptoms.

Either these protein aggregates or changes in OB neurogenesis might contribute to olfactory deficits in PD. The numbers of proliferating cells in the subventricular zone and neural precursors in the OB are reduced [7], and some animal PD models exhibit OB neurogenesis changes [8,9].

Rasagiline (N-propargyl-1-(R)-aminoindan) is an irreversible monoamine oxidase (MAO)-B inhibitor, prescribed as monotherapy in early-stage PD and as an adjunct to levodopa in moderate to advanced PD [10]. It reduces motor deficits and ameliorates motor fluctuations [11–13]. A double–blind, delayed-start trial (ADAGIO) indicated that early rasagiline treatment provides benefits consistent with a possible disease-modifying effect [14]. Rasagiline is reported to be neuroprotective in different animal models of neurodegeneration [15–17]. Interestingly, preliminary evidence suggests that rasagiline improves olfaction in PD [18,19] and ongoing clinical studies address this possibility [NCT00902941, NCT01007630].

To investigate the effect of an accumulation of wild-type α-synuclein, we studied a transgenic mouse model of PD expressing human wild-type α-synuclein under the control of the mouse α-synuclein promoter, which is likely to lead to an expression pattern of the human α-synuclein that is similar to the pattern of endogenous mouse α-synuclein expression. We first characterized olfactory deficits in a transgenic mouse model of PD expressing

A

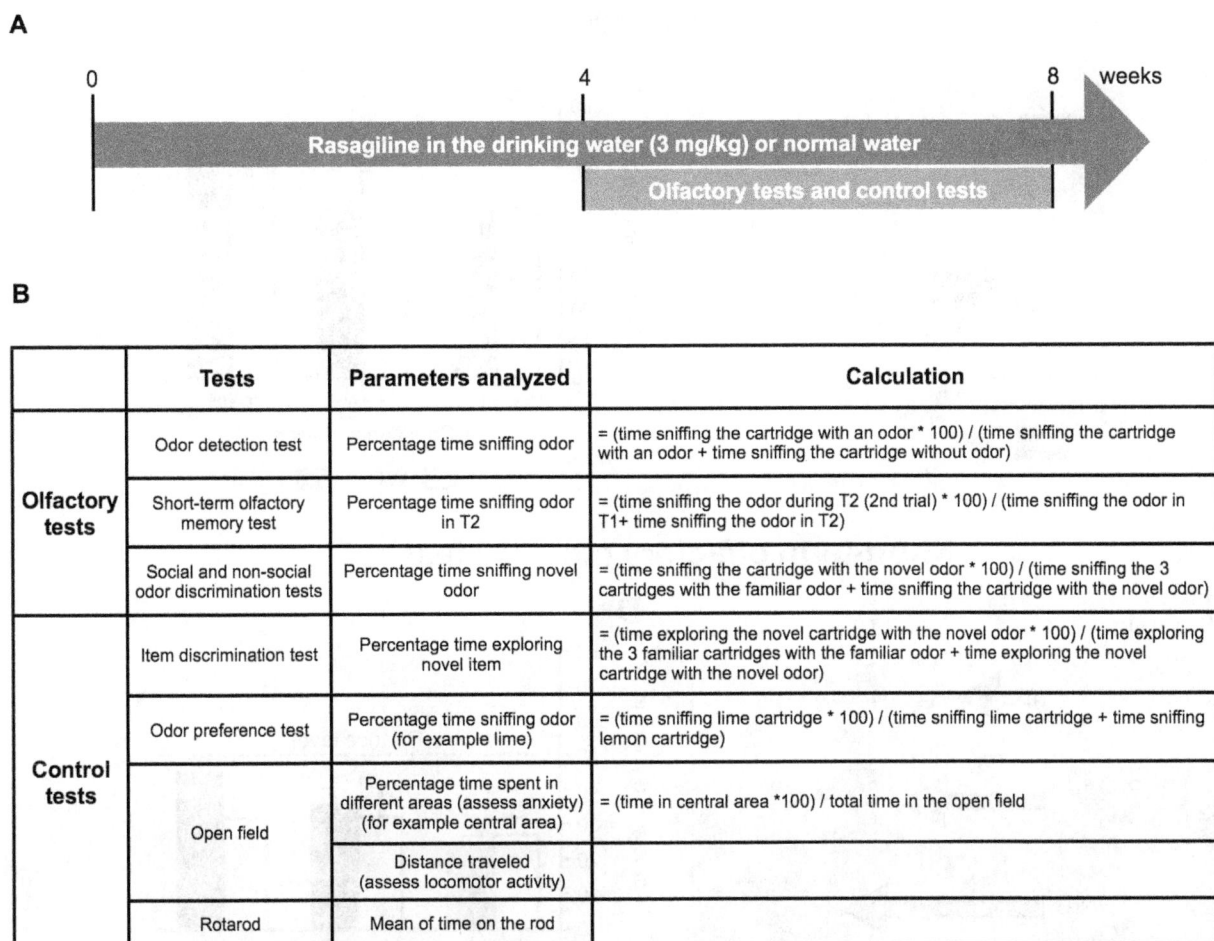

B

	Tests	Parameters analyzed	Calculation
Olfactory tests	Odor detection test	Percentage time sniffing odor	= (time sniffing the cartridge with an odor * 100) / (time sniffing the cartridge with an odor + time sniffing the cartridge without odor)
	Short-term olfactory memory test	Percentage time sniffing odor in T2	= (time sniffing the odor during T2 (2nd trial) * 100) / (time sniffing the odor in T1+ time sniffing the odor in T2)
	Social and non-social odor discrimination tests	Percentage time sniffing novel odor	= (time sniffing the cartridge with the novel odor * 100) / (time sniffing the 3 cartridges with the familiar odor + time sniffing the cartridge with the novel odor)
Control tests	Item discrimination test	Percentage time exploring novel item	= (time exploring the novel cartridge with the novel odor * 100) / (time exploring the 3 familiar cartridges with the familiar odor + time exploring the novel cartridge with the novel odor)
	Odor preference test	Percentage time sniffing odor (for example lime)	= (time sniffing lime cartridge * 100) / (time sniffing lime cartridge + time sniffing lemon cartridge)
	Open field	Percentage time spent in different areas (assess anxiety) (for example central area)	= (time in central area *100) / total time in the open field
		Distance traveled (assess locomotor activity)	
	Rotarod	Mean of time on the rod	

Figure 1. Behavioral experiments: design and parameters analyzed. A. Experimental design of the behavioral study. **B**. Olfactory- and control tests and the parameters analyzed from these experiments.

human wild-type α-synuclein. Subsequently, we monitored the effects of rasagiline on these deficits and OB neurogenesis.

Materials and Methods

Ethics Statement

This study was carried out in strict accordance with the recommendations in the Guide for the Care and Use of Laboratory Animals of the National Institutes of Health. The Malmö-Lund Animal Ethical Committee approved all experimental procedures (permit number: M55-09). All efforts were made to minimize suffering.

Transgenic mice and rasagiline treatment

We studied 3, 10–11 and 18 month-old F28 α-synuclein mice (α-syn mice, provided by H. Lundbeck A/S, Denmark) overexpressing wild-type human α-synuclein under the control of the partial mouse α-synuclein promoter [20]. Western blot analysis showed that the level of α-synuclein in total brain lysates of the α-syn mice is slightly less than 3 fold compared to the wild-type level (data not shown). Moreover, real-time PCR quantification showed an increase of α-synuclein in the striatum of approximately 3 fold in α-syn mice compared to wild-type [20]. Previous immunostaining of human α-synuclein protein indicated a cytoplasmic accumulation of α-synuclein in cell bodies of the hippocampal

CA1 region, striatum, thalamus, amygdala and in several cortical layers [21].

The 10–11 month-old mice received either rasagiline in the drinking water (3 mg/kg) or normal drinking water. The liquid intake was carefully monitored for each mouse before the experiment and then every week during the experiment. We did not see any difference in water consumption between wild-type and α-synuclein mice (wild-type: 3.23±0.08 ml/day; α-synuclein mice 3.08 ± 0.08 ml/day; Unpaired t-test p = 0.192). The weight of both wild-type and α-synuclein mice was also similar (wild-type mice: 36.1±0.8 g; α-synuclein mice: 35.6±0.7 g; unpaired t-test p = 0.615). Rasagiline concentration was individually adapted throughout the treatment period according to the water intake and weight for each mouse. The drinking water was changed twice a week. During rasagiline treatment and behavioral testing, animals were kept in individual cages (12 h light/dark cycle), with access to food and water *ad libitum*.

Behavioral tests

The experimental design of the behavioral study and parameters analyzed for these experiments are described in Figure 1A-B. We began treating mice with rasagiline 4 weeks prior to behavioral testing and continued treatment throughout the period of behavioral testing. To investigate different aspects of olfactory function, we performed a set of olfactory tests (odor detection test,

Odor detection test

A

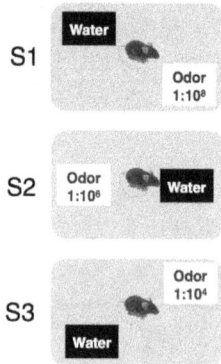

B

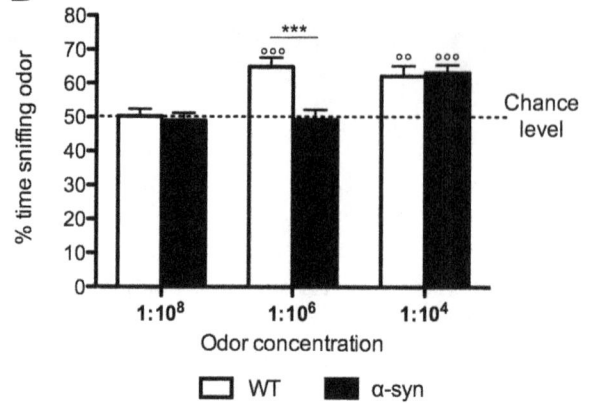

Short-term olfactory memory test

C

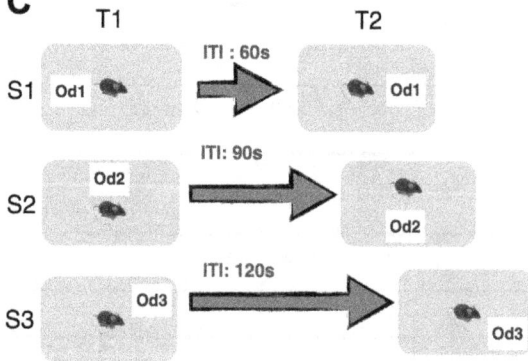

D

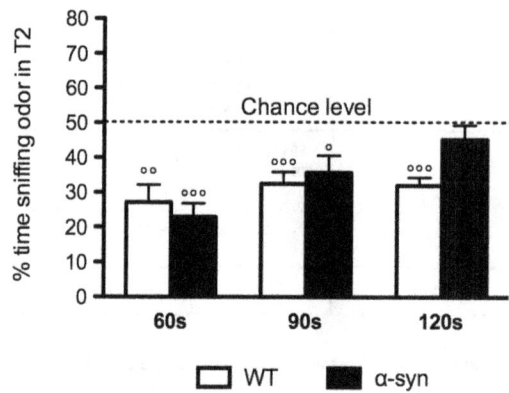

Odor discrimination test

E Social odor discrimination test

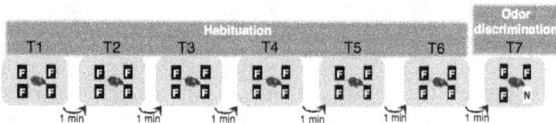

F Non-social odor discrimination test

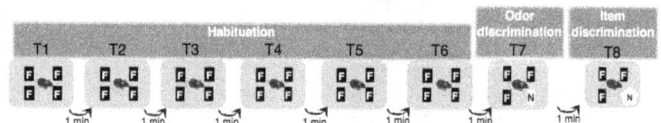

G

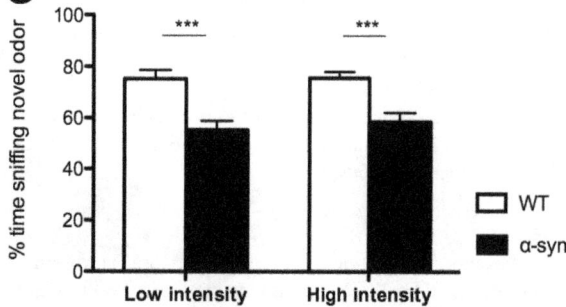

H

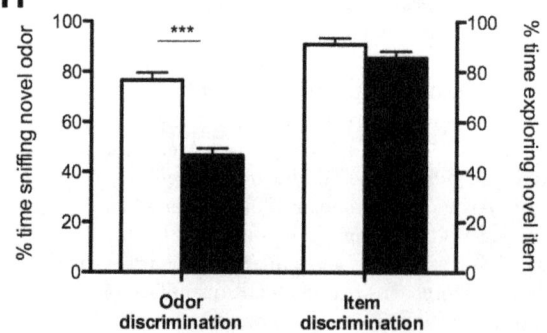

Figure 2. Olfactory deficits in the α-syn mouse model of PD. A-B: Odor detection test. A. Description of the protocol composed of 3 sessions (S). In each 5-min session, mice were exposed to 2 cartridges, one filled with water, the other with increasing odor concentrations from $1:10^8$ to $1:10^4$. **B.** Percentage of time sniffing the odor for the different concentrations. WT mice start detecting the odor at the concentration $1:10^6$ when percentage of time sniffing the odor is significantly different from the chance level (50%, where mice spent same time sniffing water and odor cartridges) ($^{\circ\circ\circ}$p<0.001, $^{\circ\circ}$p<0.01, one-sample t-test). α-Syn mice can detect the odor only at $1:10^4$ ($^{\circ\circ\circ}$p<0.001, one-sample t-test). At the concentration $1:10^6$, α-syn mice are significantly impaired compared to WT. N = 10 for each group aged 10–11 months. Statistics: One-sample t-test to compare each value to chance level (50%), ($^{\circ\circ\circ}$p<0.001, $^{\circ\circ}$p<0.01). Two-way RM ANOVA: odor concentration, p = 0.0001, F(2,36) = 11.96; genotype, p = 0.015, F(1,36) = 7.2; odor concentration×genotype, p = 0.0073, F(2,36) = 5.65; Bonferroni post-hoc (***p<0.001). **C-D: Short-term olfactory memory test. C.** Description of the protocol composed of 3 sessions (S). Each session consisted of two 5 min-trials (T) where mice are exposed to a novel odor separated by an increasing inter-trial time (ITI) from 60 s to 120 s. **D.** Percentage of time sniffing the odor during T2 compared to the total time spent sniffing during both trials. WT mice remember the odor during the 2nd exposure for the 3 ITI tested and their percentage of time sniffing the odor during T2 was significantly different from the chance level (50%, where mice spent same time sniffing the odor during T1 and T2) ($^{\circ\circ}$p<0.01, $^{\circ\circ\circ}$p<0.001, one-sample t-test). By contrast, short-term olfactory memory of α-syn mice was impaired from an ITI of 120 s (p>0.05, one-sample t-test). However, it was significantly different from chance level at 60 s and 90 s ($^{\circ\circ\circ}$p<0.001 and $^{\circ}$p<0.05 respectively, one-sample t-tests). N = 10 for each group aged 10–11 months. Statistics: One-sample t-test to compare each value to chance level (50%), ($^{\circ\circ\circ}$p<0.001, $^{\circ\circ}$p<0.01, $^{\circ}$p<0.05). **E-H: Odor discrimination test. E and G:** Social odor discrimination test. **E.** Description of the protocol composed of 6 habituation trials where mice are exposed to a familiar odor (F, odor of the tested mouse); and one odor discrimination trial, where one familiar odor is replaced by a novel odor (N, another mouse's odor). This test was performed with low or high odor intensities (wood blocks impregnated with mouse's odor for 2 or 7 days respectively). Each trial lasted 2 min and was separated by 1 min. **G.** Percentage of time sniffing novel odor. For both low and high odor intensities, α-syn mice have impaired odor discrimination with the percentage of time sniffing the odor significantly lower than WT. N = 19–21 for each group aged 10–11 months. Statistics: Two-way RM ANOVA: odor intensity, p = 0.55, F(1,38) = 0.37; genotype, p<0.0001, F(1,38) = 27.1; odor intensity×genotype, p = 0.63, F(1,38) = 0.23; Bonferroni post-hoc (***p<0.001). **F and H:** Non-social odor discrimination test. **F.** Description of the protocol based on the same principle of the social odor discrimination test but using non-social odors (lemon and lime). In the 8th 2 min-trial, an item discrimination trial was added where the usual cartridge, with the novel odor (lime), was replaced by a novel item (a novel type of cartridge associated with the same novel odor, lime). **H.** Percentage of time sniffing the novel odor during the odor discrimination trial and percentage of time exploring the novel item in the item discrimination trial. α-syn mice had significantly impaired odor discrimination of the social odor. By contrast, the ability to discriminate the novel item was similar between WT and α-syn mice suggesting that the discrimination deficit is specific to olfaction. Statistics: unpaired t-test, non-social odor discrimination p<0.0001, N = 19–21 for each group aged 10–11 months; item discrimination p = 0.16, N = 10 for each group aged 10-11 months (***p<0.001).

short-term olfactory memory test, social and non-social odor discrimination tests, odor preference test) and control tests (item discrimination test, open field, rotarod) on mice aged 10–11 months. We investigated 21 non-treated wild-type (WT), 18 rasagiline-treated WT, 19 non-treated α-syn mice and 20 rasagiline-treated α-syn mice. The item discrimination, open field, short-term olfactory memory and odor detection tests were performed on only 10 non-treated WT, 9 treated WT, 10 non-treated α-syn mice and 9 treated α-syn mice. In order to assess the progression of olfactory deficits with age we investigated 14 wild-type and α-syn mice aged 3 months, 10 WT and α-syn mice aged 11 months and 14 WT and α-syn mice aged 18 months in the odor detection test, short-term olfactory memory test and the non-social odor discrimination test. All experiments were performed blinded to group identity. Olfactory tests were performed in the mouse's home cage. Two days before commencing the first olfactory test that required the use of plastic cartridges, we placed four cartridges without specific odor in the mouse cage allowing the mice to habituate to the object. Cages were cleared of cartridges or wood blocks and nests 2 h before testing to allow mice to habituate to the experimental environment. The non-social odors used in the olfactory tests (vanilla, lemon, lime, cinnamon, black pepper and anise) were prepared from pure essential oil (Aroma Creative AB, Sweden). Odors were diluted from the pure essential oil stock to the following concentrations $1:10^8$, $1:10^6$ and $1:10^4$. As a social odor, we used wood blocks impregnated with mouse odors for either 7 days (high odor intensity) or 2 days (low odor intensity). The wood blocks were placed in clean cages of individually housed mice and beddings were not changed during the time of odor impregnation (Figure S1 for more details about cartridges and wood blocks). During the olfactory tests, we measured the time spent sniffing cartridges or wood blocks when the following criteria were fulfilled: The nose of the mouse was oriented towards the object and the mouse moved its nose/whiskers (as observed when sniffing). Physical contact with the object was not necessary, however the nose of the mouse had to be close to the object (about

max 2 cm away the object). Measurements were made during the tests but all tests were videotaped in order to be able to verify measurements if necessary.

Odor detection test. To determine the threshold of odor detection, we adapted an odor detection test from Breton-Provencher et al. (2009). The test was composed of two or three sessions. In each 5-min session, mice were exposed to two cartridges, one filled with water, the other with increasing novel odor concentrations (vanilla, concentration: $1:10^8$, $1:10^6$ or $1:10^4$, Figure 2A). Normal mice instinctively spend more time sniffing new odors. The test determines if animals can detect the novel odor by comparing the time they spend sniffing the two cartridges.

Short-term olfactory memory test. We assessed short-term olfactory memory according to Breton-Provencher et al. (2009). During each of the two or three sessions (S), we exposed mice to a novel odor for two 5 min-trials (T1 and T2) separated by 60, 90 or 120 s inter-trial intervals (ITI) (Figure 2C). The novel odors we used for the ITIs of 60, 90 and 120 s were cinnamon, anise and black pepper (concentration on $1:10^4$) respectively. If mice remembered the odor from the first trial, they were expected to spend less time sniffing it during the second.

Social odor discrimination test. We assessed the ability of the mice to discriminate between social odors and we used two levels of odor intensities (high or low) [22,23]. First, mice were subjected to six 2-min habituation trials (separated by 1 min ITI) when they were exposed to four wood blocks with a "familiar social odor". Thereafter, during in the odor discrimination trial, they had to detect that one wood block had been replaced by a block impregnated with a "novel social odor" (Figure 2E). Mice that were able to discriminate between the familiar and novel odors spent more time sniffing the novel odor.

Non-social odor discrimination test. The non-social odor discrimination test was identical to the social odor discrimination test with the exception that we used cartridges filled with non-social odors (lemon or lime) instead of wooden blocks (Figure 2F).

Odor preference test

A

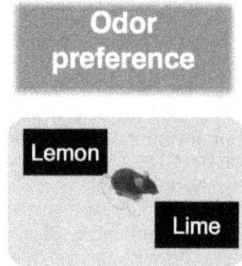

B

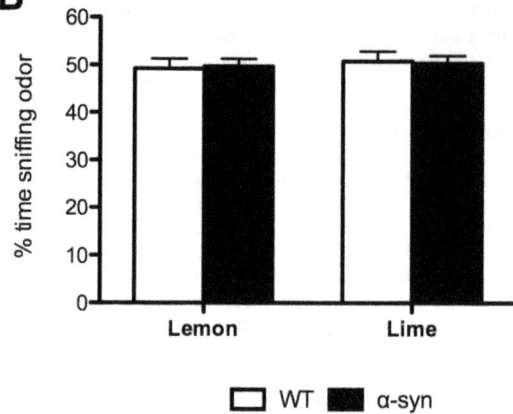

Open field

C Distance traveled

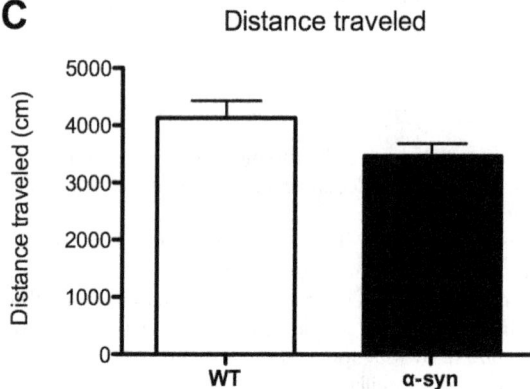

D Percentage time spent in different areas

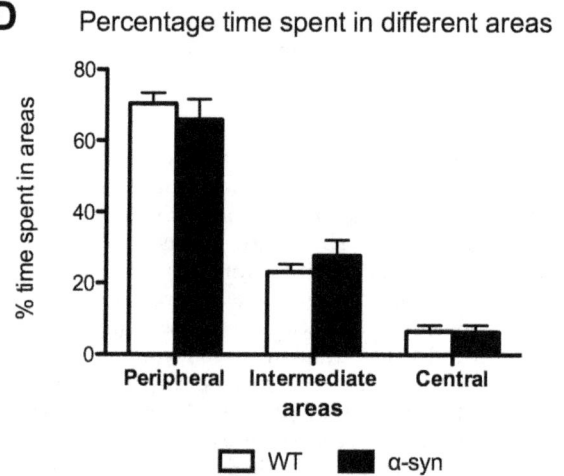

Rotarod

E

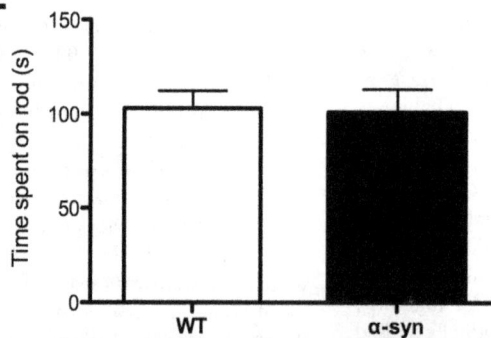

Figure 3. Specificity of the olfactory deficits in α-syn mice. A–B: Odor preference test. A. Description of the protocol. Two cartridges, filled with either lemon or lime are placed in the cage for 5 min. If mice do not have any odor preference they spend a similar time sniffing either cartridge. **B.** Percentage of time sniffing lemon and lime showing no difference between control and α-syn mice and no difference between lime and lemon odors. N = 19–21 for each group aged 10–11 months. Statistics: two-way ANOVA, odor, p = 0.57, $F_{(1,76)} = 0.33$; genotype, p = 1.00, $F_{(1,76)} \approx 0$; odors×genotype, p = 0.80, $F_{(1,76)} = 0.06$). **C-D Open field test. C.** Distance traveled in the open field. No significant difference was observed between WT and α-syn mice. α-syn mice show similar locomotor activity to control mice. N = 10 for each group aged 10–11 months. Statistics: unpaired t-test, p = 0.088. **D.** Percentage of time spent in different areas. No significant difference between WT and α-syn mice (p>0.05, two-way ANOVA), suggesting each type of mouse exhibited the same level of anxiety. N = 10 for each group aged 10–11 months. Statistics: two-way ANOVA,

genotype, p = 1, F(1,54) = 0; areas, p<0.0001, F(1,54) = 165.1; genotype×areas, p = 0.42, F(2,54) = 0.87; Bonferroni post-hoc between WT and α-syn mice, p>0.05). **E. Rotarod test**. Time spent on the rod was similar between both groups of mice. N = 10 for each group aged 10–11 months. Statistics: unpaired t-test, p = 0.9.

Item discrimination test. One min after the non-social discrimination test, we performed an 8th 2-min trial aimed at determining whether the mice could discriminate a novel item. Thus, we replaced the cartridge with the novel (lime) odor with a different cartridge type ("novel item") that contained the same lime odor (Figure 2F). We assessed item discrimination by noting an increase in time exploring the novel item. Exploration time includes all different types of exploratory behaviors and interactions with the object. This includes sniffing time as described previously but also time spent touching, manipulating, moving, turning around, or biting the object.

Odor preference test. The test consisted of a single 5 min-trial during which we exposed mice to two cartridges (one with a lemon and one with a lime odor) and monitored how much time they spent sniffing each odor (Figure 3A). If mice preferred an odor, they spent more time sniffing it.

Open field test. Using a video tracking system, we monitored general activity and anxiety status in an open field (42 cm^2, Noldus, Ethovision, Holland) for 10 min. To assess locomotor activity, we recorded distance traveled. We evaluated anxiety levels based on the time spent in the peripheral, intermediate and central areas [24].

Accelerating rotarod test. We assessed motor ability using a rotarod (Rotamex 4/8, Columbus Instruments, USA; 3.8 cm in rod diameter, 4.5 cm in wide section). After a training phase, during which mice had to stay on the rod for 30 s while it was turning at a constant speed (5 rotations per min (rpm)), we tested mice in 4 trials during which the speed of the rotation increased gradually from 4 to 40 rpm over a 5 min period. We averaged the time spent on the rotarod for the four trials.

BrdU (5-bromo-2′-deoxyuridine) experiment

We studied neurogenesis in four independent groups of mice (6 non-treated WT; 4 rasagiline-treated WT, 4 non-treated α-syn, 4 rasagiline-treated α-syn) that had not undergone behavioral testing. At 10 months of age, we gave them rasagiline in the drinking water for seven weeks. Four weeks prior to sacrificing, we injected them with BrdU (80 mg/kg, i.p., in PBS, pH 7.4) twice daily for 6 consecutive days.

Histological analysis

Immunohistochemistry. We perfused the 12 month-old mice transcardially with 0.9% saline followed by 4% paraformaldehyde and prepared 40 μm thick coronal sections for immunohistochemistry. Free-floating sections were treated with 10% H_2O_2 in PBS for 20 min. Specifically for BrdU staining, we treated sections with 2 N HCl in water for 30 min at 37 °C. We used the following primary antibodies: mouse anti-human α-synuclein (1:2000, Ab36615, Abcam), rat anti-BrdU (1:100, Oxford Biotechnology OBT0030) and/or mouse anti-NeuN (neuronal nuclei, 1:100, AB MAB377, Millipore). For detection of human α-synuclein and BrdU antibodies with the chromogen 3,3′diaminobenzidine (DAB), sections were incubated in biotinlylated horse anti-mouse (1:200, BA-2001, Vector Laboratories) or rabbit anti-rat secondary antibodies (1:200, E0468, Dako) respectively and then processed using a standard peroxidase-based method (Vectastain ABC kit and DAB kit; Vector Laboratories). For immunofluorescence staining, we used Alexa 488 anti-mouse and Alexa 568 anti-rat secondary antibodies respectively (raised in

goat, 1:200, A11029 or A11077, Invitrogen). Specimen analyses were performed either with a conventional light microscope (Eclipse 80i microscope; Nikon), a confocal laser microscope (Leica TCS SL) or with a stereological setting (Olympus BX50 microscope with a Marzhauser X–Y–Z step motor stage and the Visiopharm Integrator System software, Visiopharm A/S).

Cell Counting. We counted the number of BrdU positive cells in the granule cell layer of the OB using a systematic, random counting procedure optical dissector (section interval: 240 μm, counting frame: 100 μm×100 μm; counting grid: 300 μm×300 μm), [25,26]. We also determined the frequency of newborn neurons, by using confocal microscopy (focal plane of 1 μm) to quantify cells double-stained for NeuN and BrdU staining. On average, we analyzed 100 BrdU-positive cells in each animal (3 animals/group).

Statistical analysis

We expressed data as means±SEM. Statistical tests are described in figure legends. We performed statistical analysis using GraphPad Prism (version 5.0c, USA). We used the Bonferroni post-hoc test when the one-way ANOVA (analysis of variance), two-way ANOVA or two-way repeated measures (RM) ANOVA revealed significant differences.

Results

α-syn mice overexpress WT α-synuclein in the olfactory bulb

We observed significantly higher levels of α-synuclein immunoreactivity in the OB in the transgenic mice (Figure 4A-B). In different layers of the OB, including the glomerular and granule cell layers, we observed few large (approximately 4 μm diameter) and several smaller α-synuclein immunoreactive profiles (Figure 4C-D).

Olfactory deficits in the mouse model of PD

α-Syn mice show odor detection impairment. In the odor detection test, control mice detected an odor at the dilution 1:10^6, whereas α-syn mice needed a higher concentration (dilution 1:10^4) indicating that their ability to detect odor is significantly impaired. Neither WT nor α-syn mice were able to detect the lowest odor concentration (1:10^8) (Figure 2B).

α-Syn mice have short-term olfactory memory deficit. In the short-term olfactory memory test, we found that control mice spent less time sniffing the novel odor during the 2nd exposure, for all the three ITI tested, indicating that they remembered the odor (Figure 2D). α-Syn mice also remembered the odor after ITIs of 60 s and 90 s, but after an ITI of 120 s, they behaved as if they could not remember that they had been exposed to the odor before.

α-Syn mice have impaired odor discrimination. In the social odor discrimination test, we found that the α-syn mice were impaired compared to control mice at both odor intensity levels (Figure 2G). The non-social odor discrimination test yielded similar results (Figure 2H). By contrast, control and α-syn mice spent similar time exploring the novel item, meaning that α-syn mice using visual input can discriminate between novel and familiar items (Figure 2H). We also examined if mice prefer either of the two non-social odors we used (Figure 3A). This was not the

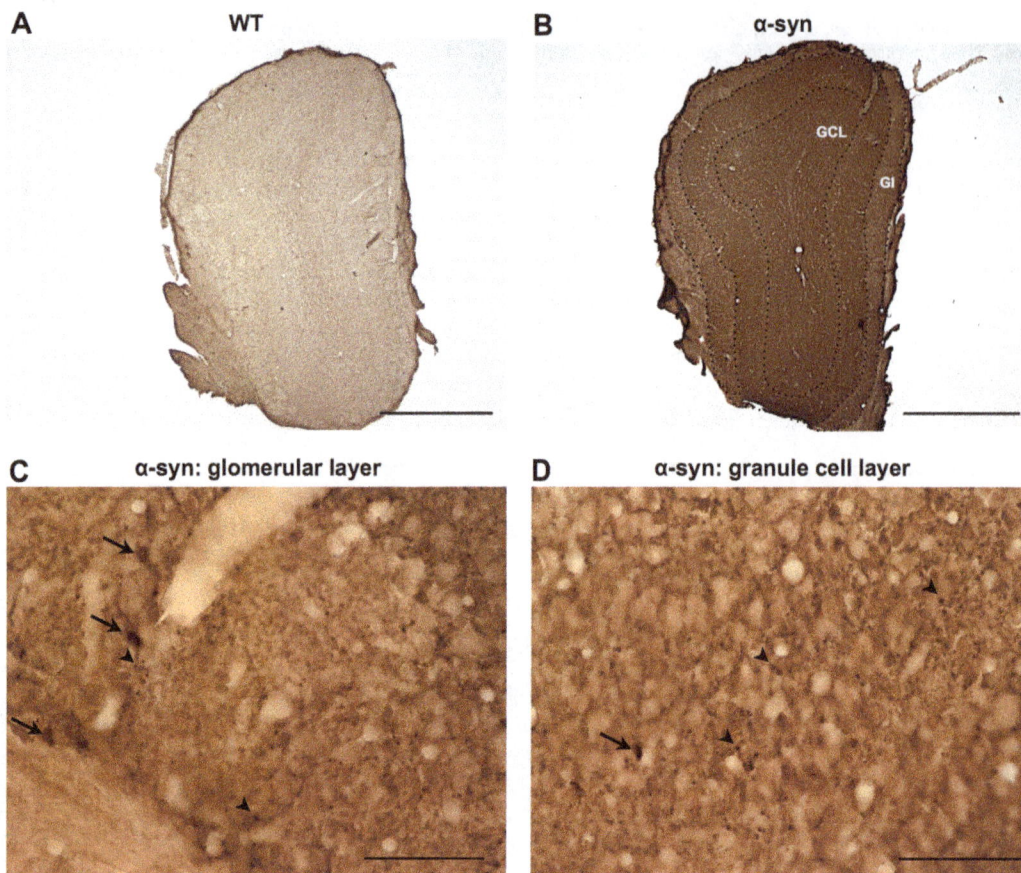

Figure 4. Overexpression of α-synuclein in the olfactory bulb of the α-syn transgenic mice. Immunostaining of human wild-type α-synuclein in OB of **A.** WT mice and **B-D.** α-syn mice aged 12 months. **A-B.** Scale bars: 500 μm. **C-D.** High magnification of **C.** the glomerular layer (Gl) and **D.** the granule cell layer (GCL). Scale bars: 50 μm. α-Syn mice exhibit high expression of human α-synuclein in the different layers of the OB. α-Synuclein immunoreactivity indicates large profiles (arrows) as well as numerous small α-synuclein immunoreactive puncta (arrow heads).

case, as when they were exposed to lime and lemon odors, both WT and α-syn mice spent equal time sniffing the odors (Figure 3B).

α-Syn mice show similar activity and motor ability compared to WT mice. In the open field, α-syn mice traveled a similar distance to control mice (Figure 3C). Likewise, α-syn mice spent similar time compared to control mice in the different areas of the open field (Figure 3D). Finally, control and α-syn mice spent a similar time on the rotarod (Figure 3E). Thus the α-syn mice did not exhibit any signs of anxiety or deficits in locomotor activity and motor function, which could have interfered with the interpretation of odor tests.

Olfactory deficits are age-dependent

To better characterize the olfactory deficits in this transgenic model and to determine if they are progressive with age, we assessed these deficits in animals aged 3, 11 and 18 months.

α-Syn mice exhibit an age-dependent odor detection impairment. In the odor detection test (Figure 5A) using the dilution $1:10^6$, α-syn mice were impaired at 11 and 18 months, whereas at 3 months the performances of WT and α-syn mice were similar (Figure 5B). Using the lower dilution ($1:10^4$), for the 3 ages studied (3, 11 and 18 months) both WT and α-syn mice were able to detect the odor (Figure 5C). Thus, α-syn mice became progressively impaired at detecting the odor at the concentration $1:10^6$ with aging.

With age, α-syn mice show a progressive short-term olfactory memory deficit. In the short-term olfactory memory test (Figure 5D) using the shorter ITI (60 s), α-syn mice aged 3 and 11 months remembered the odor and performed similarly to WT mice whereas they became impaired when they reached 18 months of age (Figure 5E). When mice were tested with the longer ITI (120 s), all α-syn mice failed to remember the odor whatever their age (3, 11 or 18 months) (Figure 5F). Taken together, the short-term olfactory memory deficit of α-syn mice progressively increases with age.

α-Syn mice show an age-dependent non-social odor discrimination deficit. In the non-social odor discrimination test (Figure 5G), α-syn mice aged 3, 11 and 18 months were impaired compared to WT mice. Interestingly, two-way ANOVA analysis indicated that age had a significant effect on the short-term olfactory memory performance (Figure 5H).

Rasagiline improves olfaction in α-syn mice

We evaluated whether 4–8-week rasagiline treatment ameliorates the olfactory deficit exhibited in the α-syn mouse model of PD aged 10–11 months.

Rasagiline improves odor detection in α-syn mice. We found that rasagiline treatment normalizes the ability of α-syn mice to detect odors at a concentration ($1:10^6$) when they otherwise are impaired compared to control mice. Whereas the untreated α-syn mice spent a short time (close to chance level)

Odor detection test

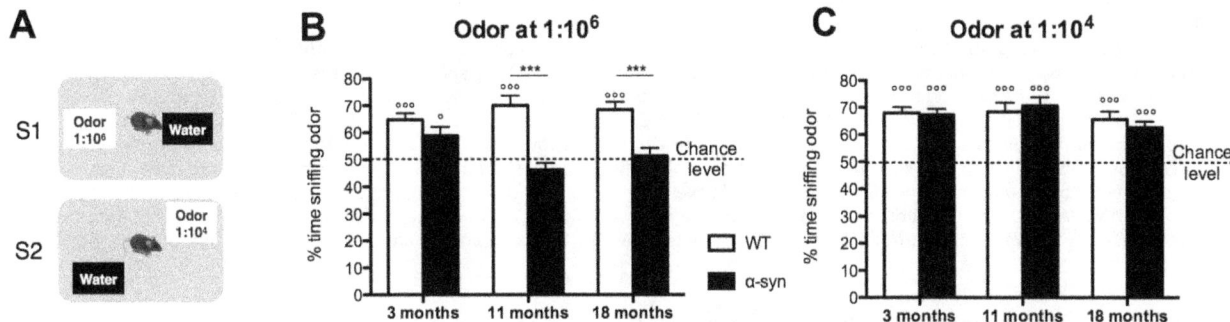

Short-term olfactory memory test

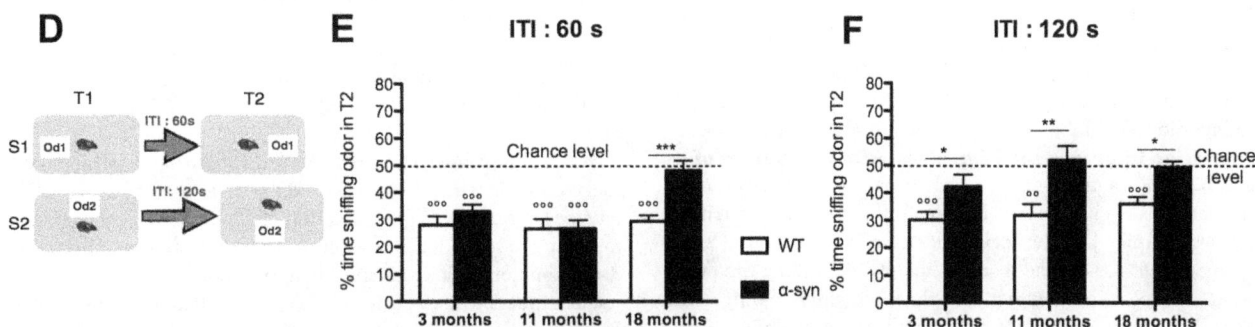

Non-social odor discrimination test

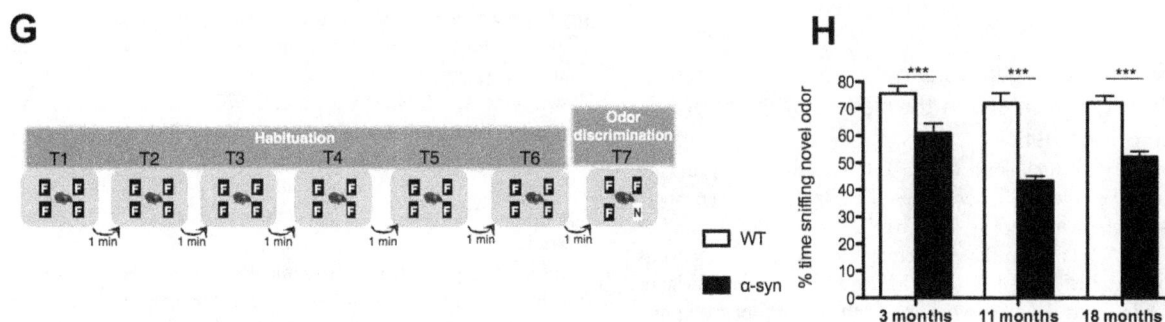

Figure 5. Olfactory deficits are age-dependent. A-C: Odor detection test. Description of the protocol consisting of 2 sessions (S). **B.** Percentage of time spent sniffing the odor at the concentration of 1:10^6 (session 1). WT mice aged 3, 11 and 18 months could detect the odor and the percentage of time sniffing the odor was significantly different from the chance level (50%) (^{ooo}p<0.001). On the contrary, α-syn mice are progressively impaired in detecting the odor. Whereas at 3 months transgenic mice spent more time sniffing the odor compared to the chance level (p<0.05), from 11 months of age their scores no longer differed from the chance level (p>0.05) and the percentage of time spent by α-syn mice to sniff the odor is significantly different from WT mice (two way ANOVA). Statistics: One-sample t-tests to compare each value to chance level (50%) (^{o}p<0.05, ^{ooo}p<0.001). Two-way ANOVA: age, p = 0.49, F(2,70) = 0.71; genotype, p<0.0001, F(1,70) = 40.21; age×genotype, p = 0.016, F(2,70) = 4.42; Bonferroni post-hoc (***p<0.001). **C.** Percentage of time spent sniffing the odor at the concentration of 1:10^4 (session 2). Both WT and α-syn mice aged 3, 11 and 18 months can detect the odor at the concentration of 1:10^4 and their percentage of time sniffing the odor is significantly different from the chance level (^{ooo}p<0.001). Moreover, there is no significant difference between the genotypes (two-way ANOVA p>0.05). Statistics: One-sample t-tests to compare each value to chance level (50%) (^{ooo}p<0.001). Two-way ANOVA: age, p = 0.12, F(2,70) = 2.15; genotype, p = 0.83, F(1,70) = 0.045; age×genotype, p = 0.64, F(2,70) = 0.45. **D-F: Short-term olfactory memory test. D.** Description of the protocol consisting of 2 sessions (S). **E.** Session 1 with an inter-trial interval of 60 s. Percentage of time spent sniffing the odor during T2 (trial 2) compared to the total time spent sniffing during both trials. All groups of WT mice aged 3, 11 and 18 months as well as α-syn mice aged 3 and 11 months remember the odor during the 2nd exposure and their percentage of time sniffing the odor during T2 is significantly different from the chance level (50%) (^{ooo}p<0.001). However, from 18 months of age, α-syn mice are impaired in remembering the odor during the 2nd exposure (one-sample t-test, p>0.05) and the percentage of time spent sniffing the odor during T2 is significantly higher compared to 18 month-old WT mice (two-way ANOVA, p<0.001).

Statistics: One-sample t-test to compare each value to chance level (50%) (^{ooo}p<0.001). Two-way ANOVA: age, p=0.0010, $F_{(2,70)}$=7.63; genotype, p=0.0032, $F_{(1,70)}$=9.32, age×genotype, p=0.011, $F_{(2,70)}$=4.78; Bonferroni post-hoc (***p<0.001). **F.** Session 2 with an inter-trial interval of 120 s. Percentage of time spent sniffing the odor during T2 compared to the total time spent sniffing during both trials. All groups of WT mice aged 3, 11 and 18 months remember the odor during the 2nd exposure and their percentage of time spent sniffing the odor during T2 is significantly different from the chance level (one-sample t-tests, ^{ooo}p<0.001). On the contrary, α-syn mice aged 3, 11 and 18 months, are all impaired in remembering the odor during the 2nd exposure (one-sample t-tests, p>0.05) and the percentage of time spent sniffing the odor during T2 is significantly higher compared to WT mice of the same age (two way ANOVA, *p<0.05, **p<0.01). Statistics: One-sample t-tests to compare each value to chance level (50%) (^{oo}p<0.01, ^{ooo}p<0.001). Two-way ANOVA: age, p=0.13, $F_{(2,70)}$=2.12; genotype, p<0.0001, $F_{(1,70)}$=26.86; age×genotype, p=0.53, $F_{(2,70)}$=0.64; Bonferroni post-hoc (*p<0.05, **p<0.01). **G-H: Odor discrimination test. G.** Description of the protocol consisting of 6 habituation trials and one odor discrimination trial. **H.** Percentage of time spent sniffing the novel odor. At 3, 11 and 18 months, α-syn mice spend significantly less time compared to age-matched control mice to sniff the novel odor suggesting that they are impaired in their ability to discriminate the novel odor (two-way ANOVA, ***p<0.001). Statistics: Two-way ANOVA: age, p=0.0028, $F_{(2,70)}$=6.42; genotype, p<0.0001, $F_{(1,70)}$=77.78; age×genotype, p=0.077, $F_{(2,70)}$=2.66; Bonferroni post-hoc (***p<0.001). For all tests, N=14 for group aged 3 and 18 months; N=10 for group aged 11 months.

sniffing the novel odor, the time the rasagiline-treated α-syn mice spent sniffing the odor was significantly greater than chance level. Moreover, the α-syn mice treated with rasagiline spent a similar amount of time sniffing a novel odor as the control mice. Thus, rasagiline restores the odor detection ability in α-syn mice to the level of control mice (Figure 6A).

Rasagiline does not ameliorate the short-term olfactory memory deficit in α-syn mice. We found that α-syn mice were impaired in their short-term olfactory memory when exposed to a test involving a 120 s ITI. The α-syn mice treated with rasagiline did not exhibit any improvement (Figure 6B).

Rasagiline improves the odor discrimination ability of α-syn mice. Rasagiline rescued the social or non-social odor discrimination deficits observed in α-syn mice. Thus, for both odor intensities examined in the social odor discrimination test, α-syn mice treated with rasagiline spent a similar time sniffing the novel odor as control mice and significantly more time sniffing than untreated α-syn mice (Figure 6C). Similarly, α-syn mice were impaired in the non-social odor discrimination test. We found that the rasagiline-treated α-syn mice significantly improved their ability to discriminate non-social odors. Thus, they spent significantly more time sniffing the novel odor compared to non-treated α-syn mice (Figure 6C) and behaved like normal control mice.

Olfactory bulb neurogenesis is not involved in olfactory deficits in α-syn mice and in the beneficial effect of rasagiline on the deficits

The number of newborn cells (BrdU-positive cells) in the granule cell layer did not differ between control and α-syn mice, regardless of whether they had been treated with rasagiline or not (Figure 7A-B). Likewise, the percentage of newborn neurons (NeuN-positive/BrdU-positive cells) in the OB granule cell layer was similar in control and α-syn mice, with or without rasagiline treatment (Figure 7C-D).

Discussion

The F28 transgenic mouse overexpressing human wild-type α-synuclein displays age-dependent olfactory impairments that are manifest as deficits in odor detection, discrimination and short-term memory. Control behavioral tests confirmed that these deficits are specifically due to alterations in olfaction. We found no changes in OB neurogenesis that could explain the olfactory deficits. Importantly, rasagiline improved the ability of α-syn mice to detect and discriminate odors, whereas olfactory short-term memory was unchanged.

Olfactory deficits in PD mouse models

We found that α-syn mice are impaired in their ability to detect an odor compared to control mice. They also exhibited impaired recollection of an odor after an inter-trial interval of 120 s. Interestingly, odor detection and short-term olfactory deficits have been described in mice after disruption of OB neurogenesis by treatment with the mitosis inhibitor AraC [27], suggesting a role for newborn olfactory interneurons in these functions. Even though OB neurogenesis was not impaired in the α-syn mice, it is interesting to note that these olfactory functions require olfactory interneurons.

We demonstrated impaired discrimination of social and non-social odors in α-syn mice. Mice exhibited no preference for either lime or lemon odors, indicating that the greater time spent sniffing the lime ("novel odor") was due it being perceived as novel. The discrimination deficit was specific to olfaction because when using visual input α-syn mice were able to discriminate a novel item from familiar ones. Our results are consistent with odor discrimination deficits previously described in a different α-syn mouse model overexpressing human wild-type α-syn under the Thy1 promoter [23].

Although the OB plays a crucial role in odor detection, odor discrimination and short-term olfactory memory, others brain structures, such as the olfactory cortex, could also be involved in olfactory deficits observed in our model. The piriform cortex, for example, is involved in the identification, categorization and discrimination of olfactory stimuli [28]. In the same way, the olfactory tubercle contributes to the odor perception, odor discrimination and higher–order olfactory functions [29]. Moreover, these structures are also affected in PD patients, exhibiting alpha-synucleinopathy [6] and could play a role in the hyposmia related to PD.

Interestingly, in this transgenic model, the deficits to detect, discriminate and to remember an odor during a short time interval are age-dependent, which emphasizes the relevance of this model for the neurodegenerative Parkinson's disease.

Specific olfaction deficits relevant to PD

Our behavioral studies show clearly that α-syn mice were actually capable of performing each test per se, and that the impairments they exhibited were specifically due to reduced olfactory functions. Thus, although α-syn mice could not detect an odor at a concentration of 1:10^6, they were able to detect the same odor at a concentration of 1:10^4. In the test of olfactory memory, they failed to remember an odor presented 120 s earlier, but were successful in doing so when the inter-trial interval was as short as 60 s or 90 s. Moreover, while they were impaired in discriminating a novel odor, α-syn overexpressing mice were normal when it came to discriminating novel objects. Finally, in the test for motor function and anxiety, the α-syn mice did not differ from control

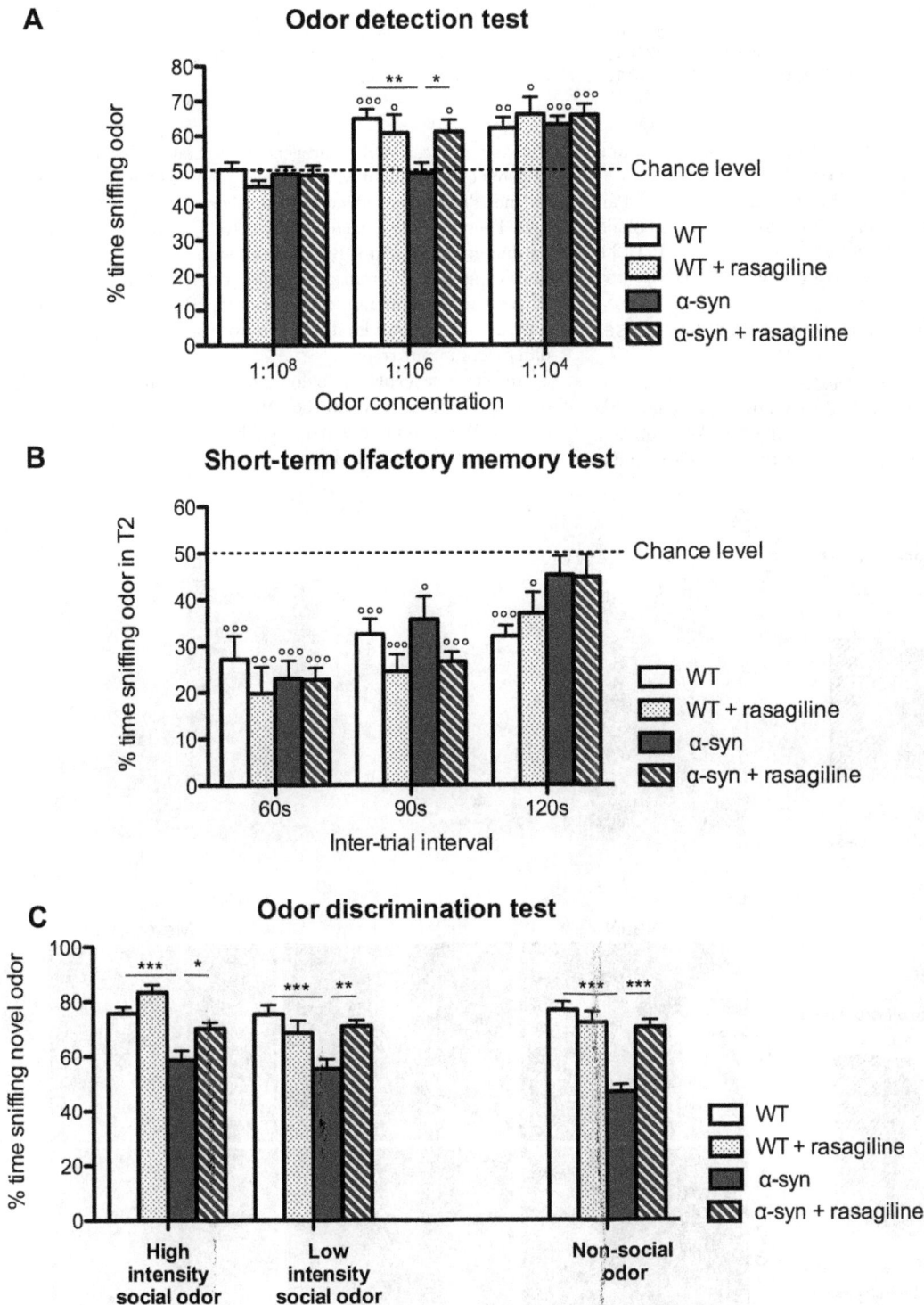

Figure 6. Rasagiline improved some aspects of olfaction in α-syn mice. A. Effect of rasagiline on odor detection deficit in α-syn mice. Rasagiline rescued the odor detection deficit in α-syn mice. At a concentration of 1:10^6, non-treated α-syn mice do not detect the odor and the percentage of time spent sniffing the odor was close to chance level, whereas rasagiline treated mice were significantly higher than the chance level. Moreover, rasagiline treated mice spent a similar time sniffing the odor compared to control mice. N = 9–10 for each group aged 10–11 months. Statistics: One-sample t-test to compare each value to chance level (50%), ($^{\circ}p<0.05$, $^{\circ\circ}p<0.01$, $^{\circ\circ\circ}p<0.001$). Two-way RM ANOVA: odor concentration, $p<0.0001$, $F_{(2,66)}=29$; group, $p=0.19$, $F_{(3,66)}=1.67$; odor concentration×group, $p=0.06$, $F_{(6,66)}=2.12$; Bonferroni post-hoc ($^*p<0.05$, $^{**}p<0.001$). **B. Effect of rasagiline on short-term olfactory memory impairment in α-syn mice.** For the 120 s-ITI, percentage of time spent sniffing the odor in T2 was not different from chance level for both α-syn mice groups, treated or not treated with rasagiline. Rasagiline did not improve the short-term olfactory memory in α-syn mice. N = 9–10 for each group aged 10–11 months. Statistics: One-sample t-test compare to chance level (50%), ($^{\circ}p<0.05$ and $^{\circ\circ\circ}p<0.001$). Two-way RM ANOVA: ITI, $p<0.0001$, $F_{(2,68)}=15.65$; group, $p=0.13$, $F_{(3,68)}=2.04$; ITI×group, $p=0.23$, $F_{(6,68)}=1.39$; Bonferroni post-hoc. **C. Effect of rasagiline on odor discrimination deficit in α-syn mice.** Percentage of time spent sniffing the novel odor of α-syn mice was increased by rasagiline treatment for both intensities of the social odor as well as for the non-social odor. α-

Syn mice treated with rasagiline were similar to control mice (p>0.05). Rasagiline rescued the odor discrimination deficit of α-syn mice. N = 18–21 for each group aged 10–11 months. Statistics for social odor discrimination: Two-way RM ANOVA, odor intensity, p = 0.032, F(1,74) = 4.78; group, p<0.0001, F(3,74) = 13.3; odor intensity×group, p = 0.034, F(3,74) = 3.04; Bonferroni post-hoc (*p<0.05, **p<0.01, ***p<0.001). Statistics for non-social odor discrimination: one-way ANOVA, p<0.001, F(3,73) = 18.16; Bonferroni post-hoc (***p<0.001).

mice, indicating that changes in these behavioral parameters were unlikely to be involved in the observed olfactory impairments.

The olfactory deficits we observed in the mouse PD model are consistent with clinical observations of impairments in the abilities to detect, discriminate and identify odors in idiopathic PD [30–34]. Therefore, our animal model is relevant to the clinical setting.

Do changes in neurogenesis or neuronal activity cause olfactory deficits?

We hypothesized that the olfactory deficits were due to alterations in OB neurogenesis. Reduced OB neurogenesis has previously been associated with deficits in odor detection and short-term olfactory memory [27]. Contrary to previous studies

showing reduced OB neurogenesis in transgenic mice overexpressing either human wild-type or mutant α-synuclein [8,35], we did not detect any reduction of newborn cells or neurons in the OB. Therefore, the olfactory deficits we observed are not likely to be due to changes in OB neurogenesis. Two earlier studies used mice expressing the transgene under a different promoter (PDGF whereas we used mouse α-synuclein) [8,35], which may lead to a different pattern and level of α-syn overexpression, explaining the differences in our results.

An alternate explanation for the olfactory impairment is that overexpression of α-syn in the OB directly affects local neuronal activity. We found that α-syn is highly overexpressed in the OB of our transgenic model, in particular in the glomerular layer

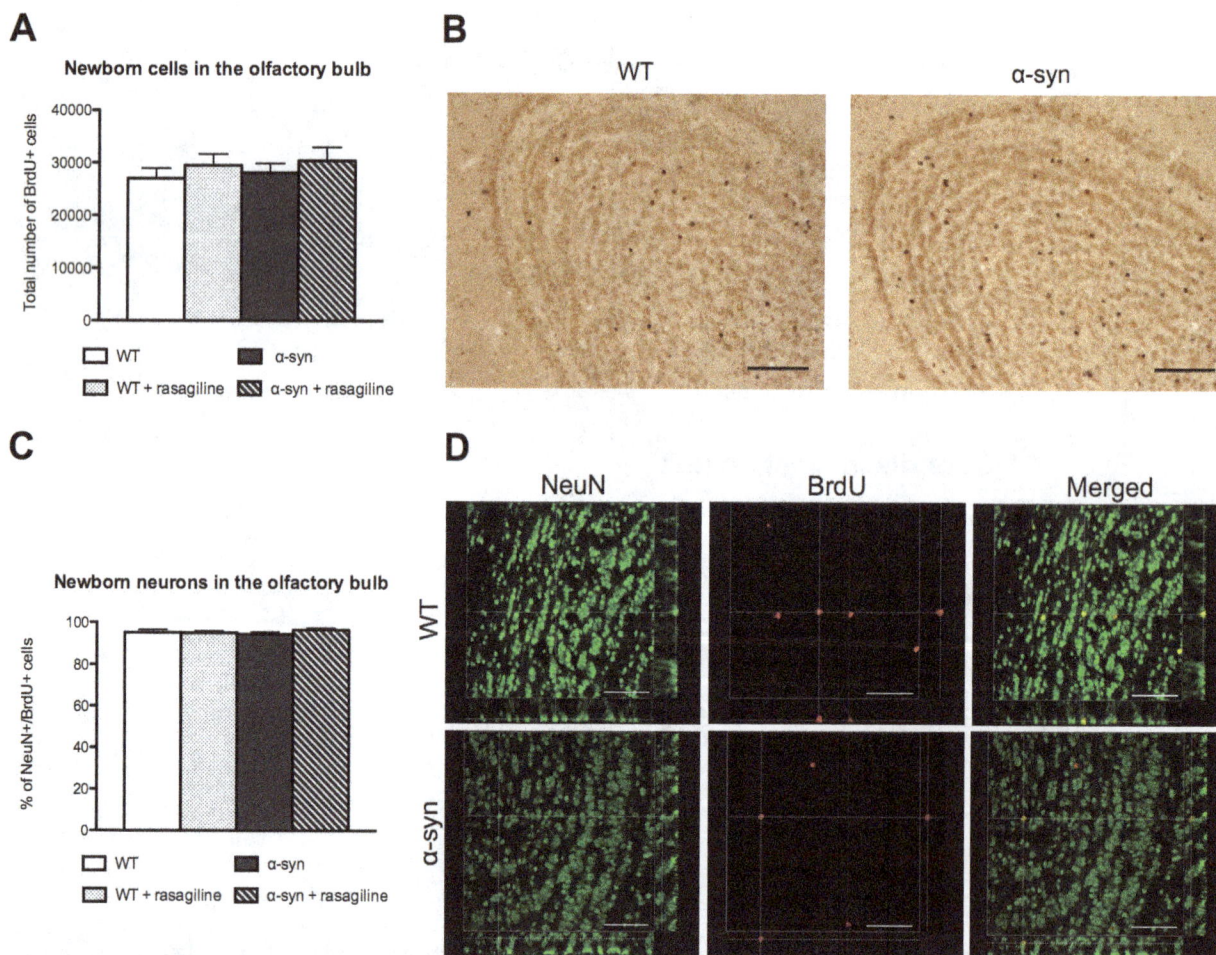

Figure 7. Neurogenesis changes are not involved in the olfactory deficit of α-syn mice and rasagiline-induced improvement. A. Quantification of newborn cells in the granule cell layer of the OB. Total number of BrdU positive cells was assessed every sixth section by stereology (counting frame 100 μm×100 μm; counting grid: 300 μm×300 μm). No difference between control and α-syn mice as well as no effect of rasagiline treatment was observed. N = 4-6 for each group aged 12 months. Statistic: one-way ANOVA, p = 0.66, F(3,14) = 0.54. **B.** BrdU staining in the olfactory bulb of WT and α-syn mice. Scale bars: 100 μm. **C.** Quantification of newborn neurons in the granule cell layer of the OB. The proportion of BrdU positive cells, which are also NeuN positive, was assessed by confocal microscopy. No difference between control and α-syn mice as well as no effect of rasagiline treatment was observed. On average, we analyzed 100 BrdU-positive cells in each animal, N = 3 mice in each group aged 12 months. Statistic: one-way ANOVA, p = 0.61, F(3,8) = 0.65. **D.** NeuN (green) and BrdU (red) double staining in the OB. Examples of NeuN-positive/BrdU positive-cells observed in WT and α-syn mice. Scale bars: 55.5 μm.

including dopaminergic periglomerular interneurons, in the mitral cell layer and the granule cell layer. Moreover, these cells and layers of the OB are clearly involved in the olfactory functions that we found to be impaired in the α-syn mice [27,36].

Rasagiline reverses certain olfactory deficits

We found that rasagiline treatment improved olfaction of α-syn mice and rescued odor detection and odor discrimination deficits. Rasagiline did not, however, ameliorate short-term olfactory deficits. Since we did not observe any reduction of neurogenesis in α-syn mice, nor any positive effect of rasagiline on neurogenesis, it is highly unlikely that the improved olfaction following rasagiline treatment is related to enhanced neurogenesis.

The rasagiline dose used (3 mg/kg) has previously been found to be efficacious in models of cerebral ischemia [37], vitamin deficiency [38] and PD [16]. We chose a long-term treatment (4–8 weeks) because we were not only interested in the MAO-B inhibitory activity of rasagiline [39], but also in its potential neuroprotective effects [16,40].

The rasagiline metabolite aminoindan is reported to be neuroprotective in several models of neuronal damage. This effect appears to be independent of MAO-B inhibition [41]. One potential mode of action of rasagiline is the stabilization of the mitochondrial membrane potential [42]. Interestingly, α-synuclein interacts directly with mitochondrial membranes [43], inhibits complex I [44], and thereby reduces the mitochondrial membrane potential [45]. Moreover, mitochondria in transgenic mice overexpressing mutant α-syn have been reported to display abnormal structure and function [46,47]. Therefore it is plausible that overexpression of α-syn directly affects mitochondria and thereby impairs neuronal function, and that rasagiline could potentially mitigate these effects.

Another option is that rasagiline, which is known to inhibit MAO-B, could improve endogenous dopamine concentrations and transmission [39]. The rasagiline dose (3 mg/kg/day) can reduce the residual MAO-B activity in the brain from 2% to 0.08% compared to untreated controls [38,48]. Knowing that MAO-B is expressed in the olfactory bulb [49,50], it is likely that rasagiline therapy could affect dopamine transmission in the olfactory bulb especially between interneurons and olfactory receptor neurons or mitral/tufted cells in the glomerular layer. These cells are involved in odor detection and discrimination,

functions, which are both improved by rasagiline treatment. Interestingly, dopamine receptor (D1 or D2) agonists or antagonists, affect odor discrimination learning as well as odor detection threshold [51–53]. In the same vein, transgenic mice lacking either dopamine transporters or D2 dopamine receptors exhibit odor discrimination impairment [22] suggesting that D2 dopamine receptor activation is important for odor discrimination. Therefore, the MAO-B inhibitory activity of rasagiline might underlie the beneficial effects on odor discrimination and detection.

In conclusion, our study shows a robust positive effect of rasagiline treatment on olfactory deficits in a transgenic mouse model of PD. The underlying mechanisms require further elucidation. Meanwhile it would be valuable to systematically examine if rasagiline improves olfaction in PD patients.

Supporting Information

Figure S1 Plastic cartridge and wood block used in the olfactory tests. A. The cartridge is a plastic tube (eppendorf), open at the two extremities, filled with a piece of compress. The compress is not accessible to the mice. During olfactory tests, odor solutions are prepared daily and we apply 400 µl of the solution (200 µl each side) to the compress. As both ends of the tubes are open, the odor can easily diffuse during the tests. **B.** The wood block is approximately 3 cm^3. During the impregnation time, wood blocks will get mouse odors mainly coming from mouse' body fluids.

Acknowledgments

We thank Birgit Haraldsson, Michael Sparrenius, Britt Lindberg, Alicja Flasch and Nikki Jane Damsgaard for excellent technical assistance and Sonia George for providing language help. GHP and PB are part of the Strong Research Environment Multipark (Multidisciplinary research in Parkinson's disease at Lund University) and Neurofortis.

Author Contributions

Conceived and designed the experiments: GHP EB MH PK KF CFA PB. Performed the experiments: GHP. Analyzed the data: GHP. Contributed reagents/materials/analysis tools: GHP EB PK KF CFA PB. Wrote the paper: GHP EB MH PK KF CFA PB.

References

1. Chaudhuri KR, Naidu Y (2008) Early Parkinson's disease and non-motor issues. J Neurol 255: S33–38.
2. Kranick SM, Duda JE (2008) Olfactory dysfunction in Parkinson's disease. Neurosignals 16: 35–40.
3. Haehner A, Boesveldt S, Berendse HW, Mackay-Sim A, Fleischmann J, et al. (2009) Prevalence of smell loss in Parkinson's disease--a multicenter study. Parkinsonism Relat Disord 15: 490–494.
4. Ross GW, Petrovitch H, Abbott RD, Tanner CM, Popper J, et al. (2008) Association of olfactory dysfunction with risk for future Parkinson's disease. Ann Neurol 63: 167–173.
5. Haehner A, Hummel T, Reichmann H (2010) Olfactory Function in Parkinson's Disease. Eur Neurol Rev 5: 26–29.
6. Ubeda-Bañon I, Saiz-Sanchez D, de la Rosa-Prieto C, Argandoña-Palacios L, Garcia-Muñozguren S, et al. (2010) alpha-Synucleinopathy in the human olfactory system in Parkinson's disease: involvement of calcium-binding protein- and substance P-positive cells. Acta Neuropathol 119: 723–735.
7. Höglinger GU, Rizk P, Muriel MP, Duyckaerts C, Oertel WH, et al. (2004) Dopamine depletion impairs precursor cell proliferation in Parkinson disease. Nat Neurosci 7: 726–735.
8. Winner B, Rockenstein E, Lie DC, Aigner R, Mante M, et al. (2008) Mutant alpha-synuclein exacerbates age-related decrease of neurogenesis. Neurobiol Aging 29: 913–925.
9. Marxreiter F, Nuber S, Kandasamy M, Klucken J, Aigner R, et al. (2009) Changes in adult olfactory bulb neurogenesis in mice expressing the A30P mutant form of alpha-synuclein. Eur J Neurosci 29: 879–890.
10. Stocchi F (2006) Rasagiline: defining the role of a novel therapy in the treatment of Parkinson's disease. Int J Clin Pract 60: 215–221.
11. Parkinson Study Group (2002) A controlled trial of rasagiline in early Parkinson disease: the TEMPO Study. Arch Neurol 59: 1937–1943.
12. Parkinson Study Group (2005) A randomized placebo-controlled trial of rasagiline in levodopa-treated patients with Parkinson disease and motor fluctuations: the PRESTO study. Arch Neurol 62: 241–248.
13. Rascol O, Brooks DJ, Melamed E, Oertel W, Poewe W, et al. (2005) Rasagiline as an adjunct to levodopa in patients with Parkinson's disease and motor fluctuations (LARGO, Lasting effect in Adjunct therapy with Rasagiline Given Once daily, study): a randomised, double-blind, parallel-group trial. Lancet 365: 947–954.
14. Olanow CW, Rascol O, Hauser R, Feigin PD, Jankovic J, et al. (2009) A double-blind, delayed-start trial of rasagiline in Parkinson's disease. N Engl J Med 361: 1268–1278.
15. Youdim MBH, Amit T, Falach-Yogev M, Bar-Am O, Maruyama W, et al. (2003) The essentiality of Bcl-2, PKC and proteasome-ubiquitin complex activations in the neuroprotective-antiapoptotic action of the anti-Parkinson drug, rasagiline. Biochem Pharmacol 66: 1635–1641.
16. Blandini F, Armentero MT, Fancellu R, Blaugrund E, Nappi G (2004) Neuroprotective effect of rasagiline in a rodent model of Parkinson's disease. Exp Neurol 187: 455–459.
17. Olanow CW (2006) Rationale for considering that propargylamines might be neuroprotective in Parkinson's disease. Neurology 66: S69–79.

18. Alvarez M, Grogan P (2011) Olfaction in Parkinson Disease patients taking Rasagiline: a case series. 10th International Conference on Alzheimer's & Parkinson's Diseases (AD/PD) 8: 927.

19. Alvarez M, Grogan P (2010) Can hyposmia recover with treatment of Parkinson's disease? 14th International Congress of Movment Disorder Society (MDS) 25: S255.

20. Westerlund M, Ran C, Borgkvist A, Sterky FH, Lindqvist E, et al. (2008) Lrrk2 and α-synuclein are co-regulated in rodent striatum. Mol Cell Neurosci 39: 586–591.

21. Oksman M, Kallunki P, Iivonen H, Miettinen P, Tanila H (2007) Cognitive deficits and cortical pathology in alpha-synuclein transgenic mice. 8th International Conference on Alzheimer's & Parkinson's Diseases (AD/PD) 4: S234.

22. Tillerson JL, Caudle WM, Parent JM, Gong C, Schallert T, et al. (2006) Olfactory discrimination deficits in mice lacking the dopamine transporter or the D2 dopamine receptor. Behav Brain Res 172: 97–105.

23. Fleming SM, Tetreault NA, Mulligan CK, Hutson CB, Masliah E, et al. (2008) Olfactory deficits in mice overexpressing human wildtype alpha-synuclein. Eur J Neurosci 28: 247–256.

24. Crawley JN, Belknap JK, Collins A, Crabbe JC, Frankel W, et al. (1997) Behavioral phenotypes of inbred mouse strains: implications and recommendations for molecular studies. Psychopharmacology (Berl) 132: 107–124.

25. Gundersen HJ, Bendtsen TF, Korbo L, Marcussen N, Møller A, et al. (1988) Some new, simple and efficient stereological methods and their use in pathological research and diagnosis. APMIS 96: 379–394.

26. Williams RW, Rakic P (1988) Three-dimensional counting: an accurate and direct method to estimate numbers of cells in sectioned material. J Comp Neurol 278: 344–352.

27. Breton-Provencher V, Lemasson M, Peralta MR, Saghatelyan A (2009) Interneurons produced in adulthood are required for the normal functioning of the olfactory bulb network and for the execution of selected olfactory behaviors. J Neurosci 29: 15245–15257.

28. Gottfried JA (2010) Central mechanisms of odour object perception. Nat Rev Neurosci 11: 628-641.

29. Wesson DW, Wilson DA (2011) Sniffing out the contributions of the olfactory tubercle to the sense of smell: hedonics, sensory integration, and more? Neurosci Biobehav Rev 35: 655–668.

30. Mesholam RI, Moberg PJ, Mahr RN, Doty RL (1998) Olfaction in neurodegenerative disease: a meta-analysis of olfactory functioning in Alzheimer's and Parkinson's diseases. Arch Neurol 55: 84–90.

31. Meusel T, Westermann B, Fuhr P, Hummel T, Welge-Lüssen A (2010) The course of olfactory deficits in patients with Parkinson's disease--a study based on psychophysical and electrophysiological measures. Neurosci Lett 486: 166–170.

32. Boesveldt S, Verbaan D, Knol DL, Visser M, van Rooden SM, et al. (2008) A comparative study of odor identification and odor discrimination deficits in Parkinson's disease. Mov Disord 23: 1984–1990.

33. Ansari KA, Johnson A (1975) Olfactory function in patients with Parkinson's disease. J Chronic Dis 28: 493–497.

34. Doty RL, Stern MB, Pfeiffer C, Gollomp SM, Hurtig HI (1992) Bilateral olfactory dysfunction in early stage treated and untreated idiopathic Parkinson's disease. J Neurol Neurosurg Psychiatry 55: 138–142.

35. Winner B, Lie DC, Rockenstein E, Aigner R, Aigner L, et al. (2004) Human wild-type alpha-synuclein impairs neurogenesis. J Neuropathol Exp Neurol 63: 1155–1166.

36. Abraham NM, Egger V, Shimshek DR, Renden R, Fukunaga I, et al. (2010) Synaptic inhibition in the olfactory bulb accelerates odor discrimination in mice. Neuron 65: 399–411.

37. Speiser Z, Mayk A, Litinetsky L, Fine T, Nyska A, et al. (2007) Rasagiline is neuroprotective in an experimental model of brain ischemia in the rat. J Neural Transm 114: 595–605.

38. Eliash S, Dror V, Cohen S, Rehavi M (2009) Neuroprotection by rasagiline in thiamine deficient rats. Brain Res 1256: 138–148.

39. Guay DRP (2006) Rasagiline (TVP-1012): a new selective monoamine oxidase inhibitor for Parkinson's disease. Am J Geriatr Pharmacother 4: 330–346.

40. Stefanova N, Poewe W, Wenning GK (2008) Rasagiline is neuroprotective in a transgenic model of multiple system atrophy. Exp Neurol 210: 421–427.

41. Youdim MBH, Bar-Am O, Yogev-Falach M, Weinreb O, Maruyama W, et al. (2005) Rasagiline: neurodegeneration, neuroprotection, and mitochondrial permeability transition. J Neurosci Res 79: 172–179.

42. Naoi M, Maruyama W, Akao Y, Yi H (2002) Mitochondria determine the survival and death in apoptosis by an endogenous neurotoxin, N-methyl(R)salsolinol, and neuroprotection by propargylamines. J Neural Transm 109: 607–621.

43. Nakamura K, Nemani VM, Wallender EK, Kaehlcke K, Ott M, et al. (2008) Optical reporters for the conformation of alpha-synuclein reveal a specific interaction with mitochondria. J Neurosci 28: 12305–12317.

44. Devi L, Raghavendran V, Prabhu BM, Avadhani NG, Anandatheerthavarada HK (2008) Mitochondrial import and accumulation of alpha-synuclein impair complex I in human dopaminergic neuronal cultures and Parkinson disease brain. J Biol Chem 283: 9089–9100.

45. Banerjee K, Sinha M, Pham Cle L, Jana S, Chanda D, et al. (2010) Alpha-synuclein induced membrane depolarization and loss of phosphorylation capacity of isolated rat brain mitochondria: implications in Parkinson's disease. FEBS Lett 584: 1571–1576.

46. Martin LJ, Pan Y, Price AC, Sterling W, Copeland NG, et al. (2006) Parkinson's disease alpha-synuclein transgenic mice develop neuronal mitochondrial degeneration and cell death. J Neurosci 26: 41–50.

47. Schapira AHV, Gegg M (2011) Mitochondrial contribution to Parkinson's disease pathogenesis. Parkinsons Dis doi: 10.4061/2011/159160.

48. Eliash S, Speiser Z, Cohen S (2001) Rasagiline and its (S) enantiomer increase survival and prevent stroke in salt-loaded stroke-prone spontaneously hypertensive rats. J Neural Transm 108: 909–923.

49. Büyüköztürk A, Kanit L, Ersöz B, Menteş G, Hariri NI (1995) The effects of hydergine on the MAO activity of the aged and adult rat brain. Eur Neuropsychopharmacol 5: 527–529.

50. Hashizume C, Suzuki M, Masuda K, Momozawa Y, Kikusui T, et al. (2003) Molecular cloning of canine monoamine oxidase subtypes A (MAOA) and B (MAOB) cDNAs and their expression in the brain. J Vet Med Sci 65: 893–898.

51. Yue EL, Cleland TA, Pavlis M, Linster C (2004) Opposing effects of D₁ and D₂ receptor activation on odor discrimination learning. Behav Neurosci 118: 184–190.

52. Wei CJ, Linster C, Cleland TA (2006) Dopamine D(2) receptor activation modulates perceived odor intensity. Behav Neurosci 120: 393–400.

53. Escanilla O, Yuhas C, Marzan D, Linster C (2009) Dopaminergic modulation of olfactory bulb processing affects odor discrimination learning in rats. Behav Neurosci 123: 828–833.

12

Stem and Progenitor Cell Subsets Are Affected by JAK2 Signaling and Can Be Monitored by Flow Cytometry

Ryuji Iida[1], Robert S. Welner[2], Wanke Zhao[3], José Alberola-Ila[1], Kay L. Medina[4], Zhizhuang Joe Zhao[3], Paul W. Kincade[1]*

1 Immunobiology and Cancer Program, Oklahoma Medical Research Foundation, Oklahoma City, Oklahoma, United States of America, **2** Beth Israel Deaconess Medical Center, Boston, Massachusetts, United States of America, **3** Department of Pathology, University of Oklahoma Health Sciences Center, Oklahoma City, Oklahoma, United States of America, **4** Department of Immunology, Mayo Clinic, Rochester, Minnesota, United States of America

Abstract

Although extremely rare, hematopoietic stem cells (HSCs) are divisible into subsets that differ with respect to differentiation potential and cell surface marker expression. For example, we recently found that CD86− CD150+ CD48− HSCs have limited potential for lymphocyte production. This could be an important new tool for studying hematological abnormalities. Here, we analyzed HSC subsets with a series of stem cell markers in JAK2V617F transgenic (Tg) mice, where the mutation is sufficient to cause myeloproliferative neoplasia with lymphocyte deficiency. Total numbers of HSC were elevated 3 to 20 fold in bone marrow of JAK2V617F mice. Careful analysis suggested the accumulation involved multiple HSC subsets, but particularly those characterized as CD150^{HI} CD86− CD18^{Lo}CD41+ and excluding Hoechst dye. Real-Time PCR analysis of their HSC revealed that the erythropoiesis associated gene transcripts Gata1, Klf1 and Epor were particularly high. Flow cytometry analyses based on two differentiation schemes for multipotent progenitors (MPP) also suggested alteration by JAK2 signals. The low CD86 on HSC and multipotent progenitors paralleled the large reductions we found in lymphoid progenitors, but the few that were produced functioned normally when sorted and placed in culture. Either of two HSC subsets conferred disease when transplanted. Thus, flow cytometry can be used to observe the influence of abnormal JAK2 signaling on stem and progenitor subsets. Markers that similarly distinguish categories of human HSCs might be very valuable for monitoring such conditions. They could also serve as indicators of HSC fitness and suitability for transplantation.

Editor: Connie J. Eaves, B.C. Cancer Agency, Canada

Funding: This work was supported by grants AI020069, and HL107138 (PWK); R01 HL096108 (KLM); HL079441 (ZJZ) from the National Institutes of Health and 4340-04-09-0 from the Oklahoma Center for Adult Stem Cell Research. PWK holds the William H. and Rita Bell Endowed Chair in Biomedical Research and is Scientific Director, Oklahoma Center for Adult Stem Cell Research. The funders had no role in study design, data collection and analysis, decision to publish, or preparation of the manuscript.

Competing Interests: The authors have declared that no competing interests exist.

* E-mail: kincade@omrf.org

Introduction

Hematopoietic stem cells (HSCs) normally replace blood cells according to need, but particular lineages are disproportionally expanded in myeloproliferative neoplasia (MPNs). The transition from normal, steady-state to disease likely involves HSC or very primitive hematopoietic progenitors, but it has been difficult to pinpoint such early changes [1,2].

Though extremely rare within bone marrow, HSC are heterogeneous and divisible with recently developed methods [3,4]. For example, single cell transplantation experiments revealed that some HSC preferentially generate myeloid or lymphoid lineage cells, while others are "balanced" with respect to blood cell formation [5–7]. New and less tedious barcoding approaches yielded essentially the same information [8,9]. Importantly, only the myeloid-biased and balanced HSC have durable self-renewal properties. Individual HSC also differ with respect to time spent in a quiescent state and ability to produce blood cells for prolonged periods [10,11]. At least some of these functional characteristics remain stable through multiple cycles of serial transplantation.

HSC subsets are also divisible on the basis of cell surface marker expression and fluorescent dye efflux [12–14]. For example, we found that unique CD150^{Hi} CD86− HSCs from normal animals are poor at replenishing the adaptive immune system [15]. HSCs with those characteristics accumulate with age and in animals repeatedly exposed to small amounts of lipopolysaccharide [16].

The JAK2V617F mutation is found in more than 95% of polycythemia vera (PV) patients as well as in many others with MPNs [1]. The consequences of this abnormality have been extensively studied with experimental models where single or multiple copies of JAK2V617F were introduced to mice [17–28]. In all of these circumstances, JAK2V617F causes progressive erythro-megakaryocytic abnormalities, and myelofibrosis has occasionally been observed. Transplantation experiments suggest that abnormal JAK2 signaling affects HSCs in the lineage marker negative, Sca-1 antigen positive, c-Kit+ (LSK) fraction. In contrast, the disease is not transferrable by multipotent progenitor (MPP), megaryocyte-erythroid progenitor (MEP), common myeloid progenitor (CMP) and granulocyte-macrophage progenitor (GMP) subsets [23,24,29].

Much has been learned about intracellular signaling pathways that involve JAK2 and participate in disease. For example, total protein phosphorylation was increased in PV patients [30]. Also, deletion of Stat 1 and 5 in JAK2 knock-in mice blocked disease progression [31,32]. The thrombopoietin receptor (TpoR or MPL) is known to be important for HSC self-renewal. Expression of mutant JAK2V617F in cultured cells reduced TpoR protein, inhibited apotosis and promoted cell division [33].

A potential therapy was suggested by the fact that IFNα selectively modulates the JAK2V617F burden in HSC in mice [2,34]. However, pegylated interferon alpha-2 may have increased numbers of somatic mutations outside of the JAK-STAT pathway in a clinical trial [35]. JAK2 phosphorylates the arginine methyltransferase PRMT5, leading to increased genetic instability [36].

It should be possible to use HSC characteristics to monitor early changes in such hematologic diseases, and we have now tested that with a model system. We selected a JAK2V617F transgenic model, in part because there was previous evidence for lymphopenia [21]. Overgrowth of HSC predicted to have poor lymphoid potential correlated with reduced lymphocyte progenitor numbers, but not their ability to differentiate. If markers of equivalent utility can be found for human HSC, they might be valuable for predicting rebound of the immune system following chemotherapy and marrow transplantation.

Results

HSC Numbers are Increased in JAK2V617F Transgenic Mice

A variety of experimental strategies have been used to introduce single or multiple copies of the JAK2V617F mutation to mice [17–28]. Different levels and sites of expression in these models might account for different conclusions about the effects on HSC [37]. Although conditional knock-in mice have been described, PV patients can have multiple copies of the mutant JAK2V617F gene and we wished to exploit a robust model. Therefore, transgenic Line A animals [21] were studied between 10–21 weeks of age. Marrow cellularities were determined with two tibias from each mouse and significant differences were not found between any of the control and transgenic animals. However, cell numbers in the stem/progenitor rich $Lin^- Sca-1^+ c-Kit^{Hi}$ (LSK) fraction were significantly elevated (Fig. 1A). All of these animals had marked splenomegaly, and the incidences of LSK were also increased in that site (Fig. 1B). Elevations of more stringently gated $CD150^+$ $CD48^-$ within the LSK fraction [13] were also seen in both organs (Fig. 1C and 1D).

There are many phenotypic definitions of HSC. For example, Eaves and colleagues gate on $EPCR^+ CD45^+$ cells that are also $CD150^+ CD48^-$ [38]. There was a five-fold accumulation of HSC enumerated in this way in JAK2V617F mice (Fig. 1E). We conclude that the transgene expands numbers of HSC defined by two sets of inclusive criteria.

A $CD150^{HI}$ $CD86^-$ $CD18^{Lo}CD41^+$ HSC Subset Accumulates in JAK2V617F Transgenic Mice

Previous studies revealed that HSC are heterogeneous and composed of functionally specialized subpopulations [3,4]. In some cases, this corresponds to phenotypes [10,39], and the JAK2V617F transgene preferentially expands $CD150^{Hi}$ HSC (Fig. 2A). That category of HSC was reportedly less likely to generate lymphocytes than $CD150^{Lo/-}$ HSC [10]. Similarly, Goodell and colleagues found lymphopoietic potential was lowest among HSC that strongly exclude Hoechst dye [40]. These "lower side population" HSC were the most increased by JAK2V617F (Fig. 2B).

We recently found that a distinct population of $CD150^{Hi}$ $CD86^-$ HSC have poor lymphopoietic potential and accumulate with age or chronic Toll-like receptor (TLR) stimulation [15,16]. Cells with this distinctive phenotype were prominent in the JAK2V617F transgenic mice (Fig. 2C). Percentages and numbers of LSK $CD48^-$ $CD150^{Hi}$ $CD86^-$ HSC increased in response to JAK2 signaling in all of the transgenics (Fig. 2C). While most of the increases in this transgenic model involved $CD86^-$ HSC, numbers of HSC expressing CD86 also increased in some of the animals.

CD18 staining patterns in healthy mice are similar to those obtained with CD86 [16], and that was also the case for transgenic mice (compare Fig. S1A with Fig. 2C above). In addition, we reported that CD41 expression on HSC increases with chronological age [15]. Another group concluded that CD41 acquisition corresponds to loss of lymphopoietic potential [41]. A similar phenomenon resulted from the JAK2V617F transgene in young mice (Fig. S1B). Although gating for $CD41^-$ cells has been used as a strategy to enrich HSC [13,41], CD41 was not included in our lineage depletion cocktail for all analyses. Interestingly, expression of CD41 tended to be reciprocal with CD86 on $CD150^{Hi}$ $CD48^-$ HSC (Fig. S1C). In contrast to these relationships, significant shifts were not recorded in HSC bearing VCAM-1 or CD39 (Fig. S1D, E). These markers were investigated because they appeared to change in a previous study involving TLR stimulation [15].

We then sorted $CD150^+$ $CD48^-$ LSK cells from normal and JAK2V617F transgenic marrow for Real Time-PCR analyses with a small set of genes. This revealed depressed CD86 and Mpl transcripts, offset by increased expression of Klf1, Gata1 and Epor (Fig. 3A and B). In contrast, expression of Lnk, Flk2, Satb1, Ikzf1 HSC/lymphoid genes was unaltered (Fig. 3A). Stability of Mpo, Csf3r, Sfpi1 and Vwf lineage associated genes was also seen (Fig. 3B). The same PCR analyses were simultaneously performed and validated with sorted progenitors from normal mice (Fig. S2A and B). These are valuable when compared to Fig. 3B in showing that the transgene preferentially primes transcripts corresponding to primitive erythroid (CFU-E) lineage cells.

It seemed possible that the JAK2V617F transgene caused interferon pathway signaling as is the case in patients with essential thrombocythemia [30]. Also, HSC themselves are capable of Ifnγ production [42]. Stem/progenitors that normally lack Sca-1 could have acquired it as a result of such cytokine influence [43,44]. However, Real-Time PCR analyses suggested that mechanism does not account for the expansion of LSK (Fig. 3C). That is, transcripts corresponding to Ifn inducible genes were not elevated.

Thus, the $CD86^+$ HSCs that predominate in healthy young animals are present and occasionally even elevated in marrow of JAK2V617F transgenic mice. However, they are diluted by large numbers of Lin^- $Sca-1^+$ $c-Kit^+$ $CD150^{Hi}$ $CD34^-$ $CD48^-$ $CD86^-$ $CD18^-$ $CD41^+$ and/or Hoechst dye excluding cells. Earlier studies associated those properties with overlapping populations of HSC that are poor at replenishing the adaptive immune system [13–15]. Possibly related to that, JAK2 transgenic HSC had elevated expression of two transcription factors required for erythropoiesis.

Primitive Hematopoietic Progenitors are Affected by JAK2 Signaling

Multipotent progenitors with limited self-renewal capability represent the immediate progeny of HSC and subsets of them have been defined with various collections of markers. A method described by Trumpp and colleagues [45] revealed that expansion of $CD150^+$ $CD48^-$ $CD34^-$ LSK (MPP1) as well as $CD150^+$ $CD48^+$ $CD34^+$ LSK (MPP2) but not $CD150^-$ $CD48^+$ $CD34^+$

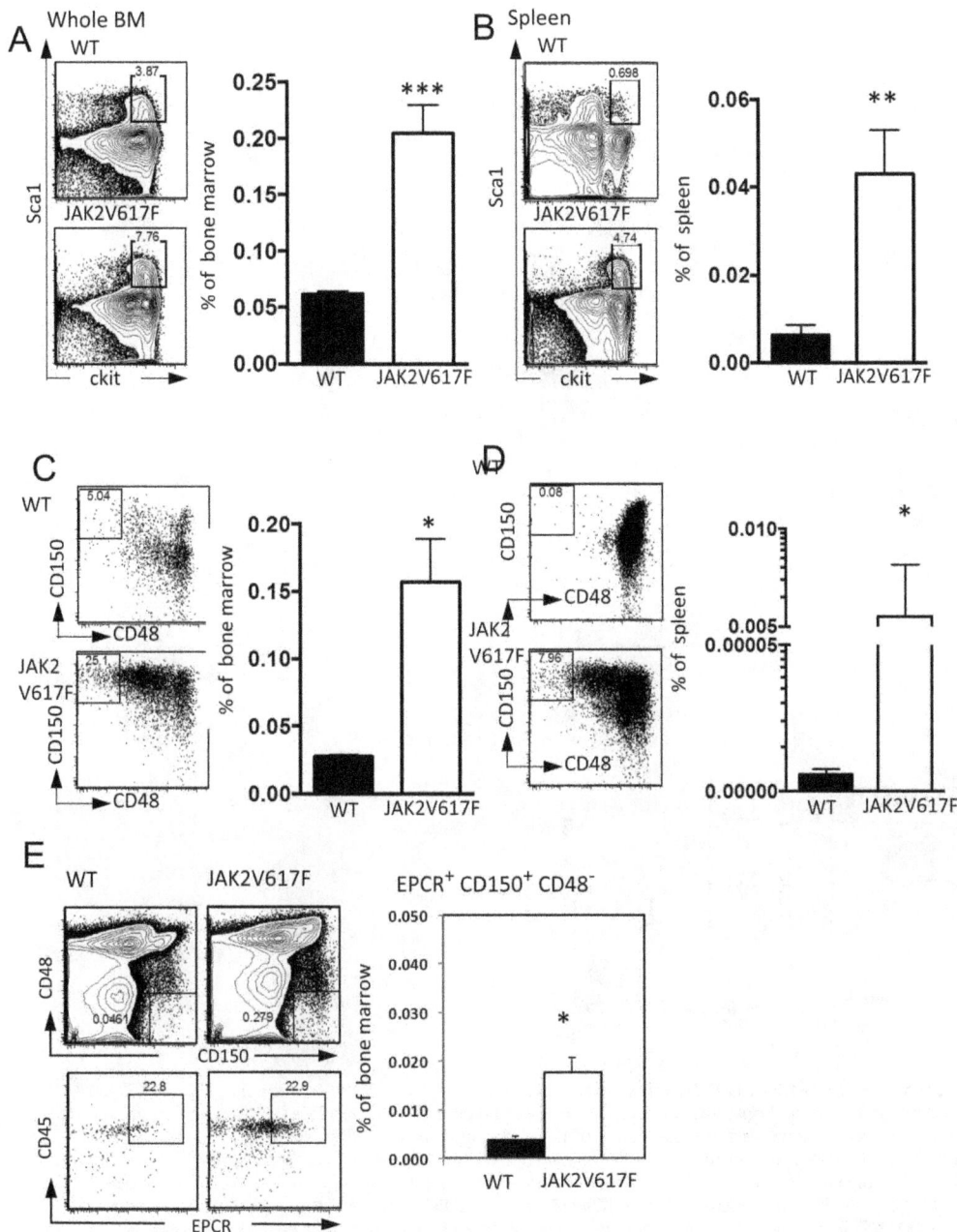

Figure 1. LSK and CD150⁺ CD48⁻ HSCs expand in JAK2V617F Tg mice. (A, B) LSK were gated with c-kit and Sca-1 in BM (**A**) and spleen (**B**). The percentages of LSK are shown(right panel in A and B). CD150⁺ CD48⁻ HSC in BM (**C**) and spleen (**D**) are given. Closed and open bars indicate WT and JAK2V617F individual mice, respectively. The data are representative of those obtained in two independent experiments (N = 8, 4) (**E**) Distinct HSC were resolved as EPCR⁺ CD45⁺ CD150⁺ and CD48⁻. The percentages of CD45⁺ EPCR⁺ CD150⁺ CD48⁻ in 10⁶ BM were increased in JAK2V617F mice. The data are representative of those obtained in two independent experiments (N = 6). $p<0.001(***)$, $p<0.01(**)$, $p<0.05(*)$.

LSK (MPP3) occurs in our transgenic mice (Figure 4A). Further resolution of those subsets showed that most of this expansion involved MPP lacking the CD86 indicator of lymphopoietic potential (Figure 4B).

Morrison and colleagues recently used a different collection of markers to define a different series of multipotent progenitors, while retaining the same MPP nomenclature [46]. We found that this approach showed significant depletion of CD150⁻ CD48⁻ CD229⁻ CD244⁻ (MPP1), CD150⁻ CD48⁻ CD229⁺ CD244⁻ (MPP2), CD150⁻ CD48⁻ CD229⁺ CD244⁺ (MPP3), and

CD150⁻ CD48⁺ (HPC) in the JAK2 mice (Figure 4C). Remarkably, this depletion primarily involved the lymphopoietic CD86⁺ progenitors (Figure 4D).

These changes suggest that multipotent cells capable of generating lymphocytes are either greatly diluted (analysis with the Trumpp marker scheme) or depleted (Morrison method) as a result of JAK2 signaling. Furthermore, they are consistent with those described above for HSC and again suggest that potential to replenish the immune system is probably diminished.

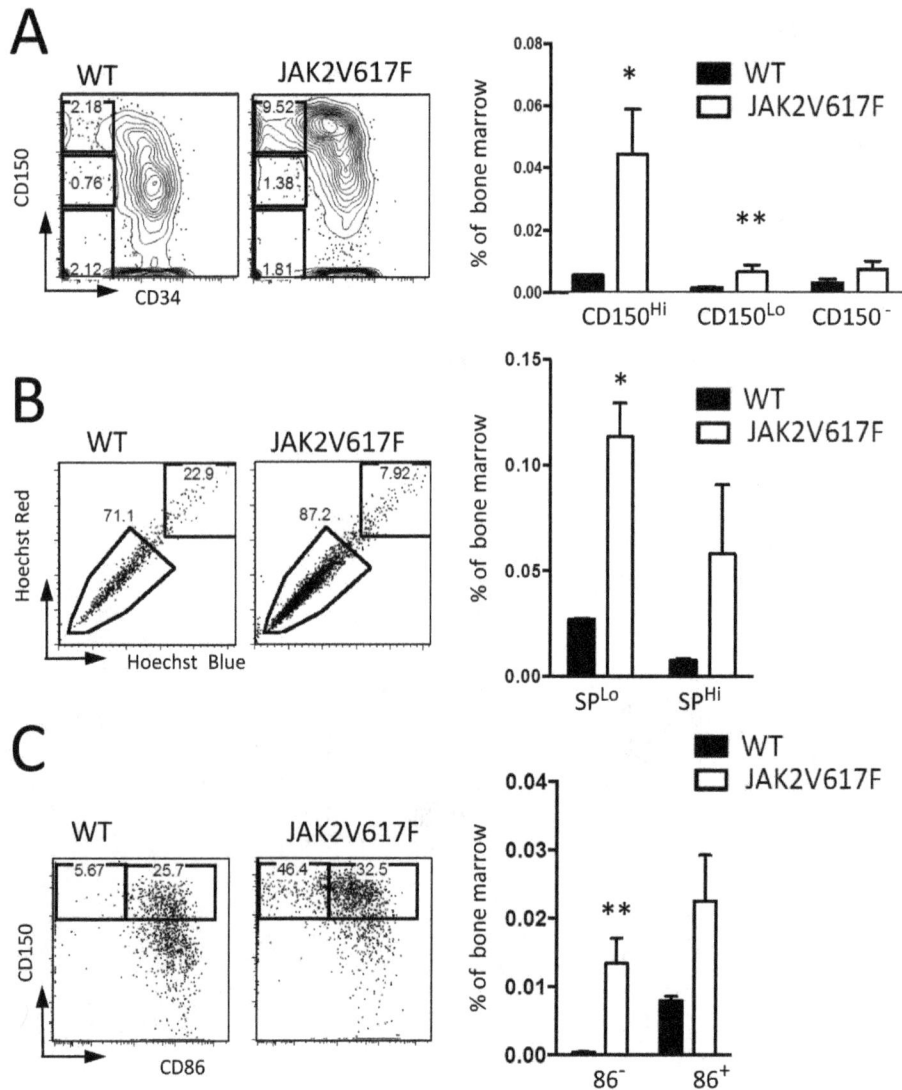

Figure 2. Particular HSC subsets expand in JAK2V617F Tg mice. (A) LSK were resolved with CD150 and CD34. CD34$^-$ CD150$^{Hi, Lo}$, and negative HSC were gated as described by Morita [10]. The percentages of CD34$^-$ HSC subsets in BM were calculated. Closed and open circles indicate individual WT and JAK2V617F mice, respectively. The data are representative of those obtained in two independent experiments (N = 6). **(B)** Lower side population cells in BM (spindle-shaped gating) are thought to include myeloid biased HSC. The "side population tip" was increased in JAK2V617F mice. Representative plots are shown in the left panels, while cell percentages of lower and upper cells are given in right panels. **(C)** Both CD86$^-$ and CD86$^+$ CD150$^+$ CD48$^-$ LSK cells in BM were increased in JAK2V617F mice. CD48$^-$ LSK were resolved with CD86 and CD150. Cell percentages of CD86$^-$ and CD86$^+$ in CD150^{Hi} were calculated. The data are representative of those obtained in three independent experiments (N = 8) $p < 0.01$(**), $p < 0.05$(*).

Lymphoid Progenitors Progressively Decline in JAK2V617F Transgenic Mice

A previous analysis of this line of transgenic mice revealed small elevations in peripheral and bone marrow leukocytes, most likely granulocytes, as well as subtle changes in CD45R/B220$^+$ lymphocytes [21]. We conducted a more thorough analysis of bone marrow and found evidence for reduced lymphopoiesis. That is, there were significantly reduced numbers of lymphoid lineage cells marked by B220 or CD19 expression coincident with expansion of Gr-1 bearing myeloid cells (Fig. 5A). As noted above, total marrow cellularities were unaffected by the transgene. Lineage marker negative (Lin$^-$) subsets were then gated for analysis of primitive hematopoietic cells (Fig. 5B). That includes Lin$^-$ c-Kit^{Lo} Flk-2^{Hi} IL-7Rα^+ common lymphoid progenitors

(CLP) that we found were depleted more than 10 fold in the transgenics (Fig. 5C, left panel). CLP are thought to derive from a fraction of Lin$^-$ Sca-1$^+$ c-Kit^{Hi} Flk2^{Hi} lymphoid primed multipotent progenitors (LMPP) [47,48]. Numbers of these primitive lymphopoietic cells were also reduced (Fig. 5C, right panel). These changes were more pronounced in older transgenic animals (data not shown).

Given that the transgene is widely expressed in hematopoietic tissue, it was possible that lymphoid progenitors were directly affected. Therefore, we sorted the small numbers of CLP present in young JAK2V617F animals and placed them in stromal cell-free cultures that support B lymphopoiesis (Fig. 5D). Progenitors enriched in this way generated normal numbers of B lineage lymphocytes marked by CD19 expression.

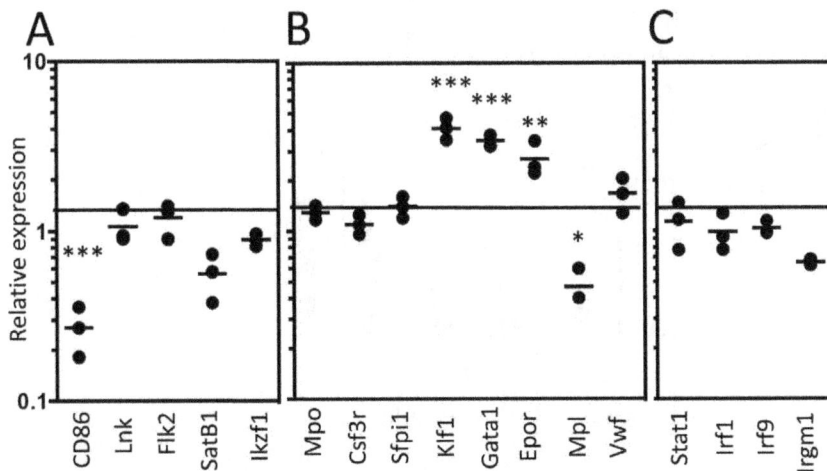

Figure 3. Erythropoiesis associated transcripts are overrepresented in JAK2V167F HSC. Gene expression analysis was performed by Real-Time PCR on a small panel of genes, normalizing with GAPDH for a reference housekeeping gene. Lymphoid genes (**A**), myeloid-, erythroid-, and megakaryocytic- genes (**B**), and IFNγ responsive genes (**C**) were indicated. The data are representative of those obtained in two independent experiments $p < 0.001(***)$, $p < 0.05(*)$.

We conclude that the JAK2 transgene affects early events in lymphopoiesis, depressing numbers of lymphoid progenitors. This is consistent with the erythroid associated gene expression in their HSC shown above. However, artificial JAK2 signaling in these animals did not interfere with the ability of residual lymphoid progenitors to differentiate. Those that escaped a JAK2 regulated checkpoint were fully able to generate B lineage lymphocytes.

JAK2 May Slightly Influence Stem/Progenitor Cell Proliferation

Increased self-renewal, prolonged survival and/or decreased export of stem/progenitor cells from the marrow could all account for their expansion in the transgenic mice. One of these parameters was assessed by staining HSC subsets with the proliferation associated Ki67 marker together with Hoechst dye (Figure 6A). While the results would be consistent with expansion of CD86⁻ HSC in JAK2 transgenics, they did not reach statistical significance and the same was true for some multipotent progenitor subsets (Figure 6B, Trumpp method, middle panel; Morrison method, far right panel). As another approach, BrdU was added to very short term (1 hr) cultures of bone marrow cells (Figure 6C, D). This revealed that there were significantly more dividing stem and progenitor cells in transgenic marrow. Note that all of these data were calculated and expressed in terms of percentages, but would reflect absolute numbers per bone because cellularities were unaffected. Even a small change in proliferation may be significant over time and it is interesting that the changes preferentially occurred in CD86⁻ cells.

Two HSC Subsets can Transfer Myeloproliferative Disease

Previous transplantation studies revealed that HSC, but not hematopoietic progenitors from JAK2V617F mice can transmit disease [2,23–25,29]. Therefore, we could now ask if normal versus disease differentiation decisions occur in a particular HSC subset. Marrow was harvested from 15 week old JAK2V617F animals that had high platelet (>4×10⁶/mm³) and RBC (> 11×10⁶/mm³) counts as well as splenomegaly (>7 fold increased weight). The CD150⁺ CD48⁻ LSK cells were then sorted according to CD86 expression (as illustrated in Figure 2C) and 100 of each HSC subset were mixed with 10⁵ whole rescue

marrow cells before transplantation into lethally irradiated mice (Figure 7). The design was such that progeny of CD45.2 marked transgenic HSC could be distinguished from rescue and recipient mouse cells identified as CD45.1.

Sampling of peripheral blood began 4 weeks later and continued for an additional 14 weeks. Engraftment was considered successful when >1.0% of total nucleated cells were persistently of donor type (Figure 7A). This occurred in five of ten CD86⁻ and nine of ten CD86⁺ HSC transplants. Animals were considered to have myeloproliferative disease when platelet counts exceeded 10³ per mm³ of blood. This occurred by 12 weeks post-transplant in three of the five (60%) CD86⁻ recipients as compared to eight of the nine (89%) animals that received CD86⁺ HSC (Figure 7B). Death tended to occur earlier in mice transplanted with the CD86⁻ HSC, and especially those with low chimerism (Figure 7C and data not shown). As described above, CD86⁻ HSC preferentially expand in response to JAK2 signaling (Figure 3C), but these transplantation results indicate that disease can initiate in either subset.

Discussion

Many questions arise from recent findings relating to HSC heterogeneity. For example, are the properties of HSC populations affected by malignancies? At least some functional characteristics, such as the ability to restore the immune system, correlate with patterns of cell surface marker expression. That technical breakthrough suggested that flow cytometry rather than lengthy transplantation assays could be used to address some of these issues. Here we have tracked HSC subset changes when hematopoiesis was perturbed in a model of myeloproferative neoplasia. Dilution or depletion of HSC predicted to be lymphopoietic correlated with declines in common lymphoid progenitors.

Our prior studies indicated that loss of CD86 and CD18 together with increases in CD41 and CD150 defines HSC that accumulate in aged or chronically LPS treated animals [15,16]. Even in untreated adult mice, such HSC are poor with respect to lymphocyte formation and probably overlap with myeloid-skewed HSC identified in single cell transplantation experiments [5–7]. While the JAK2V617F transgene caused accumulation of HSC

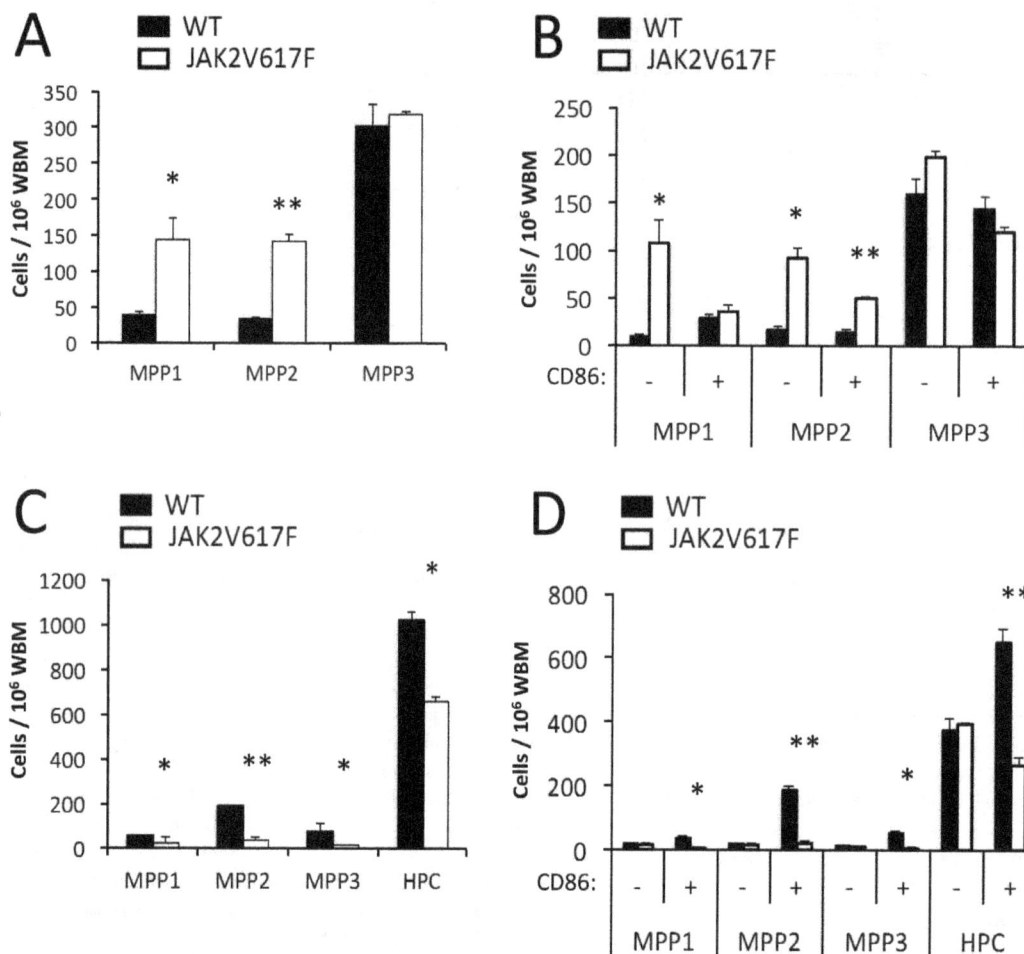

Figure 4. Multipotent progenitors are altered in JAK2V617F Tg mice. (**A**) LSK from control and Tg mice were resolved with CD150, CD48, Flk2 and CD34 staining into MPP subsets as described by Trumpp and colleagues [45]. The subset definitions are CD150$^+$ CD48$^-$ CD34$^+$ LSK (MPP1), CD150$^+$ CD48$^+$ CD34$^+$ LSK (MPP2) and CD150$^-$ CD48$^+$ CD34$^+$ LSK (MPP3). (**B**) These populations were further subdivided on the basis of CD86 expression. (**C**) Morrison and colleagues exploited CD150, CD48, CD229 and CD244 to define MPP subsets [46], and (**D**) again we subdivided them on the basis of CD86 display. The surface markers define CD150$^-$ CD48$^-$ CD229$^-$ CD244$^-$ (MPP1), CD150$^-$ CD48$^-$ CD229$^+$ CD244$^-$ (MPP2), CD150$^-$ CD48$^-$ CD229$^+$ CD244$^+$ (MPP3), and CD150$^-$ CD48$^+$ (HPC). The results are given as the percentages of BM cells and are representative of those obtained in two independent experiments (N = 3). Despite the common nomenclature used with these methods, distinct subsets and possibly separate differentiation pathways are identified (See Figure 8).

defined with five commonly used methods, closer examination revealed that it primarily involved those that lacked CD86. In fact, most of them were Lin$^-$ Sca-1$^+$ c-Kit$^+$ CD150^{Hi} CD34$^-$ CD48$^-$ CD86$^-$ CD18$^-$ CD41$^+$ and/or Hoechst dye excluding cells.

Multipotent progenitors with reduced self-renewal potential have been defined with various markers. Here we exploited two approaches that utilize the same nomenclature but describe different subsets of cells [45,46]. The JAK2 driven changes in these subsets suggest alternative differentiation pathways may exist for these very primitive cells (Figure 8). Acquisition of CD48 marks one option (MPP1 to 2, 3 in the Trumpp scheme) and it is the one promoted by JAK2 signaling. Alternatively, HSC lose CD150 (formation of MPP 1, 2, 3 in the Morrison pathway), and this route is diminished in the transgenics. It is important to stress that the changes in MPP paralleled those in HSC and mainly involved CD86$^-$ subsets. Further analysis of these distinct MPP populations might reveal if there are also differences in lymphoid versus myeloerythroid lineage priming.

We do not have a precise explanation for why particular populations expand and others contract in this model. Slightly

increased proliferation in cells that are normally quiescent could be significant over time. It is noteworthy that JAK2V617F effects on stem cell proliferation are highly dependent on the model system used and need not parallel changes in progenitors [37].

It is interesting to compare manipulations of JAK2V617F to those involving Lnk. The latter is an adaptor molecule that negatively regulates JAK2 signaling via its SH2 domain [49]. That in turn inhibits downstream signaling mediated by Epor and MPL. Lnk deficient mice have increased numbers of HSC with enhanced self-renewal potential and increased quiescence [50]. Lnk$^{-/-}$ mice also develop myeloproliferative neoplasia with thrombocytosis, splenomegaly and fibrosis [51]. The MPN phenotype develops more rapidly in Lnk$^{-/-}$ mice that also have the JAK2V617F mutation [52]. Unlike the case with JAK2V617F transgenics, Lnk$^{-/-}$ mice are characterized by B cell overproduction. Perhaps the impact of JAK2 pathway signaling in lymphopoietic cells is different in these two circumstances. Lnk can also inhibit c-kit and Flk2 signaling, raising the possibility that it influences lymphopoiesis independently of JAK2. The few CLP recovered from JAK2V617F transgenics retain normal potential to

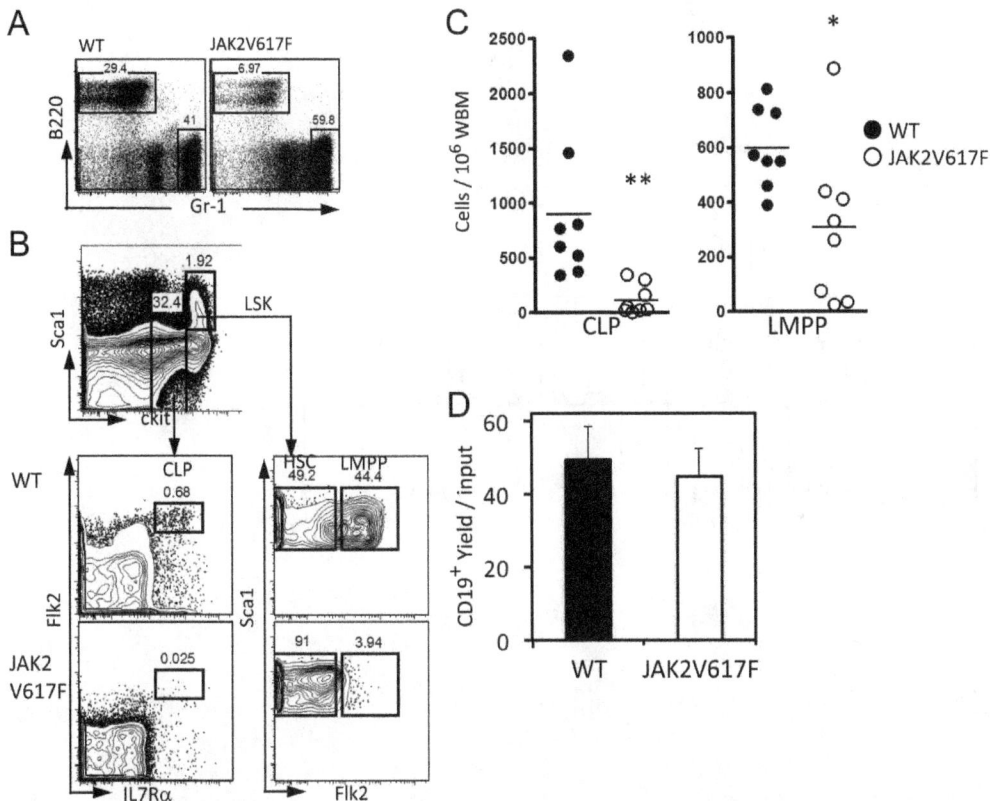

Figure 5. Lymphoid progenitors are depleted in JAK2V617F transgenic mice. (A) Whole bone marrow cells were characterized by staining with B220 (lymphoid specific) and Gr1 (myeloid specific) antibodies. (B) The c-kit^{Lo}CLP fraction was defined by staining for Flk2 and IL-7Rα (left panels). The LSK fraction was divided for enumeration of Flk2^{Hi} LMPP. (C) Percentages of CLP (left panel) and LMPP (right panel) are shown. The data are representative of those obtained in three independent experiments (N = 8). (D) Lymphoid potential is similar between WT and JAK2V617F CLP. CLP were cultured in serum- and stromal cell-free culture for 7 days. Production of differentiated CD19$^+$ cells were calculated as yields per input progenitor. The data are representative of those obtained in two independent experiments $p<0.01$(**), $p<0.05$(*).

generate lymphocytes in culture, indicating they have no signaling abnormalities. Lnk expression was increased in CD34$^+$ peripheral blood cells in MPN and PV patients with the JAK2V617 mutation [53,54]. Although JAK2 and Lnk could be mutually regulated, we did not observe altered expression of Lnk in CD150$^+$ CD48$^-$ HSC from our Tg mice. Thus, our results indicate that the Lnk gene is not a direct or secondary target of JAK2V617F. Strong expression of JAK2V617F did reduce transcripts in HSC for Mpl, a Tpo receptor known to be important for hematopoiesis [55]. Also, JAK2 partially activates normal MPL function in HSC [56].

Results from our limited PCR analyses might reflect a shift in gene expression in HSC towards erythropoiesis, consistent with reduced numbers of LMPP and CLP. Lymphopoiesis might be further suppressed because potent HSC are diluted in lympho-poietic niches by CD86$^-$ HSC. Regardless of the reason, our results are best explained in terms of an early JAK2 modulated checkpoint. Progenitors that progress beyond that point seem fully competent to generate lymphocytes.

As concluded by others, JAK2 affects HSCs, but JAK2V617F bearing ones can transfer the disease and in some models have a growth advantage over wild-type HSC [2,29]. We now show that it preferentially involves a normally rare subset and speculate that JAK2 signaling can affect lineage choice decisions prior to the diversion of myelo-erythroid-megakaryocytic and lymphoid path-ways. It is unclear if these CD86$^-$ HSCs should be considered injured or re-programmed. Their staining characteristics predicted impaired lymphopoietic potential, and we found reduced numbers

of functionally competent lymphoid progenitors. This could suggest that the immune systems of patients with JAK2 mutations are not being replenished to a normal degree.

These findings provide proof of principal that cell surface markers can be used to track HSC subsets in normal and disease circumstances. For example, we can now appreciate that selective expansion and mobilization of certain HSC occurs in MPNs. It is clear that more effort should be expended to identify and sub-divide HSC in humans.

Materials and Methods

Mice

C57BL/6 (Jackson Laboratory, Bar Harbor, ME) and JAK2 line A transgenics backcrossed for at least nine generations to C57BL/6 [21] were bred and maintained in the Laboratory Animal Resource Center at the Oklahoma Medical Research Foundation (Oklahoma City, OK) or the University of Oklahoma Health Sciences Center (Oklahoma City, OK). They were analyzed between 10 and 21 weeks of age. Experiments were performed in accordance with approved IACUC protocols.

Ethics Statement

C57BL/6 (Jackson Laboratory, Bar Harbor, ME) and JAK2 transgenics backcrossed for at least nine generations to C57BL/6 [21] were bred and maintained in the Laboratory Animal Resource Center at the Oklahoma Medical Research Foundation

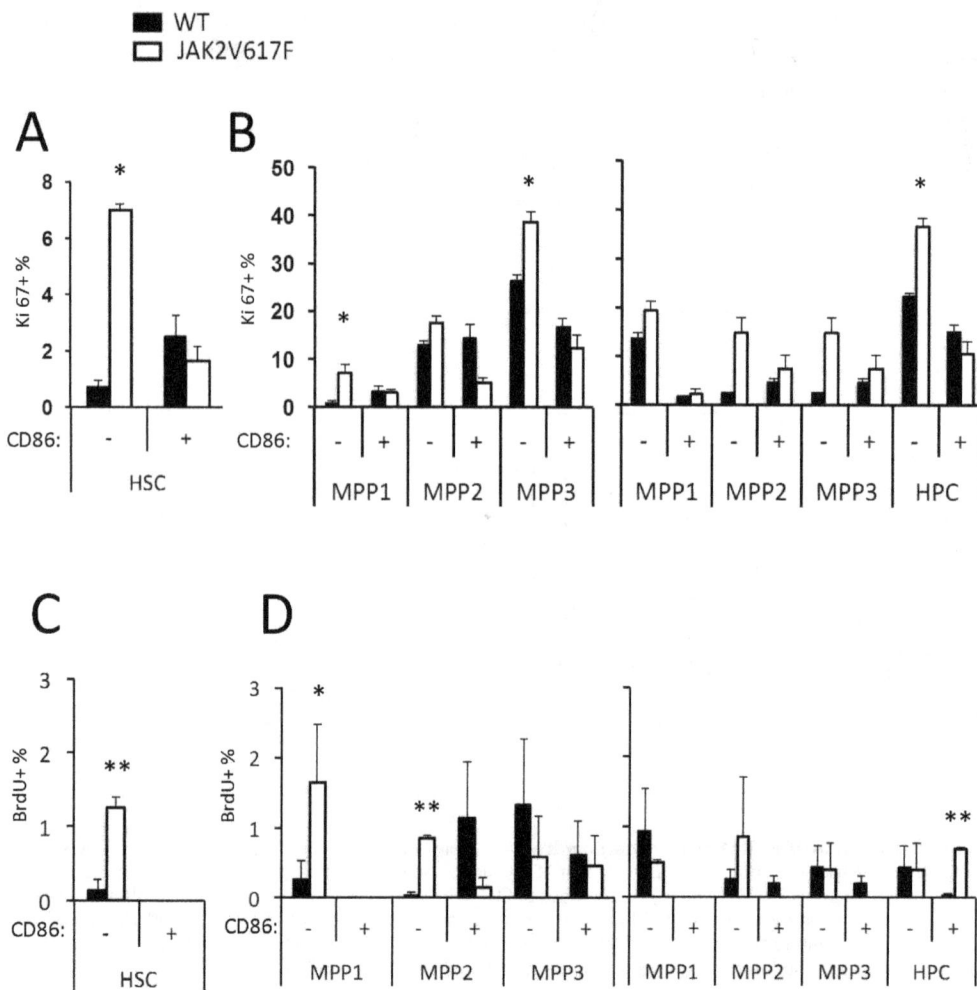

Figure 6. The JAK2V617F transgene may slightly alter proliferation of stem/progenitor cells. The CD150$^+$ CD48$^-$ HSC (**A**) and the multipotent progenitors (**B**) described Fig. 4 were stained with the proliferation associated Ki67 marker. Percentages of Ki67$^+$ cells in BM were calculated. As another approach, the same marrow samples were pulsed with BrdU in very short-term cultures before staining (**C, D**). These results are given as percentages of BrdU$^+$ cells per subset and significance is indicated by asterisks $p<0.01$(**), $p<0.05$(*). These data were obtained in a single experiment (N = 3).

(OMRF) (Oklahoma City, OK) or the University of Oklahoma Health Sciences Center (OUHSC) (Oklahoma City, OK). All experimental procedures were conducted under Institutional Animal Care and Use Committees. OMRF: Protocol KA-1251-1 (co-PI: Kincade and Alberola-Ila), approved November 15, 2012, and annually thereafter, Animal Welfare Assurance #A3127-01. OUHSC: Protocol #: 12-109-T (PI: Zhao), approved August 16, 2013, and annually thereafter, Animal Welfare Assurance #A3165-01.

Antibodies and Flow Cytometry

Tissue and cell manipulations were performed in PBS with 3% fetal calf serum (v/v). Erythrocytes were lysed in NH$_4$Cl$^-$ hypotonic solution. Bone marrow was stained for 15 min on ice. Antibodies included APC-Cy7- lineage (CD3, CD8, CD11b, TER-119, NK1.1, CD19); FITC- FcgRIII(2.4G2), CD41(MWReg30); PE- CD86 (GL1); PE-Cy5- CD135/Flt-3 (A2F10); PerCP Cy5.5- CD48(HM48-1); PE-Cy7- CD150 (TC15-12F12.2); APC- EPCR (ebio1560), IL-7Rα (A7R34), Alexa645- CD34 (RAM34); Alexa700- Gr-1(RB6-8C5); Pacific Blue- Sca-1 (D7); Brilliant violet 510^{TM}- cKit (2B8); Brilliant violet

605^{TM}-B220 (RA3-6B2) and biotin- CD105 (MJ7/18). A secondary streptavidin PE-Texas Red was used for IL-7Rα staining Dead cells were excluded by fixable viability dye eFluorR780 (ebioscience). Cells were sorted using either a MoFlo (DakoCytomation, Ft. Collins, CO) or FACS-Aria cytometer (BD Biosciences, San Diego, CA). Purification of each subset was confirmed by post-sort analysis. For side population analysis, 10^7 bone marrow cells were incubated with 5 μg/mL Hoechst 33342 in DMEM for 90 min at 37°C. Flow cytometry was performed on a BD LSRII (BD Biosciences, San Jose, CA), and FlowJo software (Treestar, San Carlos, CA) was used for data analysis.

Cell Cycle Analysis

Cell were fixed and permeablilized with BD Cytofix/Cytoperm^{TM} Fixation/Permeabilization solution after cell surface staining. Ki-67 was performed using the PE-mouse anti-human Ki-67 kit (BD Bioscience). For analysis of BrdU incorporation, bone marrow cells were incubated with 10 μM BrdU (Sigma) at 37°C for an hour. After cell surface staining, BrdU staining were performed using BD APC flow kit.

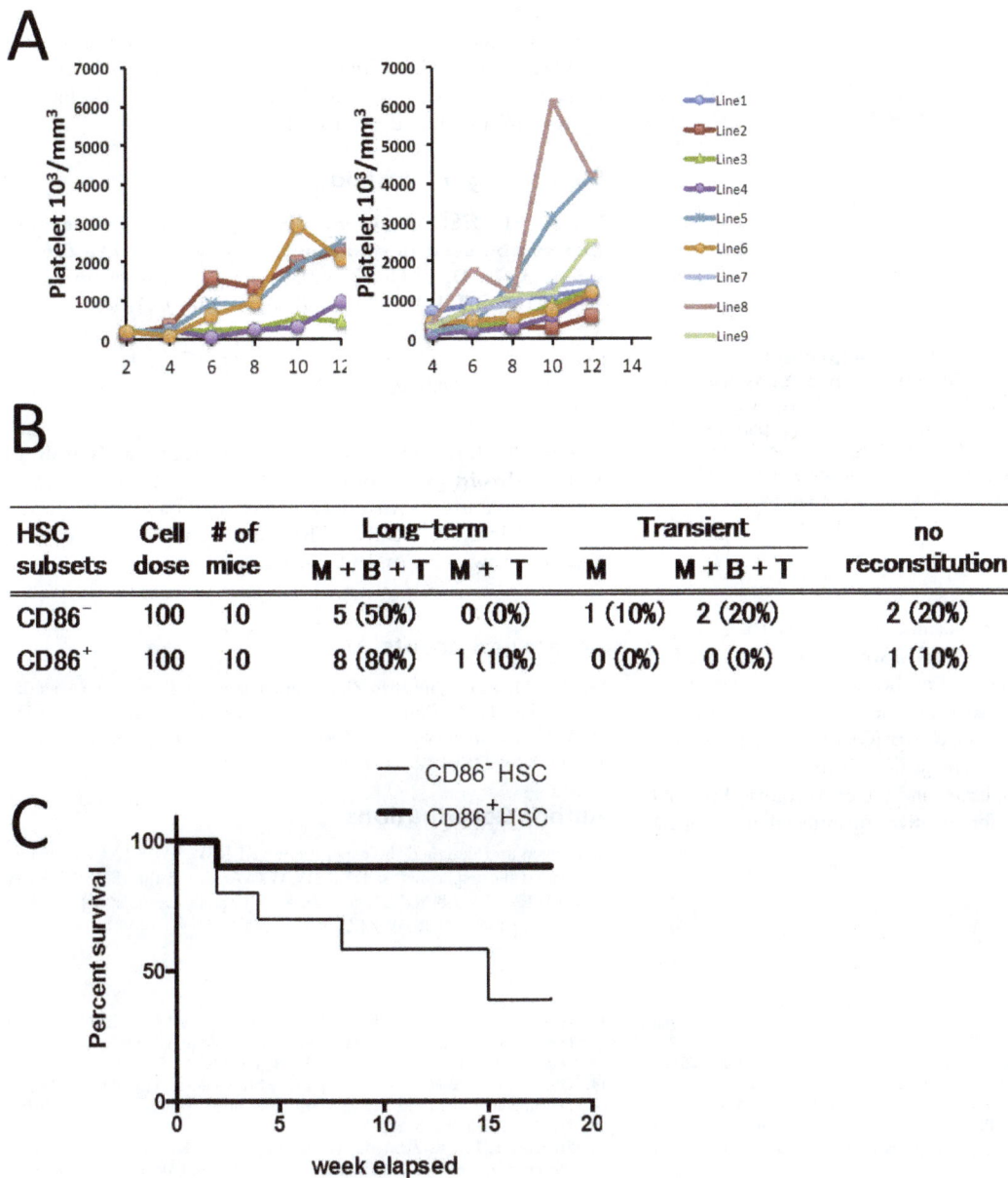

Figure 7. The JAK2V617F transgene initiates disease in both CD86+ and CD86− HSC subsets. One hundred CD86+ or CD86− CD150+ CD48− HSC from CD45.2 JAK2V617F transgenic mice were transplanted to lethally irradiated CD45.1 recipient mice. (**A**) This table summarizes recipients that survived for at least 18 weeks after transplantation. Effective chimerism was considered if the recipients had more than 1.0% donor cells in the peripheral blood. (**B**) Platelet counts are shown at the indicated intervals. (**C**) Survival curves are given for recipients of the two HSC subsets.

Cell Culture

Sorted cells were cultured in round-bottom 96-well plates (Corning, Inc.) with X-VIVO15 medium (Biowhittaker, Walkersville, MD) containing 1% detoxified bovine serum albumin (Stem Cell Technologies, Vancouver, Canada), 5×10^{-5} M 2-mercaptoethanol (2-ME), 2 mM L-glutamine, 100 U/ml penicillin, and 100 mg/ml streptomycin. Culture medium was enriched with 100 ng/mL FL, 20 ng/mL SCF and 1 ng/mL IL-7. Incubation was maintained at 37°C in a 5% CO_2 humidified atmosphere. Cells were fed by replacing half culture volume with fresh media and cytokines every three to four days. Cells were harvested at designated times and stained with monoclonal antibodies to CD19, B220, CD11c, Ly6c, CD11b/Mac-1, and NK1.1.

Transplantation

Recipient (CD45.1) mice were lethally irradiated (2×6.5 Gy [650 rad]) with a 137Cs source (Mark I irradiator; J. L. Shepard and Associates). Mice were anesthetized with isoflurane, and cells were infused intravenously by retro-orbital injection. For purified HSC transplantations, HSCs were sorted directly into 96-well plates containing rescue marrow cells (1×10^5 cells/200 µL). Competitive repopulation was assessed at 2-week intervals by peripheral blood analysis.

Real-Time PCR

The mRNAs were isolated from sorted cells with Trizol (Invitrogen). The cDNAs were then prepared by using random

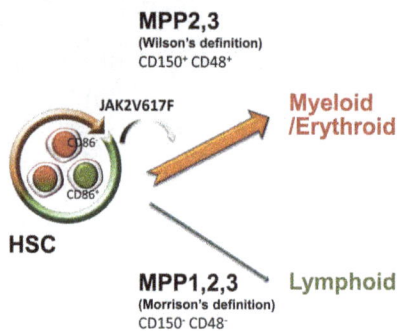

Figure 8. Two methods for resolving multipotent progenitor cells may identify alternate differentiation options for HSC. Trumpp and colleagues described a progression marked by acquisition of CD48 (MPP2 and 3) [45] and we show it as the top pathway. This option and particularly MPP that lack CD86, is promoted by the JAK2 mutant transgene. Alternatively, the Morrison lab showed that HSC can lose CD150 (formation of MPP 1, 2, 3 in the lower pathway) [46]. This route is diminished in the transgenics and most of the loss is in CD86+ MPP.

primers and Moloney murine leukemia virus reverse transcriptase (Invitrogen). Reactions were quantified with fluorescent TaqMan technology. TaqMan primers and probes specific for every genes were used in the ABI7500 sequence detection system (Applied Biosystems) using QuanTitect PCR Mix (Qiagen). Reactions were run at an annealing temperature of 60°C with 45 cycles. Each sample was measured in triplicate, and the comparative threshold cycle method was used for relative quantification of gene expression.

Statistics

Prism V5.0d software (GraphPad, San Diego, CA) was used for statistical analysis. Unpaired t-test analyses were employed, and p-values were considered significant if less than 0.05. Bars on figures depict Standard Error of the Mean.

Supporting Information

Figure S1 HSC differentially express other markers that can be used to monitor HSC subsets in JAK2V617F mice. (**A, B, D, E**) CD48$^-$ LSK were gated and characterized with CD18, CD41, VCAM-1 and CD39 in the CD150^{Hi} fraction. (**C**) Altered ratios of CD86$^-$ CD41$^+$ and CD86$^+$ CD41$^-$ were found in the CD150^{Hi} HSC fraction. CD150$^+$ CD48$^-$ LSK were resolved with CD86 and CD41.

Figure S2 Lineage associated transcripts in lymphoid and erythroid progenitors. Real-Time PCR was performed using cDNA from sorted Flk2$^+$ LSK and Lin$^-$ ckit^{Hi} Sca1$^-$ CD150$^+$ CD105$^+$ pre-CFUE. The data are representative of those obtained in two independent experiments.

Acknowledgments

We thank Doctors Qingzhao Zhang and Tomoyuki Shimazu for scientific consultation, Karla Garrett for technical assistance, Jacob Bass and Dr. Diana Hamilton for cell sorting, Beverly Hurt for graphic design and Shelli Wasson for editorial assista.

Author Contributions

Conceived and designed the experiments: RI RSW JAI KLM ZJZ PWK. Performed the experiments: RI RSW WZ. Analyzed the data: RI RSW ZJZ PWK. Contributed reagents/materials/analysis tools: ZJZ PWK. Wrote the paper: RI RSW ZJZ PWK.

References

1. Chen E, Staudt LM, Green AR (2012) Janus kinase deregulation in leukemia and lymphoma. Immunity 36: 529–541.
2. Hasan S, Lacout C, Marty C, Cuingnet M, Solary E, et al (2013) JAK2V617F expression in mice amplifies early hematopoietic cells and gives them a competitive advantage that is hampered by IFNα. Blood 122: 1464–1477.
3. Muller-Sieburg CE, Sieburg HB, Bernitz JM, Cattarossi G (2012) Stem cell heterogeneity: implications for aging and regenerative medicine. Blood 119: 3900–3907.
4. Copley MR, Beer PA, Eaves CJ (2012) Hematopoietic stem cell heterogeneity takes center stage. Cell Stem Cell 10: 690–697.
5. Müller-Sieburg CE, Cho RH, Thoman M, Adkins B, Sieburg HB (2002) Deterministic regulation of hematopoietic stem cell self-renewal and differentiation. Blood 100: 1302–1309.
6. Dykstra B, Kent D, Bowie M, McCaffrey L, Hamilton M, et al (2007) Long-term propagation of distinct hematopoietic differentiation programs in vivo. Cell Stem Cell 1: 218–229.
7. Benz C, Copley MR, Kent DG, Wohrer S, Cortes A, et al (2012) Hematopoietic Stem Cell Subtypes Expand Differentially during Development and Display Distinct Lymphopoietic Programs. Cell Stem Cell 10: 273–283.
8. Gerrits A, Dykstra B, Kalmykowa OJ, Klauke K, Verovskaya E, et al (2010) Cellular barcoding tool for clonal analysis in the hematopoietic system. Blood 115: 2610–2618.
9. Lu R, Neff NF, Quake SR, Weissman IL (2011) Tracking single hematopoietic stem cells in vivo using high-throughput sequencing in conjunction with viral genetic barcoding. Nat Biotechnol 29: 928–933.
10. Morita Y, Ema H, Nakauchi H (2010) Heterogeneity and hierarchy within the most primitive hematopoietic stem cell compartment. J Exp Med 207: 1173–1182.
11. Benveniste P, Frelin C, Janmohamed S, Barbara M, Herrington R, et al (2010) Intermediate-term hematopoietic stem cells with extended but time-limited reconstitution potential. Cell Stem Cell 6: 48–58.
12. Adolfsson J, Månsson R, Buza-Vidas N, Hultquist A, Liuba K, et al (2005) Identification of Flt3+ lympho-myeloid stem cells lacking erythro-megakaryocytic potential a revised road map for adult blood lineage commitment. Cell 121: 295–306.

13. Kiel MJ, Yilmaz OH, Iwashita T, Yilmaz OH, Terhorst C, et al (2005) SLAM family receptors distinguish hematopoietic stem and progenitor cells and reveal endothelial niches for stem cells. Cell 121: 1109–1121.
14. Challen GA, Boles NC, Chambers SM, Goodell MA (2010) Distinct hematopoietic stem cell subtypes are differentially regulated by TGF-beta1. Cell Stem Cell 6: 265–278.
15. Shimazu T, Iida M, Zhang Q, Welner RS, Medina KL, et al (2012) CD86 is expressed on murine hematopoietic stem cells and denotes lymphopoietic potential. Blood 119: 4889–4897.
16. Esplin BL, Shimazu T, Welner RS, Garrett KP, Nie L, et al (2011) Chronic Exposure to a TLR Ligand Injures Hematopoietic Stem Cells. J Immunol 186: 5367–5375.
17. Wernig G, Mercher T, Okabe R, Levine RL, Lee BH, et al (2006) Expression of Jak2V617F causes a polycythemia vera-like disease with associated myelofibrosis in a murine bone marrow transplant model. Blood 107: 4274–4281.
18. Lacout C, Pisani DF, Tulliez M, Gachelin FM, Vainchenker W, et al (2006) JAK2V617F expression in murine hematopoietic cells leads to MPD mimicking human PV with secondary myelofibrosis. Blood 108: 1652–1660.
19. Zaleskas VM, Krause DS, Lazarides K, Patel N, Hu Y, et al (2006) Molecular pathogenesis and therapy of polycythemia induced in mice by JAK2 V617F. PLoS One 1: e18.
20. Tiedt R, Hao-Shen H, Sobas MA, Looser R, Dirnhofer S, et al (2008) Ratio of mutant JAK2-V617F to wild-type Jak2 determines the MPD phenotypes in transgenic mice. Blood 111: 3931–3940.
21. Xing S, Wanting TH, Zhao W, Ma J, Wang S, et al (2008) Transgenic expression of JAK2V617F causes myeloproliferative disorders in mice. Blood 111: 5109–5117.
22. Shide K, Shimoda HK, Kumano T, Karube K, Kameda T, et al (2008) Development of ET, primary myelofibrosis and PV in mice expressing JAK2 V617F. Leukemia 22: 87–95.
23. Akada H, Yan D, Zou H, Fiering S, Hutchison RE, et al (2010) Conditional expression of heterozygous or homozygous Jak2V617F from its endogenous promoter induces a polycythemia vera-like disease. Blood 115: 3589–3597.
24. Mullally A, Lane SW, Ball B, Megerdichian C, Okabe R, et al (2010) Physiological Jak2V617F expression causes a lethal myeloproliferative neoplasm

with differential effects on hematopoietic stem and progenitor cells. Cancer Cell 17: 584–596.

25. Li J, Spensberger D, Ahn JS, Anand S, Beer PA, et al (2010) JAK2 V617F impairs hematopoietic stem cell function in a conditional knock-in mouse model of JAK2 V617F-positive essential thrombocythemia. Blood 116: 1528–1538.

26. Marty C, Lacout C, Martin A, Hasan S, Jacquot S, et al (2010) Myeloproliferative neoplasm induced by constitutive expression of JAK2V617F in knock-in mice. Blood 116: 783–787.

27. Van Etten RA, Koschmieder S, Delhommeau F, Perrotti D, Holyoake T, et al (2011) The Ph-positive and Ph-negative myeloproliferative neoplasms: some topical pre-clinical and clinical issues. Haematologica 96: 590–601.

28. Li J, Kent DG, Chen E, Green AR (2011) Mouse models of myeloproliferative neoplasms: JAK of all grades. Dis Model Mech 4: 311–317.

29. Mullally A, Poveromo L, Schneider RK, Al-Shahrour F, Lane SW, et al (2012) Distinct roles for long-term hematopoietic stem cells and erythroid precursor cells in a murine model of Jak2V617F-mediated polycythemia vera. Blood 120: 166–172.

30. Chen E, Beer PA, Godfrey AL, Ortmann CA, Li J, et al (2010) Distinct clinical phenotypes associated with JAK2V617F reflect differential STAT1 signaling. Cancer Cell 18: 524–535.

31. Walz C, Ahmed W, Lazarides K, Betancur M, Patel N, et al (2012) Essential role for Stat5a/b in myeloproliferative neoplasms induced by BCR-ABL1 and JAK2(V617F) in mice. Blood 119: 3550–3560.

32. Yan D, Hutchison RE, Mohi G (2012) Critical requirement for Stat5 in a mouse model of polycythemia vera. Blood 119: 3539–3549.

33. Pecquet C, Diaconu CC, Staerk J, Girardot M, Marty C, et al (2012) Thrombopoietin receptor down-modulation by JAK2 V617F: restoration of receptor levels by inhibitors of pathologic JAK2 signaling and of proteasomes. Blood 119: 4625–4635.

34. Mullally A, Bruedigam C, Poveromo L, Heidel FH, Purdon A, et al (2013) Depletion of Jak2V617F myeloproliferative neoplasm-propagating stem cells by interferon-α in a murine model of polycythemia vera. Blood 121: 3692–3702.

35. Quintás-Cardama A, Abdel-Wahab O, Manshouri T, Kilpivaara O, Cortes J, et al (2013) Molecular analysis of patients with polycythemia vera or essential thrombocythemia receiving pegylated interferon alpha-2a. Blood 122: 893–901.

36. Liu F, Zhao X, Perna F, Wang L, Koppikar P, et al (2011) JAK2V617F-mediated phosphorylation of PRMT5 downregulates its methyltransferase activity and promotes myeloproliferation. Cancer Cell 19: 283–294.

37. Kent DG, Li J, Tanna H, Fink J, Kirschner K, et al (2013) Self-renewal of single mouse hematopoietic stem cells is reduced by JAK2V617F without compromising progenitor cell expansion. PLoS Biol 11: e1001576.

38. Kent DG, Copley MR, Benz C, Wöhrer S, Dykstra BJ, et al (2009) Prospective isolation and molecular characterization of hematopoietic stem cells with durable self-renewal potential. Blood 113: 6342–6350.

39. Beerman I, Bhattacharya D, Zandi S, Sigvardsson M, Weissman IL, et al (2010) Functionally distinct hematopoietic stem cells modulate hematopoietic lineage potential during aging by a mechanism of clonal expansion. Proc Natl Acad Sci U S A 107: 5465–5470.

40. Weksberg DC, Chambers SM, Boles NC, Goodell MA (2008) CD150- side population cells represent a functionally distinct population of long-term hematopoietic stem cells. Blood 111: 2444–2451.

41. Gekas C, Graf T (2013) CD41 expression marks myeloid biased adult hematopoietic stem cells and increases with age. Blood 121: 4463–4472.

42. Sugimura R, He XC, Venkatraman A, Arai F, Box A, et al (2012) Noncanonical Wnt signaling maintains hematopoietic stem cells in the niche. Cell 150: 351–365.

43. Spangrude GJ, Brooks DM (1993) Mouse strain variability in the expression of the hematopoietic stem cell antigen Ly-6A/E by bone marrow cells. Blood 82: 3327–3332.

44. Essers MA, Offner S, Blanco-Bose WE, Waibler Z, Kalinke U, et al (2009) IFNalpha activates dormant haematopoietic stem cells in vivo. Nature 458: 904–908.

45. Wilson A, Laurenti E, Oser G, van der Wath RC, Blanco-Bose W, et al (2008) Hematopoietic stem cells reversibly switch from dormancy to self-renewal during homeostasis and repair. Cell 135: 1118–1129.

46. Oguro H, Ding L, Morrison SJ (2013) SLAM Family Markers Resolve Functionally Distinct Subpopulations of Hematopoietic Stem Cells and Multipotent Progenitors. Cell Stem Cell 13: 102–116.

47. Igarashi H, Gregory SC, Yokota T, Sakaguchi N, Kincade PW (2002) Transcription from the RAG1 locus marks the earliest lymphocyte progenitors in bone marrow. Immunity 17: 117–130.

48. Adolfsson J, Borge OJ, Bryder D, Theilgaard-Mönch K, Astrand-Grundström I, et al (2001) Upregulation of Flt3 expression within the bone marrow Lin(-)Sca1(+) c-kit(+) stem cell compartment is accompanied by loss of self-renewal capacity. Immunity 15: 659–669.

49. Tong W, Zhang J, Lodish HF (2005) Lnk inhibits erythropoiesis and Epo-dependent JAK2 activation and downstream signaling pathways. Blood 105: 4604–4612.

50. Takaki S, Morita H, Tezuka Y, Takatsu K (2002) Enhanced hematopoiesis by hematopoietic progenitor cells lacking intracellular adaptor protein, Lnk. J Exp Med 195: 151–160.

51. Velazquez L, Cheng AM, Fleming HE, Furlonger C, Vesely S, et al (2002) Cytokine signaling and hematopoietic homeostasis are disrupted in Lnk-deficient mice. J Exp Med 195: 1599–1611.

52. Bersenev A, Wu C, Balcerek J, Jing J, Kundu M, et al (2010) Lnk constrains myeloproliferative diseases in mice. J Clin Invest 120: 2058–2069.

53. Gery S, Cao Q, Gueller S, Xing H, Tefferi A, et al (2009) Lnk inhibits myeloproliferative disorder-associated JAK2 mutant, JAK2V617F. J Leukoc Biol 85: 957–965.

54. Baran-Marszak F, Magdoud H, Desterke C, Alvarado A, Roger C, et al (2010) Expression level and differential JAK2-V617F-binding of the adaptor protein Lnk regulates JAK2-mediated signals in myeloproliferative neoplasms. Blood 116: 5961–5971.

55. Kimura S, Roberts AW, Metcalf D, Alexander WS (1998) Hematopoietic stem cell deficiencies in mice lacking c-Mpl, the receptor for thrombopoietin. Proc Natl Acad Sci U S A 95: 1195–1200.

56. Vainchenker W, Delhommeau F, Constantinescu SN, Bernard OA (2011) New mutations and pathogenesis of myeloproliferative neoplasms. Blood 118: 1723–1735.

13

Autoimmunity-Associated LYP-W620 Does Not Impair Thymic Negative Selection of Autoreactive T Cells

Dennis J. Wu[1][9], Wenbo Zhou[2][9], Sarah Enouz[3], Valeria Orrú[4,5], Stephanie M. Stanford[1], Christian J. Maine[6], Novella Rapini[1], Kristy Sawatzke[2], Isaac Engel[7], Edoardo Fiorillo[4,5], Linda A. Sherman[6], Mitch Kronenberg[7], Dietmar Zehn[3], Erik Peterson[2*][¶], Nunzio Bottini[1,4*][¶]

1 Division of Cellular Biology, La Jolla Institute for Allergy and Immunology, La Jolla, California, United States of America, 2 Center for Immunology, Department of Medicine, University of Minnesota, Minneapolis, Minnesota, United States of America, 3 Swiss Vaccine Research Institute, Epalinges, and Division of Immunology and Allergy, Department of Medicine, Lausanne University Hospital, Lausanne, Switzerland, 4 Institute for Genetic Medicine, University of Southern California, Los Angeles, California, United States of America, 5 Istituto di Ricerca Genetica e Biomedica (IRGB), CNR, Monserrato, Italy, 6 Department of Immunology, The Scripps Research Institute, La Jolla, California, United States of America, 7 Division of Developmental Immunology, La Jolla Institute for Allergy and Immunology, La Jolla, California, United States of America

Abstract

A C1858T (R620W) variation in the *PTPN22* gene encoding the tyrosine phosphatase LYP is a major risk factor for human autoimmunity. LYP is a known negative regulator of signaling through the T cell receptor (TCR), and murine *Ptpn22* plays a role in thymic selection. However, the mechanism of action of the R620W variant in autoimmunity remains unclear. One model holds that LYP-W620 is a gain-of-function phosphatase that causes alterations in thymic negative selection and/or thymic output of regulatory T cells (T_{reg}) through inhibition of thymic TCR signaling. To test this model, we generated mice in which the human LYP-W620 variant or its phosphatase-inactive mutant are expressed in developing thymocytes under control of the proximal *Lck* promoter. We found that LYP-W620 expression results in diminished thymocyte TCR signaling, thus modeling a "gain-of-function" of LYP at the signaling level. However, LYP-W620 transgenic mice display no alterations of thymic negative selection and no anomalies in thymic output of $CD4^+Foxp3^+$ T_{reg} were detected in these mice. *Lck* promoter-directed expression of the human transgene also causes no alteration in thymic repertoire or increase in disease severity in a model of rheumatoid arthritis, which depends on skewed thymic selection of $CD4^+$ T cells. Our data suggest that a gain-of-function of LYP is unlikely to increase risk of autoimmunity through alterations of thymic selection and that LYP likely acts in the periphery perhaps selectively in regulatory T cells or in another cell type to increase risk of autoimmunity.

Editor: Song Guo Zheng, Penn State University, United States of America

Funding: This study was supported by NIH grants R21AI065643 and R01AI070544 and an Innovative Grant from the Juvenile Diabetes Research Foundation (JDRF) to N.B., grant NIH R01AR057781, the American College of Rheumatology Research and Education Foundation, the Alliance for Lupus Research, and Lupus Foundation of Minnesota to E.J.P., grant NIH R37AI71922 to M.K. Valeria Orrú and Edoardo Fiorillo were supported by Master and Back fellowships from the Sardinian government. Stephanie M. Stanford is supported by a fellowship from JDRF. DZ was supported by grants from the Swiss National Science Foundation [PP00P3_144883, PDFMP3_137128, CRSII3_141879] and funds from the Swiss Vaccine Research Foundation. This is publication #1650 from the La Jolla Institute for Allergy and Immunology. The funders had no role in study design, data collection and analysis, decision to publish, or preparation of the manuscript.

Competing Interests: The authors have declared that no competing interests exist.

* E-mail: peter899@umn.edu (EP); nunzio@lji.org (NB)

[9] These authors contributed equally to this work.

[¶] These authors also contributed equally to this work.

Introduction

The *PTPN22* gene, encoding the lymphoid tyrosine phosphatase LYP, has emerged as one of the major non-HLA risk factors for a wide range of autoimmune diseases, including type 1 diabetes, rheumatoid arthritis (RA), systemic lupus erythematosus, Graves' disease and others [1,2]. A missense *C1858T* single nucleotide polymorphism in exon 14 of the *PTPN22* gene leads to LYP-R620W substitution. The variant *PTPN22* allele confers to carriers a roughly two-fold increased risk of autoimmunity [2–5]. LYP inhibits signaling through the T cell receptor (TCR), and its substrates in T cells include the phosphorylated tyrosine residues in the activation motifs of Lck, Zap-70 and other signaling molecules [4,6–8]. Mice made deficient for *Ptpn22* (encoding Pep,

the murine LYP-homolog PEST-enriched phosphatase) display a phenotype of increased TCR signaling in effector T cells, which correlates with an expansion of the effector-memory T cell compartment [9,10]. The LYP-R620W substitution impairs the ability of the phosphatase to bind to the SH3 domain of the C-terminal Src-family kinase CSK [3,4], which is a major LYP interactor in T cells [7,11]. LYP-W620 also displays 1.5–2 fold increased intrinsic phosphatase activity compared to the common R620 variant [12–14].

Studies of the effect of the LYP-R620W substitution on immune cell signaling have not yet yielded a unifying model. We and others reported that TCR signaling is impaired in T cells from patients with autoimmune disease who carry the LYP-W620 variant [12,15–17]. Reduced signaling through antigen receptors has also been reported in B cells and peripheral blood mononuclear cells (PBMC) of both patient and healthy donor LYP-W620 carriers [13,15,18]. Together, these findings suggest that the LYP-W620 variant is a "gain-of-function" negative regulator of antigen receptor signaling.

Several models have been proposed to explain the gain-of-function phenotype, including increased phosphatase activity following reduced CSK-mediated phosphorylation of the regulatory Tyr536 residue [14], and increased recruitment of the LYP-W620 variant to lipid rafts following release from cytoplasmic sequestration by Csk [19]. However, others have proposed an opposing model wherein the R620W substitution confers "loss-of-function" effects on antigen receptor signaling. Supporting data for a LYP-W620 "loss-of-function" hypothesis come from overexpression experiments in Jurkat T cells [20]. Enhanced TCR-driven calcium mobilization was observed in human LYP-W620 carriers and in T cells from a mouse carrying a knock-in R619W mutation in mouse Pep that is homologous to the human LYP R620W variation [21]. Chang *et al.* identified a new dominant-negative isoform of LYP and proposed a model that reconciles "gain-of-function" and "loss-of-function" observations [22]. Dai *et al.* recently reported a phenotype of enhanced TCR signaling and spontaneous autoimmunity in R619W knock-in mice [23]. Analysis of the spectrum of phosphorylated molecules in TCR-stimulated Pep-R619W T cells suggested altered enzymatic specificity [23]. In line with this "altered function" model, a recent analysis of peripheral T cells from genotyped healthy subjects suggested that the LYP-R620W mutation can positively or negatively affect TCR signaling, depending on the biochemical readout assayed and on the stage of signaling [24].

A prevailing model of thymocyte selection holds that TCR affinity for MHC/peptide ligand plays a central role in shaping the peripheral TCR repertoire [25]. Deletion of autoreactive thymocytes and agonist selection of regulatory T cells (T_{reg}) are two important mechanisms for establishing T cell tolerance that depend upon high-affinity interactions between TCR and ligand.

Phenotyping of *Ptpn22* knockout mice suggested that Pep might play a role in regulating thymic selection. Increased positive thymic selection has been reported in *Ptpn22* KO mice [9] and in two independently-generated Pep R619W knock-in mouse models, one of which developed spontaneous autoimmunity [21,23]. Increased negative selection of H-Y transgenic male thymocytes mice was reported in one Pep R619W knock-in model [23], but thymic deletion was not altered in the other Pep R619W knock-in strain [21] or in *Ptpn22* KO mice [9]. We reported increased TCR signaling in *Ptpn22* KO thymocytes, correlating with increased thymic output of T_{reg} in KO mice [26]; however, another group found no difference in thymic T_{reg} percentages in an independently generated KO model [23]. The potential effect of the autoimmune-associated human LYP-W620 variant in thymic selection has not yet been examined in a controlled genetic environment (congenic mice). Yeh *et al.* recently reported that overexpression of wild type Pep under control of the distal *Lck* promoter did not result in alteration of thymocyte numbers in the NOD background [27], however the distal *Lck* promoter is expressed only in late stages of thymocyte selection [28], and the effect of the autoimmune-predisposing R620W variant was not assessed.

In this study, we addressed effects of the human LYP-W620 variant on thymocyte signaling and selection by expressing the variant phosphatase under control of the proximal *Lck* promoter. To model effects of altered enzymatic activity inherent in LYP-W620, we studied mice exhibiting low overexpression levels of the active human phosphatase in thymocytes. To detect possible

effects unrelated to the increased phosphatase activity, we also examined animals overexpressing an enzymatically-inactive mutant (C227S) of the LYP-W620 phosphatase. Mice transgenic for the active phosphatase displayed reduced thymocyte TCR signaling when compared to mice transgenic for the inactive phosphatase or to non transgenic littermates. However no significant alterations of T cell selection or of thymic T_{reg} output were observed in LYP-W620 transgenic mice. Our study suggests that the reported gain-of-function activity of LYP-W620 is insufficient to significantly alter thymic output; thus it is unlikely that its major disease-predisposing role is exerted during thymocyte selection.

Materials and Methods

Mouse Experimental Work

Generation of Prox*Lck*-driven transgenic mice. cDNAs encoding either N-terminal HA-tagged LYP-W620 (LYPW) or its inactive C227S mutant (LYPW^C227S) were cloned into the *BamHI* site of the *p*1017 vector [29](obtained from T. Mustelin, Sanford-Burnham Medical Research Institute). The transgenes were isolated from the vector backbone by digestion with *Not*I, and pronuclear oocyte injection was performed using fertilized ova from C57BL/6 (B6)D2F1 females, and pseudopregnant outbred ICR females as foster mothers. Genotyping of transgenic founders and all transgenic animals was performed by PCR, using one primer 5′-tgtgaacttggtgcttgagg-3′ on the *Lck* promoter and another 5′-tgttatggcatgcatggagt-3′ on the LYP cDNA. All founder animals were bred with C57BL/6Tac mice and germline transmission was obtained from three B6D2F1-Tg(LckproxLYP-W620)/Igm founders and two B6D2F1-Tg(LckproxLYP-W620/S227)/Igm founders. Lines were generated from each of the founders carrying the transgenes in hemizygosity. Each line was tested for expression of the transgene and activity of the transgenic phosphatase. The two founder lines B6D2F1-Tg(LckproxLYP-W620)958/Igm (here abbreviated as TgLYPW) and B6D2F1-Tg(LckproxLYP-W620/S227)963/Igm (here abbreviated as TgLYPW^C227S) were selected for further studies and extensively backcrossed onto the BALB/c background for >15 generations before breeding with Skg mice (see below).

Generation of BAC-transgenic mice. Recombineering (homologous recombination in bacteria) [30] was used to introduce the autoimmune-associated variant (*PTPN22-C1858T*) into the human *PTPN22* locus. A 200 kb Bacterial Artificial Chromosome (BAC; Invitrogen) fragment of human genomic DNA containing the entire *PTPN22* locus, including 60 kb located 5′ to the *PTPN22* transcriptional start site, was used as a template. Following recombineering, 116 kb *Not*I restriction fragments encoding the disease variant LYP-W620 protein was used for pronuclear B6 oocyte injection and generation of founder mice (BACLYPW) directly on B6 background as described above. LYP expression in BACLYPW mice has been described elsewhere [31].

Skg mice and arthritis scoring. Skg mice were kindly provided by S. Sakaguchi (Kyoto University, Kyoto, Japan). Skg/Skg mice are homozygous for a loss-of-function W163C mutation of Zap-70 that leads to decreased thymic TCR signaling and altered selection of self-reactive CD4^+ T cells and of T_{reg} [35,36]. These mice display alterations of the Vβ thymic repertoire and develop spontaneous arthritis on the BALB/c background [35,37,38], whose frequency and clinical course is dependent on microbial colonization and on cleanliness of the environment. However severe arthritis can be induced in all Skg/Skg mice, independent on their microbial colonization, by intraperitoneal injection with fungal wall polysaccharides, including zymosan and mannan [39,40]. In the La Jolla Institute for Allergy & Immunology (LJI) animal facility, Skg/Skg animals developed spontaneous arthritis with very low frequency and the severity of arthritis was variable among cages. However all Skg/Skg mice developed severe arthritis after intraperitoneal injection of 20 mg mannan (purchased from Sigma) dissolved in 200 µl PBS. Skg/WT heterozygous mice display altered thymic selection [36,37] and alterations of the Vβ thymic repertoire [37] qualitatively similar to the ones found in Skg/Skg mice, however they do not develop spontaneous [35] or polysaccharide-induced (our unpublished observation) arthritis. Clinical scoring of arthritis was carried out as described [35].

Other mouse models. Foxp3^{GFP} mice were kindly provided by A. Rudensky (Sloan-Kettering Institute, NY) [32]. HY-TCR mice [33] were kindly provided by H. Cheroutre (La Jolla Institute for Allergy and Immunology, CA), or by K. Hogquist (University of Minnesota; for the crosses with BACLYPW mice), and Vβ5 TCRβonly transgenic mice by P. Fink (University of Washington, WA). BALB/c and DO11.10 mice [34] were purchased from Taconic and Rip-mOva and OT-1 mice from Jackson laboratories. Ptpn22-deficient mice were kindly provided by A. Chan (Genentech).

Bone marrow chimeras and infections with Lm-Ova and VSV-Ova. Recipient mice were lethally irradiated with 900 rad and one day later engrafted with bone marrow cells isolated from the femur and tibia. Prior to injection, the bone marrow was T cell depleted by incubating cells with biotinylated anti-CD3 antibody (eBiociences), anti-biotin microbeads followed by magnetic separation in LS columns (all Miltenyi Biotec, Bergisch-Gladbach, Germany). Mice were treated with antibiotics (Bactrim, Roche) until 3 weeks post-irradiation.

Frozen stocks of recombinant, ovalbumin expressing *Listeria monocytogenes* [Lm-Ova] were grown in brain-heart infusion broth (Oxoid, UK) to mid log phase. Bacterial numbers were determined by measuring the OD at 600 nm. 1000–3000 cfu were injected in PBS intravenously. *Vesicular stomatitis virus* expressing Ova (VSV-Ova) [41] were grown and titered on BHK cells. Frozen stocks were diluted in PBS to 2×10^6 pfu and injected intravenously.

Ex vivo Thymocyte or T cell Stimulation

Spleen and thymus cell suspensions were obtained by mashing the organs through a 100 µm nylon cell strainer (BD Falcon). Red blood cells were lysed with a hypotonic ACK lysis buffer. 5×10^5 CD45.1 congenic thymocytes were co-cultured in 96 well plates for 18–22 hours with 7.5×10^4 irradiated RMA cells and titrated doses of SIINFEKL, SIITFEKL, or SIIVFEKL peptide (EMC Microcolelctions, Tübingen, Germany). Cultures were stained with anti-CD8 (53–6.7, PerCP, eBioscience), CD45.1 (clone A20), CD4 (GK1.4) purchased from BioXcell (West Lebanon, NH) and conjugated to FITC or A647 using labeling reagents from Invitrogen.

To identify K^b/Ova reactive T cells in infected mice, up to 4×10^6 total splenocytes were seeded into 96 well plates. The cells were stimulated with titrated doses of SIINFEKL peptide for 30 min at 37°C, then 7 µM Brefeldin A (Sigma) was added and the cells were returned to 37°C for another 4.5 hours.

Afterwards, the cells were washed, incubated for 30 min at 4°C in PBS supplemented with 2% FBS and 0.01% azide (FACS buffer) and then stained with the following antibodies: anti-CD8 PerCP-Cy5.5 (clone 53–6.7, eBioscience), anti-CD4 and anti-IFNγ (BioXcell, self-conjugated to FITC, Invitrogen). Then the cells were fixed and permeabilized using the Cytofix/Cytoperm Kit (BD, Franklin Lakes NJ, USA) and stained with anti-IFNγ (BioXCell, self-conjugated to Alexa Fluor 647, Invitrogen). Data

were analyzed with FlowJo software (Tree Star Inc). Graphs were prepared and EC$_{50}$ concentrations were determined with GraphPad Prism.

Stimulation for the CD69 expression assay was performed by precoating 96-well plates with 10 µg/ml and 25 µg/ml anti-CD3 overnight at 4°C and then culturing thymocytes at 2×10^5 cells/well for 18 h at 37°C.

Phospho-Erk (pErk) ELISA

pErk levels were detected by ELISA using the PathScan® Phospho-p44 MAPK (Thr202/Tyr204) Sandwich ELISA antibody pair (Cell Signaling Technology) according to manufacturer's protocols.

FACS Analyses and Intracellular Staining (ICS)

Abs (conjugated to eFluor450, Pacific Blue, APC, PE, FITC, PE-Cy7, PerCP-Cy5.5, PerCP-eFluor710, V500 and Alexa-Fluor700) directed against the following surface markers were obtained from eBioscience, BD PharMingen, BioLegend, Proimmune and used at a 1/100 dilution to analyze immune cell subsets in single cell suspensions of thymi and lymph nodes of LYP Tg mice: CD3ε (145-2C11), CD4 (GK1.5), CD8α (53-6.7), CD25 (PC61.5), CD44 (IM7), CD69 (H1.2F3), B220/CD45R (RA3–6B2), Vβ3 (KJ25), Vβ5.1/5.2 (MR 9-4), Vβ8.1, 8.2 (KJ16), DO11.10 TCR (KJ1-26), HY TCR (T3.70), CD1d tetramers. For intracellular cytokine staining [40], lymph node (LN) cells were stimulated with 20 ng/ml PMA and 1 µM ionomycin in the presence of GolgiStop (BD Biosciences) for 5 h. Cells were then stained for surface antigens, fixed, and permeabilized using Cytofix/Cytoperm buffer (BD Biosciences), followed by anti-IL-17 (BD Pharmingen) staining. To detect low level expression of HA-tagged LYP by ICS indirect immunofluorescence was used to increase sensitivity. Thymocytes were first stained for surface markers followed by fixation and permeablization. Cells were then stained with an anti-HA Ab (Cell Signaling Technology) for 1 hour followed by washing and staining with FITC-conjugated anti-rabbit Ab (BD PharMingen). For pErk ICS, cells were stimulated with 10 µg/ml anti-CD3 for 20 min, followed by fixation in 4% paraformaldehyde. After washing, cells were permeabilized with 90% methanol at 4°C, then stained with anti-phospho-p44/42 MAPK (Erk1/2) (Cell Signaling Technology) followed by washing and staining with FITC-conjugated anti-rabbit Ab (BD Pharmingen). All flow cytometry experiments were performed using an LSRII (BD Bioscience) and data was analyzed using FlowJo Software (Tree Star).

Immunoprecipitations

For immunoprecipitations (IPs), cells were lysed in 20 mM Tris/HCl, pH 7.4, 150 mM NaCl, 5 mM EDTA with 1% Nonidet P-40, 1 mM phenylmethylsulfonyl fluoride, 10 µg/ml aprotinin, 10 µg/ml leupeptin and 10 µg/ml soybean trypsin inhibitor. Lysates were probed with the indicated antibodies for 2 h at 4°C, followed by incubation with PGSepharose beads (GE Healthcare) for 1 h at 4°C. Beads were then washed in lysis buffer, and proteins were eluted in Laemmli sample buffer (Bio-Rad).

Quantitative PCR (qPCR)

Total human *PTPN22* and mouse *Ptpn22* mRNA expression was measured by quantitative real-time RT-PCR (qPCR) using specific primers designed to amplify the human and mouse mRNAs. Total RNA was isolated using the RNeasy Plus Micro Kit (Qiagen). After DNase treatment of the RNA with Ambion® TURBO™ DNase (Life Technologies), cDNA was reverse transcribed using the SuperScript® III First-Strand Synthesis System (Life Technologies). Gene expression was quantified using SYBR Green reaction mixture (SaBiosciences/Qiagen) on a LightCycler 480 (Roche) with a cycling program of 94°C for 3 minutes, 35 cycles of (94°C for 45 seconds, 62°C for 30 seconds, 72°C for 1 minute), 72°C for 10 minutes. Reactions were performed in triplicate and normalized to the mouse housekeeping gene RNA Polymerase II (Polr2a). Primer sequences were:

PTPN22 forward 5'-GAATTTCTGAAGCTGAAAAGGCA-3'

PTPN22 reverse 5'-TCCCAAATCATCCTCCAGAAGTC-3'

Polr2a forward 5' CCAGGAAACACATTGCGTC-3'

Polr2a reverse 5'-GGAAGAAGAACTCAGTGGGTG-3'.

Phosphatase Assays

Phosphatase assays were performed on HA-tagged LYP following anti-HA-IPs from thymocytes from the indicated mice. Cells were lysed in 50 mM HEPES, pH 7.4, 150 mM NaCl, 5 mM EDTA with 1% Triton-X 100, 1 mM phenylmethylsulfonyl fluoride, 10 μg/ml aprotinin, 10 μl/ml leupeptin, and 10 μg/ml soybean trypsin inhibitor. The IPs were washed in 50 mM Bis-Tris, pH 6.0, and then resuspended in phosphatase buffer (50 mM Bis-Tris, pH 6.0, 5 mM DTT, 0.005% Tween-20, 5 mM sodium fluoride, 2 mM sodium pyrophosphate). After the addition of 0.1 mM 6,8-Difluoro-4-Methylumbelliferyl Phosphate (DiFMUP, Life Technologies), the reaction was monitored continuously by measuring the increase in fluorescence ($\lambda_{ex} = 340$ nm and $\lambda_{em} = 460$ nm) at 60 s intervals for 30 min using a Tecan Infinite M1000 plate reader (Tecan) [42]. The activity measured in triplicate was corrected for the nonspecific signal of identical reactions performed also in triplicate without the addition of enzyme.

Ethics statement. This study was carried out in strict accordance with the recommendations in the Guide for the Care and Use of Laboratory Animals of the National Institutes of Health. The protocol was approved by the institutional Animal Use Committees of the University of Southern California (Protocols #19853 and #10714), the La Jolla Institute for Allergy and Immunology (Protocol #AP140-NB4), the University of Minnesota (Protocol #1106A00402) or the veterinarian authorities of the Swiss Canton Vaud (Protocols 2253 and 2643). All efforts were made to minimize animal suffering.

Results

Generation of Mice Carrying Thymic Overexpression of LYP-W620

In order to assess the effect of LYP-W620 overexpression at the thymic level, we generated transgenic mouse lines overexpressing N-terminal HA-tagged LYPW (abbreviated as TgLYPW) or its inactive mutant LYPW^{C227S} (abbreviated as TgLYPW^{C227S}) under the control of the proximal Lck promoter. The Lck proximal promoter is highly active in thymocytes, including immature double negatives (DN), double positives (DP), and CD4$^+$ and CD8$^+$ single positives (SP) [29], but is repressed after thymocyte emigration. **Fig. 1A** shows that the two transgenic constructs were expressed in thymocytes, although at different levels. Progeny of an additional independent founder expressing TgLYPW^{C227S} all displayed overexpression levels several orders of magnitude higher than those observed in the TgLYPW line expressing the active phosphatase (data not shown). A single TgLYPW^{C227S} founder line was selected for all further analyses. In line with known thymus-specific activity of the proximal Lck promoter, no expression of the transgenes was detectable in the spleen and

lymph nodes of the transgenic mice (data not shown). The transgene-encoded LYP-W620 protein was active in an in vitro phosphatase assay, while as expected, LYP-W620/S227 was completely inactive (**Fig. 1B**). The highest expression of the transgenic protein was achieved in DP thymocytes, although overexpression was evident in SP and DN cells (**Fig. 1C** and **Fig. 1D**). An analysis at the mRNA level using primers that cross-react with endogenous Ptpn22 suggested that thymocytes from the TgLYPW line carry an average two-fold overexpression of PTPN22 transcript (**Fig. 1E**).

Thymic Overexpression of LYPW Leads to Reduced Thymocyte TCR Signaling

The gain-of-function model predicts that LYP-W620 will exert augmented negative regulatory function in antigen receptor signaling. We therefore assessed whether overexpression of LYPW results in decreased thymocyte TCR signaling. **Fig. 2A** and **Fig. 2B** show that Erk phosphorylation after engagement of the TCR was reduced in thymocytes from TgLYPW transgenic animals compared to cells from non transgenic littermates. Notably, despite exhibiting significantly higher expression of LYP protein compared to TgLYPW mice (**Fig. 1A**), TgLYPW^{C227S} transgenic thymocytes displayed no alteration in Erk phosphorylation compared to non transgenic cells. Similarly, TCR-dependent CD69 upregulation was reduced on DP thymocytes from TgLYPW, but not from TgLYPW^{C227S}, transgenic mice, compared to their respective littermates (**Fig. 2C**). Overall, these data indicate that thymocyte-expressed LYP-W620 can repress TCR signaling in total thymocytes and in DP thymocytes in particular, and suggest that modulation of TCR signaling by LYP-W620 is phosphatase-activity dependent.

Thymic Overexpression of LYPW does not Affect Positive or Negative Selection or the Output of Thymic CD4$^+$Foxp3$^+$ T$_{reg}$

Loss of Pep results in increased numbers of both CD4SP and CD8SP thymocytes, but no defects in negative selection [9,21]. We asked whether thymocyte expression of LYP-W620 could modulate positive or negative selection. We found no significant alterations in the total thymocyte numbers and numbers of DN, DP, CD4SP, CD8SP, NKT or gamma-delta T cells associated with TgLYPW or TgLYPW^{C227S} expression (**Table 1**). Further, we observed no differences in the percentages of CD4$^+$Foxp3$^+$ T$_{reg}$ present in the thymi of TgLYPW or TgLYPW^{C227S} mice (data not shown). To further determine whether CD4$^+$Foxp3$^+$ T$_{reg}$ thymic output is affected by the expression of the LYPW transgene, we bred the transgenic mice with GFP-Foxp3 mice carrying a GFP cassette knocked into the Foxp3 locus [32]. **Fig. 3A** shows that the frequency of GFP-Foxp3$^+$ thymocytes was unaffected by the expression of the LYPW transgene. To address the roles of LYP-W620 and of LYP enzymatic activity in TCR-driven thymocyte development, we bred the transgenic mice to DO11.10 and H-Y TCR transgenic strains that served as models of positive and positive/negative selection, respectively. **Fig. 3B** shows that the numbers of DO11.10 TCR transgenic cells were not affected by expression of the LYPW or LYPW^{C227S} transgenes pointing to the absence of anomalies of positive selection in these mice. As shown in **Fig. 3C**, total numbers of transgenic HY-TCR thymocytes were unaffected by the expression of the active phosphatase in F1 B6xBALB/c male mice. In this model, deletion of the HY-TCR transgene was minimal, likely due to the B6xBALB/c genetic background, however no skewing of the percentages of DP, CD8SP or CD4SP could be found, suggesting that negative

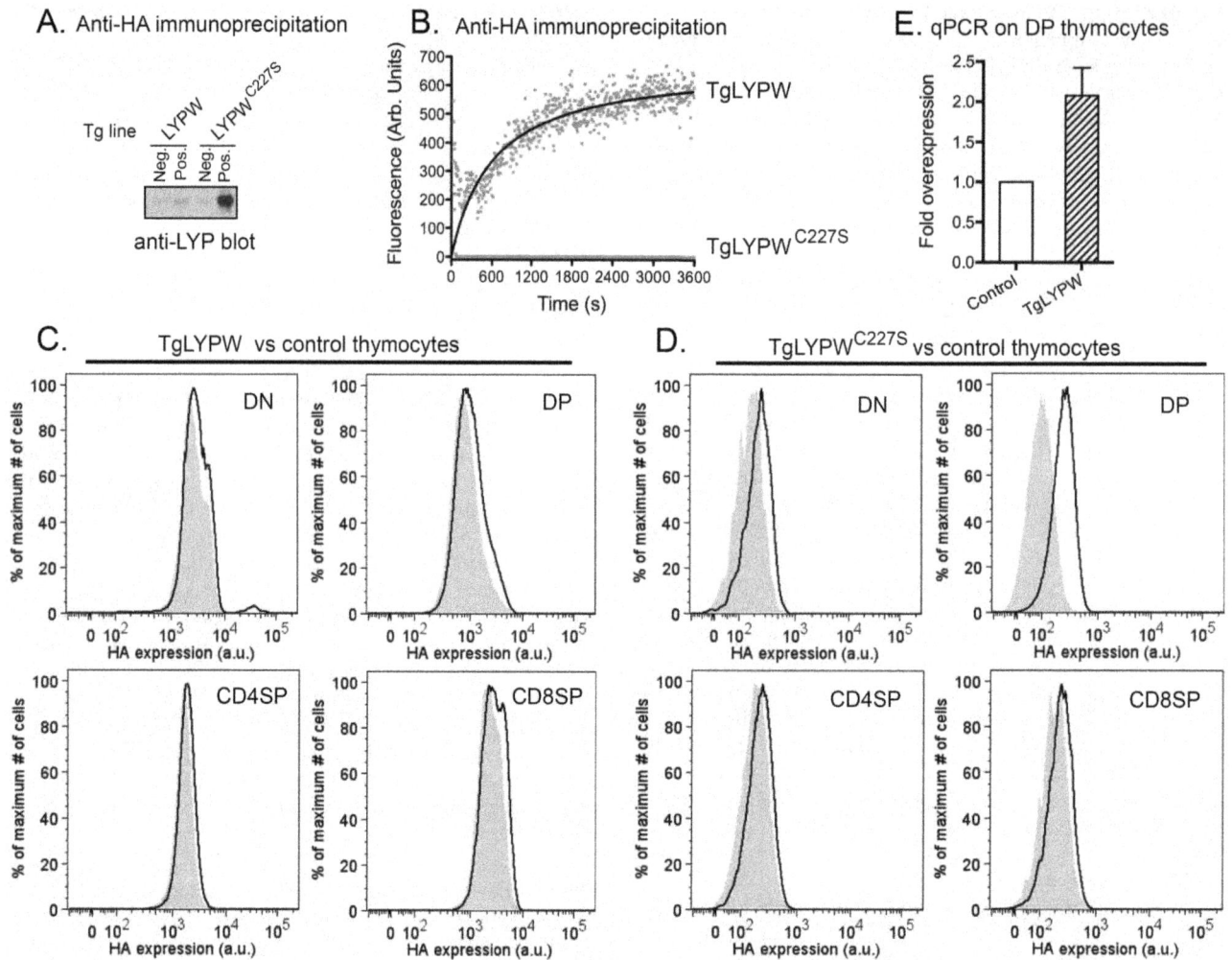

Figure 1. Mice transgenic for LYP-W620 (TgLYPW) show overexpression of LYP in DP thymocytes. A, LYP protein expression in TgLYPW thymocytes. Total thymocytes from TgLYPW (lane 2) or a non-Tg littermate (lane 1) and TgLYPW^C227S (lane 4) or a non-Tg littermate (lane 3) were lysed and subjected to immunoprecipitation (IP) using an anti-HA antibody (Ab). Panel shows Western blotting using an anti-LYP Ab. Data are representative of 3 independent experiments with similar results. B, LYP phosphatase activity in TgLYPW thymocytes. Anti-HA IPs were performed from lysates of total thymocytes from TgLYPW and TgLYPW^C227S mice. Graphs show the phosphatase activity of LYP as assessed by dephosphorylation of the fluorescent substrate DiFMUP. Data are representative of 2 independent experiments with similar results. C–D, LYP-W620 transgene expression in thymocyte subpopulations. Expression of LYPW (C) or LYPW^C227S (D) was assessed by intracellular staining using a fluorophore-conjugated anti-HA Ab in DN (upper left panel), DP (upper right panel) CD4SP (lower left panel) and CD8SP (lower right panel) thymocytes from Tg mice (black graphs) and control non-Tg littermates (grey filled graphs). E, Quantification of overexpression of LYPW relative to endogenous Pep in DP thymocytes of TgLYPW mice. mRNA encoding LYP and Pep was quantified by qPCR from sorted DP thymocytes from control BALB/c (white bar) and TgLYPW (striped bar) mice, using a primer pair that amplifies both human *PTPN22* and mouse *Ptpn22* mRNAs. Graph shows relative expression levels of total *PTPN22* after normalization to the mouse housekeeping gene *Polr2a*. Data are average and SE of 3 biological replicates.

selection of CD8+ cells operates normally in these mice (**Fig. 3C**). Similarly, we did not see any differences between non transgenic and TgLYPW OT-1 TCR transgenic mice in the efficiency of CD8+ negative selection upon crossing the animals to cognate antigen-expressing Rip-mOva mice (data not shown).

The above studies of negative selection were carried out in F1 B6xBALB/c mice in which overexpression or mis-expression of *Lck* promoter-directed LYP-W620 might obscure phosphatase-dependent changes in TCR-dependent thymocyte development. Therefore, we sought independent confirmation of our findings by carrying out studies of animals made transgenic for human *PTPN22* on a pure B6 background. Mice bearing an entire *PTPN22*-*T1858 human regulon encoding a LYP-W620 protein were generated on a B6 background under the supposition that

endogenous *PTPN22* transcriptional regulatory elements would induce physiologic spatio-temporal expression patterns. To mimic the LYP-W620 phosphatase allele "load" observed in the most common human carrier state (heterozygosity for one allele each of LYP-R620 and LYP-W620), we bred the animals with *Ptpn22*-deficient mice to generate Pep+/−.LYP-W620 (abbreviated as BACLYPW) mice. A non-significant trend toward a decrease in the total number of thymocytes was observed in BACLYPW mice compared to non transgenic littermate mice. We observed no differences between Pep+/− littermates and BACLYPW transgenic mice in thymocyte number or distribution (**Fig. 4A** and data not shown). These findings provided further evidence that LYP-W620 expression does not cause grossly altered thymic development.

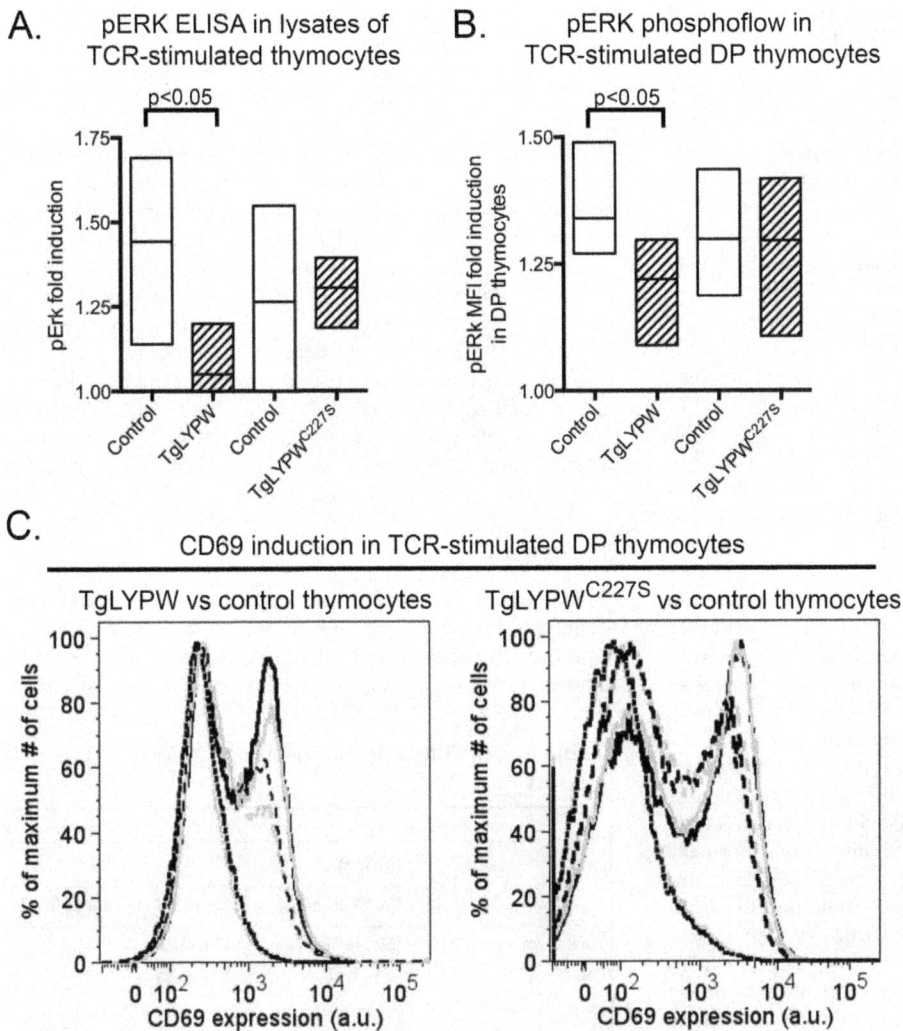

Figure 2. Overexpression of LYPW inhibits TCR signaling in DP thymocytes. A–B, Overexpression of LYPW causes reduced activation of Erk in thymocytes. Total thymocytes from TgLYPW or TgLYPW^{C227S} (striped bars) or their respective non-Tg littermates (white bars) were stimulated with 20 µg/ml anti-CD3 and 10 µg/ml anti-Armenian Hamster IgG1 crosslinker for 2.5 minutes. A, Graph shows phosphorylation of Erk in total thymocyte lysates assessed using the PathScan® phospho-p44 MAPK (Thr202/Tyr204) sandwich ELISA kit. Histogram shows mean and range of fold induction of at least 3 biological replicates. B, Phosphorylation of Erk in DP thymocytes was assessed by phosphoflow analysis after intracellular staining with an anti-pErk Ab. Fold induction of Erk phosphorylation was normalized within each experiment relative to the sample with the highest induction. Histogram shows mean and range of at least 3 biological replicates. C, Overexpression of LYPW causes reduced T cell activation in DP thymocytes. Thymocytes from TgLYPW or TgLYPW^{C227S} (grey graphs) and their respective non-Tg littermates (black graphs) were cultured in the presence of 10 µg/ml (long dashed graphs) or 25 µg/ml (solid graphs) anti-CD3, or media alone (dotted graphs), for 18 hours. Graphs shows expression of CD69 in DP thymocytes as assessed by flow cytometry analysis after staining with an anti-CD69 Ab. Median fluorescence intensity (MFI) values are indicated on each graph. Graphs are representative of at least 3 biological replicates with identical results.

We also asked whether endogenous promoter-driven LYP-W620 expression affects thymocyte positive or negative selection. We bred BACLYPW mice to H-Y TCR transgenic mice. H-Y TCR transgenic female thymi displayed no significant differences between BACLYPW and non transgenic Pep$^{+/-}$ in total thymocyte numbers or the distribution of H-Y TCR-high cells (**Fig. 4B** and data not shown). A small decrease in the total number of thymocytes was noticed in BACLYPW vs control male HY-TCR transgenic mice (**Fig. 4C**). However, the distribution of the major thymocyte subpopulations remained unaffected in BACLYPWxHY-TCR male mice, compared with age-matched controls (**Fig. 4C**). Together, these data supported the suggestion, first raised by findings in TgLYPW animals (**Fig. 3A**), that LYP-W620 does not differentially regulate thymocyte positive selection.

Significantly, the data from H-Y males suggest that if LYP-W620 does modulate negative selection, it does not impair that process. Rather, the presence of human promoter-controlled LYP-W620 may serve to promote deletion of autoreactive thymocytes.

We reasoned that the DO11.10 and HY TCR transgenic models might be inadequate to assess the effect of the R620W mutation on thymic selection if a sufficiently intense stimulation of the TCR is able to overcome the inhibitory effect of the R620W gain-of-function on signaling. We therefore also utilized altered peptide ligands to test how the R620W variant impacts negative selection in a qualitative and quantitative manner. We co-cultured non transgenic, TgLYPW, and TgLYPW^{C227S} OT-1 thymocytes with irradiated target cells and increasing concentrations of either native OT-1 peptide (SIINFEKL, N4) or altered peptide ligands

(SIITFEKL, T4 and SIIVFEKL, V4). The latter peptides provide a much lower level of stimulation to OT-1 T cells than SIINFEKL [43]. Exposing double positive thymocytes to a stimulating peptide causes a strong down-regulation of CD4 and CD8 expression [44]. As expected, compared to the wild-type ligand N4, higher concentrations of T4 and V4 were needed to induce co-receptor down-regulation but interestingly, identical dose-response curves were obtained when non transgenic, TgLYPW or TgLYPW^{C227S} OT-1 thymocytes were exposed to the different peptides. We concluded that expression of the two different forms of LYP does not change antigen-sensitivity of OT-1 thymocytes (**Fig. 5**).

Next, we assessed the effect of the LYP-W620 transgene on *in vivo* negative selection of polyclonal antigen-specific T cells using a previously established Rip-mOva and Vβ5 transgenic mouse model system [45]. Vβ5 mice express the same TCRβ chain as OT-1 T cells while they re-arrange endogenous TCRα chains [46]. Fixing TCRβ strongly elevates the frequency of K^b/Ova-specific T cells; however, these T cells cover a wide range from low to high avidity recognition for the Ova-antigen. Crossing Vβ5 to Rip-mOva mice causes the thymic deletion of high avidity Ova-reactive T cells while low avidity T cells are spared [45]. Similar results can be obtained when Vβ5 bone marrow is transferred into Rip-mOva hosts (DZ, unpublished observations). We asked whether TgLYPW could modulate negative selection in Rip-mOva mice and if this manipulation might lead to an escape of higher affinity Vβ5 T cells. Irradiated Rip-mOVA mice were reconstituted with bone marrow from either non-Tg or TgLYPW Vβ5 mice. After infecting these bone marrow chimeric mice with *Listeria monocytogenes*-Ova, we found that T cells with comparable functional avidity for K^b/Ova could be found in non-Tg or TgLYPW mice (**Fig. 6A**). Since higher avidity T cells normally overgrow lower avidity T cell clones [43], we concluded from our data that there is no change in the negative selection threshold caused by the presence of human LYP-W620. We also assessed whether LYP-W620 expression would result in enhanced expansion of peripheral autoreactive T cell numbers that are normally subject to negative selection in Rip-mOva mice. As previously described, we applied two consecutive infections to expand low avidity clones in Rip-mOva mice [47]. We observed that both normal and LYPW Rip-mOva mice contained similar numbers of low avidity K^b/Ova-reactive T cells (**Fig. 6B**). Together, the data suggest that active LYP expression does not affect negative selection of thymocytes or the frequency of autoreactive thymic emigrants.

No Alterations of CD4$^+$ Thymocyte Repertoire and Severity of Arthritis in SkgxTgLYPW Mice

In order to assess the effect of LYPW overexpression on the thymic repertoire of CD4$^+$ T cells, we crossed TgLYPW mice with Skg mice. Skg mice carry a mutation in *Zap70* (W163C) that attenuates T cell receptor signaling [35]. However, to some extent, the reduced function *Zap70* mutant can be compensated such that CD4$^+$ and CD8$^+$ T cells can be found in the periphery of Skg mice. Interestingly, a chronic form of arthritis develops spontaneously in these mice. Moreover, disease can be induced by injecting fungal-derived polysaccharides into animals that are homozygous for the Skg mutation and express at least one copy of the H2^d haplotype ([35,39,40] and our unpublished data). It has also been shown that Skg/Skg and heterozygote Skg/WT mice display alterations in the Vβ repertoire usage by CD4$^+$Foxp3$^-$ thymocytes and even more markedly in CD4$^+$Foxp3$^+$ thymocytes, which are supposedly skewed toward a more self-reactive repertoire. These alterations include decreased deletion of Vβ3$^+$ and Vβ5$^+$ T cells and decreased positive selection of Vβ8$^+$ T cells [37]. Qualitatively

similar but quantitatively more marked alterations of the CD4$^+$ Vβ repertoire, correlating with a more severe decrease in thymocyte TCR signaling and thymocyte selection, were found in homozygous Skg/Skg animals. Thus, we reasoned that Skg/WT and Skg/Skg animals would be a sensitive model to detect subtle alterations in CD4$^+$ T cell selection caused by decreased thymocyte TCR signaling [36,37]. **Fig. 7A and 7B** show that no alterations of the frequencies of Vβ3$^+$, Vβ5$^+$, or Vβ8$^+$ thymocytes were found among CD4$^+$Foxp3$^-$ or CD4$^+$Foxp3$^+$ thymocytes in Skg/WT or Skg/Skg TgLYPW mice versus non transgenic littermates, suggesting that the decrease in TCR signaling caused by the active phosphatase transgene is insufficient to cause significant alterations in CD4$^+$ T cell selection. Accordingly, the LYPW transgene did not cause increased severity of mannan-induced arthritis in Skg/Skg mice (**Fig. 7C**), and equal numbers of lymph node Th17 cells were found in Skg/Skg TgLYPW and littermate Skg/Skg mice one month after induction of arthritis by intraperitoneal injection of mannan (**Fig. 7C**).

Discussion

With respect to genetic effect potency, the LYP-R620W polymorphism currently ranks in second and third position, respectively, as a risk factor for RA and for type 1 diabetes [2,48]. Although it is evident that the R620W variation is one of the

Table 1. Cell distribution in thymus of TgLYPW and TgLYPW^{C227S}.

	TgLYPW	
	TgLYPW (n = 5)	Non Tg littermates (n = 6)
Total thymocyte (x10^6)	127±46.77	154±32.21
DN	3.12±1.97	2.49±1.07
DP	75.18±1.95	75.13±5.25
CD4SP	13.2±1.84	13.45±2.24
CD8SP	3.45±0.8	3.7±0.69
CD25$^-$CD44$^+$ % of DN	15.04±3.59	18.22±5.40
CD25$^+$CD44$^+$ % of DN	4.68±2.05	4.56±2.19
CD25$^+$CD44$^-$ % of DN	42.16±4.41	39.9±4.03
CD25$^-$CD44$^-$ % of DN	38.1±7.49	37.35±3.63
NKT cells	1.3±0.19	1.36±0.06
gammadelta T cells	0.27±0.04	0.31±0.03

	TgLYPW^{C227S}	
	TgLYPW^{C227S} (n = 5)	Non Tg littermates (n = 5)
Total thymocyte (x10^6)	160.78±23.99	131.23±21.36
DN	4.32±2.51	3.53±1.26
DP	71.78±6.41	74.78±2.37
CD4SP	17.74±5.02	15.18±2.51
CD8SP	4.13±0.75	4.60±0.73
CD25$^-$CD44$^+$ % of DN	16.32±9.46	17.16±8.71
CD25$^+$CD44$^+$ % of DN	5.31±1.65	4.50±2.37
CD25$^+$CD44$^-$ % of DN	41.96±6.21	37.80±3.33
CD25$^-$CD44$^-$ % of DN	37.02±3.34	40.60±5.95

Table shows the basic statistics (average±SD) of cell counts in transgenic animals and non transgenic littermates.

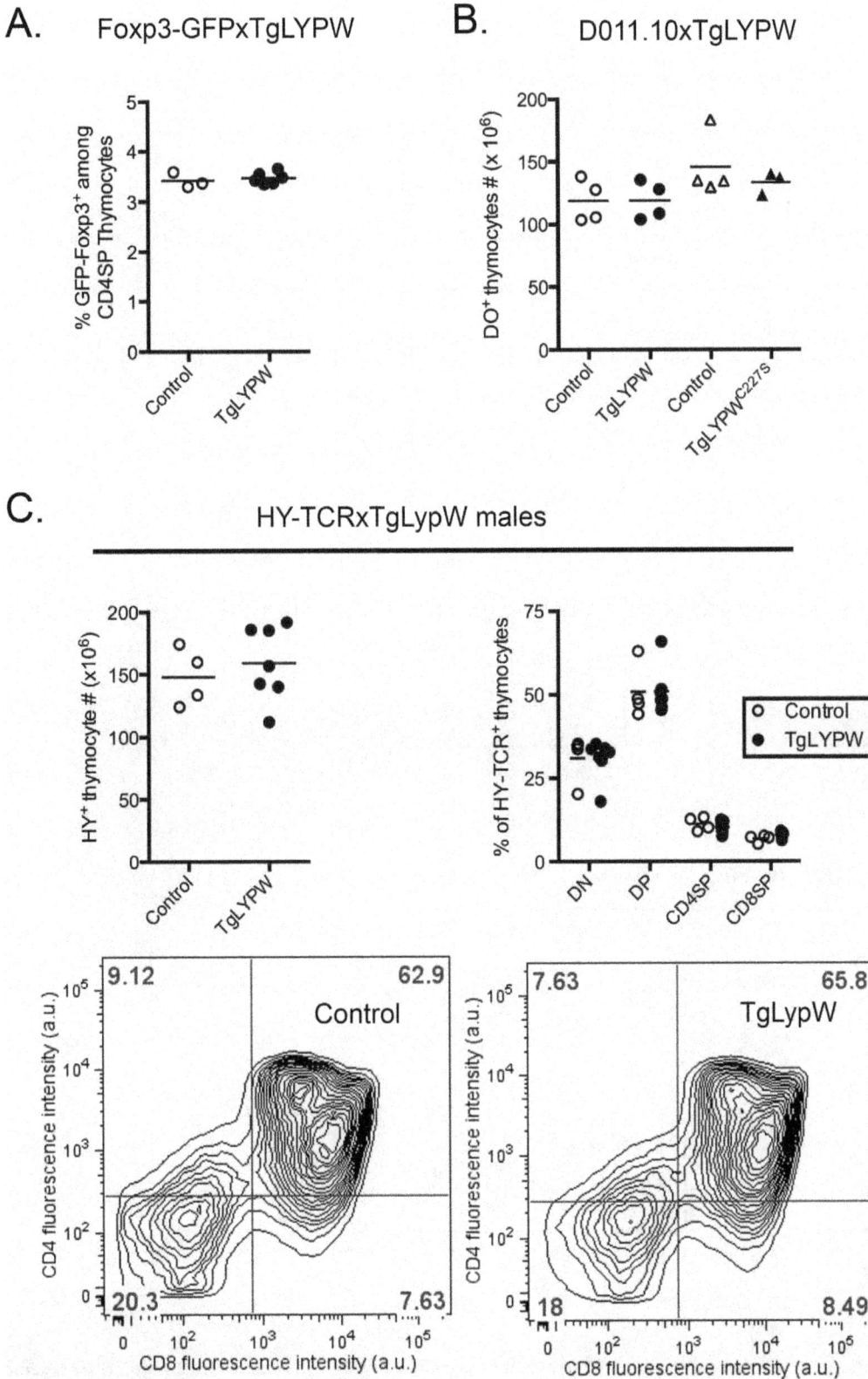

Figure 3. No decrease in thymic output of T_reg and positive or negative selection of HY-TCR-transgenic thymocytes in human LYP-W620 transgenic (TgLYPW) and control mice. A, LYP-W620 overexpression does not alter the numbers of CD4⁺Foxp3⁺ thymocytes. Panel shows % of GFP⁺ thymocytes among CD4SP from Foxp3-GFPxTgLYPW mice (black circles, n = 6) or non-Tg littermates (white circles, n = 3). B, No alterations of positive selection of DO11.10⁺ thymocytes in TgLYPW mice. Panel shows numbers of DO11.10⁺ thymocytes in DO11.10xTgLYP^WT mice or DO11.10xTgLYP^C227S mice (black circles and triangles, n = 4 and n = 3 respectively) or their respective non-Tg littermates (white circles and triangles, n = 4 and n = 4 respectively). C, Overexpression of LYP does not alter negative selection in male F1 HY-TCRx.TgLYP mice. Upper left panel shows numbers of HY-TCR⁺ thymocytes from male F1 HY.TgLYP mice (black circles, n = 7) or non-Tg littermates (white circles, n = 4). Upper right panel shows

% of CD4SP, DP, CD8SP and DN thymocytes after gating on the HY$^+$ population of thymocytes from male HY-TCRxTgLYPW mice (black circles, n = 4) or non-Tg littermates (white circles, n = 4). Bottom panels shows representative contour plot of CD4$^+$ vs CD8$^+$ expression of HY-TCR$^+$ thymocytes from HY-TCRxTgLYPW or non-Tg littermates. Transversal bars in each panel indicate mean value.

major non-HLA genetic risk factors for autoimmunity, its mechanism of action at the molecular level and its role in the immunopathogenesis of disease remain unclear.

Our experiments in primary T cells from type 1 diabetes subjects and in Jurkat and primary human T cells overexpressing LYPW lead us to propose a model that LYP-W620 acts as a gain-of-function variant in phosphatase activity and in negative regulation of TCR signaling [12]. Accordingly, an initial hypothesis as to how the R620W variation might promote autoimmunity was that variant LYP augments inhibition of

thymocyte TCR signaling and allows the escape of higher numbers of auto-reactive T cells or of T cells exhibiting a higher functional avidity (higher strength of self-pMHC and TCR interaction) [12,49]. Since formulation of the hypothesis, additional data variably supporting a "gain-of-function"[13,15–18], "loss-of-function" [20,21,50] or "altered-function" [23,24] phenotype of LYP-W620 in TCR signaling have been published.

Ptpn22-deficient mice exhibit abnormal accumulation of effector-like and memory T cells and heightened lymphoproliferation capacity without clear-cut breaches in peripheral tolerance [9], but

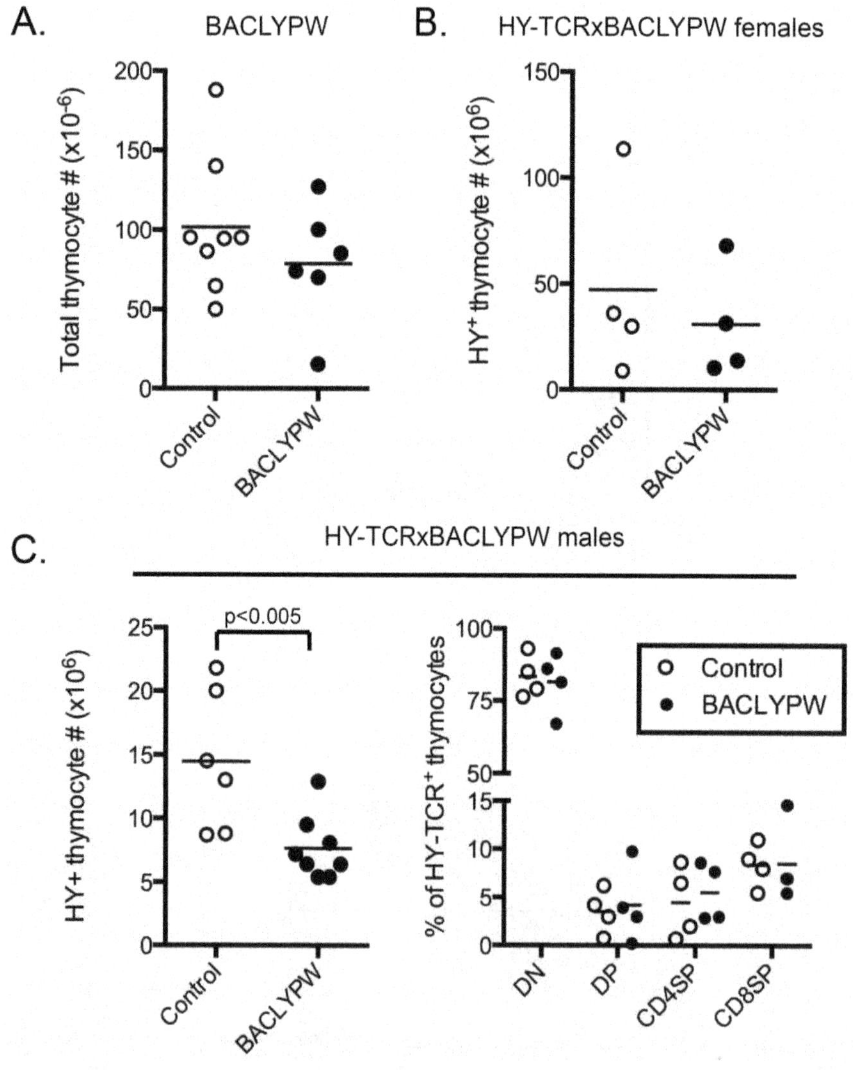

Figure 4. **No decrease in positive or negative selection of HY-TCR-transgenic thymocytes in human *PTPN22* regulon transgenic (BACLYPW) and control mice.** A, Overexpression of LYP-W620 under control of the *PTPN22* physiological promoter does not affect total thymocyte number in BAC transgenic mice. Panel shows total numbers of thymocytes in BACLYPW mice (black circles, n = 6) or their respective non-Tg littermates (white circles, n = 8). B–C, overexpression of LYP-W620 does not alter positive or negative selection and the subpopulation distribution of HY-TCR$^+$ thymocytes in BACLYPW mice. B, panel shows numbers of HY-TCR$^+$ thymocytes from female HY-TCRxBACLYPW mice (black circles, n = 4) or non-Tg littermates (white circles, n = 4). C, Left panel shows numbers of HY-TCR$^+$ thymocytes from male HY-TCRxBACLYPW mice (black circles, n = 8) or non-Tg littermates (white circles, n = 6). Right panel shows % of CD4SP, DP, CD8SP and DN thymocytes after gating on the HY$^+$ population of thymocytes from male HY-TCRxBACLYPW mice (black circles, n = 4) or non-Tg littermates (white circles, n = 4). Transversal bars in each panel indicate mean value.

Ex vivo thymocytes

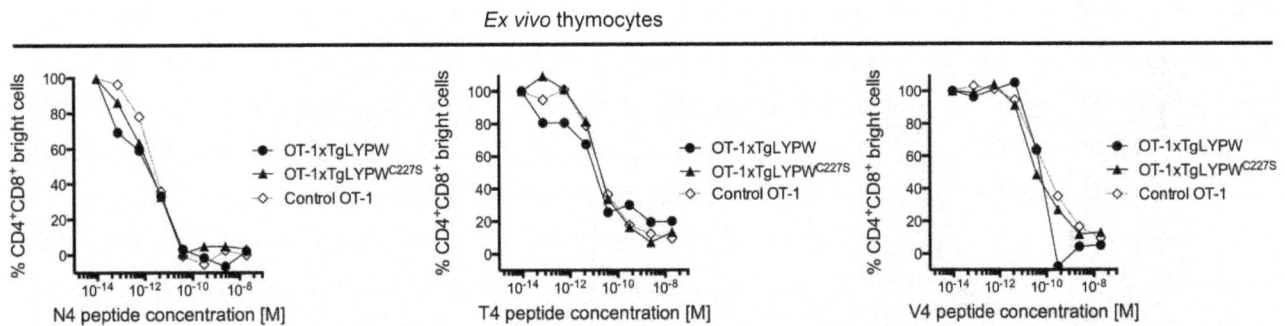

Figure 5. OT-1 thymocytes overexpressing LYP-W620 show similar antigen sensitivity as non-Tg thymocytes. Freshly isolated OT-1 transgenic thymocytes from TgLYP (black circles), TgLYP^{C227S} (black triangles) or control non-Tg mice (white diamonds) were co-cultured overnight together with RMA cells and the indicated doses of SIINFKL peptide (N4, left panel), of the low and very low affinity altered peptide ligands SIITFEKL (T4, middle panel) and SIIVFEKL (V4, right panel). Graphs show normalized dose-response values of peptide concentration vs the fraction of maximum numbers of residual DP (CD4 and CD8 bright) thymocytes.

it has remained unclear whether these phenotypes solely relied on a lack of peripheral or thymic Pep expression. Two independent studies have found evidence of increased qualitative positive thymocyte selection in *Ptpn22* KO mice transgenic for the DO11.10 [9] and OT-II TCRs [10]. Another group recently overexpressed WT Pep under control of the T lineage specific distal *Lck* promoter [27], and found no alterations in thymocyte subpopulation numbers in transgenic vs non transgenic mice. However, based on the published *Ptpn22* KO, knock-in, or overexpression studies, no prediction could be made how a presumed gain of function R620W variant of human LYP might impact negative selection of auto-reactive T cells. Our study was designed to test the hypothesis that the gain-of-function inhibition of TCR signaling by LYP-W620 in thymocytes is sufficient to alter thymic output in a manner that could increase predisposition to autoimmunity. We studied the human phosphatase variant rather than its mouse homolog W619 because it is currently unclear whether regulation by Csk is conserved between the human and mouse phosphatase, or whether the gain-of-function nature of the R620W variant is fully replicated by the R619W mutation of mouse Pep. To detect possible enzymatic activity-independent effects of the phosphatase, we also generated animals transgenic for the inactive C227S mutants of LYP-W620. In the presence of LYP-W620, we observed increased phosphatase activity of LYP-W620 in thymocytes, associated with decreased TCR signaling in total thymocytes and DP thymocytes. No anomalies in thymic TCR signaling were detected in animals carrying the inactive form of LYP-W620, despite supraphysiological expression of LYP directed by the transgene in these animals. Results with LYP-W620 C227S variant-bearing mice thus suggest that effects of phosphatase overexpression on thymocyte TCR signaling critically depend upon enzymatic function of LYP.

Overexpression of active LYP-W620 did not lead to impaired positive or negative selection of HY-TCR transgenic thymocytes in male mice or in OT-1 TCR-transgenic thymocytes crossed to Rip-mOva mice. Indeed, the finding of modestly-reduced autoreactive thymocyte numbers in BACLYP (human *PTPN22* promoter-driven) transgenic H-Y males suggests that LYP-W620 may actually promote negative selection of CD8$^+$ T cells. Such a phenotype is in line with the report by Dai *et al.*, in which Pep-W619-knockin autoreactive thymocytes exhibit augmented clonal deletion [23]. We speculate that such subtle positive effects of LYP-W620 on thymocyte selection are obliterated in our TgLYP models by the combination of supraphysiological expression levels and the use of an exogenous promoter.

Next, we considered the hypothesis that overexpression of the LYP-W620 variant would only impact negative selection at an intermediate TCR affinity range and thus might particularly impact those clones bearing TCR with affinity for self-antigen close to the threshold of negative selection. We addressed this hypothesis by comparing the affinity of the bulk population of K^b/Ova-reactive T cells that survive negative selection in Rip-mOva mice. Normally, in such mice only T cells with low affinity for K^b/Ova can be detected in the periphery [45]. We reasoned that if LYP-W620 selectively reduces negative selection of T cells bearing antigen receptors with affinity at or just above the threshold of negative selection, then expression of TgLYPW would confer increased in the overall functional avidity of K^b/Ova-reactive T cells in Rip-mOva or Rip-mOvaxVβ5 mice. Our negative data clearly argue against this hypothesis.

We also considered whether LYP-W620 might selectively impact CD4$^+$ T cell selection. The pathogenesis of disease in Skg mice depends on reduced TCR signaling in thymocytes, which results in reduced positive and negative selection of CD4$^+$ thymocytes, but the repertoire of Skg thymic emigrants contains T cells with significant auto-reactive potential [36,37]. When we overexpressed LYP-W620 in the thymus of Skg mice, we did not observe alterations of CD4$^+$ thymocyte repertoire or a gain in disease severity. Higher affinity interactions with thymocyte TCR by autoantigens are thought to be required for the differentiation of T$_{reg}$ in the thymus, thus suggesting that alterations of TCR signaling in the thymus might shape the amount and repertoire of T$_{reg}$ [51]. However, we did not see changes in the output of regulatory T cells from the TgLYPW Skg thymus, further arguing against the notion that LYP-W620 selectively affects high-affinity TCR interactions important for development.

Importantly, in the TgLYPW model, expression of the transgene is controlled by the *Lck* proximal promoter and is only expressed in the thymus but not in the periphery. Therefore, we can exclude that we created a balanced situation such that we might have augmented the thymic threshold but, at the same time, caused similar increases in the peripheral T cell activation threshold. Altogether, our observations imply that LYP-W620 does not have a major negative impact on selection of T cells in general or on the threshold of negative selection in mice. This is well in line with observations that complete absence of Pep had no impact on negative selection [9].

Given the lack of impact on thymic selection, we suggest that LYP-W620 might primarily act on peripheral T cells. A selective effect of LYP-W620 on signaling in peripheral T cells would be

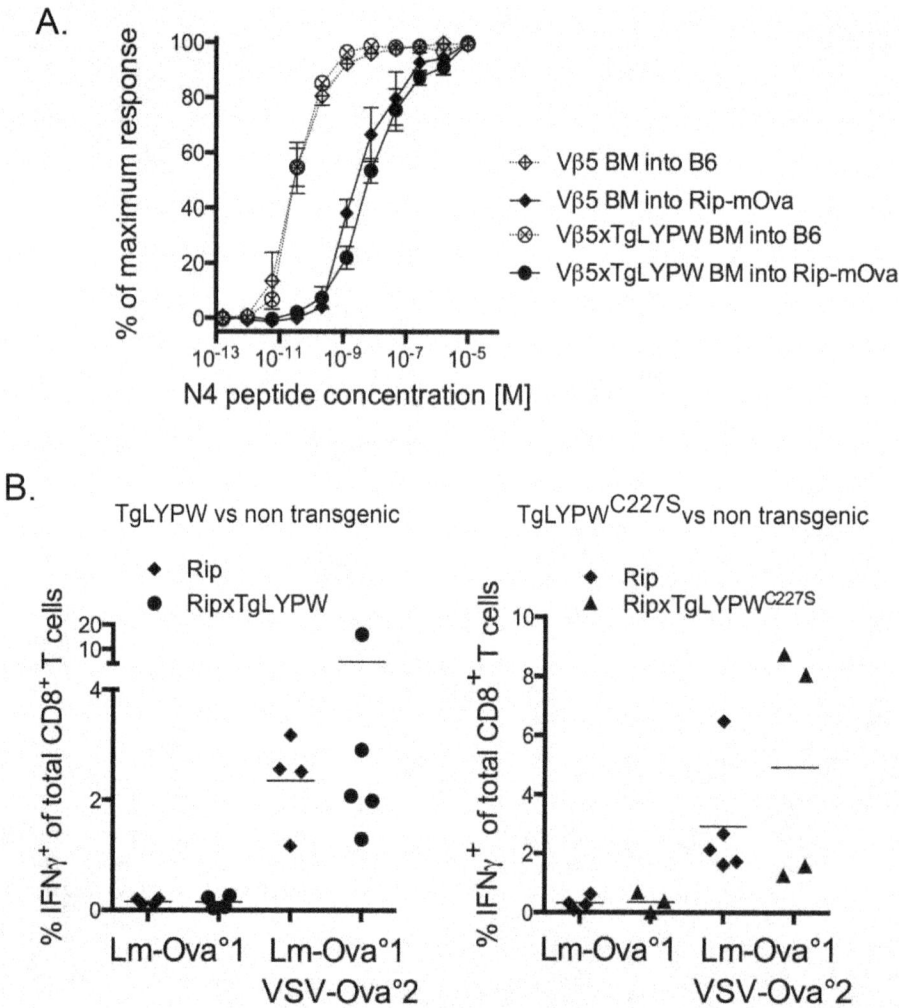

Figure 6. Polyclonal thymocytes undergo similar levels of negative selection in the presence of absence of transgenic LYP-W620. A, Lethally irradiated Rip-mOva mice (continuous graphs and black symbols) or C57BL/6 mice (dotted graphs and crossed symbols) were reconstituted with bone marrow harvested from Vβ5xLYPW (circles) or Vβ5 control (diamonds) mice. 10 weeks after the reconstitution the mice were infected with a strain of *Listeria monocytogenes* expressing Ovalbumin (Lm-Ova). Splenocytes were harvested at 8 days after the infection and briefly *in vitro* re-stimulated with titrated doses of SIINFEKL peptide. Afterwards, the cells were intracellularly stained for IFNγ. Peptide-dose response curves showing the frequency of IFNγ producing CD8⁺ T cells as fraction of maximum response are presented. B, TgLYPW, TgLYPW^{C227S} and control non-Tg mice contain similar numbers of low avidity auto-reactive T cells. RipxTgLYPW (left panel, black circles), RipxTgLYPW^{C227S} (right panel, black triangles) and control Rip mice (left and right panels, black diamonds) were infected with Lm-Ova and 4 weeks later challenged by a strain of *Vesicular stomatitis virus* expressing Ova (VSV-Ova). On day 6 after the primary (left side of each panel) or the secondary infection (right side of each panel) blood was drawn from the mice and PBMC were briefly re-stimulated with SIINFEKL peptide. The number of IFNγ producing CD8⁺ T cells was determined by intracellular cytokine staining. Panels show the frequency of Ova-specific T cells.

consistent with the observed increase of effector/memory T cells in Pep KO mice. However, a gain of function mutant that increases the activation threshold of effector T cells is difficult to reconcile with auto-immunity. We therefore consider that a peripheral model of promoting auto-immunity through the LYP-R620W mutation could involve an impaired T_{reg} function. However, others have not documented diminished T_{reg} function in 619W knock-in mice [23].

Our findings lead us to the conclusion that the mechanism by which LYP-W620 impinges on the immunopathogenesis of T cell-modulated human disease is still uncertain and does not involve major effects on thymocyte-intrinsic processes that establish central tolerance. LYP-W620 might involve yet unknown effects on T cells or it might cause a rather subtle impact that we failed to detect in our experimental systems. One study reported that

subjects carrying LYP-W620 have increased numbers of effector T cells [15]. A recent report also suggests that the R620W variation might have unknown "change-in-function" effects [24] on T cells.

The LYP-R620W variation might impact the function of other cell types including B cells. It has been suggested that an increased burden of autoreactive B cells –perhaps secondary to weakened B cell negative selection– underlies the predisposition of LYP-W620-carrying subjects to autoimmune diseases [52]. Zhang *et al.* found increased activation of dendritic cells in mice carrying a Pep-R619W knock-in mutation, suggesting that hyperactive myeloid cells also might also contribute to the mechanism of action of the autoimmune predisposing variant [21]. Notably, the complexity and subtlety of LYP-W620 function in myeloid cells likely rivals that of its action in lymphocytes. Our recent work revealed selective loss of capacity to produce type 1 interferon among

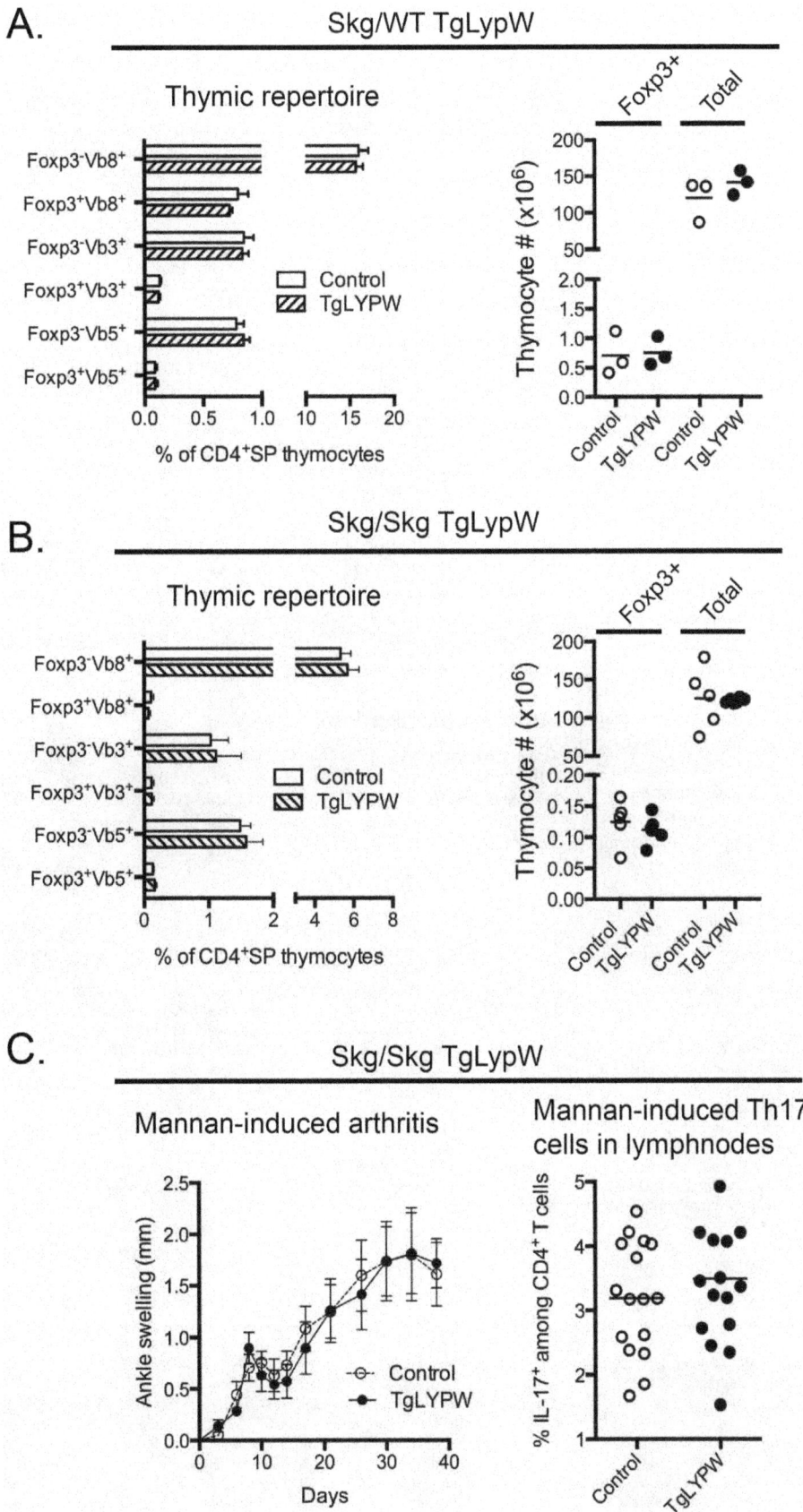

Figure 7. LYPW overexpression does not alter thymic repertoire and autoimmune phenotype of Skg mice. A–B, Overexpression of LYPW does not alter Vβ repertoire or numbers of CD4+Foxp3+ thymocytes in Skg/WT or Skg/Skg mice. Left panel shows average and SE % Vβ positive CD4+Foxp3+ and CD4+Foxp3− thymocytes from Skg/WT (A) or Skg/Skg (B) TgLYPW (striped bars, n = 3 for Skg/WT and n = 6 for Skg/Skg) and control

non-Tg littermates (white bars, n = 3 for Skg/WT and n = 5 for Skg/Skg) as assessed by flow cytometry analysis after staining with anti-Vβ3, -Vβ5 and - Vβ8 antibodies. Right panel shows mean and range of CD4$^+$Foxp3$^+$ (first and second bar) and total (third and fourth bars) thymocytes from the same Skg/WT (A) or Skg/Skg (B) TgLYPW mice (striped bars) or non-Tg littermates (white bars). C, Overexpression of LYPW does not alter the course of mannan-induced arthritis and the frequency of Th17 cells in peripheral lymph node (LN) of arthritic Skg mice. Left panel shows arthritis score (measured as ankle swelling in mm) of Skg/Skg TgLYPW mice (black circles, n = 6) and littermates non-Tg Skg/Skg mice (white circles, n = 5) followed-up for 40 days after a single i.p. injection of 20 mg mannan dissolved in 200 µl PBS. One month following mannan-injection, LN cells from Skg/Skg TgLYPW mice (black circles, n = 15) or non-Tg Skg/Skg littermates (white circles, n = 16) were stimulated with 20 ng/ml PMA and 2 mM ionomycin for 5 hours. Right panel shows % Th17$^+$ cells of the CD4$^+$ T cell population as assessed by flow cytometry analysis after intracellular staining with an anti-IL17 antibody.

dendritic cells derived from BACLYPW transgenic mice [31]. Such a myeloid cell defect could contribute to cell-extrinsic aberrations in effector T cells responding to infection or inflammation, yet be consistent with absence of a T cell developmental phenotype for LYP-W620.

Our results are supported by experiments in two different mouse transgenic systems and multiple monoclonal and polyclonal TCR transgenic models. However, there are few limitations, including 1) the use of an overexpression system, which is amenable to positional effects, 3) although knock-down and deletion experiments suggest that the function of the phosphatase in TCR signaling is conserved in human and mouse cells, the expression of a human phosphatase in a mouse context can be viewed as another limitation of our model and 4) the use of hybrid mouse strains in some experiments might have decreased the sensitivity to small effects. Despite the above-mentioned limitations, our data strongly suggest that the increased activity of LYP, which –according to several groups– is conferred by the R620W variation, is insufficient to cause anomalies in thymic selection that could underlie the increased risk of autoimmunity conferred by carriage of the variant. TgLYPW mice transgenic for the active phosphatase variant carried an average calculated phosphatase activity between those reported for heterozygous or homozygous human carriers of the LYP-W620 variant. Since the overall

overexpression of the phosphatase was low, the most likely explanation for the lack of a phenotype is that the decrease in thymocyte TCR signaling caused by the overexpression of the phosphatase was too small to significantly affect thymic selection. An additional or alternative explanation to explain the lack of phenotype in SkgxTgLYPW mice is that after the threshold of signaling inhibition necessary to trigger thymic signaling anomalies has been trespassed, quantitative reductions in signaling are required in order to see further decreases of selection and an increase in severity of arthritis. Since no studies to date have uncovered profound effects of LYP-W620 on thymic selection, an alternative explanation is that TCR signaling in thymocytes might be controlled by multiple redundant phosphatases. LYP-dependent effects might not be detectable in the presence of a dominant regulator such as CD45, which plays a major role in both positive and negative regulation of TCR signaling in double positive and single positive thymocytes [53].

Author Contributions

Conceived and designed the experiments: NB EJP DZ WZ SE IE LAS MK. Performed the experiments: DJW WZ SE VO SMS CJM NR KS EF NB. Analyzed the data: DJW WZ SE VO SMS NR KS EF DZ EJP NB. Contributed reagents/materials/analysis tools: IE MK. Wrote the paper: DZ EP NB.

References

1. Veillette A, Rhee I, Souza CM, Davidson D (2009) PEST family phosphatases in immunity, autoimmunity, and autoinflammatory disorders. Immunol Rev 228: 312–324.
2. Stanford SM, Mustelin TM, Bottini N (2010) Lymphoid tyrosine phosphatase and autoimmunity: human genetics rediscovers tyrosine phosphatases. Semin Immunopathol 32: 127–136.
3. Bottini N, Musumeci L, Alonso A, Rahmouni S, Nika K, et al. (2004) A functional variant of lymphoid tyrosine phosphatase is associated with type I diabetes. Nat Genet 36: 337–338.
4. Begovich AB, Carlton VE, Honigberg LA, Schrodi SJ, Chokkalingam AP, et al. (2004) A missense single-nucleotide polymorphism in a gene encoding a protein tyrosine phosphatase (PTPN22) is associated with rheumatoid arthritis. Am J Hum Genet 75: 330–337.
5. Kyogoku C, Langefeld CD, Ortmann WA, Lee A, Selby S, et al. (2004) Genetic association of the R620W polymorphism of protein tyrosine phosphatase PTPN22 with human SLE. Am J Hum Genet 75: 504–507.
6. Cloutier JF, Veillette A (1999) Cooperative inhibition of T-cell antigen receptor signaling by a complex between a kinase and a phosphatase. J Exp Med 189: 111–121.
7. Cohen S, Dadi H, Shaoul E, Sharfe N, Roifman CM (1999) Cloning and characterization of a lymphoid-specific, inducible human protein tyrosine phosphatase, Lyp. Blood 93: 2013–2024.
8. Wu J, Katrekar A, Honigberg LA, Smith AM, Conn MT, et al. (2006) Identification of substrates of human protein-tyrosine phosphatase PTPN22. J Biol Chem 281: 11002–11010.
9. Hasegawa K, Martin F, Huang G, Tumas D, Diehl L, et al. (2004) PEST domain-enriched tyrosine phosphatase (PEP) regulation of effector/memory T cells. Science 303: 685–689.
10. Brownlie RJ, Miosge LA, Vassilakos D, Svensson LM, Cope A, et al. (2012) Lack of the phosphatase PTPN22 increases adhesion of murine regulatory T cells to improve their immunosuppressive function. Sci Signal 5: ra87.
11. Cloutier JF, Veillette A (1996) Association of inhibitory tyrosine protein kinase p50csk with protein tyrosine phosphatase PEP in T cells and other hemopoietic cells. EMBO J 15: 4909–4918.
12. Vang T, Congia M, Macis MD, Musumeci L, Orru V, et al. (2005) Autoimmune-associated lymphoid tyrosine phosphatase is a gain-of-function variant. Nat Genet 37: 1317–1319.
13. Cao Y, Yang J, Colby K, Hogan SL, Hu Y, et al. (2012) High basal activity of the PTPN22 gain-of-function variant blunts leukocyte responsiveness negatively affecting IL-10 production in ANCA vasculitis. PLoS One 7: e42783.
14. Fiorillo E, Orru V, Stanford SM, Liu Y, Salek M, et al. (2010) Autoimmune-associated PTPN22 R620W variation reduces phosphorylation of lymphoid phosphatase on an inhibitory tyrosine residue. J Biol Chem 285: 26506–26518.
15. Rieck M, Arechiga A, Onengut-Gumuscu S, Greenbaum C, Concannon P, et al. (2007) Genetic variation in PTPN22 corresponds to altered function of T and B lymphocytes. J Immunol 179: 4704–4710.
16. Aarnisalo J, Treszl A, Svec P, Marttila J, Oling V, et al. (2008) Reduced CD4(+)T cell activation in children with type 1 diabetes carrying the PTPN22/ Lyp 620Trp variant. J Autoimmun 31: 13–21.
17. Chuang WY, Strobel P, Belharazem D, Rieckmann P, Toyka KV, et al. (2009) The PTPN22(gain-of-function)+1858T(+) genotypes correlate with low IL-2 expression in thymomas and predispose to myasthenia gravis. Genes Immun 10: 667–672.
18. Habib T, Funk A, Rieck M, Brahmandam A, Dai X, et al. (2012) Altered B cell homeostasis is associated with type I diabetes and carriers of the PTPN22 allelic variant. Journal of immunology 188: 487–496.
19. Vang T, Liu WH, Delacroix L, Wu S, Vasile S, et al. (2012) LYP inhibits T-cell activation when dissociated from CSK. Nat Chem Biol 8: 437–446.
20. Zikherman J, Hermiston M, Steiner D, Hasegawa K, Chan A, et al. (2009) PTPN22 deficiency cooperates with the CD45 E613R allele to break tolerance on a non-autoimmune background. J Immunol 182: 4093–4106.
21. Zhang J, Zahir N, Jiang Q, Miliotis H, Heyraud S, et al. (2011) The autoimmune disease-associated PTPN22 variant promotes calpain-mediated Lyp/Pep degradation associated with lymphocyte and dendritic cell hyperresponsiveness. Nat Genet 43: 902–907.
22. Chang HH, Tai TS, Lu B, Iannaccone C, Cernadas M, et al. (2012) PTPN22.6, a dominant negative isoform of PTPN22 and potential biomarker of rheumatoid arthritis. PLoS One 7: e33067.

23. Dai X, James RG, Habib T, Singh S, Jackson S, et al. (2013) A disease-associated PTPN22 variant promotes systemic autoimmunity in murine models. J Clin Invest 123: 2024–2036.

24. Vang T, Landskron J, Viken MK, Oberprieler N, Torgersen KM, et al. (2013) The autoimmune-predisposing variant of lymphoid tyrosine phosphatase favors T helper 1 responses. Hum Immunol 74: 574–585.

25. Stritesky GL, Jameson SC, Hogquist KA (2012) Selection of self-reactive T cells in the thymus. Annu Rev Immunol 30: 95–114.

26. Maine CJ, Hamilton-Williams EE, Cheung J, Stanford SM, Bottini N, et al. (2012) PTPN22 alters the development of regulatory T cells in the thymus. J Immunol 188: 5267–5275.

27. Yeh LT, Miaw SC, Lin MH, Chou FC, Shieh SJ, et al. (2013) Different Modulation of Ptpn22 in Effector and Regulatory T Cells Leads to Attenuation of Autoimmune Diabetes in Transgenic Nonobese Diabetic Mice. Journal of immunology.

28. Zhang DJ, Wang Q, Wei J, Baimukanova G, Buchholz F, et al. (2005) Selective expression of the Cre recombinase in late-stage thymocytes using the distal promoter of the Lck gene. Journal of immunology 174: 6725–6731.

29. Chaffin KE, Beals CR, Wilkie TM, Forbush KA, Simon MI, et al. (1990) Dissection of thymocyte signaling pathways by in vivo expression of pertussis toxin ADP-ribosyltransferase. Embo J 9: 3821–3829.

30. Warming S, Costantino N, Court DL, Jenkins NA, Copeland NG (2005) Simple and highly efficient BAC recombineering using galK selection. Nucleic Acids Res 33: e36.

31. Wang Y, Shaked I, Stanford SM, Zhou W, Curtsinger JM, et al. (2013) The Autoimmunity-Associated Gene PTPN22 Potentiates Toll-like Receptor-Driven, Type 1 Interferon-Dependent Immunity. Immunity 39: 111–122.

32. Fontenot JD, Rasmussen JP, Williams LM, Dooley JL, Farr AG, et al. (2005) Regulatory T cell lineage specification by the forkhead transcription factor foxp3. Immunity 22: 329–341.

33. Teh HS, Kishi H, Scott B, Borgulya P, von Boehmer H, et al. (1990) Early deletion and late positive selection of T cells expressing a male-specific receptor in T-cell receptor transgenic mice. Developmental immunology 1: 1–10.

34. Murphy KM, Heimberger AB, Loh DY (1990) Induction by antigen of intrathymic apoptosis of CD4+CD8+TCRlo thymocytes in vivo. Science 250: 1720–1723.

35. Sakaguchi N, Takahashi T, Hata H, Nomura T, Tagami T, et al. (2003) Altered thymic T-cell selection due to a mutation of the ZAP-70 gene causes autoimmune arthritis in mice. Nature 426: 454–460.

36. Sakaguchi S, Sakaguchi N, Yoshitomi H, Hata H, Takahashi T, et al. (2006) Spontaneous development of autoimmune arthritis due to genetic anomaly of T cell signal transduction: Part 1. Semin Immunol 18: 199–206.

37. Tanaka S, Maeda S, Hashimoto M, Fujimori C, Ito Y, et al. (2010) Graded attenuation of TCR signaling elicits distinct autoimmune diseases by altering thymic T cell selection and regulatory T cell function. Journal of immunology 185: 2295–2305.

38. Sakaguchi S, Tanaka S, Tanaka A, Ito Y, Maeda S, et al. (2011) Thymus, innate immunity and autoimmune arthritis: interplay of gene and environment. FEBS Lett 585: 3633–3639.

39. Yoshitomi H, Sakaguchi N, Kobayashi K, Brown GD, Tagami T, et al. (2005) A role for fungal {beta}-glucans and their receptor Dectin-1 in the induction of autoimmune arthritis in genetically susceptible mice. J Exp Med 201: 949–960.

40. Hashimoto M, Hirota K, Yoshitomi H, Maeda S, Teradaira S, et al. (2010) Complement drives Th17 cell differentiation and triggers autoimmune arthritis. J Exp Med 207: 1135–1143.

41. Kim SK, Reed DS, Olson S, Schnell MJ, Rose JK, et al. (1998) Generation of mucosal cytotoxic T cells against soluble protein by tissue-specific environmental and costimulatory signals. Proc Natl Acad Sci U S A 95: 10814–10819.

42. Montalibet J, Skorey KI, Kennedy BP (2005) Protein tyrosine phosphatase: enzymatic assays. Methods 35: 2–8.

43. Zehn D, Lee SY, Bevan MJ (2009) Complete but curtailed T-cell response to very low-affinity antigen. Nature 458: 211–214.

44. Hogquist KA (2001) Assays of thymic selection. Fetal thymus organ culture and in vitro thymocyte dulling assay. Methods in molecular biology 156: 219–232.

45. Zehn D, Bevan MJ (2006) T cells with low avidity for a tissue-restricted antigen routinely evade central and peripheral tolerance and cause autoimmunity. Immunity 25: 261–270.

46. Dillon SR, Jameson SC, Fink PJ (1994) V beta 5+ T cell receptors skew toward OVA+H-2Kb recognition. Journal of immunology 152: 1790–1801.

47. Enouz S, Carrie L, Merkler D, Bevan MJ, Zehn D (2012) Autoreactive T cells bypass negative selection and respond to self-antigen stimulation during infection. J Exp Med 209: 1769–1779.

48. Todd JA, Walker NM, Cooper JD, Smyth DJ, Downes K, et al. (2007) Robust associations of four new chromosome regions from genome-wide analyses of type 1 diabetes. Nat Genet 39: 857–864.

49. Gregersen PK, Lee HS, Batliwalla F, Begovich AB (2006) PTPN22: setting thresholds for autoimmunity. Semin Immunol 18: 214–223.

50. Lefvert AK, Zhao Y, Ramanujam R, Yu S, Pirskanen R, et al. (2008) PTPN22 R620W promotes production of anti-AChR autoantibodies and IL-2 in myasthenia gravis. J Neuroimmunol 197: 110–113.

51. Lee HM, Bautista JL, Scott-Browne J, Mohan JF, Hsieh CS (2012) A broad range of self-reactivity drives thymic regulatory T cell selection to limit responses to self. Immunity 37: 475–486.

52. Menard L, Saadoun D, Isnardi I, Ng YS, Meyers G, et al. (2011) The PTPN22 allele encoding an R620W variant interferes with the removal of developing autoreactive B cells in humans. J Clin Invest 121: 3635–3644.

53. Zikherman J, Jenne C, Watson S, Doan K, Raschke W, et al. CD45-Csk phosphatase-kinase titration uncouples basal and inducible T cell receptor signaling during thymic development. Immunity 32: 342–354.

14

Divergent Phenotypes in Mutant TDP-43 Transgenic Mice Highlight Potential Confounds in TDP-43 Transgenic Modeling

Simon D'Alton[1,2]*, Marcelle Altshuler[1], Ashley Cannon[2], Dennis W. Dickson[2], Leonard Petrucelli[2], Jada Lewis[1,2]*

1 Department of Neuroscience and Center for Translational Research in Neurodegenerative Disease, University of Florida, Gainesville, Florida, United States of America, **2** Department of Neuroscience, Mayo Clinic, Jacksonville, Florida, United States of America

Abstract

The majority of cases of frontotemporal lobar degeneration and amyotrophic lateral sclerosis are pathologically defined by the cleavage, cytoplasmic redistribution and aggregation of TAR DNA binding protein of 43 kDa (TDP-43). To examine the contribution of these potentially toxic mechanisms *in vivo*, we generated transgenic mice expressing human TDP-43 containing the familial amyotrophic lateral sclerosis-linked M337V mutation and identified two lines that developed neurological phenotypes of differing severity and progression. The first developed a rapid cortical neurodegenerative phenotype in the early postnatal period, characterized by fragmentation of TDP-43 and loss of endogenous murine Tdp-43, but entirely lacking aggregates of ubiquitin or TDP-43. A second, low expressing line was aged to 25 months without a severe neurodegenerative phenotype, despite a 30% loss of mouse Tdp-43 and accumulation of lower molecular weight TDP-43 species. Furthermore, TDP-43 fragments generated during neurodegeneration were not C-terminal, but rather were derived from a central portion of human TDP-43. Thus we find that aggregation is not required for cell loss, loss of murine Tdp-43 is not necessarily sufficient in order to develop a severe neurodegenerative phenotype and lower molecular weight TDP-43 positive species in mouse models should not be inherently assumed to be representative of human disease. Our findings are significant for the interpretation of other transgenic studies of TDP-43 proteinopathy.

Editor: Emanuele Buratti, International Centre for Genetic Engineering and Biotechnology, Italy

Funding: DoD (USAMRMC PR080354) to LP, JL (http://cdmrp.army.mil/funding/). ALSA to LP, JL (http://www.alsa.org/). NIH/NIA [P50AG16574 (DWD); R01AG026251 and R01AG026251–03A2 (LP); and P01-AG17216-08 (LP, DWD)], NIH/NINDS [R01 NS 063964-01 (LP), 5R21NS071097-02 (JL)]. (http://grants.nih.gov/grants/oer.htm). Mayo Clinic to DWD, LP, and JL (http://www.mayoclinic.org/jacksonville/). UF to JL. (http://www.ufl.edu/). The funders had no role in study design, data collection and analysis, decision to publish, or preparation of the manuscript.

Competing Interests: Dr. Petrucelli and Dr. Lewis are inventors of the iTDP-43 mice detailed; however, no revenue has been generated.

* E-mail: sdalton82@ufl.edu (SD); jada.lewis@ufl.edu (JL)

Introduction

TAR DNA binding protein of 43 kilodaltons (TDP-43) is the major pathological substrate in frontotemporal lobar degeneration with TDP-43 pathology (FTLD-TDP) and most cases of amyotrophic lateral sclerosis (ALS) [1]. TDP-43 is a member of the hnRNP family, containing two RNA Recognition Motifs (RRMs) that bind RNA and a C-terminal, glycine rich domain that mediates interactions with functional binding partners to coordinate the metabolism of a wide range of RNA substrates [2,3]. In disease, normal nuclear localization of TDP-43 is lost, and ubiquitinated and hyperphosphorylated inclusions are deposited in the form of neuronal intranuclear inclusions or in the cytoplasm as juxtanuclear aggregates or dystrophic neurites [4]. TDP-43 also undergoes processing to produce C-terminal fragments that range in size from 15–35 kDa [1,5,6]. Evidence for definitive pathological involvement in disease is additionally derived from causal familial mutations in the gene encoding TDP-43, *TARDBP*, in ALS, which are principally though not exclusively found in the glycine rich C-terminus [7].

It is unclear if one or several of these pathological TDP-43 alterations is a dominant driving mechanism in disease, and there is evidence to support the toxicity of each in FTLD-TDP. Pathological C-terminal fragments of the protein are more aggregate-prone, undergoing phosphorylation, ubiquitination and ultimately demonstrating cytotoxicity in cells [8,9,10]. Expression of TDP-43 containing disease linked mutations found in ALS appears to increase the production of C-terminal fragments, aggregation and toxicity [9,11,12]. Loss of TDP-43 also appears to be deleterious to cell health; knockdown *in vitro* confers toxicity and knockout results in lethality *in vivo*, both *in utero* in constitutive $Tardbp^{-/-}$ mice and in conditional knockout animals in which deletion is postponed until adulthood [13,14,15,16]. Loss of TDP-43 specifically in motor neurons results in cell death and an ALS-like phenotype in mice [17] and reduced TDP-43 expression in zebrafish and drosophila results in motor deficits [18,19].

To determine if one mechanism is more dominant than others *in vivo*, a host of constitutive and inducible transgenic rodents expressing human or mouse (wild type and mutant) TDP-43 under the control of various promoters have been created, which recapitulate some of the hallmark pathology of FTLD-TDP/ALS including inclusions of ubiquitinated, phosphorylated TDP-43 and TDP-43 immunoreactive lower weight molecular species [20,21,22,23,24]. Here we describe the pathological and biochemical characterization of two novel lines of mice conditionally expressing human mutant M337V TDP-43, providing further insight into the nature of neurodegenerative phenotypes in TDP-43 expressing animal models.

Methods

Ethics Statement for Animal Care

All procedures were conducted according to the National Institutes of Health guide for animal care and approved by the Institutional and Animal Care and Use Committee at University of Florida or Mayo Clinic.

Generation of Transgenic iTDP-43 Animals

iTDP-43 mice were generated similarly to a previously described protocol [25]. Full length, untagged, M337V human TDP-43 cDNA was created using the Quickchange II site directed mutagenesis kit (Stratagene) using a TDP-43-myc plasmid as a template [26], and was inserted into the inducible expression vector pUHD 10–3 containing five tetracycline operator sequences. The construct was confirmed by restriction enzyme digest and direct sequencing. The transgenic fragment was obtained by *BsrBI* digestion, gel purified followed by β–agarase digestion (NEB), filtration and concentration. The modified TDP-43 transgene was injected into the pronuclei of donor FVB/NCr embryos (Charles River). 14 founders were positive for the TDP-43 responder transgene. These were then bred with 129S6 mice (Taconic) with the tetracycline transactivator (tTA) transgene downstream of calcium calmodulin kinase II alpha (CaMKIIα) promoter elements [27] to produce the iTDP-43 transgenic mice with forebrain hTDP-43 expression. 8 founders transmitted and expressed the TDP-43 transgene and we subsequently chose the two founder lines with the highest transgenic TDP-43 expression at 2 months of age. All experimental mice used in this study were F1 hybrids produced from a breeding of TDP-43 monogenic mice on a congenic FVB/Ncr and tTA monogenic mice on a 129S6 background.

Antibodies

Antibodies used in this work are described in table s1. Epitopes recognized by TDP-43 antibodies are detailed in figure s1.

Immunohistochemistry

After euthanasia via cervical dislocation, brains were harvested and divided along the midline. The right hemisphere was flash-frozen on dry ice, while the left hemisphere was drop fixed in 10% neutral buffered formalin for histological analyses. Brains were embedded in paraffin and cut into 5 μm sagittal sections. For hTDP-43, caspase 3 and ubiquitin antibodies, tissues were immunostained using the DAKO Autostainer (DAKO Auto Machine Corporation, Carpinteria, CA) with DAKO Envision HRP System. For p403/404 detection, sections were deparaffinzed and hydrated through a graded alcohol series prior to antigen retrieval in citrate buffer (10 mM sodium citrate, 0.05% Tween, pH 6.0) for 30 minutes in a steamer with water at a rolling boil. Blocking was performed using DAKO Protein Block for 1 h followed by incubation overnight with primary antibody at 4°C. Peroxidase conjugated secondary antibody was visualized with diaminobenzidine (Vector Labs, Burlingame, CA). Hematoxylin and eosin (H&E) staining was performed by standard procedures.

Immunofluorescence

Sections were prepared as above, except visualization of primary antibody was performed using Alexa Fluor 488 secondary antibody (Invitrogen). Slides were dipped in Sudan Black to reduce background autofluorescence and vectashield with DAPI (Vector Laboratories) was used to stain nuclei. Images were captured using an Olympus BX60 microscope (Olympus).

Protein Isolation, Fractionation and Western Blotting

To isolate SDS-soluble fractions, brain or dissected cortex was homogenized in lysis buffer (50 mM Tris, 300 mM NaCl, 1% Triton X-100, 1 mM EDTA with protease inhibitors and phosphatase inhibitors (Sigma)) at 6 ml/g tissue, and aliquots of homogenate stored at −80°C. Lysate was prepared via brief sonication, addition of SDS to 1% and centrifugation at $40K \times g$ for 20 minutes at 4°C. For experiments using HEK293T cells, cells were washed once with and then harvested in ice cold PBS. Cells were then pelleted by brief centrifugation at $500 \times g$ and lysed in lysis buffer as described above. 30–50 μg protein was loaded onto 10% or 15% Tris–glycine polyacrylamide gel (Novex). Following electrophoresis and transfer, nitrocellulose membranes were blocked with 5% milk in TBS-T (Tris-buffered saline with 0.1% Triton X-100), incubated with appropriate primary and HRP-conjugated secondary antibodies and visualized using ECL reagent (Perkin-Elmer). Images were captured on the ProteinSimple FluorChem E (ProteinSimple, Santa Clara, California). In all cases where blotting was to be quantified, lysate prepared fresh from homogenate was used, and following probing with antibody of interest the blot was stripped (70 mM SDS in Tris HCl (pH 6.8) with 0.7% BME for 30 minutes at 55°C) and re-probed with loading antibody. Densitometry was performed using AlphaView (ProteinSimple), and following normalization to loading controls, fold changes calculated and unpaired *t*-test used to assess significance.

To analyze solubility of TDP-43 fragments, cortex from P5 animals was homogenized in 5 ml/g of high salt buffer (10 mM Tris pH 7.5, 5 mM EDTA, 10% sucrose, 1% triton, 0.5 M NaCl, with protease and phosphatase inhibitors), incubated at 4°C for 15 minutes and clarified by centrifugation at $110K \times g$. The supernatant containing the high salt fraction was snap frozen on dry ice and the pellet washed and re-homogenized in 5 ml/g myelin floatation buffer (high salt buffer with 30% sucrose). This homogenization, centrifugation and fraction collection sequence was repeated with sarkosyl buffer (high salt buffer substituting 1% triton for 1% N-lauroyl-sarcosine, incubation 1 hour at room temperature), and finally in 1 ml/g urea buffer (30 mM Tris pH 8.5, 4% CHAPS, 7 M urea, 2 M thiourea, brief sonication). 50 ug of high salt extract and equivalent volumes of all other fractions underwent Western blotting as described above.

Quantitative Real Time PCR

Total RNA was isolated from dissected hippocampus or cortex using TRIzol reagent (Life Technologies) and Pure Link RNA Mini Kit (Life Technologies). During this step, DNA was removed using on-column DNase digestion (Life Technologies) and the resulting RNA purity and integrity was assessed on spectrophotometer (Nanodrop, Wilmington, USA) and agarose gel electrophoresis. 2 μg of RNA was used to synthesize cDNA using the High Capacity cDNA Reverse Transcription Kit (Applied Biosystems). All samples were run in triplicate on the ABI 7900 HT Real-Time PCR Detection System using SYBR green PCR master mix (Applied Biosystems). Primer efficiencies (E) were calculated using $E = 10^{(-1/slope)}$, where slope was

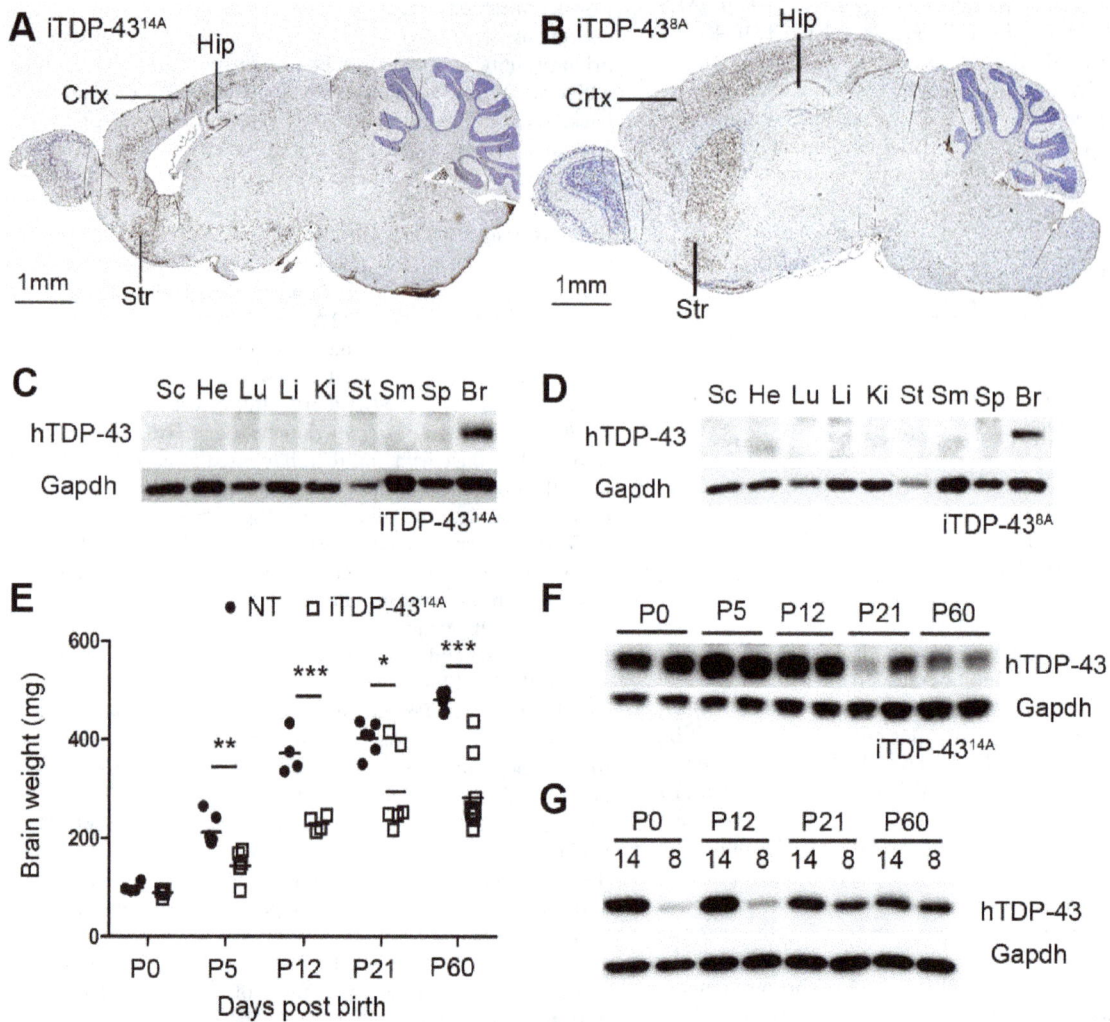

Figure 1. Expression of human TDP-43 in iTDP-43¹⁴ᴬ and iTDP-43⁸ᴬ mice in the postnatal period. Immunohistochemical detection of hTDP-43 expression in cortex (CTX), hippocampus (HIP) and striatum (STR) in iTDP-43¹⁴ᴬ (A) and iTDP-43⁸ᴬ (B). Western analysis of organs demonstrated specificity of hTDP-43 expression to the brain in both iTDP-43¹⁴ᴬ (C) and iTDP-43⁸ᴬ (D) (SC = spinal cord, He = heart, Lu = lung, Li = liver, Ki = kidney, St = stomach, SM = skeletal muscle, Sp = spleen, Br = brain). (E) Brain weight measurement of non-transgenic (NT) and iTDP-43¹⁴ᴬ mice at postnatal stages until 2 months of age (P60) (*$p<0.05$, **$p<0.01$, *** $p<0.001$, unpaired two tailed *T-test*). (F) Expression of hTDP-43 at indicated postnatal time points for iTDP-43¹⁴ᴬ. (G) Expression of hTDP-43 at indicated postnatal time points for iTDP-43¹⁴ᴬ (14) compared to iTDP-43⁸ᴬ (8).

determined by plotting the Cq (quantification cycle) values against the $\log_{10}$ input of a cDNA dilution series. A 1/50 dilution of each experimental cDNA sample was run. From this, the relative quantities (RQ) were calculated for each primer pair in each sample and normalization of *Tardbp* performed using the geometric mean of the two reference genes *Gapdh* and *β-actin*, giving the Normalized Relative Quantity (NRQ) [28]. Primer sequences for qPCR were *Tardbp*1 F (5′AAAGGTGTTTGTTGGACGTTGTACAG 3′), *Tardbp*1 R (5′ AAAGCTCTGAATGGTTTGGGAATG 3′), *Tardbp*2 F (5′ GATTGGTTTGTTCAGTGTGGAGTATATTCA 3′), *Tardbp*2 R (5′ ACAGCAGTTCACTTTCACCCACTCA 3′), *Tardbp*3 F (5′ GGTGGTTAGTAGGTTGGTTATTAGGTTAGGTA 3′), *Tardbp*3 R (5′ AAATACTGCTGAATATACTCCACACT-GAACA 3′), *Gapdh* F (5′ CATGGCCTTCCGTGTTCCTA 3′), *Gapdh* R (5′ CCTGCTTCACCACCTTCTTGAT 3′), *β-actin* F (5′ GATGACCCAGATCATGTTTGAGACCTT 3′) and *β-actin* R (5′ CCATCACAATGCCTGTGGTACGA 3′).

Single PCR products were verified by melt curve analysis. Statistical significance was assessed using unpaired t-test.

TDP-43 Plasmid Generation

Full-length human TDP-43 complementary cDNA in plasmid pEGFP-C1 [10] was used as PCR template to generate N-terminally myc tagged TDP-43²⁰⁸⁻⁴¹⁴ and TDP-43¹⁻²⁸⁰. The primers used were TDP-43²⁰⁸⁻⁴¹⁴ F (5′-TACGGATCCCAC-CATGGAACAAAAACTCATCTCAGAAGAGGATCTGCGG-GAGTTCTTCTCTCAGTACGG-3′); TDP-43²⁰⁸⁻⁴¹⁴ R (5′-GAATCGCGGCCGCCTACATTCCCCAGCCAGAAGACT-3′); TDP-43¹⁻²⁸⁰ F (5′-TACGGATCCCACCATGGAA-CAAAAACTCATCTCAGAAGAGGATCTGATGTCTGAA-TATATTCGGGTAACCGAA-3′); TDP-43¹⁻²⁸⁰ R (5′-GAATCGCGGCCGCTCATGGATTACCACCAAATCTTC-CAC-3′). Products were cloned into pcDNA3.1 (Life Technologies) using BamH1 and Not1 sites and plasmids were verified by sequence analysis.

Cell Culture and Transfection

Human Embryonic Kidney 293T cells were maintained in Dulbecco's Modified Eagle's Medium (Lonza) supplemented with 10% Fetal Calf Serum (Sigma) and penicillin-streptomycin (Life Technologies). Transfection in 6 well plates was performed for 48 hours using 2.0 ug of plasmid and Lipofectamine 2000 (Life Technologies) following the manufacturer's guidelines.

Results

Early, Rapid Postnatal Cell Loss Induced by hTDP-43^{M337V} Prevents Brain Development without Causing TDP-43 Aggregation

Fourteen monogenic, transgenic M337V hTDP-43 founder lines were bred to animals expressing the tetracycline transactivator (tTA) to produce bigenic iTDP-43 mice expressing human mutant TDP-43. We initially screened iTDP-43 mice at two months of age for expression of human TDP-43 and selected the two highest expressing lines (14A and 8A) for further analysis. Transgene expression in bigenic animals from both lines was limited exclusively to the brain, predominantly in the cortex, hippocampus and striatum (figure 1A–D), consistent with what has been previously reported with this conditional system [24,27]. Phenotypically these animals did not display any premature death or overt signs of neurological dysfunction as we and others have reported in TDP-43 transgenic animals [22,24,29]. However, while iTDP-43^{8A} animals developed normal brain structure, there was obvious reduction in the cortical volume of iTDP-43^{14A} brains compared to non-transgenic (NT) littermates (figure 1A).

To fully characterize the progression of cortical degeneration in iTDP-43^{14A}, we examined postnatal ages from P0 to P60 to determine the time point of initial phenotypic onset. During this period, brain weights of NT mice experienced a rapid phase of growth between P0 and P12, followed by more modest increases into adulthood (figure 1E). iTDP-43^{14A} mice however demonstrate striking abnormalities in brain weight during postnatal development. There was no difference in gross brain weight of iTDP-43^{14A} mice compared to NT mice at P0; however, iTDP-43^{14A} brain weight was reduced by 33% (NT = 212 mg±11, iTDP-43 = 143 mg±11, $p = 0.002$) by P5 and iTDP-43^{14A} brain weight never reached that of non-transgenic litter mates (figure 1E). This phenotype was not observed in either monogenic tTA or monogenic hTDP-43^{14A} mice, and thus is a result of hTDP-43 expression in bigenic animals (figure s2A). Analysis using a human specific TDP-43 antibody revealed large changes in observed transgene expression over this time period. Peak transgene expression at P5 coincided with age of phenotypic onset before gradual reduction to the levels observed at P60 (figure 1F). Although iTDP-43^{14A} and iTDP-43^{8A} mice expressed similar levels of human M337V TDP-43 in the initial screen at P60, transgene expression in iTDP-43^{8A} mice at earlier time points was far lower than in iTDP-43^{14A} mice, likely explaining the phenotypic difference between the two lines (figure 1G).

To examine the potential underlying mechanisms of neurodegeneration in iTDP-43^{14A}, we screened for neuropathology consistent with FTLD-TDP at P5, the earliest neurodegenerative time point studied. Immunohistochemical detection of hTDP-43^{M337V} expression using human-specific antibody revealed abundant expression in cortex, hippocampus and striatum (figure 2A). To confirm active cell death, we used antibodies to activated caspase 3. Western blotting of brain lysate revealed low level activated caspase-3 in NT mice, which is compatible with programmed apoptosis during developmental stages [30]; however, this was markedly increased in iTDP-43^{14A} animals (figure 2B).

This was confirmed by immunohistochemical detection of extensive activated caspase 3 immunoreactivity in the cortex of iTDP-43^{14A} mice (figure 2C), suggesting extensive cell death at P5 as a result of the expression of M337V TDP-43. We detected no increase in activated caspase-3 by western blot in monogenic tTA mice compared to NT mice, ruling out tTA expression as the cause of this phenotype (figure s2B).

The end-stage neuropathology of FTLD-TDP is characterized by the presence of ubiquitinated, phosphorylated TDP-43 aggregates in the cytoplasm of affected neurons and glia. We used antibodies to ubiquitin, human TDP-43 (hTDP-43), total TDP-43 (tTDP-43 Ab1) and TDP-43 phosphorylated at residues 403/404 (p403/404) and 409/410 (p409/410) to investigate the contribution of aggregation to the cell death observed in P5 iTDP-43^{14A} mice. We qualitatively observed widespread, increased ubiquitin immunoreactivity in the upper layers of the developing cortex (figure 2C). Despite being exclusively cytoplasmic, the staining was diffuse in nature and we found no evidence of ubiquitin positive aggregates characteristic of FTLD-TDP. Furthermore, we found no evidence of TDP-43 aggregation using hTDP-43, tTDP-43, p403/404 or p409/410 antibodies (figure 2D). We did frequently detect cytoplasmic human TDP-43 staining in iTDP-43^{14A} mice within the cortex; however, no aggregates were observed. Both tTDP-43 and p409/410 antibodies showed abundant cytoplasmic reactivity in NT mice, suggesting that this is a spatiotemporally normal distribution (arrowheads in figure 2D). Therefore, cell death at the earliest stage investigated occurred in the complete absence of cytoplasmic or nuclear TDP-43 aggregates of any kind.

Misprocessing of Human TDP-43 is a Feature of Early Neurodegeneration

The cleavage of TDP-43 into a variety of detergent-insoluble, urea soluble fragments ranging from 17–35 kDa is a hallmark of FTLD-TDP and these fragments are a potential source of neuronal toxicity [1,5,6]. Transgenic mice expressing wild type or familial ALS-linked mutant TDP-43 are characterized by the presence of 35 kDa and 25 kDa TDP-43 immunoreactive species [7,21,29,31,32,33,34,35]. However, the nature of these species is largely unknown. We used a panel of antibodies raised to the N-terminus (residues 3–12), C-terminus (405–414), and central RRM domains of TDP-43 (tTDP-43Ab1 and tTDP-43 Ab2) to determine the extent and nature of TDP-43 fragmentation in P5 iTDP-43^{14A} mice. To validate the antibodies used, we overexpressed either a C-terminal pathological fragment identified in human cases of FTLD-TDP (TDP-43$^{208-414}$) [36] or the first 280 amino acids of human TDP-43 (TDP-43$^{1-280}$), both with an N-terminal myc tag, in HEK293T cells. Subsequent western analysis of lysate verified that the C-terminal antibody detected TDP-43$^{208-414}$, but not TDP-43$^{1-280}$. Conversely, tTDP-43 Ab2, which is raised to residues 181–198, detected TDP-43$^{1-280}$ but not TDP-43$^{208-414}$. Human specific antibody and myc antibody detected both. TDP-43 3–12 detected TDP-43$^{1-280}$ but not the C-terminal fragment (figure s3). These findings are consistent with the antigenic epitopes of these antibodies and epitope mapping studies previously published [37,38].

In SDS soluble fractions of P5 brain lysate from iTDP-43^{14A} mice, we detected species of approximately 35 kDa (TDP-35) and 25 kDa (TDP-25) using total TDP-43 Ab1 and total TDP-43 Ab2 (figure 3A). Both of these antibodies recognize epitopes in the second RNA recognition motif of mTdp-43 and hTDP-43, which is located in the center of TDP-43. However, we failed to detect either of these low molecular weight species using antibodies raised

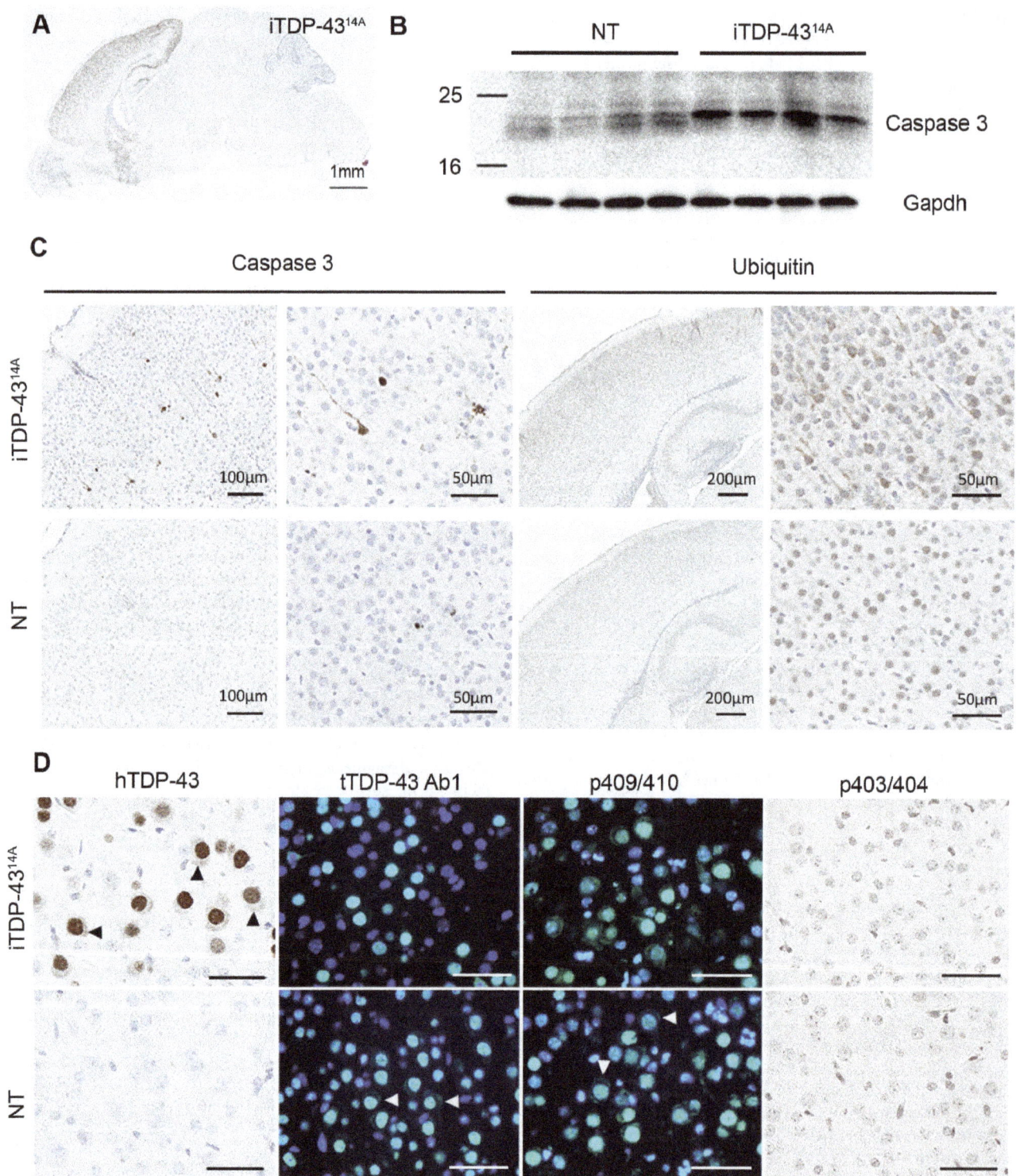

Figure 2. Early degenerative phenotype in iTDP-43^{14A} mice at P5 in the absence of FTLD-like TDP-43 aggregation. (A) Monoclonal antibody to human TDP-43 showed expression at P5 remained restricted to previously characterized regions of hippocampus, cortex and striatum. (B) Western blotting of brain lysate of P5 non-transgenic (NT) and iTDP-43^{14A} demonstrated increased levels of activated caspase 3 in iTDP-43^{14A} mice. (C) Abundant caspase 3 immunoreactivity in the cortex of iTDP-43^{14A} mice that was virtually absent in NT mice, suggestive of elevated cell death in iTDP43^{14A} compared to NT mice. iTDP-43^{14A} mice were also characterized by increased ubiquitin staining in the upper layers of the cortex compared to NT mice, which upon higher magnification appeared to be completely diffuse and cytoplasmic. (D) Immunohistochemistry for hTDP-43 and p403/404 and immunofluorescence using antibodies to total TDP-43 and p409/410 TDP-43. Significant amounts of cytoplasmic hTDP-43 were observed in iTDP-43 mice (arrowheads). Note that this cytoplasmic staining was also observed in NT mice (arrowheads) with antibodies to total TDP-43 (tTDP-43 Ab1) and TDP-43 phosphorylated at 409/410 (p409/410). Scale bars in D = 50 μm.

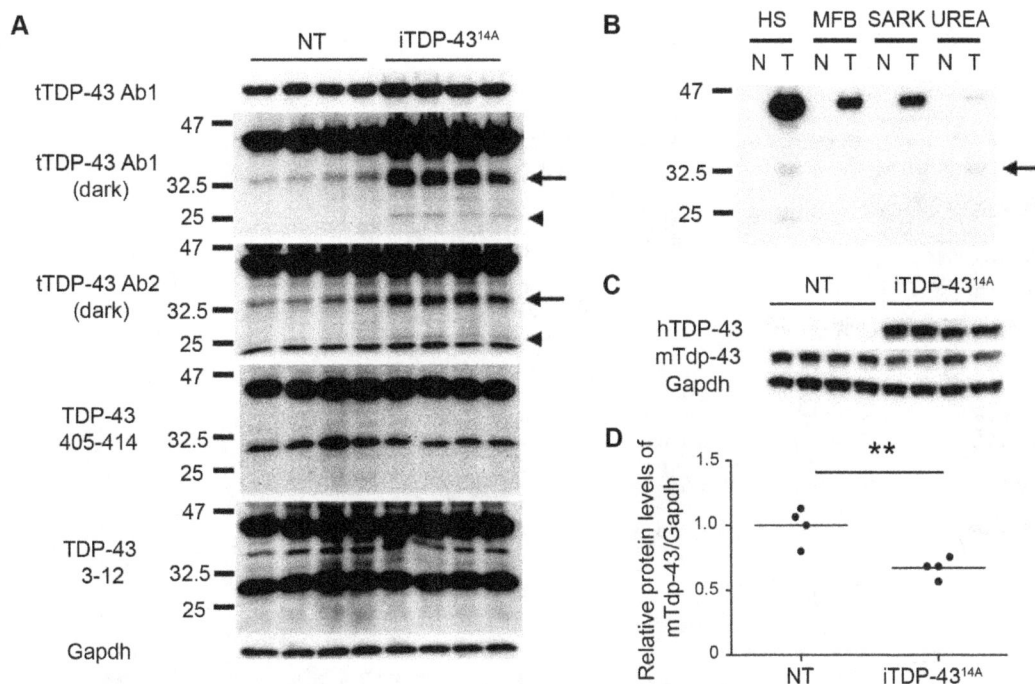

Figure 3. Biochemistry of iTDP-43[14A] brain lysates at P5. (A) Western blotting using two antibodies to total TDP-43 (tTDP-43 Ab1 and tTDP-43 Ab2) demonstrated increased levels of low molecular weight species at 35 kDa (arrow) and 25 kDa (arrowhead) in iTDP-43[14A] mice relative to NT mice. These species were not observed using antibodies to the C-terminus (405–414) or N-terminus (3–12) of TDP-43. (B) Western blot analysis of high salt (HS), myelin floatation buffer (MFB), sarkosyl (SARK) and urea fractions using antibody to human TDP-43. Note that human TDP-35 (arrow) is present in the urea fraction but is absent from MFB and SARK fractions, N = non-transgenic, T = iTDP-43[14A]. (C) Antibody to murine Tdp-43 demonstrated reduction of mTdp-43 in brain compared to NT mice. (D) Quantification of blot in (C), **$p < 0.01$, unpaired two tailed t-test.

to the C-terminus (residues 405–414) or N-terminus (residues 3–12) of TDP-43 (figure 3A).

To assess the solubility of TDP-43, TDP-35 and TDP-25, we used a series of increasingly stringent buffers to sequentially extract protein from cortical tissue of p5 iTDP-43[14A] mice. Western blotting of the resulting fractions and analysis using human specific TDP-43 antibody demonstrated that although a proportion of TDP-35 was detergent insoluble but urea soluble, the majority of TDP-35 and TDP-25 are soluble in high salt buffer (figure 3b). Consequently, during early neurodegeneration in iTDP-43[14A], TDP-43 fragments are derived from a central portion of human TDP-43 and are largely soluble.

Another cardinal feature of FTLD-TDP is loss of nuclear TDP-43, which may contribute to disease via aberrant transcriptional control and/or misprocessing of RNA substrates bound by TDP-43 [2]. In a previously published transgenic line expressing human wild type TDP-43, loss of murine Tdp-43 (mTdp-43) occurred upon human TDP-43 transgene expression and correlated with phenotypic progression, suggesting that loss of mTdp-43 may be analogous to loss of nuclear TDP-43 in human disease [23] despite human transgene expression in the nucleus. Using an antibody specific to mTdp-43, we detected a 33% decrease in the expression of mTdp-43 in iTDP-43[14A] brain extract compared to NT animals (0.67 ± 0.03, $p = 0.007$ unpaired two tailed t-test, figure 3C and D).

Low Level Misprocessing of TDP-43 is Well Tolerated *in vivo*

Given the lack of an overt neurodegenerative phenotype in the second iTDP-43[8A] line, we aged iTDP-43[8A] cohorts to 10 and 25 months. In both cohorts, brain weights of both tTA monogenic

and iTDP-43[8A] mice were reduced 5–7% compared to NT (10 M: NT = 496 ± 6 mg $N = 5$,   iTDP-43[8A] = 462 ± 4 mg $N = 7$, tTA = 464.9 ± 9 mg $N = 7$, $p < 0.05$ one way ANOVA Bonferroni *post-hoc* analysis; 25 M: NT = 505 ± 8 mg $N = 7$,   iTDP-43[8A] = 470 ± 7 mg, tTA = 473 ± 7 mg, $p < 0.05$ one way ANOVA Bonferroni *post-hoc* analysis, figure s4) and there was decreased volume of the dentate gyri, a previously documented observation that likely reflects a phenotype driven by this tTA line rather than the expression of the TDP-43 transgene [32]. With this exception, we observed no phenotype in these animals as assessed by gross weight or brain morphology using hematoxylin and eosin staining (figure s4). We verified that expression of tTA did not alter the levels of endogenous mTdp-43 or the abundance of lower molecular weight TDP species (figure s4). Therefore, although tTA expression results in reduced brain weight in older mice, it does not alter TDP-43 biochemistry.

Due to the lack of an overt phenotype, we hypothesized that any TDP-43 biochemical alterations in the iTDP-43[8A] line may be small in magnitude. In order to increase detection of low level alterations over background, we used lysate of dissected cortex as opposed to brain lysate, thus enriching hTDP-43[M337V] expressing regions and reducing noise from regions not expressing the transgene. Expression levels of the hTDP-43[M337V] transgene were exceptionally low in both the 10 and 25 month iTDP-43[8A] cohorts. Western blotting and densitometry analysis of cortical extracts using tTDP-43 Ab2 demonstrated ~ 1.25 fold overexpression of TDP-43 compared to NT mice (10 M = 1.22 ± 0.07, 25 M = 1.33 ± 0.05, figure 4A, quantification in 4B). In both cohorts, increased levels of TDP-35 but not TDP-25 was observed relative to control mice (10 M TDP-35 = 1.32 ± 0.09, $p = 0.049$; 10 M   TDP-25 = 1.08 ± 0.05, $p = 0.396$;   25 M   TDP-

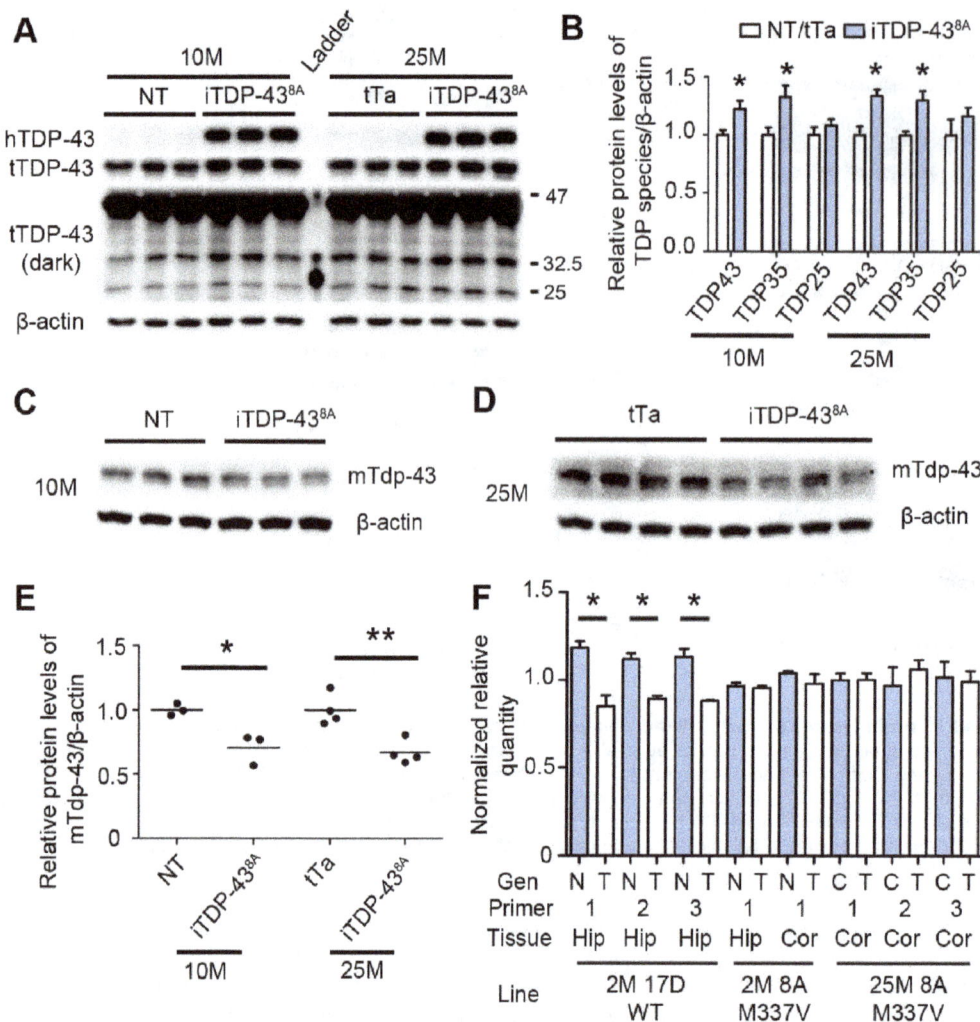

Figure 4. Loss of murine Tdp-43 and increased TDP-35 in 25 month (25 M) iTDP-43^8A. (A) Analysis of transgene expression in 10 M and 25 M cohorts using hTDP-43 and tTDP-43 Ab2 antibodies, β-actin for loading. (B) Densitometric quantitation of TDP-43, TDP-35 and TDP-25 from blot in (A), represented as relative levels of TDP species following sample normalization to β-actin loading controls. (C) mTdp-43 expression in 10 M iTDP-43^8A, β-actin loading control. (D) mTdp-43 expression in 25 M iTDP-43^8A, β-actin loading control. (E) Densitometric quantitation of reduction in mTdp-43 protein levels from blots in (C) and (D), represented as relative levels of mTdp-43 following sample normalization to β-actin loading controls. (F) Quantitative PCR analysis of *Tardbp* transcript in 2 month old transgenic (T) iTDP-43^17D mice hippocampus (HIP) using *Tardbp* primers 1, 2 and 3 detected significant reduction in murine Tdp-43 mRNA levels compared to nontransgenic (N) mice and acted as a positive control for our methodology ($N = 3$ per genotype). No reduction in *Tardbp* mRNA was detected in the cortex (COR) and hippocampus of 2 month (2 M) iTDP-43^8A mice relative to nontransgenic mice using murine-specific *Tardbp* primer pair 1 ($N = 3$ per genotype). No downregulation of *Tardbp* was observed in the cortex of 25 M iTDP-43^8A ($N = 3$) relative to control (C) mice ($N = 4$), using *Tardbp* primers 1, 2 or 3. Values shown as Normalized Relative Quantity, see Materials and Methods for further details. Control at 25 M consisted of 2 non-transgenic and 2 tTA mice. In graphs (B),(E) and (F): *$p < 0.05$, **$p < 0.01$, unpaired *t*-test, error bars are SEM.

35 = 1.30±0.08, $p = 0.026$; 25 M TDP-25 = 1.16±0.16 $p = 0.353$, figure 4A and B). We also measured levels of mTdp-43 protein in cortical extracts of both cohorts and found it to be downregulated in iTDP-43^8A compared to controls (10 M = 0.71±0.07, $p = 0.017$; 25 M = 0.67±0.05, $p = 0.005$; figure 4C, D and E). Given that TDP-43 autoregulates the abundance of *Tardbp* mRNA *in vivo* and *in vitro* [2,39], we used quantitative PCR to examine levels of murine *Tardbp* transcript. Using three different primer pairs to *Tardbp*, we found no evidence in the cortex of 25 month old animals for the downregulation of murine *Tardbp* transcript in iTDP-43^8A mice compared to control animals (figure 4F). Additionally, we failed to detect downregulation in the cortex and hippocampus of 2 month old iTDP-43^8A animals (figure 4F). Seeking to validate our methodology, we analyzed hippocampal

RNA from a line of transgenic mice overexpressing human wild type TDP-43, iTDP-43^17D, that have previously been reported to express reduced *Tardbp* mRNA [24]. We observed a statistically significant reduction in *Tardbp* using all three primer pairs (figure 4F). These data suggest that additional mechanisms may exist *in vivo* that regulate Tdp-43 levels independent of *Tardbp* total transcript levels.

To determine the neuropathological consequences of low level M337V TDP-43 expression in the older 25 M cohort, we again examined markers of FTLD-TDP pathology. The cortex of 25 M old iTDP-43^8A animals was characterized by rare cells bearing intense ubiquitin reactivity, predominantly in the lower cortical layers (figure 5A and B). This increased staining for ubiquitin was both nuclear and cytoplasmic (inset, figure 5A), which contrasts

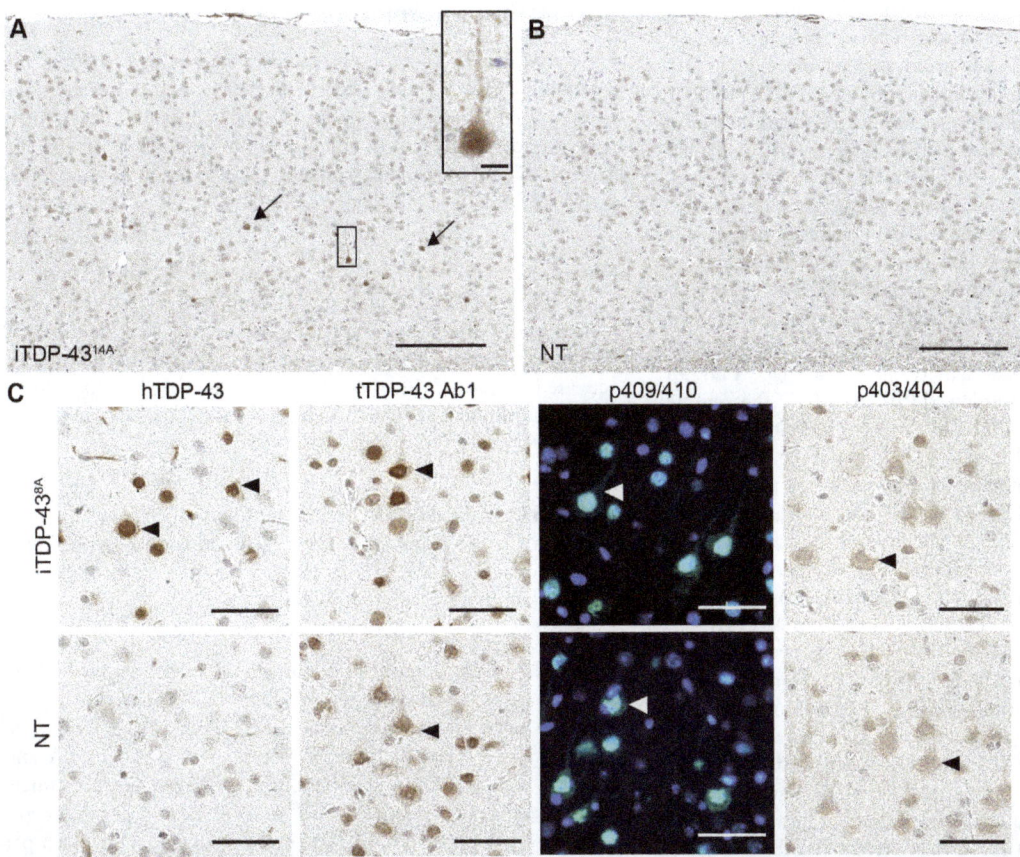

Figure 5. Neuropathology of 25 month old iTDP-43^{8A} mice. Immunohistochemical detection of ubiquitin revealed rare cells bearing increased ubiquitin staining in the cortex of iTDP-43^{8A} mice (arrows, A) that was absent in NT animals (B, scale bar = 200 µm). Staining was detected in both nucleus and cytoplasm of affected cells (inset in A, scale bar = 10 µm). (C) In iTDP-43^{8A} animals hTDP-43 was predominantly nuclear, some cells displaying cytoplasmic localization without aggregation. Cytoplasmic localization was observed in NT and iTDP-43^{8A} mice using antibodies to total TDP-43 (tTDP-43 Ab1) and phosphorylated forms of TDP-43 (p403/404, p409/410).

with the exclusively cytoplasmic distribution in P5 iTDP-43^{14A} mice. As in P5 iTDP-43^{14A} mice, we found no evidence of TDP-43 aggregation using hTDP-43, tTDP-43, p403/404 or p409/410 antibodies (figure 5C). Although we observed neurons in iTDP-43 mice bearing hTDP-43 reactivity in the cytoplasm, numerous cells were observed in both NT and iTDP-43 mice with cytoplasmic tTDP-43, p409/410 and p403/404 (arrowheads in figure 5C). In conjunction with our biochemical analysis, this evidence suggests that an approximately 30% reduction in mTdp-43 in the presence of low level expression of hTDP-43 and simultaneous low level increases in TDP-35 result in relatively minor neuropathological abnormalities and are well tolerated even over long time spans.

Discussion

TDP-43 plays a major role in the pathogenesis of FTLD-TDP. Mutations in *TARDBP* cause some cases of TDP-43-proteinopathy in amyotrophic lateral sclerosis. Numerous transgenic rodents overexpressing wild type or mutant TDP-43 have been reported that have developed profound neurodegenerative phenotypes, early lethality and gait abnormalities. Some features of FTLD-TDP and/or ALS have been recapitulated in these transgenic lines including aggregated, hyperphosphorylated TDP-43, and mis-processing of TDP-43 into lower molecular weight species.

In this paper, we generated two lines of mice expressing human TDP-43 containing the familial M337V mutation driven by the CaMKIIα promoter to determine the effects of mutant TDP-43 expression in the forebrain. Of note, neither iTDP-43 line exhibited overt phenotype comparable to previous reports (early lethality, weight loss, limb weakness or gait disturbances). As transgenic lines driven by the CaMKIIα promoter or bearing the M337V mutation have already been reported that do exhibit these phenotypes [21,24,33], the absence here of early lethality or gait disturbance cannot be entirely explained by these variables alone, suggesting that a combination of these and other factors such as timing of transgene expression and background strain govern ALS-like phenotypes.

The human neuropathology of FTLD-TDP - consisting of loss of nuclear TDP-43, TDP-43 aggregation, phosphorylation and fragmentation to C-terminal lower molecular weight species - has driven hypotheses as to TDP-43 toxicity both in human disease and in mice expressing transgenic TDP-43. Herein, we found that early neurodegeneration in the iTDP-43^{14A} mouse line occurred in the complete absence of aggregation. This observation is consistent with previous research demonstrating that aggregation does not appear to be a major driver of cell death in TDP-43 transgenic animals [22,23]. However, we cannot rule out that aggregates may be neurotoxic, particularly in lines in which they feature more prominently [21,24]. In contrast, downregulation of endogenous mTdp-43 is a feature of early degeneration in iTDP-

43^{14A}, and this has been implicated elsewhere as a potential driver of neurodegeneration where it correlates with phenotypic progression [23]. However, data from our second line iTDP-43^{8A} – in which we observed downregulation of mTdp-43 of 30% until 25 months of age, the oldest age examined – indicates that if loss of murine Tdp-43 is capable of driving aggressive neurodegenerative phenotypes, a critical threshold must be reached. Interestingly, we failed to detect downregulation of murine *Tardbp* transcript in either young or aged iTDP-43^{8A} cortex despite the reduction in mTdp-43 protein. This contrasts with findings from our lab and elsewhere demonstrating *in vivo*, hTDP-43 autoregulated reduction in *Tardbp* as the underlying mechanism of reduced Tdp-43 [2,20,33]. This finding may be indicative of other post-transcriptional autoregulatory activity (for example, TDP-43 mediated translational repression [40]) or that the subtle phenotype observed in aged iTDP-43^{8A} mice is sufficient to cause a measurable reduction in Tdp-43.

A potential additional source of neurotoxicity in iTDP-43^{14A} may be fragmentation of TDP-43 to the low molecular weight species TDP-35 and TDP-25. At the earliest phenotypic time point studied here, we observed accumulation of these species at least partially derived from the human TDP-43 transgene and found evidence of detergent insoluble, urea soluble TDP-35. Although these species may be inherently neurotoxic and contribute to cell death in iTDP-43^{14A}, we also determined that they are derived from a central portion of TDP-43 and are thus may not be biochemically identical to the C-terminally derived fragments of human disease. Consequently, we cannot be certain that any neurotoxicity these species invoke is relevant to human pathology. More study is required to determine if these fragments are identical to those almost universally present in other TDP-43 transgenic animals. Elsewhere, these fragments have been described as C-terminal in nature, based on the use of antibodies raised to the C-terminal 154 residues of TDP-43 [20]. To our knowledge the epitope of this polyclonal antibody has not been mapped. However, as a large proportion of the immunogen corresponds to the center of TDP-43, it is possible this 'C-terminal' antibody detects centrally derived fragments also.

Consequently, our data here add to prior studies suggesting that neurodegenerative phenotypes in existing TDP-43 transgenic animals might be influenced by a number of factors depending on the transgenic line, and that it may prove difficult to separate toxicity of potentially disease relevant phenomena from one another or from global RNA dysregulation that occurs upon mutant TDP-43 overexpression [41]. Additionally, the phenotype in iTDP-43^{14A} is aggressive and occurs prior to full development of the brain, an effect that we have documented in a previous line constitutively expressing wild type TDP-43 and that may be the case in several other high expressing transgenic lines with early phenotypes [20,22,24,29,33]. This clearly makes for a poor model of an age-related neurodegenerative condition. While existing TDP-43 transgenic mice will undoubtedly provide insight into TDP-43 function and in some instances can be used as investigative tools, the field still lacks *in vivo* models that faithfully recapitulate age dependent accumulation of disease relevant TDP-43 pathology.

References

1. Neumann M, Sampathu DM, Kwong LK, Truax AC, Micsenyi MC, et al. (2006) Ubiquitinated TDP-43 in frontotemporal lobar degeneration and amyotrophic lateral sclerosis. Science 314: 130–133.
2. Polymenidou M, Lagier-Tourenne C, Hutt KR, Huelga SC, Moran J, et al. (2011) Long pre-mRNA depletion and RNA missplicing contribute to neuronal vulnerability from loss of TDP-43. Nature neuroscience 14: 459–468.

Supporting Information

Figure S1 TDP-43 antibodies used in this study. N-terminal, C-terminal, mTDP-43 and tTDP-43 Ab2 are raised to the indicated epitopes. hTDP-43 and tTDP-43 Ab1 antibodies have been mapped to the indicated epitopes in previous studies [37,38]. Murine Tdp-43 specific antibody was raised as previously described [42].

Figure S2 tTA expression is not responsible for iTDP-43^{14A} phenotypes. (A) Brain weights of NT, monogenic hTDP-43^{14A} and monogenic tTA mice at P5 and 2 months of age were identical. (B) Western analysis of P5 brain lysate demonstrated no increase in activated caspase 3 in tTA only mice compared to NT mice. However, elevated activated caspase 3 was observed in iTDP-43^{14A}.

Figure S3 Validation of TDP-43 antibody specificity. Western analysis of lysate from HEK-293T cells transfected with empty vector, myc-hTDP-43^{1-280} or myc-hTDP-$43^{208-414}$. Blots were probed with the indicated antibodies.

Figure S4 Expression of tTA trangene does not affect TDP-43 metabolism. (A) Brain weights of iTDP-43^{8A} mice covering the postnatal period and aged cohorts. No significant difference in weight was observed in the time period to 2 months; tTA and iTDP-43 mice in 10 month and 25 month cohorts showed a small, significant decrease relative to NT mice (N(P0) = 4 per genotype; N(P12) = 3 per genotype; N(P21) = 4 per genotype; N(2 M) = 5 per genotype; N(10 M) = 5 NT, 8 iTDP-43, 7 tTA; N(25 M) = 7 NT, 7 iTDP-43, 5 tTA. Error bars are SD, *p<0.05, one way ANOVA, Bonferroni *post hoc*). (B) Hematoxylin and eosin staining of 25 M cohorts from line iTDP-43^{8A} show no gross morphological differences between tTA and iTDP-43^{8A} cortex. There was a trend toward smaller dentate gyri in tTA mice compared to NT mice. (C) Long exposure of western blot of 11 M NT and tTA cortical lysates using antibody to total TDP-43 (tTDP-43 Ab2), mouse specific TDP-43 antibody was used to probe same lysates, β-actin for loading. (D) Densitometric quantitation of blots in (C) confirmed no change in TDP-43, TDP-35 or TDP-25 in tTA mice versus NT mice; values are represented as fold change relative to NT mice following sample normalization to β-actin loading controls.

Table S1 Primary antibodies used in this study.

Author Contributions

Conceived and designed the experiments: SD LP JL. Performed the experiments: SD MA AC DWD. Analyzed the data: SD MA. Contributed reagents/materials/analysis tools: DWD LP JL. Wrote the paper: SD JL. Revision of drafts of the article and final approval: DWD LP JL MA AC.

3. Buratti E, Brindisi A, Giombi M, Tisminetzky S, Ayala YM, et al. (2005) TDP-43 binds heterogeneous nuclear ribonucleoprotein A/B through its C-terminal tail: an important region for the inhibition of cystic fibrosis transmembrane conductance regulator exon 9 splicing. The Journal of biological chemistry 280: 37572–37584.

4. Mackenzie IR, Neumann M, Baborie A, Sampathu DM, Du Plessis D, et al. (2011) A harmonized classification system for FTLD-TDP pathology. Acta neuropathologica 122: 111–113.

5. Yamashita T, Hideyama T, Hachiga K, Teramoto S, Takano J, et al. (2012) A role for calpain-dependent cleavage of TDP-43 in amyotrophic lateral sclerosis pathology. Nature communications 3: 1307.

6. Hasegawa M, Arai T, Nonaka T, Kametani F, Yoshida M, et al. (2008) Phosphorylated TDP-43 in frontotemporal lobar degeneration and amyotrophic lateral sclerosis. Annals of neurology 64: 60–70.

7. Lagier-Tourenne C, Cleveland DW (2009) Rethinking ALS: the FUS about TDP-43. Cell 136: 1001–1004.

8. Igaz LM, Kwong LK, Chen-Plotkin A, Winton MJ, Unger TL, et al. (2009) Expression of TDP-43 C-terminal Fragments in Vitro Recapitulates Pathological Features of TDP-43 Proteinopathies. J Biol Chem 284: 8516–8524.

9. Nonaka T, Kametani F, Arai T, Akiyama H, Hasegawa M (2009) Truncation and pathogenic mutations facilitate the formation of intracellular aggregates of TDP-43. Hum Mol Genet 18: 3353–3364.

10. Zhang YJ, Xu YF, Cook C, Gendron TF, Roettges P, et al. (2009) Aberrant cleavage of TDP-43 enhances aggregation and cellular toxicity. Proc Natl Acad Sci U S A 106: 7607–7612.

11. Barmada SJ, Skibinski G, Korb E, Rao EJ, Wu JY, et al. (2010) Cytoplasmic mislocalization of TDP-43 is toxic to neurons and enhanced by a mutation associated with familial amyotrophic lateral sclerosis. J Neurosci 30: 639–649.

12. Johnson BS, Snead D, Lee JJ, McCaffery JM, Shorter J, et al. (2009) TDP-43 is intrinsically aggregation-prone, and amyotrophic lateral sclerosis-linked mutations accelerate aggregation and increase toxicity. J Biol Chem 284: 20329–20339.

13. Suzuki H, Lee K, Matsuoka M (2011) TDP-43-induced Death Is Associated with Altered Regulation of BIM and Bcl-xL and Attenuated by Caspase-mediated TDP-43 Cleavage. J Biol Chem 286: 13171–13183.

14. Ayala YM, Misteli T, Baralle FE (2008) TDP-43 regulates retinoblastoma protein phosphorylation through the repression of cyclin-dependent kinase 6 expression. Proc Natl Acad Sci U S A 105: 3785–3789.

15. Wu LS, Cheng WC, Hou SC, Yan YT, Jiang ST, et al. (2010) TDP-43, a neuro-pathosignature factor, is essential for early mouse embryogenesis. Genesis 48: 56–62.

16. Chiang PM, Ling J, Jeong YH, Price DL, Aja SM, et al. (2010) Deletion of TDP-43 down-regulates Tbc1d1, a gene linked to obesity, and alters body fat metabolism. Proc Natl Acad Sci U S A 107: 16320–16324.

17. Wu LS, Cheng WC, Shen CK (2012) Targeted depletion of TDP-43 expression in the spinal cord motor neurons leads to the development of amyotrophic lateral sclerosis-like phenotypes in mice. The Journal of biological chemistry 287: 27335–27344.

18. Feiguin F, Godena VK, Romano G, D'Ambrogio A, Klima R, et al. (2009) Depletion of TDP-43 affects Drosophila motoneurons terminal synapsis and locomotive behavior. FEBS Lett 583: 1586–1592.

19. Kabashi E, Lin L, Tradewell ML, Dion PA, Bercier V, et al. (2010) Gain and loss of function of ALS-related mutations of TARDBP (TDP-43) cause motor deficits in vivo. Hum Mol Genet 19: 671–683.

20. Xu YF, Gendron TF, Zhang YJ, Lin WL, D'Alton S, et al. (2010) Wild-type human TDP-43 expression causes TDP-43 phosphorylation, mitochondrial aggregation, motor deficits, and early mortality in transgenic mice. The Journal of neuroscience : the official journal of the Society for Neuroscience 30: 10851–10859.

21. Tsai KJ, Yang CH, Fang YH, Cho KH, Chien WL, et al. (2010) Elevated expression of TDP-43 in the forebrain of mice is sufficient to cause neurological and pathological phenotypes mimicking FTLD-U. The Journal of experimental medicine 207: 1661–1673.

22. Janssens J, Wils H, Kleinberger G, Joris G, Cuijt I, et al. (2013) Overexpression of ALS-Associated p.M337V Human TDP-43 in Mice Worsens Disease Features Compared to Wild-type Human TDP-43 Mice. Molecular neurobiology.

23. Igaz LM, Kwong LK, Lee EB, Chen-Plotkin A, Swanson E, et al. (2011) Dysregulation of the ALS-associated gene TDP-43 leads to neuronal death and degeneration in mice. The Journal of clinical investigation 121: 726–738.

24. Cannon A, Yang B, Knight J, Farnham IM, Zhang Y, et al. (2012) Neuronal sensitivity to TDP-43 overexpression is dependent on timing of induction. Acta neuropathologica 123: 807–823.

25. Santacruz K, Lewis J, Spires T, Paulson J, Kotilinek L, et al. (2005) Tau suppression in a neurodegenerative mouse model improves memory function. Science 309: 476–481.

26. Zhang YJ, Xu YF, Dickey CA, Buratti E, Baralle F, et al. (2007) Progranulin mediates caspase-dependent cleavage of TAR DNA binding protein-43. The Journal of neuroscience : the official journal of the Society for Neuroscience 27: 10530–10534.

27. Mayford M, Bach ME, Huang YY, Wang L, Hawkins RD, et al. (1996) Control of memory formation through regulated expression of a CaMKII transgene. Science 274: 1678–1683.

28. Hellemans J, Mortier G, De Paepe A, Speleman F, Vandesompele J (2007) qBase relative quantification framework and software for management and automated analysis of real-time quantitative PCR data. Genome biology 8: R19.

29. Wils H, Kleinberger G, Janssens J, Pereson S, Joris G, et al. (2010) TDP-43 transgenic mice develop spastic paralysis and neuronal inclusions characteristic of ALS and frontotemporal lobar degeneration. Proceedings of the National Academy of Sciences of the United States of America 107: 3858–3863.

30. Srinivasan A, Roth KA, Sayers RO, Shindler KS, Wong AM, et al. (1998) In situ immunodetection of activated caspase-3 in apoptotic neurons in the developing nervous system. Cell death and differentiation 5: 1004–1016.

31. Wegorzewska I, Bell S, Cairns NJ, Miller TM, Baloh RH (2009) TDP-43 mutant transgenic mice develop features of ALS and frontotemporal lobar degeneration. Proceedings of the National Academy of Sciences of the United States of America 106: 18809–18814.

32. Han HJ, Allen CC, Buchovecky CM, Yetman MJ, Born HA, et al. (2012) Strain background influences neurotoxicity and behavioral abnormalities in mice expressing the tetracycline transactivator. The Journal of neuroscience : the official journal of the Society for Neuroscience 32: 10574–10586.

33. Xu YF, Zhang YJ, Lin WL, Cao X, Stetler C, et al. (2011) Expression of mutant TDP-43 induces neuronal dysfunction in transgenic mice. Molecular neurodegeneration 6: 73.

34. Swarup V, Phaneuf D, Bareil C, Robertson J, Rouleau GA, et al. (2011) Pathological hallmarks of amyotrophic lateral sclerosis/frontotemporal lobar degeneration in transgenic mice produced with TDP-43 genomic fragments. Brain : a journal of neurology 134: 2610–2626.

35. Stallings NR, Puttaparthi K, Luther CM, Burns DK, Elliott JL (2010) Progressive motor weakness in transgenic mice expressing human TDP-43. Neurobiology of disease 40: 404–414.

36. Igaz LM, Kwong LK, Chen-Plotkin A, Winton MJ, Unger TL, et al. (2009) Expression of TDP-43 C-terminal Fragments in Vitro Recapitulates Pathological Features of TDP-43 Proteinopathies. The Journal of biological chemistry 284: 8516–8524.

37. Tsuji H, Nonaka T, Yamashita M, Masuda-Suzukake M, Kametani F, et al. (2012) Epitope mapping of antibodies against TDP-43 and detection of protease-resistant fragments of pathological TDP-43 in amyotrophic lateral sclerosis and frontotemporal lobar degeneration. Biochemical and biophysical research communications 417: 116–121.

38. Zhang HX, Tanji K, Mori F, Wakabayashi K (2008) Epitope mapping of 2E2-D3, a monoclonal antibody directed against human TDP-43. Neuroscience letters 434: 170–174.

39. Avendano-Vazquez SE, Dhir A, Bembich S, Buratti E, Proudfoot N, et al. (2012) Autoregulation of TDP-43 mRNA levels involves interplay between transcription, splicing, and alternative polyA site selection. Genes & development 26: 1679–1684.

40. Wang IF, Wu LS, Chang HY, Shen CK (2008) TDP-43, the signature protein of FTLD-U, is a neuronal activity-responsive factor. Journal of neurochemistry 105: 797–806.

41. Arnold ES, Ling SC, Huelga SC, Lagier-Tourenne C, Polymenidou M, et al. (2013) ALS-linked TDP-43 mutations produce aberrant RNA splicing and adult-onset motor neuron disease without aggregation or loss of nuclear TDP-43. Proceedings of the National Academy of Sciences of the United States of America 110: E736–745.

42. Xu YF, Prudencio M, Hubbard JM, Tong J, Whitelaw EC, et al. (2013) The Pathological Phenotypes of Human TDP-43 Transgenic Mouse Models Are Independent of Downregulation of Mouse Tdp-43. PloS one 8: e69864.

15

Xmrk, *Kras* and *Myc* Transgenic Zebrafish Liver Cancer Models Share Molecular Signatures with Subsets of Human Hepatocellular Carcinoma

Weiling Zheng[1], Zhen Li[1], Anh Tuan Nguyen[1], Caixia Li[1], Alexander Emelyanov[2], Zhiyuan Gong[1]*

1 Department of Biological Sciences, National University of Singapore, Singapore, Singapore, **2** Institute of Cell and Molecular Biology, Singapore, Singapore

Abstract

Previously three oncogene transgenic zebrafish lines with inducible expression of *xmrk*, *kras* or *Myc* in the liver have been generated and these transgenic lines develop oncogene-addicted liver tumors upon chemical induction. In the current study, comparative transcriptomic approaches were used to examine the correlation of the three induced transgenic liver cancers with human liver cancers. RNA profiles from the three zebrafish tumors indicated relatively small overlaps of significantly deregulated genes and biological pathways. Nevertheless, the three transgenic tumor signatures all showed significant correlation with advanced or very advanced human hepatocellular carcinoma (HCC). Interestingly, molecular signature from each oncogene-induced zebrafish liver tumor correlated with only a small subset of human HCC samples (24–29%) and there were conserved up-regulated pathways between the zebrafish and correlated human HCC subgroup. The three zebrafish liver cancer models together represented nearly half (47.2%) of human HCCs while some human HCCs showed significant correlation with more than one signature defined from the three oncogene-addicted zebrafish tumors. In contrast, commonly deregulated genes (21 up and 16 down) in the three zebrafish tumor models generally showed accordant deregulation in the majority of human HCCs, suggesting that these genes might be more consistently deregulated in a broad range of human HCCs with different molecular mechanisms and thus serve as common diagnosis markers and therapeutic targets. Thus, these transgenic zebrafish models with well-defined oncogene-induced tumors are valuable tools for molecular classification of human HCCs and for understanding of molecular drivers in hepatocarcinogenesis in each human HCC subgroup.

Editor: Xiaolei Xu, Mayo Clinic, United States of America

Funding: This work was supported by a grant from National Medical Research Council of Singapore. The funders had no role in study design, data collection and analysis, decision to publish, or preparation of the manuscript.

* E-mail: dbsgzy@nus.edu.sg

Introduction

Human hepatocellular carcinoma (HCC) is known to be a very heterogeneous disease, especially at intermediate and advanced stages [1]. Due to the diverse and complex etiologies contributing to HCC incidence, different genetic mutations or altered molecular pathways could be responsible for hepatocarcinogenesis. So far, several carcinogenic pathways have been identified to be involved in the development and progression of HCC, including the VEGFR, EGFR, and mTOR pathways [2]. In effort to decipher the role of different oncogenic pathways, a number of transgenic mouse models have been established [3,4] and comparative transcriptomic analyses have been used to identify the best transgenic mouse models for human HCCs [4].

The zebrafish has been increasingly recognized as a valuable experimental model for human diseases, particularly for cancers [5] including liver cancers [6–11]. It has been shown that the zebrafish tumors had striking similarities with human cancer histologically [12,13]. Transcriptomic and epigenetic analyses have also indicated conserved features of carcinogen-induced zebrafish HCC with human HCC [14–16].

Previously we have generated several liver tumor models by transgenic expression of three different oncogenes (*kras*, *xmrk* or *Myc*) specifically in the zebrafish liver and these transgenic zebrafish usually produce liver tumors with variable degrees of severity from hepatocellular adenoma (HCA) to HCC [6–9]. The three oncogenes we used in the zebrafish have all been shown to be involved in hepatocarcinogenesis. *KRAS* is mutated in ~7% of liver cancers in human [17] but Ras signaling is ubiquitously activated in HCC [18]. *Xmrk* is a naturally occurring variant of the EGFR in fish of the genus *Xiphophorus* (platyfish and swordtails) with constitutive autophosphorylation and activation of downstream signals [19]. Activation of EGFR signal is correlated with poor prognosis of HCC patients [20]. *MYC* is commonly amplified in many cancers including HCC and higher expression level of MYC is associated with more advanced status of HCC [21]. We have shown that overexpression of *kras* and *xmrk* in the zebrafish liver could induce HCC [7–9], while overexpression of *Myc* induced mostly HCA [6].

Comparative transcriptomic analyses of animal models and human clinical samples provide a powerful tool for identification of conserved molecular pathways and biomarker genes for diagnosis and therapy [15,22]. Our existing oncogene transgenic

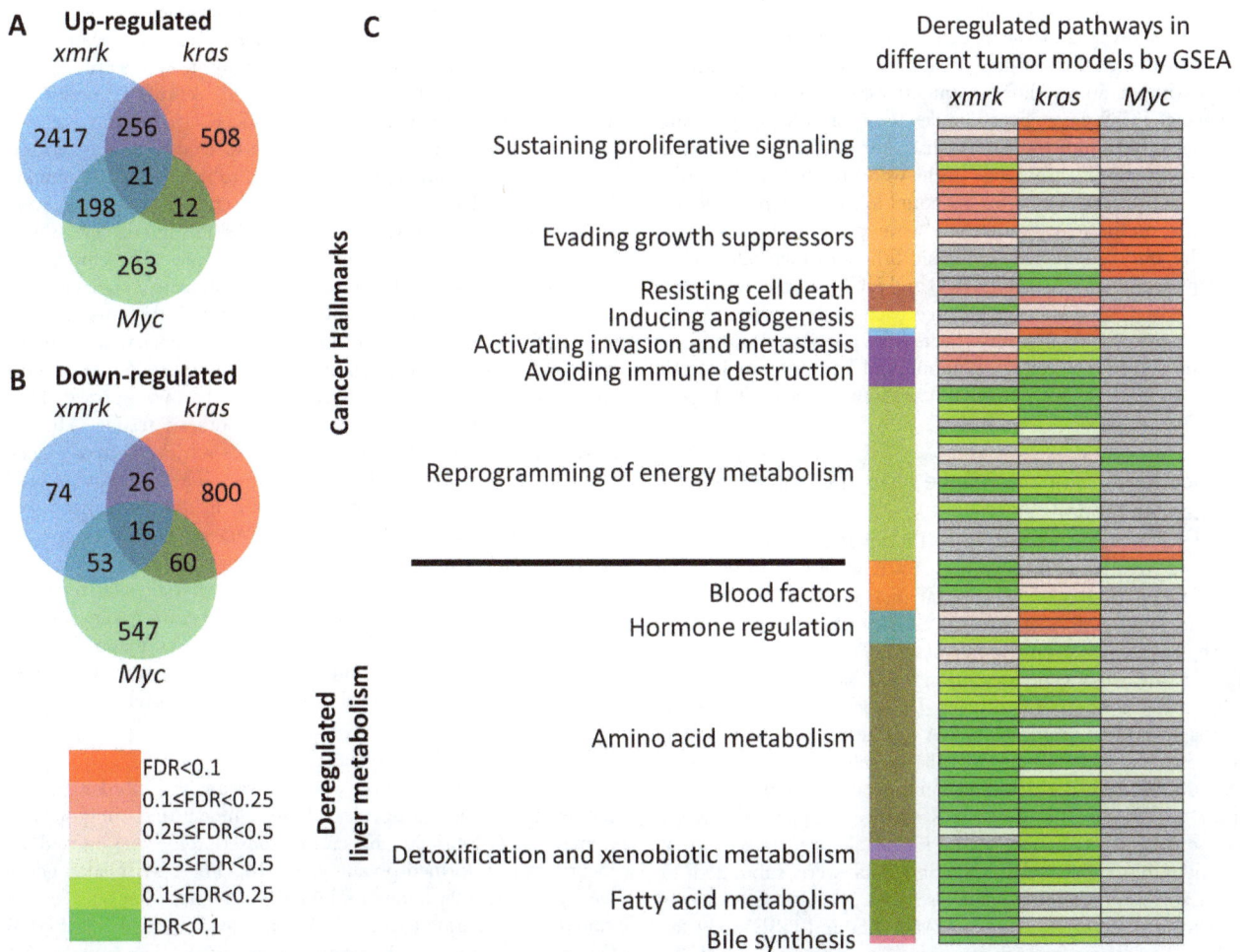

Figure 1. Deregulated genes and pathways in *xmrk, kras* and *Myc* transgenic liver tumors. (A) Venn diagram of up-regulated genes in the three zebrafish transgenic HCC models. (B) Venn diagram of down-regulated genes in the three zebrafish transgenic HCC models. (C) Deregulated pathways in the three oncogene transgenic liver tumors. Up- and down-regulated pathways in each zebrafish transgenic HCC model were analyzed by GSEA pre-ranked analysis. Red and green indicate up-and down-regulation respectively and the color code based on FDR is shown on the left. Pathways correlating with grey cells were not detected with any change. The up-regulated pathways were assigned into seven cancer hallmark categories (excluding enabling replicative immortality) according to Hanahan and Weinberg [27] and down-regulated pathways was classified based on different aspects of the liver metabolism (see Table S1 for details).

zebrafish models have well defined up-regulation of driver oncogene and this may provide a valuable tool to identify the molecular driving forces in human carcinogenesis by comparative transcriptomic analyses. Thus, in this study, we employed RNA sequencing technology to compare the transcriptomic profiles of the three oncogene-induced liver tumors in transgenic zebrafish. By comparative analyses with human liver transcriptomes from cirrhotic livers to very advanced HCC, we found that they all showed strong molecular correlation with advanced or very advanced human HCCs. Nevertheless, there are quite distinct deregulated biological pathways based on deregulated genes in the three oncogene transgenic models. Interestingly, each zebrafish liver tumor model correlated with a subset of human HCCs and each subset has some distinct molecular features. We showed that the transgenic zebrafish models with well-defined driver-gene activity should be valuable for classification of human HCCs and for understanding the molecular mechanisms behind each HCC subtype.

Materials and Methods

Treatment and Induction of Liver Cancer in the Three Zebrafish Transgenic Models

Zebrafish were maintained following the approved protocol by Institutional Animal Care and Use Committee of National University of Singapore (Protocol 079/07). The generation of *xmrk* and *Myc* transgenic zebrafish was previously described and they were termed as *TO(xmrk)* [9] and *TO(Myc)* [6] respectively in the previous publications. The two transgenic lines were constructed by using a tetracycline-inducible transgenic system and the oncogene expression were induced by doxycycline. The *kras*^{V12} transgenic line used in the present study was newly generated by using a mifepristone-inducible system [7] in combination with a Cre-loxP system (unpublished). For the *xmrk* and *Myc* transgenic lines, transgenic fish and their non-transgenic siblings were treated with 60 µg/ml doxycycline (Sigma, USA) starting from 3.5 mpf (month post fertilization) for 6 weeks. All *xmrk* fish developed HCC and all *Myc* fish developed HCA. In total, for each transgenic line, one tumor sample (transgenic fish treated with doxycycline) and three control samples (non-transgenic siblings similarly treated with doxycycline, transgenic siblings without doxycycline treatment, and non-transgenic siblings without doxycycline treatment) were collected for RNA sequencing. In all cases, liver samples used for RNA sequencing were pooled from four to five male fish. For the $kras^{V12}$ transgenic fish, one-month-old transgenic fish were treated with 1 µM mifepristone (Sigma, USA) for 36 hours to induce Cre-mediated recombination for activation of $kras^{V12}$ transgene expression in the liver. The $kras^{V12}$ activated transgenic fish were then allowed to grow for six months to develop HCC and HCC samples were then collected for RNA sequencing. Three liver tumors from induced transgenic fish and three normal livers from uninduced transgenic fish were pooled separately as tumor and control samples. All samples used were from male fish and two sets of biological replicates were used.

Identification of Signature Gene Lists in Each Zebrafish Liver Cancer Model

Total RNA was extracted using TRIzol Reagent (Invitrogen, USA) and treated with DNase I to remove genomic DNA contamination. 3′ RNA-SAGE (serial analysis of gene expression) sequencing was performed on ABI SOLiD platform by Mission Biotech (Taiwan) according to manufacturer's protocol and 10–23 million reads were generated from each sampler (Table S1). Briefly, mRNA was purified using Dynabeads Oligo(dT) EcoP (Invitrogen) and subjected to cDNA synthesis. Resultant cDNA was digested by NlaIII and EcoP15I to result in a 27 nucleotides cDNA tag between the two sequencing adapters. The tags were mapped to the NCBI RefSeq (Reference Sequence) mRNA database for zebrafish with a criterion of allowing maximum 2 nucleotide mismatches. All RNA-Seq data were submitted to Gene Expression Omnibus database with the following access numbers: GSE53342 for $xmrk$ and Myc data and GSE53630 for $kras$ data. Tag counts for each transcript were normalized to TPM (transcripts per million) to facilitate comparison among different samples. The differentially regulated genes in the Myc- and $xmrk$-induced liver cancers were identified using one sample t-test as previously described [23], and the differentially expressed genes in the $kras$-induced liver cancer was identified using two-tailed Student's t-test.

To facilitate functional implications of zebrafish transcriptome, all zebrafish genes were mapped to annotated human genes in order to use existing online software developed in human genes. Thus, human homology mapping of zebrafish Unigene clusters were retrieved from the Genome Institute of Singapore Zebrafish Annotation Database (http://123.136.65.67/). For Unigene clusters mapped by more than one transcript entries, the highest TPM was used to represent the expression level of the Unigene cluster. Some zebrafish Unigene clusters were mapped to more than one human Unigene clusters, which usually came from the same gene family. To remove redundancy and avoid causing bias in functional analyses, only the first human Unigene cluster in the list was selected to represent the zebrafish Unigene clusters. The lists of significantly up-regulated zebrafish genes that were mapped with human homologs and used for comparative analyse with human HCC data are shown in Table S2.

Gene Set Enrichment Analysis (GSEA)

GSEA was used to establish the relatedness between the transgenic zebrafish models and human liver cancers [24]. GSEA is a computational method that determines whether a *priori* defined set of genes shows statistically significant, concordant differences between two biological samples; it calculates an enrichment score

using a running-sum statistic through a ranked list of gene expression data set. Human homologs of the significantly up-regulated genes from the zebrafish tumor tissues were used as cancer signatures for each transgenic zebrafish model for transcriptomic comparison with human HCC data. Each chosen phenotype of human HCCs (either one specific HCC stage or one particular HCC sample) was compared to the rest of the samples in the same dataset. All genes in the chosen phenotype were ranked by t-test to determine expression differences among different HCC stages or different HCC samples. The enrichment score of the pre-defined transgenic zebrafish cancer signature was calculated using a running-sum statistic through the ranked genes. The statistical significance of the enrichment score was estimated by using an empirical phenotype-based permutation test procedure. An FDR (false discovery rate) value was provided by introducing adjustment of multiple hypothesis testing. Human liver cancer transcriptome data were retrieved from Gene Expression Omnibus (GEO) database. The human dataset including different stages of HCCs used in the comparison was GSE6764 [25]. The ten human HCC datasets used for examining the representation of zebrafish liver cancer gene signatures are summarized in Table S3. Annotation information was retrieved from SOURCE (http://smd.stanford.edu/cgi-bin/source/sourceSearch). For multiple probes which can be mapped to one Unigene cluster, the maximum signal intensity was selected to represent the expression level of the Unigene cluster.

GSEA Pre-ranked Analysis

GSEA pre-ranked option was used to analyze the deregulated pathways in each transgenic zebrafish model and subgroups of human HCCs. Briefly, the entire transcriptome was ranked by logarithm transformed p-value (base 10). The up-regulated genes were given positive values, and the down-regulated genes negative values. The curated canonical pathways from the MSigDB (Molecular Signature Database) were used. The number of permutation used was 1000.

RT-qPCR Validation

Total RNA were reverse-transcribed using the SuperScript II cDNA Synthesis Kit (Invitrogen). RT-qPCR was performed with same sets of cDNAs used for SAGE sequencing using the LightCycler 480 SYBR Green I Master system (Roche). Reactions were conducted in triplicate for each cDNA sample and primer sequences are shown in the Table S4. Gene expression levels in each control or transgenic liver sample were normalized with the level of β-actin mRNA as the internal control. The $\log_2$ fold changes between tumor and control samples were calculated using the CT method according to the formula: $\log_2$ fold changes $= -\Delta\Delta CT = -[(CT$ gene of interest–CT β-actin) transgenic sample–(CT gene of interest–CT β-actin) control sample]. Two-tailed heteroscedastic t test was performed using normalized CT values (CT gene - CT β-actin) and changes with p<0.05 are considered to be significant.

Results

Identification of Differentially Expressed Genes in the Three Transgenic Zebrafish Liver Cancer Models

The three oncogene transgenic lines ($xmrk$, Myc and $kras^{V12}$) were induced to develop liver tumors (Figure S1) and these tumor samples were subjected to RNA-SAGE sequencing. By a selection criteria of fold change>1.5, p value<0.05 and TPM>10 (in either control or tumor samples), differentially expressed genes were selected from the three tumor sets. There were 2,892, 797 and 494

Table 1. Commonly deregulated genes in the three transgenic zebrafish liver cancer models.

	Common	Gene Symbol	Human homolog Gene Symbol	xmrk		kras		Myc	
				FC	P value	FC	P value	FC	P value
Up	18858688	foxa3	FOXA3	3.58	6.66E-03	1.78	4.16E-02	1.55	9.18E-03
	18858728	gart	GART	3.32	1.42E-03	2.87	5.87E-03	7.69	6.93E-04
	18859502	tp53	TP53	5.09	1.43E-03	1.92	1.09E-02	3.61	3.30E-03
	41054118	c20orf24	C20orf24	4.92	2.13E-03	1.93	2.83E-03	3.14	5.22E-03
	41393080	itm1	STT3A	1.98	7.59E-03	1.56	3.02E-02	2.01	8.07E-04
	47086630	rhot1a	RHOT1	1.62	3.11E-02	1.87	3.00E-02	1.58	3.39E-02
	47087206	abce1	ABCE1	4.12	1.97E-03	6.79	1.25E-02	3.01	4.40E-03
	50344865	cdkrap3	CDK5RAP3	2.09	4.98E-03	1.85	2.06E-02	2.42	1.43E-03
	50345019	mgat4b	MGAT4B	4.09	1.62E-02	20.55	2.92E-02	1.70	3.03E-03
	50540209	srprb	SRPRB	1.85	2.00E-02	4.47	6.89E-04	2.65	5.84E-03
	55742596	eif5a2	EIF5A2	4.83	1.09E-03	1.45	4.07E-03	2.98	2.14E-04
	62955566	cirbp	CIRBP	6.11	1.63E-03	2.00	3.17E-02	2.16	1.44E-02
	66773145	noc4l	NOC4L	2.67	4.20E-02	1.63	3.58E-02	4.34	1.05E-02
	71834591	reep2	REEP2	4.38	3.51E-02	7.17	3.26E-03	8.38	7.95E-03
	76253887	srp14	SRP14	4.17	1.62E-03	1.71	1.61E-02	1.91	1.15E-04
	94536632	stmn1a	STMN1	13.83	2.30E-03	3.05	4.93E-03	4.30	1.56E-02
	113679133	mrps9	MRPS9	1.71	2.85E-02	1.56	3.71E-02	2.40	1.59E-02
	121583749	hmgcra	HMGCR	7.42	5.42E-03	30.45	7.30E-04	8.71	2.41E-03
	126723627	ubap2	UBAP2	1.94	3.33E-02	1.88	4.99E-03	1.58	9.64E-03
	148225559	fam162a	FAM162A	2.15	2.76E-03	2.16	4.72E-02	1.97	2.22E-02
	169646807	rrp9	RRP9	4.20	1.01E-02	2.99	3.32E-02	2.71	3.48E-02
Down	23308680	cyp2ad2	–	4.35	1.99E-02	9.09	8.07E-03	2.94	2.73E-02
	41053663	scp2	SCP2	5.00	9.46E-03	2.94	2.02E-02	1.96	2.17E-02
	41055025	hsd17b3	HSD17B3	10.00	3.20E-02	100.00	7.61E-03	33.33	1.66E-02
	41152446	hsdl2	HSDL2	3.85	4.21E-02	2.13	3.55E-02	2.56	6.38E-03
	47085884	fbp1b	FBP1	9.09	3.91E-04	8.33	1.86E-02	2.38	2.87E-02
	47086066	tdh	TDH	6.67	3.28E-02	2.86	1.50E-02	3.45	3.93E-02
	47086178	itgb1b.2	ITGB1	7.14	4.19E-02	7.14	6.40E-03	4.35	3.30E-03
	47086928	ak3l1	AK3L1	2.17	2.31E-02	10.00	1.56E-03	2.00	2.17E-02
	54400637	ech1	ECH1	1.52	2.29E-02	2.17	3.78E-03	2.00	1.51E-02
	55925455	gpx4a	GPX4	3.57	6.79E-03	2.27	3.32E-02	5.26	7.31E-03
	56790261	sod1	SOD1	2.94	4.18E-02	1.85	5.32E-03	2.27	2.48E-02
	70778900	slc27a2	SLC27A2	9.09	3.26E-02	5.00	5.13E-03	2.78	1.08E-02
	71834285	apobl	APOB	5.26	4.21E-02	1.79	1.20E-02	1.75	1.46E-02

Table 1. Cont.

Common	Gene Symbol	Human homolog Gene Symbol	xmrk		kras		Myc	
			FC	P value	FC	P value	FC	P value
71834671	miox	MIOX	3.13	3.05E-02	9.09	8.07E-03	4.55	4.30E-02
121583789	nrxn1b	NRXN1	4.76	4.87E-03	1.96	2.18E-02	6.67	1.51E-02
148230211	slco1d1	-	3.70	6.22E-03	4.17	1.00E-04	2.70	1.11E-03

genes up-regulated and 169, 902 and 676 genes down-regulated in the *xmrk-*, *kras-* and *Myc-* induced zebrafish liver cancer, respectively (Figure 1A,B). Deregulated genes from the three transgenic models showed relatively small overlaps, indicating that the three oncogenes regulated quite distinct sets of genes. This is consistent with the report that morphologically uniform cancer type is frequently classified into different subgroups based on their distinct gene expression patterns [26]. Interestingly, there were 21 up-regulated and 16 down-regulated genes commonly found in all the three tumor models (Figure 1A, B, Table 1).

Distinct Pathways Regulated in the Three Zebrafish Liver Tumor Models

Pathway analysis using GSEA showed that the three transgenic liver cancer models have different pathways deregulated (Figure 1C, Table S5). It has been widely accepted that there are eight cancer hallmarks for multistep tumorigenesis and the complexities of neoplasms [27,28]. We found that the GSEA-identified pathways fallen into at least seven cancer hallmarks (except for enabling replicative immortality). *Xmrk* mainly up-regulated pathways involved in evading growth suppressors and avoiding immune destruction, which included activating cell cycle, promoting RNA transcription, up-regulating proteasome and altering immune properties. *Kras* provided the tumor cells with the ability of self-sustaining proliferative signals by up-regulating EGFR, Raf-MEK-ERK, PI3K-AKT-mTOR and GSK3 signaling pathways. Specifically, it also altered the focal adhesive characters of tumor cells which could activate invasion and metastasis. *Myc* mainly up-regulated translation and proteolysis to assist tumor cells to evade growth suppressors, and it also up-regulated VEGF pathway, thus potentially inducing angiogenesis. While there was no single pathway significantly up-regulated in all three tumor models, there were some pathways up-regulated in two models, such as proteasome in the *xmrk* and *Myc* models, and IGF1 pathway, mTOR pathway, tRNA biosynthesis and focal adhesion in the *xmrk* and *kras* models. In contrast, pathways in reprogramming of energy metabolism were generally down-regulated in all three models though the down-regulation in the *Myc* model was less apparent than the other two models. Meanwhile, many other pathways involved in normal liver function such as blood factors, amino acid metabolism, detoxification and xenobiotic metabolism, fatty acid metabolism, and bile synthesis were uniformly down-regulated in all three tumor models. However, one exception was hormone regulation that was apparently up-regulated in the *kras* tumors.

Correlation of the Three Gene Signatures of Zebrafish Liver Tumors with Different Stages of Human HCCs

The up-regulated genes, 2,892, 797 and 494 from *xmrk*, *kras* and *Myc* tumors respectively, were used as the signature genes for each model. These up-regulated genes were converted to 1,362, 490, and 146 human Unigenes respectively in order to compare with available transcriptomic data from human studies. We then compared the three signature gene sets with a set of human transcriptomic data (GSE6764) from different stages of human liver conditions: cirrhotic liver, low grade dysplastic nodules (LGDN), high grade dysplastic nodules (HGDN), and very early, early, advanced, and very advanced HCC (veHCC, eHCC, aHCC and vaHCC), in which the pathological HCC stages have been defined by tumor size, differentiation status and metastasis level [25]. The three signature gene sets were all up-regulated as the disease progresses, but they started to be up-regulated at different stages of tumorigenesis (Figure 2). The *xmrk* signature

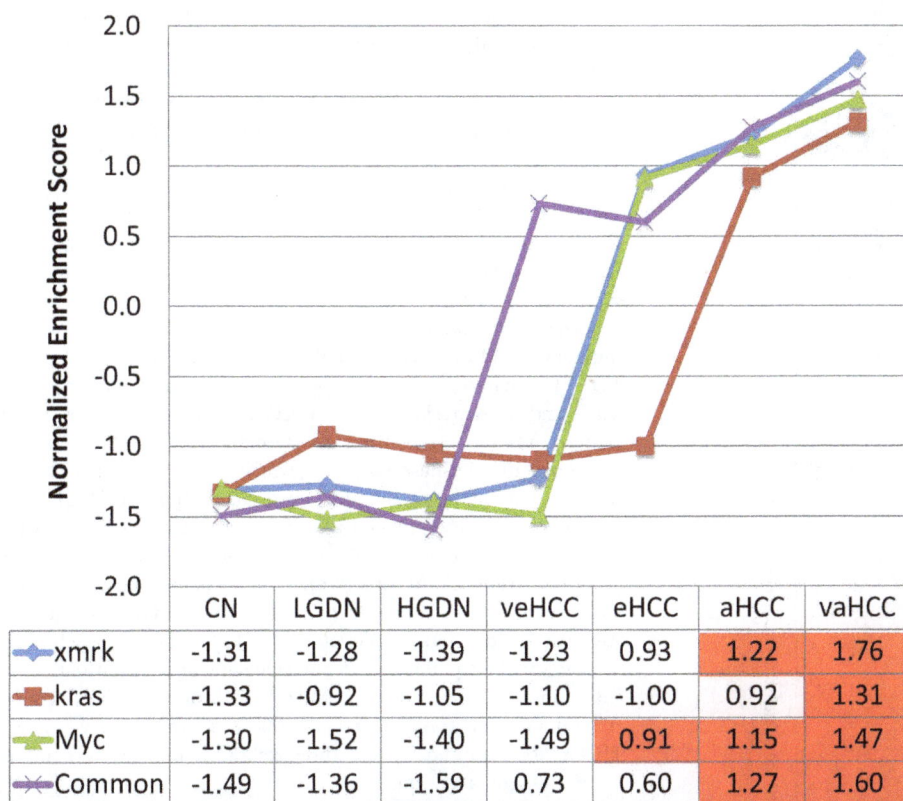

	CN	LGDN	HGDN	veHCC	eHCC	aHCC	vaHCC
xmrk	-1.31	-1.28	-1.39	-1.23	0.93	1.22	1.76
kras	-1.33	-0.92	-1.05	-1.10	-1.00	0.92	1.31
Myc	-1.30	-1.52	-1.40	-1.49	0.91	1.15	1.47
Common	-1.49	-1.36	-1.59	0.73	0.60	1.27	1.60

Figure 2. Correlation of the zebrafish liver tumor signatures with different stages of human hepatocarcinogenesis. The up-regulated gene signatures from the three transgenic tumors (xmrk, kras and Myc) and 21 commonly up-regulated genes (Common) were used for GSEA and NESs (normalized enrichment scores (y-axis) are plotted against different stages of human HCCs (x-axis). NES is listed in the table and asterisk indicate statistical significance: FDR<0.25. Abbreviations: CL, cirrhotic livers; LGDN, low grade dysplastic nodules; HGDN, high grade dysplastic nodules; veHCC, very early HCC; eHCC, early HCC; aHCC, advanced HCC; vaHCC, very advanced HCC.

showed positive correlation with HCCs starting from eHCC, and it was significantly correlated with aHCC and vaHCC. The *kras* signature positively correlated with aHCC and vaHCC, and it was only significantly correlated with vaHCC. The *Myc* signature showed significantly positive correlation with HCCs starting from eHCC, similar with our previous result using an independent set of RNA-seq data from the same transgenic line [6]. Interestingly, the common 21 up-regulated genes in all three tumor models also showed up-regulation even from the very early stage of HCC, indicating these genes are correlated with the entire neoplastic process.

Representation of the Three Gene Signatures of Zebrafish Liver Tumors in Human HCC Samples

Given the distinctive pathways altered in the three zebrafish liver tumor models and the high level of heterogeneity in human HCC patients, we sought to examine the degree of representation of the three zebrafish liver cancer models in human HCC samples. We examined ten sets of published clinical HCC microarray data, each of which included at least 80 clinical samples. The ten datasets contain a total of 1,272 samples covering different ethnic groups and risk factors (Table S2). We found that the *xmrk*, *kras* and *Myc* gene signatures were significantly correlated with 30.8%, 24.8% and 25.6% of all clinical samples, respectively. 47.2% of the human HCC samples had significant correlation with at least one of the three zebrafish gene signatures (Table 2, Figure S2). In different human HCC datasets, the percentage ranges from 22.0%

to 60.4%. Moreover, some of the clinical samples were correlated with two or even three zebrafish gene signatures. Co-correlation of two gene signatures, namely *xmrk/kras*, *kras/Myc* and *kras/Myc*, accounted for 17.5%, 13.5% and 12.4% of the human HCC samples. 9.3% of the human samples showed co-correlation of all the three gene signatures. Thus, it appears that the three transgenic zebrafish liver tumor models represent molecular mechanisms of hepatocarcinogenesis in almost half of the human HCC cases and the other half of human HCC may be due to different molecular mechanisms.

We further demonstrated that the subsets of human HCCs which showed significantly positive correlation with the same gene signature also shared similar up-regulated pathways (Figure 3, Table S6). For this analysis, each of the human HCC sets was separated into two subgroups: those showing significant correlation with one of the zebrafish signatures, and the rest. Differentially expressed genes between the two groups were identified by two-tailed t-test, and pathway analyses were performed by GSEA pre-ranked analysis. The pathways were subjected to hierarchical clustering by logarithm-transformed FDR values using MeV [29,30]. As shown in Figure 3, the pathways differentially regulated in human HCCs which showed significant correlation with one of the zebrafish signatures had distinct patterns. The pathway cluster A is consisted of pathways up-regulated in all the three subgroups, including proteasome, tRNA biosynthesis and oxidative phosphorylation. Ribosome, transcription and translation, and cell cycle were also highly up-regulated, suggesting that the human HCC samples significantly correlated with any of the

Table 2. Degree of representation of the three transgenic zebrafish liver cancer signatures in human HCCs.

	GEO accession	xmrk	kras	Myc	Total*	xmrk/kras	xmrk/Myc	kras/Myc	xmrk/kras/Myc
A	GSE364	12.6%	17.2%	19.5%	31.0%	8.0%	6.9%	6.9%	3.4%
B	GSE1898	9.9%	11.0%	16.5%	22.0%	8.8%	7.7%	7.7%	6.6%
C	GSE10141	28.8%	25.0%	17.5%	46.3%	16.3%	7.5%	5.0%	3.8%
D	GSE9843	35.2%	23.1%	25.3%	54.9%	14.3%	9.9%	8.8%	4.4%
E	GSE19977	37.8%	21.3%	25.0%	45.1%	20.1%	18.9%	13.4%	13.4%
F	GSE10186	32.2%	27.1%	12.7%	44.9%	19.5%	5.1%	6.8%	4.2%
G	GSE20017	44.4%	23.7%	30.4%	50.4%	23.0%	24.4%	17.8%	17.0%
H	GSE25097	34.7%	31.0%	34.7%	60.4%	20.1%	14.9%	15.3%	10.4%
I	GSE5975	26.9%	32.4%	32.4%	46.2%	22.3%	17.2%	21.4%	15.5%
J	GSE14520	30.5%	20.0%	20.5%	47.6%	11.4%	10.0%	6.2%	3.3%
	Total	29.2%	23.5%	23.8%	47.2%	16.9%	12.5%	11.5%	8.8%

Percentages indicate the percentages of human HCC samples which showed significant positive correlation (FDR<0.25) with the zebrafish liver cancer signatures.
*Total: The total percentage of human HCCs which showed significant correlation with any one or more of the zebrafish signatures.

zebrafish signatures were probably more proliferative than those not significantly correlated. The pathway cluster B contains pathways down-regulated in most of the subgroups associated with xmrk but up-regulated in the subgroups associated with kras and Myc. Interestingly, theses pathways were consistent with highly down-regulated pathways in the xmrk-induced zebrafish HCC, including energy metabolism, amino acid metabolism, fatty acid metabolism, bile acid biosynthesis, complement pathway, biosynthesis of steroid, and N-glycan biosynthesis. The pathway cluster C is more up-regulated in the kras-associated HCC subgroups, including glutamate metabolism and glycolysis. It has been reported that Kras could increase the conversion of glucose to glutamate, and this was essential for Kras-mediated tumorigenicity [31]. The pathway cluster D was generally up-regulated in the xmrk- and kras-correlated human HCCs, but showed a disparate pattern in the Myc-correlated human HCCs. This cluster contained many kinase pathways which were deregulated in the xmrk- and kras-induced zebrafish liver cancer, but not significantly changed in the Myc-induced zebrafish liver cancer. The pathway cluster E was quite heterogeneous. The pathway cluster F was well separated from all the others and it contained pathways which were mostly down-regulated in all the human HCC subgroups associated with zebrafish signatures, including hematopoietic cell lineage, cytokine pathway, calcium signaling, and GPCR pathway. Since most of these down-regulated pathways are involved in inflammatory response, it is likely that the subgroups of human HCCs not captured by the three zebrafish models have more severe inflammatory status.

The Commonly Up- and Down-regulated Genes in Zebrafish Liver Tumors are also Consistently Up- and Down-regulated in Human HCCs

While the three gene signatures represent up-regulation of distinct pathways and correlate with different subgroups of human HCCs, we sought to investigate whether the 21 commonly up-regulated and 16 commonly down-regulated genes in all three transgenic models would be similarly regulated in human HCCs. Among the ten human HCC datasets we used, we were able to examine two datasets which included both HCCs and their corresponding non-tumor tissues (GSE14520 and GSE25097). However, only GSE14520 could be compared with the zebrafish data as most of the human genes homologous to the common zebrafish genes could be identified from the microarray platform used: 19 of the 21 up-regulated genes (except FOXA3 and MRPS9) and 12 of the 16 down-regulated genes (except for cyp2ad2, slco1d1, TDH and MIOX). GSE14520 dataset contains 229 primary HCCs and the corresponding paired non-tumor hepatic tissues [32,33]. The xmrk, kras and Myc signatures were significantly correlated with 30.5%, 20.0%, and 20.5% of the HCC samples, and in total 47.6% of the HCC samples showed significant up-regulation of at least one of the zebrafish liver cancer signatures (Table 2). Among the 19 up-regulated genes, 16 of them were up-regulated in more than half of the HCC patients, and 9 of them were up-regulated in more than 75% of the patients (Figure 4). STMN1 was up-regulated in 96.9% of the human HCCs from the dataset, which is the most ubiquitously up-regulated. STT3A and SRP14 were up-regulated in 93.0% of the human HCCs. Among the 12 down-regulated genes, 11 of them were down-regulated in more than half of the HCC patients. FBP1 was down-regulated in 96.9% of the examined human HCCs. FBP1 is a gluconeogenesis regulatory enzyme and it functions to antagonize glycolysis in gastric cancer [34]. FBP1 is down-regulated in majority of human HCCs by methylation [35]. Restoration of FBP1 expression in human HCC

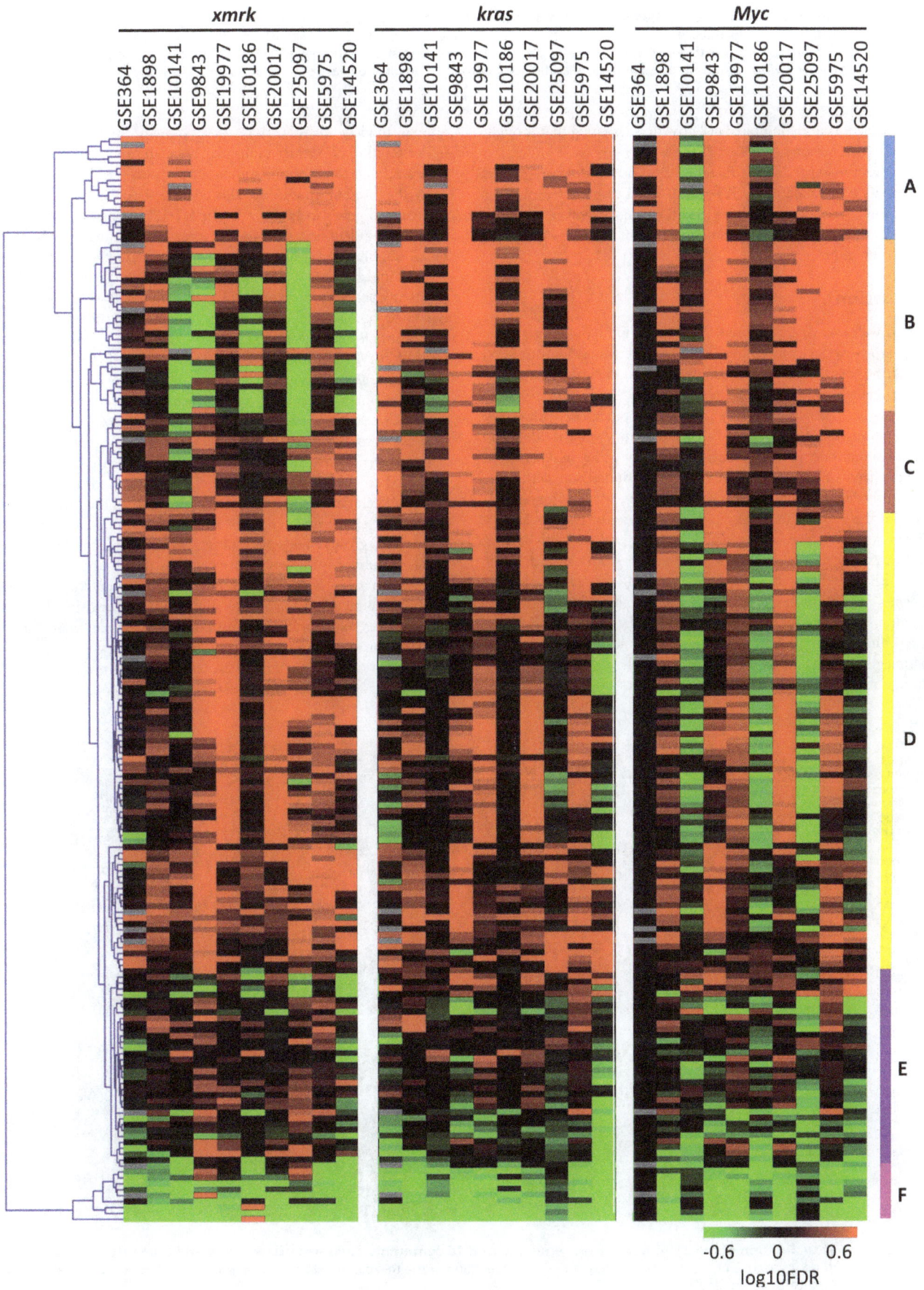

Figure 3. Hierarchical clustering of pathways differentially expressed in the subgroups of human HCCs significantly correlated with the three zebrafish tumor signatures. The color bars represent the logarithm-transformed FDR values. Up-regulated pathways are given positive values (red) and down-regulated pathways negative values (green). Pathways correlating with grey cells are not detected in the pathway analyses. Pathways with either FDR = 1 or not detected in more than five of the 30 combinations are pre-excluded from the analysis.

cell lines significantly inhibited cell growth, suggesting that it might function as a tumor suppressor [35].

Finally, the expression of these 19 up- and 16 down-regulated genes was also validated by RT-qPCR in all three different types of tumors. As shown in Figure S3, the majority of gene in the majority of tests (~90%) confirmed consistent trend of changes in the tumor samples.

Discussion

It is well known that human HCCs are highly heterogeneous; thus, cross-species comparative studies at the transcriptomic level should be valuable to identify conserved and critical pathways in carcinogenesis in vertebrate species. Here we first determined deregulated pathways from each of the oncogene transgenic zebrafish model. Although the three transgenic tumor models had quite distinct deregulated biological pathways, they all correlated to advanced or very advanced HCCs by comparison with gene signatures from human HCCs. Furthermore, we also found that each of the zebrafish model represent a subset of human HCCs. Since our oncogene transgenic lines have well-defined driving pathways in carcinogenesis, the information from the transgenic zebrafish should be valuable for understanding the main molecular mechanisms of each HCC subgroup, which is imperative for developing more effective therapeutics specific for each subgroup. Interestingly, our three oncogene transgenic zebrafish models significantly represent only less than half of the human HCCs and there is a need to develop more and different oncogene transgenic

animal models for covering more human HCCs for further understanding of distinct molecular mechanisms in hepatocarcinogenesis, with focus on inflammatory pathways.

In the present study, we also identified a list of commonly deregulated genes in liver tumors induced by different oncogenic signals was identified and their expressional changes in human HCC samples were validated *in silico* (Table 1, Figure 4). These genes could be served as potential therapeutic targets since they were independent of individual oncogenic pathways. The up-regulated genes includes those in protein translation and processing (*eif5a2, abce1, rrp9, srp14, itm1, srprb*), pro-apoptosis (*rhot1a, c20orf24, fam162a*) anti-apoptosis (*tp53*), cell cycle regulation (*stmn1a, cdkrap3*), purine synthesis (*gart*), rRNA processing (*noc4l*), G protein-coupled receptors signaling (*reep2*), n-glycan biosynthesis (*mgat4b*), stress response (*cirbp*), ubiquitination (*ubap2*), peroxisome (*hmgcra*), mitochondrial function (*mrps9*) and transcription (*foxa3*). Some of them have been implicated in hepatocarcinogenesis or identified as therapeutic targets. For example, *hmgcra*, a top up-regulated gene in all three tumor models, encodes the rate-limiting enzyme for cholesterol synthesis. Inhibition of Hmgcr could block tumor cell growth and metastasis [36], but clinical trials with Hmgcr inhibitor (pravastatin) have shown discrepant results [37–39], which may be attributed to the genetic heterogeneity of HCCs. The fact that *hmgcra* was highly up-regulated in all of the three transgenic zebrafish liver tumor models and these tumor models represent about half of human HCCs may suggest that it should be a therapeutic target in a broad, though not all, range of HCCs. Another top up-regulated gene, *mgat4b*, is one of

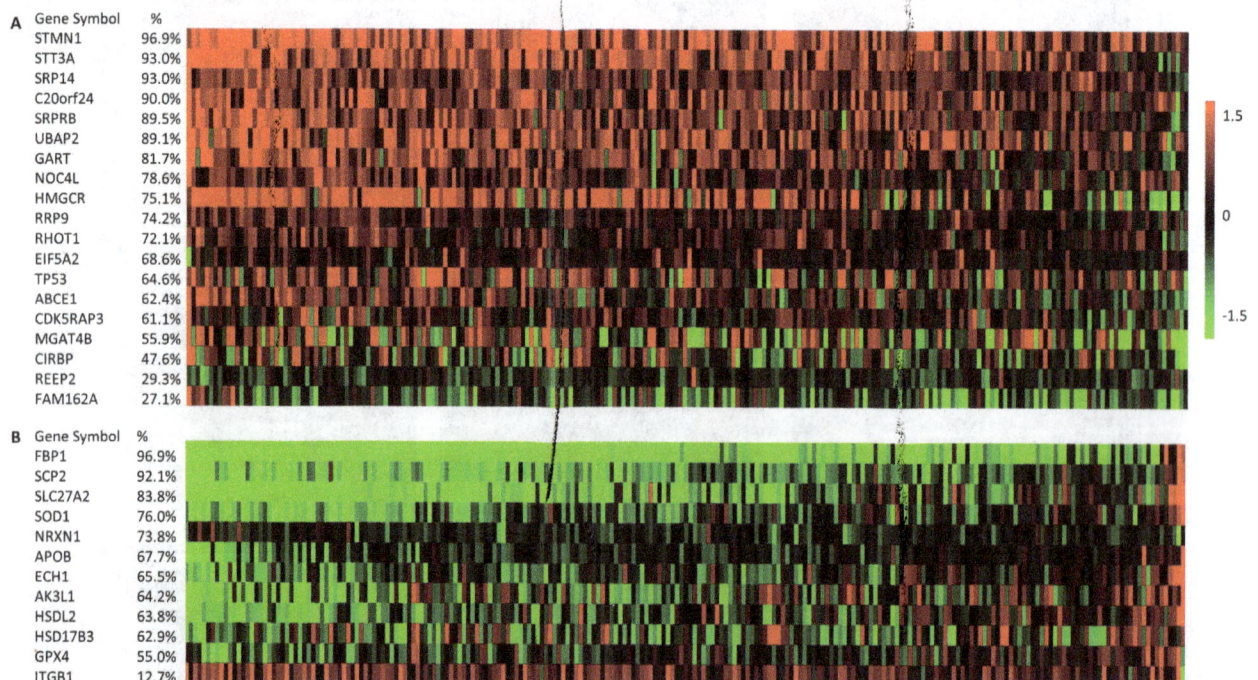

Figure 4. *In silico* validation of the 21 commonly up-regulated and 16 commonly down-regulated genes in human HCC dataset. 19 of the up-regulated genes and 12 of the down-regulated genes were identified in the microarray platform. The red color indicates up-regulation, and the green color indicates down-regulation.

the important enzymes in the biosynthetic pathway of N-glycans. N-glycan is up-regulated in human HCC [40] and is associated with drug resistance [41]. Another gene which was highly up-regulated is *stmn1a*. The human homolog of *stmn1a*, *STATHMIN1*, is over-expressed and is associated with polyploidy, metastasis, early recurrence, and poor prognosis in hepatocarcinogenesis [42–44]. It has also been identified as a major molecular target of an anticancer drug [45].

Most of the 16 genes commonly down-regulated (Table 1) are apparently involved in metabolism for normal liver function, including fatty acid metabolism (*cyp2ad2*, *hsdl2*, *ech1*), intracellular fatty acid and lipid transport (*scp2*, *slc27a2*, *apobl*), androgen metabolism (*hsd17b3*), glucose metabolism (*fbp1b*), mitochondria function (*tdh*, *ak4*), integrin complex (*itgb1b.2*), antioxidation (*gpx4a*, *sod1*), inositol catabolism (*miox*), bile acid metabolism (*slco1d1*) and neuron cell adhesion (*nrxn1b*). Several of them may have a direct connection with hepatocarcinogenesis. For example, *Sod1* deficient mice showed extensive cellular oxidative damage and majority of them developed HCC [46]. Moreover, SOD1 has also been markedly down-regulated in human HCCs induced from different etiological factors [47]. Gpx4 has also been reported to be associated with multiple types of cancers, including breast cancer [48], colorectal cancer [49], and aggressive prostate cancer [50]; however, no study has presented its correlation with HCC. Both Sod1 and Gpx4 are important components of the cellular antioxidant mechanisms. The down-regulation of these two genes in the three transgenic zebrafish liver cancer models suggested that oxidative damage is a common and conserved part of hepatocarcinogenesis.

Supporting Information

Figure S1 Induction of liver tumors in the three zebrafish transgenic liver cancer models. (A–D) Gross morphology of treated transgenic fish and the non-transgenic siblings. The treated non-transgenic siblings have normal liver size and gross morphology (A). The livers in the treated transgenic fish were obviously enlarged (B–D) compared to the treated non-transgenic siblings (A). (E–H) Histological examination of the treated transgenic fish and non-transgenic siblings. Treated *xmrk* fish developed HCC (F), treated *Myc* fish developed HCA (G), and treated *kras* fish developed heterogeneous HCC (H). The liver was circled out in white dotted lines. Scale bars: 25 mm for A–D, 50 μm for E–H.

Figure S2 Correlation of the three transgenic zebrafish liver cancer signatures with human HCC samples. The heat maps showed the positive- and negative-correlation of the three transgenic zebrafish liver cancer signatures with 9 sets of human HCC samples, including GSE364 (A), GSE1898 (B), GSE10141 (C), GSE9843 (D), GSE19977 (E), GSE10186 (F),

GSE20017 (G), GSE25097 (H) and GSE5975 (I). The color was determined by normalized enrichment score (NES) of the GSEA analysis. The red color indicates up-regulation or positive correlation, and the green color indicates down-regulation or negative correlation.

Figure S3 RT-qPCR validation of commonly up- and down-regulated genes in three transgenic zebrafish liver tumors. (A) Up-regulated genes. (B) Down-regulated genes. Gene names are indicated at the top and transgenic lines are indicaed on the left. RT-qPCR data are presented with red bars in comparison with corresponding RNA-Seq data represented by blue bars. Y-axes indicate fold changes on Log2 scale. Standard error bars are included for the RT-qPCR data and asterisks indicate statistically significance ($P<0.05$).

Table S1 Summary of RNA-SAGE data.

Table S2 Significantary up-regulated zebrafish genes with mapped human homologs.

Table S3 Summary of human HCC datasets used in the present Study.

Table S4 Sequences of PCR primers used for RT-qPCR validation of commonly up- and down-regulated genes in zebrafish liver tumors.

Table S5 Details of pathways deregulated in the three transgenic zebrafish liver cancer models as classified into the seven cancer hallmarks and different aspects of the liver metabolisms.

Table S6 Details of pathways differentially expressed in the subgroups of human HCCs which showed significantly correlation with the three zebrafish signatures.

Acknowledgments

We thank Dr. Serguei Parinov for helping in the works on generation of *kras* transgenic zebrafish.

Author Contributions

Conceived and designed the experiments: ZL ATN AE ZG. Performed the experiments: ZL ATN CL. Analyzed the data: WZ ZG. Contributed reagents/materials/analysis tools: WZ ZL ATN AE. Wrote the paper: WZ ZG.

References

1. Llovet JM, Bustamante J, Castells A, Vilana R, Ayuso Mdel C, et al. (1999) Natural history of untreated nonsurgical hepatocellular carcinoma: rationale for the design and evaluation of therapeutic trials. Hepatology 29: 62–67.

2. Siegel AB, Olsen SK, Magun A, Brown RS Jr (2010) Sorafenib: where do we go from here? Hepatology 52: 360–369.

3. Leenders MW, Nijkamp MW, Borel Rinkes IH (2008) Mouse models in liver cancer research: a review of current literature. World J Gastroenterol 14: 6915–6923.

4. Lee JS, Chu IS, Mikaelyan A, Calvisi DF, Heo J, et al. (2004) Application of comparative functional genomics to identify best-fit mouse models to study human cancer. Nat Genet 36: 1306–1311.

5. Liu S, Leach SD (2011) Zebrafish models for cancer. Annu Rev Pathol 6: 71–93.

6. Li Z, Zheng W, Wang Z, Zeng Z, Zhan H, et al. (2012) An inducible Myc zebrafish liver tumor model revealed conserved Myc signatures with mammalian liver tumors. Dis Model Mech.

7. Nguyen AT, Emelyanov A, Koh CH, Spitsbergen JM, Parinov S, et al. (2012) An inducible kras(V12) transgenic zebrafish model for liver tumorigenesis and chemical drug screening. Dis Model Mech 5: 63–72.

8. Nguyen AT, Emelyanov A, Koh CH, Spitsbergen JM, Lam SH, et al. (2011) A high level of liver-specific expression of oncogenic Kras(V12) drives robust liver tumorigenesis in transgenic zebrafish. Dis Model Mech 4: 801–813.

9. Li Z, Huang X, Zhan H, Zeng Z, Li C, et al. (2011) Inducible and repressable oncogene-addicted hepatocellular carcinoma in Tet-on xmrk transgenic zebrafish. J Hepatol 56: 419–425.

10. Lu JW, Hsia Y, Tu HC, Hsiao YC, Yang WY, et al. (2011) Liver development and cancer formation in zebrafish. Birth Defects Res C Embryo Today 93: 157–172.

11. Liu W, Chen JR, Hsu CH, Li YH, Chen YM, et al. (2012) A zebrafish model of intrahepatic cholangiocarcinoma by dual expression of hepatitis B virus X and hepatitis C virus core protein in liver. Hepatology 56: 2268–2276.

12. Amatruda JF, Shepard JL, Stern HM, Zon LI (2002) Zebrafish as a cancer model system. Cancer Cell 1: 229–231.

13. Spitsbergen JM, Tsai HW, Reddy A, Miller T, Arbogast D, et al. (2000) Neoplasia in zebrafish (Danio rerio) treated with 7,12-dimethylbenz[a]anthracene by two exposure routes at different developmental stages. Toxicol Pathol 28: 705–715.

14. Mirbahai L, Williams TD, Zhan H, Gong Z, Chipman JK (2011) Comprehensive profiling of zebrafish hepatic proximal promoter CpG island methylation and its modification during chemical carcinogenesis. BMC Genomics 12: 3.

15. Lam SH, Gong Z (2006) Modeling liver cancer using zebrafish: a comparative oncogenomics approach. Cell Cycle 5: 573–577.

16. Lam SH, Wu YL, Vega VB, Miller LD, Spitsbergen J, et al. (2006) Conservation of gene expression signatures between zebrafish and human liver tumors and tumor progression. Nat Biotechnol 24: 73–75.

17. Karnoub AE, Weinberg RA (2008) Ras oncogenes: split personalities. Nat Rev Mol Cell Biol 9: 517–531.

18. Calvisi DF, Ladu S, Gorden A, Farina M, Conner EA, et al. (2006) Ubiquitous activation of Ras and Jak/Stat pathways in human HCC. Gastroenterology 130: 1117–1128.

19. Gomez A, Wellbrock C, Gutbrod H, Dimitrijevic N, Schartl M (2001) Ligand-independent dimerization and activation of the oncogenic Xmrk receptor by two mutations in the extracellular domain. J Biol Chem 276: 3333–3340.

20. Foster J, Black J, LeVea C, Khoury T, Kuvshinoff B, et al. (2007) COX-2 expression in hepatocellular carcinoma is an initiation event; while EGF receptor expression with downstream pathway activation is a prognostic predictor of survival. Ann Surg Oncol 14: 752–758.

21. Gan FY, Gesell MS, Alousi M, Luk GD (1993) Analysis of ODC and c-myc gene expression in hepatocellular carcinoma by in situ hybridization and immuno-histochemistry. J Histochem Cytochem 41: 1185–1196.

22. Sweet-Cordero A, Mukherjee S, Subramanian A, You H, Roix JJ, et al. (2005) An oncogenic KRAS2 expression signature identified by cross-species gene-expression analysis. Nat Genet 37: 48–55.

23. Zheng W, Wang Z, Collins JE, Andrews RM, Stemple D, et al. (2011) Comparative transcriptome analyses indicate molecular homology of zebrafish swimbladder and mammalian lung. PLoS One 6: e24019.

24. Subramanian A, Tamayo P, Mootha VK, Mukherjee S, Ebert BL, et al. (2005) Gene set enrichment analysis: a knowledge-based approach for interpreting genome-wide expression profiles. Proc Natl Acad Sci U S A 102: 15545–15550.

25. Wurmbach E, Chen YB, Khitrov G, Zhang W, Roayaie S, et al. (2007) Genome-wide molecular profiles of HCV-induced dysplasia and hepatocellular carcinoma. Hepatology 45: 938–947.

26. Lee JS, Grisham JW, Thorgeirsson SS (2005) Comparative functional genomics for identifying models of human cancer. Carcinogenesis 26: 1013–1020.

27. Hanahan D, Weinberg RA (2011) Hallmarks of cancer: the next generation. Cell 144: 646–674.

28. Hanahan D, Weinberg RA (2000) The hallmarks of cancer. Cell 100: 57–70.

29. Saeed AI, Sharov V, White J, Li J, Liang W, et al. (2003) TM4: a free, open-source system for microarray data management and analysis. Biotechniques 34: 374–378.

30. Saeed AI, Bhagabati NK, Braisted JC, Liang W, Sharov V, et al. (2006) TM4 microarray software suite. Methods Enzymol 411: 134–193.

31. Weinberg F, Hamanaka R, Wheaton WW, Weinberg S, Joseph J, et al. (2010) Mitochondrial metabolism and ROS generation are essential for Kras-mediated tumorigenicity. Proc Natl Acad Sci U S A 107: 8788–8793.

32. Roessler S, Long EL, Budhu A, Chen Y, Zhao X, et al. (2012) Integrative genomic identification of genes on 8p associated with hepatocellular carcinoma progression and patient survival. Gastroenterology 142: 957–966 e912.

33. Roessler S, Jia HL, Budhu A, Forgues M, Ye QH, et al. (2010) A unique metastasis gene signature enables prediction of tumor relapse in early-stage hepatocellular carcinoma patients. Cancer Res 70: 10202–10212.

34. Liu X, Wang X, Zhang J, Lam EK, Shin VY, et al. (2010) Warburg effect revisited: an epigenetic link between glycolysis and gastric carcinogenesis. Oncogene 29: 442–450.

35. Chen M, Zhang J, Li N, Qian Z, Zhu M, et al. (2011) Promoter hypermethylation mediated downregulation of FBP1 in human hepatocellular carcinoma and colon cancer. PLoS One 6: e25564.

36. Cao Z, Fan-Minogue H, Bellovin DI, Yevtodiyenko A, Arzeno J, et al. (2011) MYC phosphorylation, activation, and tumorigenic potential in hepatocellular carcinoma are regulated by HMG-CoA reductase. Cancer Res 71: 2286–2297.

37. Lersch C, Schmelz R, Erdmann J, Hollweck R, Schulte-Frohlinde E, et al. (2004) Treatment of HCC with pravastatin, octreotide, or gemcitabine–a critical evaluation. Hepatogastroenterology 51: 1099–1103.

38. Graf H, Jungst C, Straub G, Dogan S, Hoffmann RT, et al. (2008) Chemoembolization combined with pravastatin improves survival in patients with hepatocellular carcinoma. Digestion 78: 34–38.

39. Kawata S, Yamasaki E, Nagase T, Inui Y, Ito N, et al. (2001) Effect of pravastatin on survival in patients with advanced hepatocellular carcinoma. A randomized controlled trial. Br J Cancer 84: 886–891.

40. Goldman R, Ressom HW, Varghese RS, Goldman L, Bascug G, et al. (2009) Detection of hepatocellular carcinoma using glycomic analysis. Clin Cancer Res 15: 1808–1813.

41. Kudo T, Nakagawa H, Takahashi M, Hamaguchi J, Kamiyama N, et al. (2007) N-glycan alterations are associated with drug resistance in human hepatocellular carcinoma. Mol Cancer 6: 32.

42. Hsieh SY, Huang SF, Yu MC, Yeh TS, Chen TC, et al. (2010) Stathmin1 overexpression associated with polyploidy, tumor-cell invasion, early recurrence, and poor prognosis in human hepatoma. Mol Carcinog 49: 476–487.

43. Gan L, Guo K, Li Y, Kang X, Sun L, et al. (2010) Up-regulated expression of stathmin may be associated with hepatocarcinogenesis. Oncol Rep 23: 1037–1043.

44. Yuan RH, Jeng YM, Chen HL, Lai PL, Pan HW, et al. (2006) Stathmin overexpression cooperates with p53 mutation and osteopontin overexpression, and is associated with tumour progression, early recurrence, and poor prognosis in hepatocellular carcinoma. J Pathol 209: 549–558.

45. Wang X, Chen Y, Han QB, Chan CY, Wang H, et al. (2009) Proteomic identification of molecular targets of gambogic acid: role of stathmin in hepatocellular carcinoma. Proteomics 9: 242–253.

46. Elchuri S, Oberley TD, Qi W, Eisenstein RS, Jackson Roberts L, et al. (2005) CuZnSOD deficiency leads to persistent and widespread oxidative damage and hepatocarcinogenesis later in life. Oncogene 24: 367–380.

47. Li Y, Wan D, Wei W, Su J, Cao J, et al. (2008) Candidate genes responsible for human hepatocellular carcinoma identified from differentially expressed genes in hepatocarcinogenesis of the tree shrew (Tupaia belangeri chinesis). Hepatol Res 38: 85–95.

48. Mavaddat N, Dunning AM, Ponder BA, Easton DF, Pharoah PD (2009) Common genetic variation in candidate genes and susceptibility to subtypes of breast cancer. Cancer Epidemiol Biomarkers Prev 18: 255–259.

49. Meplan C, Hughes DJ, Pardini B, Naccarati A, Soucek P, et al. (2010) Genetic variants in selenoprotein genes increase risk of colorectal cancer. Carcinogenesis 31: 1074–1079.

50. Abe M, Xie W, Regan MM, King IB, Stampfer MJ, et al. (2011) Single-nucleotide polymorphisms within the antioxidant defence system and associations with aggressive prostate cancer. BJU Int 107: 126–134.

16

Hypoxia Inducible Factor 3α Plays a Critical Role in Alveolarization and Distal Epithelial Cell Differentiation during Mouse Lung Development

Yadi Huang[1], Joshua Kapere Ochieng[1], Marjon Buscop-van Kempen[1], Anne Boerema-de Munck[1], Sigrid Swagemakers[2,3], Wilfred van IJcken[4], Frank Grosveld[5], Dick Tibboel[1], Robbert J. Rottier[1,5]*

1 Department of Pediatric Surgery, Erasmus MC-Sophia Children's Hospital, Rotterdam, The Netherlands, **2** Department of Bioinformatics, Erasmus MC, Rotterdam, The Netherlands, **3** Department of Genetics, Erasmus MC, Rotterdam, The Netherlands, **4** Department of Biomics, Erasmus MC, Rotterdam, The Netherlands, **5** Department of Cell Biology, Erasmus MC, Rotterdam, The Netherlands

Abstract

Lung development occurs under relative hypoxia and the most important oxygen-sensitive response pathway is driven by Hypoxia Inducible Factors (HIF). HIFs are heterodimeric transcription factors of an oxygen-sensitive subunit, HIFα, and a constitutively expressed subunit, HIF1β. HIF1α and HIF2α, encoded by two separate genes, contribute to the activation of hypoxia inducible genes. A third HIFα gene, *HIF3α*, is subject to alternative promoter usage and splicing, leading to three major isoforms, HIF3α, NEPAS and IPAS. HIF3α gene products add to the complexity of the hypoxia response as they function as dominant negative inhibitors (IPAS) or weak transcriptional activators (HIF3α/NEPAS). Previously, we and others have shown the importance of the Hif1α and Hif2α factors in lung development, and here we investigated the role of Hif3α during pulmonary development. Therefore, HIF3α was conditionally expressed in airway epithelial cells during gestation and although HIF3α transgenic mice were born alive and appeared normal, their lungs showed clear abnormalities, including a post-pseudoglandular branching defect and a decreased number of alveoli. The HIF3α expressing lungs displayed reduced numbers of Clara cells, alveolar epithelial type I and type II cells. As a result of HIF3α expression, the level of Hif2α was reduced, but that of Hif1α was not affected. Two regulatory genes, Rarβ, involved in alveologenesis, and Foxp2, a transcriptional repressor of the Clara cell specific Ccsp gene, were significantly upregulated in the HIF3α expressing lungs. In addition, aberrant basal cells were observed distally as determined by the expression of Sox2 and p63. We show that Hif3α binds a conserved HRE site in the Sox2 promoter and weakly transactivated a reporter construct containing the Sox2 promoter region. Moreover, Hif3α affected the expression of genes not typically involved in the hypoxia response, providing evidence for a novel function of Hif3α beyond the hypoxia response.

Editor: Oliver Eickelberg, Helmholtz Zentrum München/Ludwig-Maximilians-University Munich, Germany

Funding: This work was supported in part by the Sophia Foundation for Medical Research, SSWO nr. 482 (YH). No additional external funding was received for this study. The funders had no role in study design, data collection and analysis, decision to publish, or preparation of the manuscript.

Competing Interests: The authors have declared that no competing interests exist.

* E-mail: r.rottier@erasmusmc.nl

Introduction

The lung originates from the primitive foregut early in the development of land dwelling organisms, and through a complex interplay of signaling molecules the future airway epithelium and surrounding mesenchyme develop into the highly structured arbor-like bronchial-vascular tree (reviewed in [1,2,3]). Normal development in mammals occurs in a relative hypoxic environment, which is beneficial for lung organogenesis [4,5]. Cellular responses to different levels of oxygen are important for development and homeostasis [6], and the most important oxygen-sensing mechanism to protect cells from oxygen toxicity is the transcriptional response mediated by Hypoxia Inducible Factors (HIF), which are also expressed in the lungs [7].

HIFs are critical mediators of the hypoxic cellular response and regulate cellular adaptation by transactivating genes involved in angiogenesis, metabolism and cellular homeostasis (for recent reviews see [6,8,9]). HIFs are heterodimeric transcription factors which have two structurally related subunits, an oxygen sensitive

HIFα subunit and a constitutively expressed HIFß or ARNT subunit (Aryl hydrocarbon Receptor Nuclear Translocator). Both subunits belong to the transcription factor family containing a basic Helix-Loop-Helix (bHLH) and a Per/ARNT/Sim (PAS) domain at the N-terminus, which mediate heterodimerization and DNA binding [10,11]. HIFß is expressed ubiquitously and as such, the level and expression patterns of the HIFα proteins are mostly determining the activity of the heterodimers [12]. Currently, three genes have been identified in human and mouse that encode HIFα isoforms, *HIF1α* [10,11], *HIF2α* or *EPAS1* [13,14,15], and *HIF3α* [16,17,18,19,20]. Aside from the N-terminal bHLH/PAS domain, the HIFα subunits contain an Oxygen-Dependent Degradation Domain (ODDD) in the center of the protein, an N-terminal transactivation domain (NTAD) and a C-terminal transactivation domain (CTAD) [21,22,23,24,25]. The CTAD is absent in the HIF3α subunit, which significantly reduces the transcriptional activity of the protein [16,26]. The three α subunits are post-transcriptionally regulated by prolyl hydroxylase domain-containing enzymes (PHD1-3), which hydroxylate with different specificity the HIFα subunits at two critical prolyl residues in the ODDD under normoxic conditions [22,27]. The PHD proteins are dioxygenases which require oxygen for their function and as such are sensitive to oxygen concentrations, losing their activity under low oxygen concentration [22]. The hydroxylated HIFα proteins are poly-ubiquitinylated and targeted for 26S proteosomal degradation through the von Hippel-Lindau (pVHL)/Elongin BC/Cul2 ubiquitin-ligase complex [28,29,30,31,32,33,34]. Under low oxygen conditions, the PHD proteins are inactive, so the HIFα proteins are not hydroxylated and stable. They will translocate to the nucleus and dimerize with HIF1β, leading to the transcription of target genes, such as EPO and VEGF, through the binding to specific DNA seqences (Hypoxia Responsive Elements, HRE) [8,35,36]. Aside from the regulation of the stability of the HIFα isoforms by PHDs, additional regulatory activities are identified. The oxygen-dependent asparaginyl hydroxylase Factor Inhibiting HIF (FIH), a member of the Fe(II) and 2-oxoglutarate-dependent dioxygenase, hydroxylates a conserved asparaginyl residue in the CTAD, preventing the association of HIFα with the p300 coactivator [37,38,39]. In addition to these hydroxylation dependent regulation of HIFα isoforms, several other posttranslational modifications have been identified (for review, see [8,40,41]).

The regulation and functions of the HIF3α gene and isoforms is very complex, contrasting HIF1α and HIF2α. The *HIF3α* locus gives rise to different splice variants, resulting in three protein isoforms, HIF3α, NEPAS (neonatal and embryonic PAS) and IPAS (inhibitory PAS) [19,20,42]. HIF3α and NEPAS only differ in the first eight N-terminal amino acids due to alternative exon usage. IPAS and NEPAS are hypoxia inducible, whereas HIF3α is not because of alternative usage of promoters [43,44]. HIF3α expression is induced under hypoxia in several organs, including cortex, hippocampus, lung, heart, kidney, cerebral cortex [17,45,46]. NEPAS is almost exclusively expressed during late embryonic and neonatal stages of development, especially in the lung and heart, while HIF3α mRNA is rarely detectable during embryonic and neonatal stages [42]. HIF3α has a high homology to HIF1α and HIF2α at the N-terminus, but only a low degree of sequence similarity across the C-terminus [26]. The HIF3α/HIF1ß (HIF3) and NEPAS/HIF1ß dimers suppress basal and hypoxia induced reporter gene activation, as well as HIF1 (HIF1α/HIF1ß) or HIF2 (HIF2α/HIF1ß) driven expression [16,42]. HIF3 binds to HRE sites in promoter regions, but the transcriptional activity is much weaker than that of HIF1 and HIF2, because it lacks the CTAD [16,26,42]. Therefore, both HIF3α and NEPAS serve as competitors of HIF1 and HIF2 dependent transcription, not only by occupying identical promoter regions, but also by associating with the same HIF1ß partner [16,42]. The splice variant IPAS lacks both the NTAD and CTAD domains producing a dominant negative regulator of the HIF1α and HIF2α dependent pathway [16,18,43]. It was shown that IPAS directly associates with HIFα isoforms, thereby displacing Hif1β, and the resulting IPAS/Hifα dimer is unable to bind to DNA [18]. Both short HIF3α isoforms related to IPAS in human and the IPAS in mouse have antagonistic effects on the expression of HIF1 and HIF2 dependent hypoxia regulated target genes [47]. Thus, the *HIF3α* locus encodes isoforms generally thought to act as negative regulators of the hypoxic response.

The importance of the hypoxia response was shown by the identification of mutations in the VHL-HIF pathway in different human diseases (reviewed in [9]). Specific gene ablation studies in mice also added to the knowledge on the pleiotropic effects of the members of the hypoxia response pathway. Complete ablation of this pathway through inactivation of Hif1ß resulted in a severe lethal phenotype with defective angiogenesis of the yolk sac and branchial arches, stunted development and embryo wasting [48,49]. Hif1α knockout mice also died early during development with cardiac malformations and vascular defects [50]. Hif2α null mice displayed a pleiotropic phenotype ranging from premature death until postnatal abnormalities, depending on the background of the mouse strain [51,52,53,54]. The neonates that survived suffered from breathing problems and did not produce sufficient surfactant phospholipids and surfactant associated proteins [51]. It is interesting to note that the inactivation and ectopic activation of Hif2α showed comparable phenotypes, suggesting that type II cells require different levels of Hif2α at distinct phases of type II cell maturation [51,55]. Homozygous mutant NEPAS/Hif3α$^{-/-}$ mice were alive at birth, but displayed enlarged right ventricle and impaired lung remodelling, suggesting that NEPAS/Hif3α is important in lung and heart development during embryonic and neonatal stages [42]. Interestingly, the *Hif3α* gene contains hypoxia response elements in its promoter region and has been shown to be a transcriptional target of Hif1α [56].

In order to understand the precise role of Hif3α during pulmonary epithelium development, we generated transgenic mice with an inducible *HIF3α* gene. Mice expressing the *HIF3α* transgene in the developing airways showed a post-pseudoglandular branching defect with a reduced number of airspaces and a clear reduction in the number of alveolar type I and type II cells. Importantly, expression of the HIF3α transgene did not lead to changes in the levels of Hif1α, but affected Hif2α. The lungs of the HIF3α expressing mice showed an upregulation of genes normally expressed in the proximal parts of the lung, while genes only expressed in distal parts of the lung were downregulated. Specifically, Foxp2, a repressor of distal cell markers, and Rarß were induced in the lungs of Hif3α expressing mice, which may explain the reduction in the number of distal cell types. Furthermore, we showed that Hif3α binds a conserved HRE in the Sox2 promoter and induces the expression of a Sox2 promoter driven reporter gene, explaining the appearance of aberrant Sox2- and p63 positive cells. Collectively, our results show that Hif3α is involved in modulating correct development of the lung epithelium.

Materials and Methods

Generation of transgenic animal

The myc epitope encoding sequence (EQKLISEEDL) was cloned directly after the endogenous ATG start codon of the full length human HIF3α cDNA (GenBank: BC080551) and subcloned into a modified pTRE-Tight vector [55]. Transgenic lines were produced by pronuclear injection of FVB/N fertilized eggs, and tail tip DNA of transgenic lines was initially genotyped by Southern blot analysis, after which positive lines were routinely checked by PCR, using transgene-specific primers (sense: 5'-GTCAAGCTTATGGCGCTGGGGCTGCAGCG; antisense 5'-GCATCTAGATCAGTCAGCCTGGGCTGAGC). Three independent lines were initially analyzed, which all produced the same phenotype as described in this manuscript. Lung-specific expression of the HIF3α transgene, i-Tg-mycHIF3α, was obtained by crossing the mycHIF3α lines with the SPC-rtTA transgenic mice (A generous gift of Jeffrey Whitsett). Administration of doxycycline to pregnant mothers from gestational age 6.5 onward in the drinking water (2 mg/ml, 5% sucrose) resulted in lung epithelium-specific expression. Each experiment was performed with at least three independent litters containing double transgenic, single transgenic and wild type pups. All double transgenic animals receiving doxycycline expressed mycHIF3α in the pulmonary epithelium and showed the described phenotype. Mice were housed under standard conditions at 40–50% relative humidity and 21±1°C (12/12 hour dark/light cycle) with food and water ad libitum. All animal experiments were performed according to the Dutch and European guidelines and approved by the local ethics committee (DEC Nr 1657, 1833 and 2206).

Immunohistochemistry

Immunohistochemistry was essentially performed as previously described [57]. Briefly, lungs were dissected and fixed in formal saline (BDH) overnight at 4°C before processing for paraffin embedding according to routine protocols Antigen retrieval was performed with microwave treatment in 10 mM citric acid buffer pH 6.0 or Tris-EDTA. Sections were blocked with 5% BSA or 5% ELK in PBS for 10min and incubated with primary antibody diluted in 5% BSA or 5% ELK in PBS overnight at 4. The following antibodies were used: Myc (9E10, Roche; 4A6, Millipore), Hif3α (Ab2165, Abcam; NBP1-03155, Novus Biologicals), ß-tubulin IV (bioGenex), proSP-C (Chemicon), T1α (University of Iowa Hybridoma bank), Ttf1 (Thermo), Ccsp (seven hills), Sox2 (seven hills), Foxp2 (Abcam), Lpcat1 (Seven hills Bioreagents), α-Sma (Thermo), Ki67, cGRP Secondary antibodies against the correct IgG species were conjugated with peroxidase (Dako).

Lungs were imaged using an Olympus BX41 microscope and DP71 camera (Olympus, Zoeterwoude, The Netherlands). Subsequent airspaces counting were performed with SIS Softward Cell D (Olympus). Three independent samples of control and double-transgenic lungs of gestational age E18.5 were used to count the number of airspaces on a selected surface area (140000 μm^2) on those selected lung samples.

Microarray analysis

Lungs of three control and three double transgenic embryos were dissected at E18.5 and the middle and caudal lobes were used for total RNA isolation with Trizol reagent according to the manufacturer's instructions (Invitrogen life technologies, Carlsbad, CA, USA). RNA was purified using the RNeasy MinElute Cleanup kit. (Qiagen, Valencia, CA, USA) and cDNA was synthesized from 3 μg RNA using the GeneChip Expression 3'-Amplification Reagents One-Cycle cDNA Synthesis kit (Affymetrix, Santa Clara, CA, USA). Biotin-labelled cRNA synthesis, purification and fragmentation were performed according to standard conditions. Fragmented biotinylated cRNA was subsequently hybridized onto Affymetrix Mouse Genome 430 2.0 microarray chips. After normalization, the data were analysed with OmniViz software, version 3.6.0 (Omniviz, Inc., Maynard, MA, USA).

Functional annotation of the statistical analysis of microarrays results was done using Ingenuity Pathway Analysis (Ingenuity, Mountain View, CA) and DAVID (http://david.abcc.ncifcrf.gov). The results are shown for biological processes, which are significantly (P <0.05) enriched after multiple testing.

RT-PCR

RNA isolation and subsequent quantitative PCR analysis was essentially performed as previously described [7]. Gene-specific primer sets were Abca3: 5'-TTACGGTCCAAGTTCCTGAG-3' and 5'-TAACATCAGCACCTTAGAGCC-3'; Aqp5: 5'-GTGGTCATGAATCGGTTCAG-3' and 5'-CAAGTAGAAG-TAGAGGATTGCAG-3'; Epas1: 5'-CTGTGACGACA-GAATCTTGG-3' and 5'-GGCATGGTAGAACTCATAGG-3'; Foxp2: 5'-TGTCATCAGAGATTGCCC-3' and 5'-ATAGCCTGCCTTATGAGTG-3'; Rarß: 5'-AACTGCGT-

CATTAACAAGGTC-3' and 5'-TCATTCCTAACA-GACTCTTTGG-3'; Scd1: 5'-GAGCCACAGAACTTA-CAAGG-3' and 5'-GTACACGTCATTCTGGAACG-3'; Sftpd: 5'-GGAAGCAATCTGACATGCTG-3' and 5'-GAGGCTCTT-CATTTCTGCTC-3'. Standard deviations of the duplicates are calculated with the SPSS program (Independent-samples T test), which also generated the P values.

Luciferase reporter activity assays

HEK293T cells were transfected in duplo with Lipofectamine 2000 (Invitrogen) with a total concentration of 500 ng DNA/well, using 9*HREluc (Gift from Manuel Landazuri), pGL3-mpSox2 and pGL3-mpSox2delta (Named Sox2-Luc and ΔSox2-Luc; Gift from Victoria Moreno), Hif2α-pcDNA3, (gift from Carole Peyssonnaux), Hif3α-pcDNA3 or pcDNA3. Cells were lysed with passive lysis buffer (Promega) 24-hours after transfection and processed for lucifease analysis by the addition of the LARII reagent (Promega), which was subsequently quantified with the VICTOR luminometer. A construct containing the renilla gene (10 ng/well) was co-transfected in each well to serve as an internal control for transfection efficiency. The renila luciferase activity was quantified by addition of Stop&Glio reagent and also detected with the VICTOR luminometer. The experiment was repeated three times, and all samples were measured at least in duplo. The average luciferase activity was calculated and divided by the average of renilla activity. Standard deviations were measured with the SPSS program (Independent-samples T test), which also generated the P values.

Chromatin immunoprecipitation (ChIP)

ChIP assay was performed essentially as previously described [58], with some modifications. Chromatin-protein complexes of confluent A549 cells were fixed by adding 1% formaldehyde to the cultures. Nuclear extracts were made and chromosomal DNA was fragmented by sonication. Equal amounts of DNA was diluted 1:10 with ChIP dilution Buffer (0.01% SDS, 1.1% Triton X-100, 1.2 mM EDTA, 16.7 mM Tris-HCl pH 8.1 and 167 mM NaCl) and the samples were pre-cleared with 80 μl prot A/G agarose beads for 1 hour, after which the sample was split in equal volumes and incubated O/N with 6 μg antibody specific for HIF3α (NBP1-03155) or control IgG (rabbit). Immune complexes were subsequently purified by adding 80 μl of prot A/G beads, which were washed several times before the immune-precipitated DNA was eluted with elution buffer (1% SDS and 0.1 m NaHCO$_3$). After de-crosslinking the DNA-protein complexes by incubation at 65°C O/N with 200 mM NaCl, the eluted DNA was phenol-extracted, precipitated and qPCRs were performed to analyze the enrichment of HIF3α specific binding to the HRE in the SOX2 gene using the following primer set 5'-CAAGTGCATTTTAGC-CACAAAG-3' and 5'-CCCAAGAGGGTAATTTTAGCCG-3', while the primers for the ARRDC3 and EGLN3-D were described previously [36,59]. The data are the average of two independent ChIP assays, which were each analyzed by duplicate qPCRs, and are represented as the fold enrichment of the specific immune precipitation compared to the control IgG precipitation.

Results

Ectopic expression of mycHIF3α causes late branching defects

Previously, it was shown that homozygous NEPAS/Hif3α knockout mice were viable, but displayed an enlarged right ventricle and impaired lung remodelling, suggesting that Hif3α plays an important role during pulmonary development. However, the precise role of Hif3α during the formation of the lung is not fully understood. We first analyzed the endogenous expression of Hif3α in normal fetal lungs isolated at the end of gestation (E18.5) and in lungs of adult mice (8 weeks). Hif3α positive cells were present in the epithelium of the developing lung, as well as in the type II pneumocytes of the adult lung (arrows in Figure 1A, B). In order to determine the precise role of Hif3α in the epithelium during lung development, and more specifically in type II pneumocytes, we generated transgenic mice carrying a myc-epitope tagged HIF3α under the control of a doxycycline-inducible tet-on promoter (i-Tg-mycHif3α; Figure 1C). Expression of mycHIF3α in embryonic lung epithelium was established by crossing the i-Tg-mycHIF3α transgenic line with the established SPC-rtTA line, which drives the expression of the rtTA gene in epithelial cells of the embryonic lung [60]. Pregnant females from timed matings between SPC-rtTA and i-Tg-mycHif3α mice received doxycycline to induce the expression of the HIF3α transgene in double-transgenic fetuses. Lungs isolated from doxycycline-induced or non-induced single i-Tg-mycHif3α or SPC-rtTA transgenic mice, or lungs from non-induced double transgenic i-Tg-mycHif3α/SPC-rtTA animals appeared indistinguishable from wild type lungs. Doxycycline-induced, double-transgenic pups were born at Mendelian ratio and did not show obvious external abnormalities compared to their control litter mates.

In order to determine whether expression of mycHIF3α leads to pulmonary development defects, we analyzed lungs of double-transgenic animals and control lungs at different gestational ages. Macroscopic analysis of isolated lungs did not show clear abnormalities in double-transgenic animals at gestational ages E16.5, E17.5, E18.5 days and postnatal day 1 (PN1) (Figures 1E and F, I and J; Figure S1). Histological examinations at E16.5 did not show clear differences between control and mycHIF3α transgenic lungs (Figures S1C and D). However analysis of a series of developmental ages clearly showed aberrant alveolar airspaces in mycHIF3α expressing lungs starting at E17.5 compared to controls (Figure S1G, H). mycHIF3α expressing lungs contained significant fewer alveolar spaces compared to control ones at E18.5 and PN1 (Figures 1D). Staining with a specific antibody against the myc-epitope confirmed the expression of transgenic mycHIF3α protein in the epithelium of double-transgenic lungs (Figures 1H and L, Figure S1). The abnormal alveolar spaces remain present in the PN1 stages, but apparently, the mice do not suffer from respiratory distress, indicating that the initial requirements for life are present. So, we conclude that mycHIFα expression in epithelial cells leads to aberrant alveolar formation and affects late branching morphogenesis during pulmonary development.

This post-pseudoglandular branching defect prompted us to analyze the expression of the mycHIF3a at early embryonic stages of development. This showed that the transgene is expressed in a non-uniform manner in the epithelium of early E11.5 lungs (Figure 2A), but gradually all epithelial cells express the transgene (Figure 2B-D). Next, we analyzed whether the primary airway branches appropriately expressed some of the major branch-inducing genes [2]. Therefore, embryonic lungs of controls and double transgenic animals were isolated at gestational age 12.5. At this stage of development, the primary bronchi are already present, and these branches start to form secondary and tertiary branches. The expression of Fgf10, the growth factor with a very potent branch-inducing activity, was found in the mesenchymal compartment, alongside the epithelium that is in the process of branching (Figure 2E and I, arrows). Moreover, its receptor, FgfR2, was detected at the tips of the epithelium, in close proximity

of the Fgf10 signal (Figure 2F and J). Next, we also analyzed the expression of two genes known to be induced as a result of the Fgf10-FgfR2 signalling, Shh and Bmp4. Both genes were also expressed in the epithelium at the same location as the FgfR2, indicating that the Fgf10-FgfR2 signalling cascade is intact (Figure 2G and K; H and L). In addition, quantitative PCR analysis of embryonic lungs isolated at E12.5, E15.5 and E17.5 of controls and double transgenic mice using primers specific for FgfR2, FgfR2-IIIb, FgfR2-IIIc, Bmp4 and Spry did confirm the absence of differential expression of these important branch-inducing genes (data not shown). In conclusion, no differences in expression pattern were observed for the early branch-inducing genes between controls and double transgenics, suggesting that the initiation of the branching process occurred normally.

mycHIF3α expression inhibits Clara cells differentiation

Since we observed significant alveolar changes and aberrant branching morphogenesis, we analyzed the integrity and differentiation potential of fetal transgenic lungs by immunohistochemistry with cell-specific markers. The smooth muscle cell component of the mesenchyme (α-Sma) did not reveal striking differences between control and transgenic lungs (Figures 3A, B). Thyroid transcription factor (Ttf1) was expressed in nearly all epithelial cells in both control and transgenic lungs (Figures 3C, D). Ciliated cells (β-tubulin) and neuroendocrine cells (cGRP) were present in proximal conducting airways of control and transgenic lungs at gestational age E18.5 (Figures 3E, F and 3G, H, arrows). Moreover, both type I (T1α; Figure 4A, B) and type II pneumocytes (Lpcat1; Figures 4C, D) were present in the alveolar regions. These results indicate that differentiation into the various epithelial cell types is not hampered by Hif3α, although the total number of each cell type may be different. In addition, no differences were observed in the proliferation of epithelial or mesenchymal cells between control and transgenic lungs as indicated by Ki67 staining (Figure 4E, F).

Next, three mycHIF3α-expressing lungs and three control lungs were processed at gestational age 18.5 days for microarray analysis, to elucidate the origin of the aberrant branching morphogenesis. Hierarchical clustering of differentially expressed genes revealed large differences between controls and double transgenic lungs (Figure 5A) and the major biological processes (Figure 5B) and molecular functions (Figure 5C) are indicated. Although mycHIF3α does not prevent the differentiation of epithelial cells into Clara cells, we noticed that the number of Clara cells was significantly reduced. Both in the microarray analysis as well as the qPCR validation showed downregulation of the Ccsp gene in mycHIF3α transgenic mice. These gene expression results were confirmed by immunohistochemistry, showing that Ccsp positive cells were less prominent in the proximal airways of the Hif3α expressing lungs compared to control lungs (Figures 6A-D). Quantification of the total number of Clara cells revealed a significant reduction in the double transgenic mice (Figures 6H). So, our data show that mycHIF3α expression inhibits Clara cells differentiation during pulmonary development.

mycHIF3α induces airway epithelial cells to differentiate into proximal cell types

Analysis of the microarray data revealed that genes associated with proximal cell types of the lung appeared to be upregulated, whereas genes specifically expressed in distal epithelial cells were downregulated (Table 1 and Table 2). The induction of proximal markers is reflected by the significant downregulation of genes specific for the distal lung epithelium. The type 1 pneumocyte cell

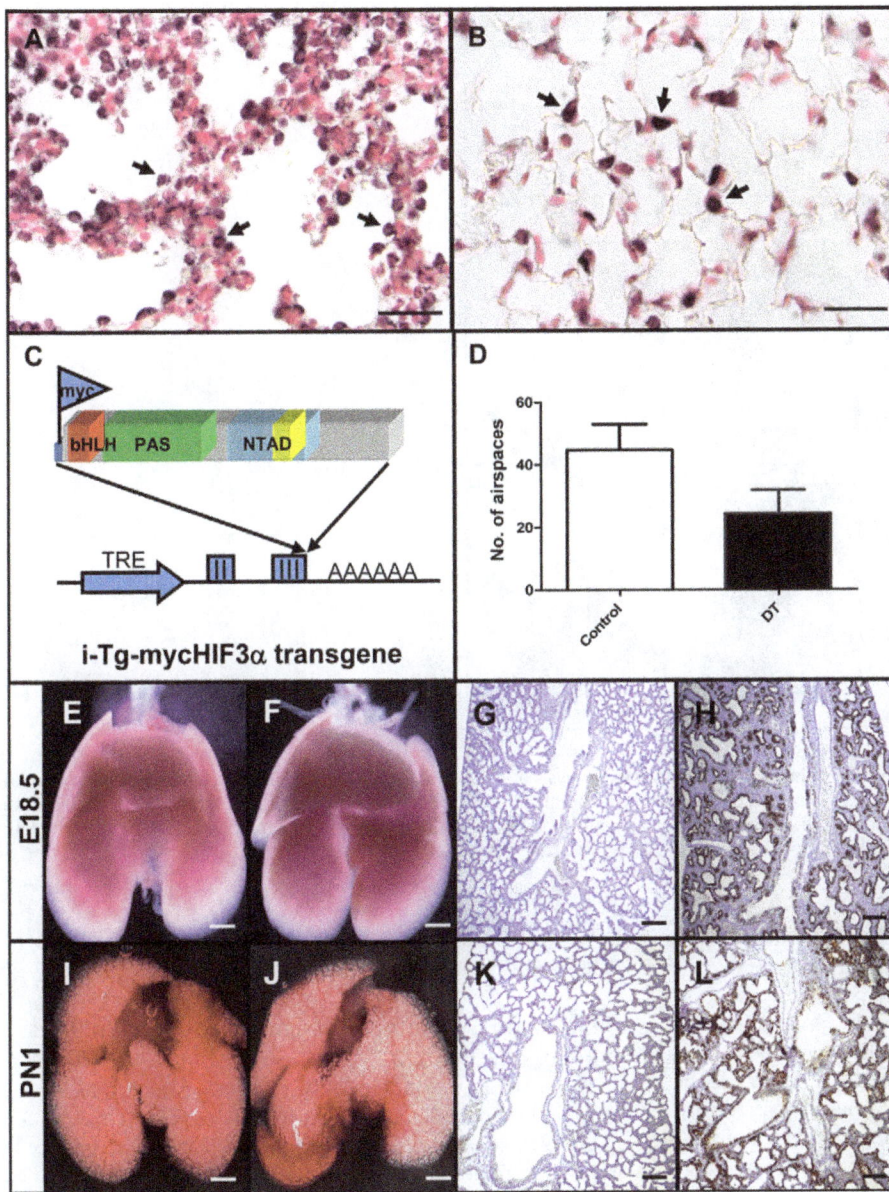

Figure 1. Enhanced expression of HIF3α results in late branching defect. Endogenous expression of Hif3α was detected in epithelial cells at gestational age E18.5 (A, arrows) and in type II pneumocytes in adult mice (B, arrows). (C) Graphic representation of the tet-inducible Hif3α/NEPAS cDNA construct used to generate transgenic mice. TRE is the Tet-responsive element containing minimal promoter, II and III refer to exon 2 and 3 of the ß-blobin gene and AAAAAA is the poly-adenylation signal. Indicated are the position of the myc-epitope, and the bHLH, PAS and NTAD domains (see text) (D) Quantification of the number of airspaces in the lung. Three independent samples of control and double-transgenic lungs at gestational age E18,5 were used to count the number of airspaces. External appearances of control (E, I) and double transgenic mycHIF3α (F, J) lungs at E18.5 days of gestation (E18.5) and post natal age 1 (PN1) do not show apparent differences. Histological analysis of control (G, K) and double transgenic lungs (H, L) showed decreased number of alveolar spaces and reduced branching in the double transgenic lungs (H and L). Anti-Myc epitope staining confirmed the expression of the mycHIF3α transgene in double transgenic lungs (H and L), which is absent in control lungs (G and K). Scale bars: Scale bars: 25 μm (A, B), 2 mm (E, F, I, J) or 200 μm (G, H, K, L).

marker Aquaporin 5 (*Aqp5*) was dowregulated in the Hif3α expressing mice, as were three genes specifically expressed in type II pneumocytes, stearoyl-coenzyme A desaturase (*Scd1*), surfactant associated protein D (*Sftpd*) and ATP-binding cassette (ABC) subfamily A3 (*Abca3*) (Figure 6E) [61,62,63]. Quantification of the number of type II pneumocytes present in the Hif3α expressing lungs using *Sftpd* in reference to *Ttf1* confirmed a significant reduction in these cells (Figure 6G). Since we are inducing the Hif3α family member of hypoxia inducible genes, we analyzed the

expression of Hif1α and Hif2α in the transgenic lungs. Although no apparent difference could be detected for Hif1α (Figure 6F), but we did notice a significant downregulation of Hif2α (*Epas1*) (Figure 6E). Previously, we showed that Hif2α is involved in maturation of type II pneumocytes, so the reduction of *Epas1* expression could be directly related to the loss of type II cells.

Among the upregulated genes are two transcription factors known to play important functions during lung development, *Foxp2* and *Sox2* [57,64]. Foxp2 is important during lung

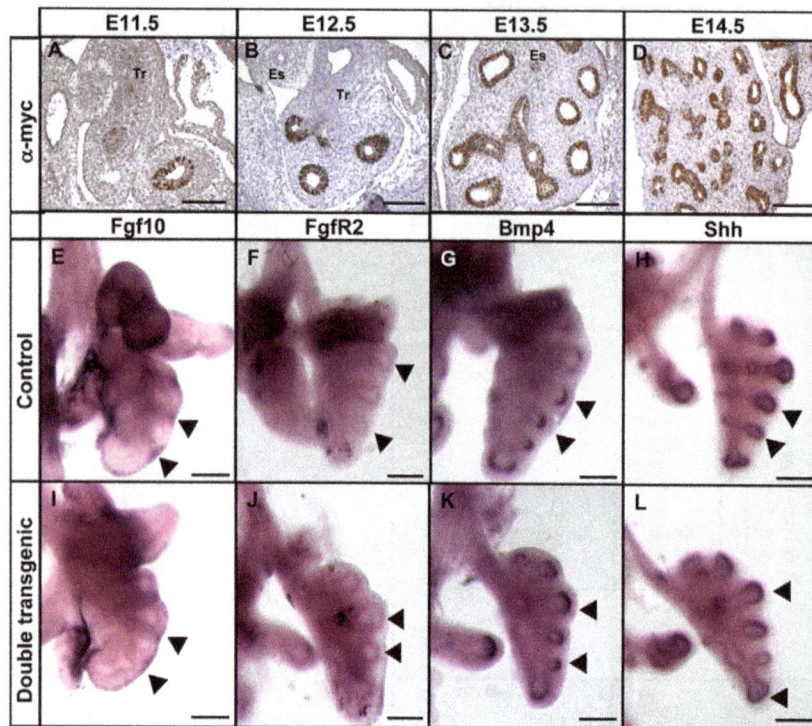

Figure 2. Expression of genes involved in branching morphogenesis. Analysis of the distribution of mycHIF3α early in lung development in double transgenic animals at E11.5 (A), E12.5 (B), E13.5 (C) and E14.5 (D). Whole mount in situ hybridization to detect the expression and localization of Fgf10 (E and I), FgfR2 (F and J), Bmp4 (G and K) and Shh (H and L) in lungs isolated at gestational age E12.5 from control (E–H) and mycHIF3α double transgenic animals (I–L). Tr: Trachea; Es: Esophagus. Scale bars: 200 μm.

development and is expressed in the distal parts of the lung. It represses the transcription of several distal cell markers, such as T1α, Spc, and Ccsp [65]. In our microarray analysis, *Foxp2* was significantly upregulated, which we validated by quantitative PCR (Table 1 and Figure 7G). Staining with a Foxp2 antibody show that the distribution of Foxp2 positive cells in Hif3α double transgenic lungs was expanded compared to control lungs (Figures 7A, D), suggesting that Hif3α suppressed the transcription of genes specific for alveolar epithelial cells through the induction of Foxp2. In addition, *Rarβ*, which is expressed at proximal sites in the lung from embryonic day 11 to 12 and not in the distal epithelium of the lung [66,67], was significantly induced in Hif3α transgenic mice (Figure 7G), confirming the expansion of proximal cell makers in these lungs [64,65].

Sox2 is important for pulmonary branching morphogenesis, epithelial cell differentiation and is exclusively expressed in the proximal parts of the lung [57]. However, in mycHIF3α expressing lungs, Sox2 is present in epithelial cells of both proximal airways and certain alveoli at postnatal day 1, suggesting that Hif3α is able to induce proximal cell fate (Figures 7B, E, arrows). The basal cell marker p63 is expressed in the esophagal and tracheal epithelium, and previously we showed that ectopic Sox2 expression induced the appearance of p63 positive cells in the epithelium of the bronchioles and enlarged distal airspaces [57]. Therefore, we analysed the distribution of basal cells in the mycHIF3α expressing lungs and found that p63 is abnormally expressed in the alveolar epithelial cells of mycHIF3α expressing lungs, contrasting the unique expression in the trachea (Figures 7C insert, *arrows* F). Collectively, our data indicate that mycHIF3α expression leads to the induction of crucial genes, such as Sox2,

Foxp2 and Rarß, which cause airway epithelial cells to differentiate into proximal cell types.

Hif3α binds the promoter region of Sox2 and induces transcription of Sox2

The promoter region of the *Sox2* gene contains two functional HREs, which are bound by Hif2α [68]. Since Sox2 is upregulated in Hif3α transgenic lungs, we analyzed whether Hif3α can directly induce the transcription of Sox2. Therefore, we first performed transcription reporter assays using a luciferase reporter construct under the influence of the Sox2 promoter containing two HREs, or two mutated HREs (Sox2-Luc and ΔSox2-Luc [68]). Hif3α induced the expression of the Sox2-Luc promoter about 2 fold, whereas the ΔSox2-Luc promoter was hardly induced compared to controls (Figure 7H). The positive control, HRE, was considerably induced by Hif2α, but only mildly by Hif3α, corresponding with the weak transcriptional activity of Hif3α [16,42]. Under hypoxia-mimicking conditions, induced by adding $CoCl_2$ to the medium, which inhibits prolyl hydroxylases by displacement of Fe(II) from their catalytic center [22], Hif3α could induce the 9*HRE-Luc considerably, and the difference with the Hif2α induced expression was much reduced (10 times versus 2 times). Moreover, the induction of the Sox2-Luc construct by Hif3α was 4 times higher than under normoxic conditions, and was comparable between Hif2α and Hif3α (Figure 7H). Subsequent analysis of the 1 kilobase region immediately upstream of the Sox2 transcriptional start site revealed that the most upstream of the two putative HRE sites was highly conserved between mice and human [68]. In order to investigate whether Hif3α could directly bind this conserved HRE site, we performed a chromatin immunoprecipitation of chromatin-protein complexes isolated

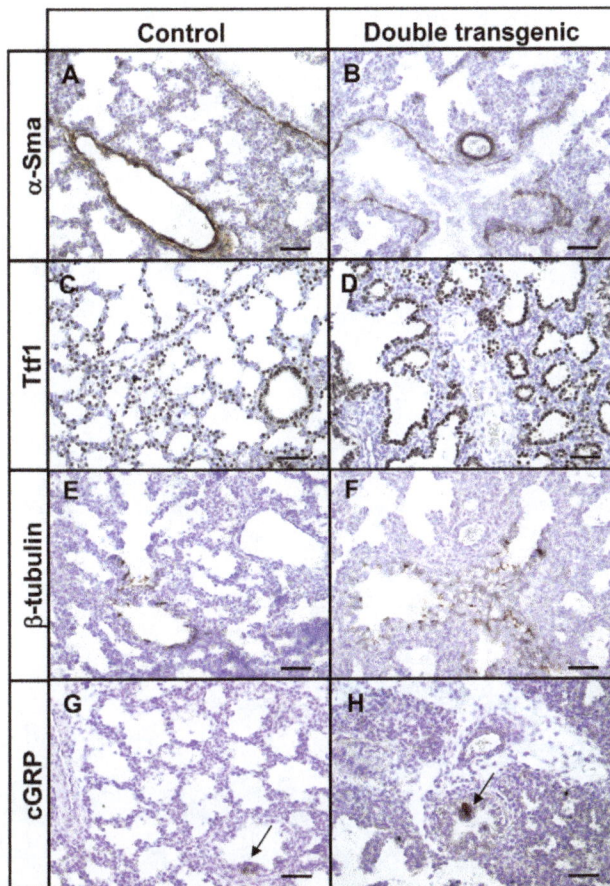

Figure 3. Normal differentiation of proximal epithelial cells in mycHIF3α transgenic lungs. The site and expression pattern of α-Sma (A and B), Ttf1 (C and D), β-tubulin (E and F) and cGRP (arrows in G and H) are comparable between control and mycHIF3α double transgenic lungs at gestational age E18.5. Scale bars: 100 μm.

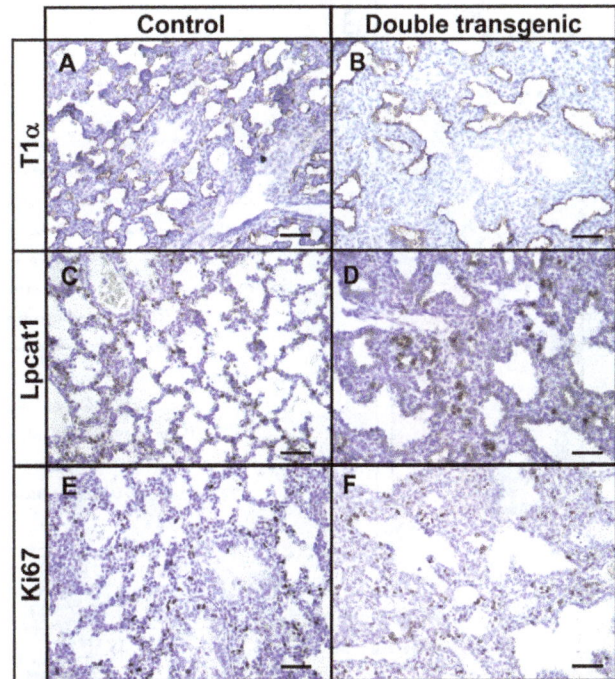

Figure 4. Normal differentiation of distal epithelial cells in mycHIF3α transgenic lungs. The site and expression pattern of T1α (A and B), Lpcat1 (C and D) and Ki67 (E and F) are comparable between control and mycHIF3α double transgenic lungs at gestational age E18.5. Scale bars: 100 μm

from human A549 cells. Analysis of the HIF3α precipitated chromatin showed that the region containing the conserved HRE site in the SOX2 promoter region was indeed preferentially enriched compared to the IgG fraction (Figure 7I). ARRDC3 was used as a potential positive control, as it is bound by both HIF1α and HIF2α, and the enhancer region D of the EGLN3 gene served as negative control [36,59]. Indeed, HIF3α did not bind to the EGLN3-D region, but did bind to the ARDDC3-HRE. This indicated that HIF3α could bind the HRE site present in the Sox2 promoter, suggesting a potential direct regulatory role of Hif3α in the transcription of Sox2.

So, Hif3α binds to the conserved HRE in the Sox2 promoter and weakly induces Sox2 expression, resulting in an abnormal Sox2 expression in airway epithelial cells of HIF3α transgenic lungs.

Discussion

Hypoxia inducible factors are an important family of proteins involved in the regulation of the cellular response to hypoxia. Its functions are required from the earliest steps of mammalian life to the correct development of multiple organs and tissues, like the placenta, trophoblast formation, bone development, heart and vascular development (reviewed in [6,8]). The importance of the hypoxia response was shown by the identification of human

mutations in the VHL-HIF pathway in different diseases [9]. Gene ablation studies in mice have revealed in more detail the specific and important roles of the different subunits of the Hifα/Hifß heterodimers. Inactivation of the stable subunit, Hif1ß, resulted in severe embryonic defects and premature death [48,49]. The disruption of the different Hifα genes identified specific roles for the individual Hifα isoforms. Hif1α knockout mice die early at gestation, have multiple developmental defects in neural tube-forrmation, vascularization, heart development, neural crest migration [69,70,71], whereas depending on the genetic back-ground of the mouse strain, Hif2α knockout out mice ranging from early embryonic lethality to adulthood [51,52,53,54].

Hif genes and lung development

The lung is under continuous exposure of external oxygen and several (patho)-physiologic conditions trigger global or local hypoxia in the lung, resulting in pulmonary abnormalities to which HIFs contribute, such as lung cancer, acute lung injury and pulmonary hypertension (reviewed in [72]). Long term changes in oxygen levels, as experienced at high altitude gives rise to lung damage as a result of chronic mountain sickness. Recently, the *EPAS1* gene, encoding for HIF2α, was shown to be associated with adaptation of living at high altitude [73,74,75,76].

Inactivation of Hif2α in mice resulted in respiratory distress and surfactant deficiency in newborns on a mixed genetic background [51]. Remarkably, heterozygous Hif1α$^{+/-}$ or Hif2α$^{+/-}$ mice showed a reduced increase in pulmonary arterial pressure and right ventricular hypertrophy upon exposure to chronic hypoxia in comparison with wild type mice [77,78]. Ectopic expression of an oxygen-insensitive Hif1α transgene in lung epithelial cells during development resulted in defective branching, impaired epithelial maturation and respiratory distress. Moreover, increased expres-

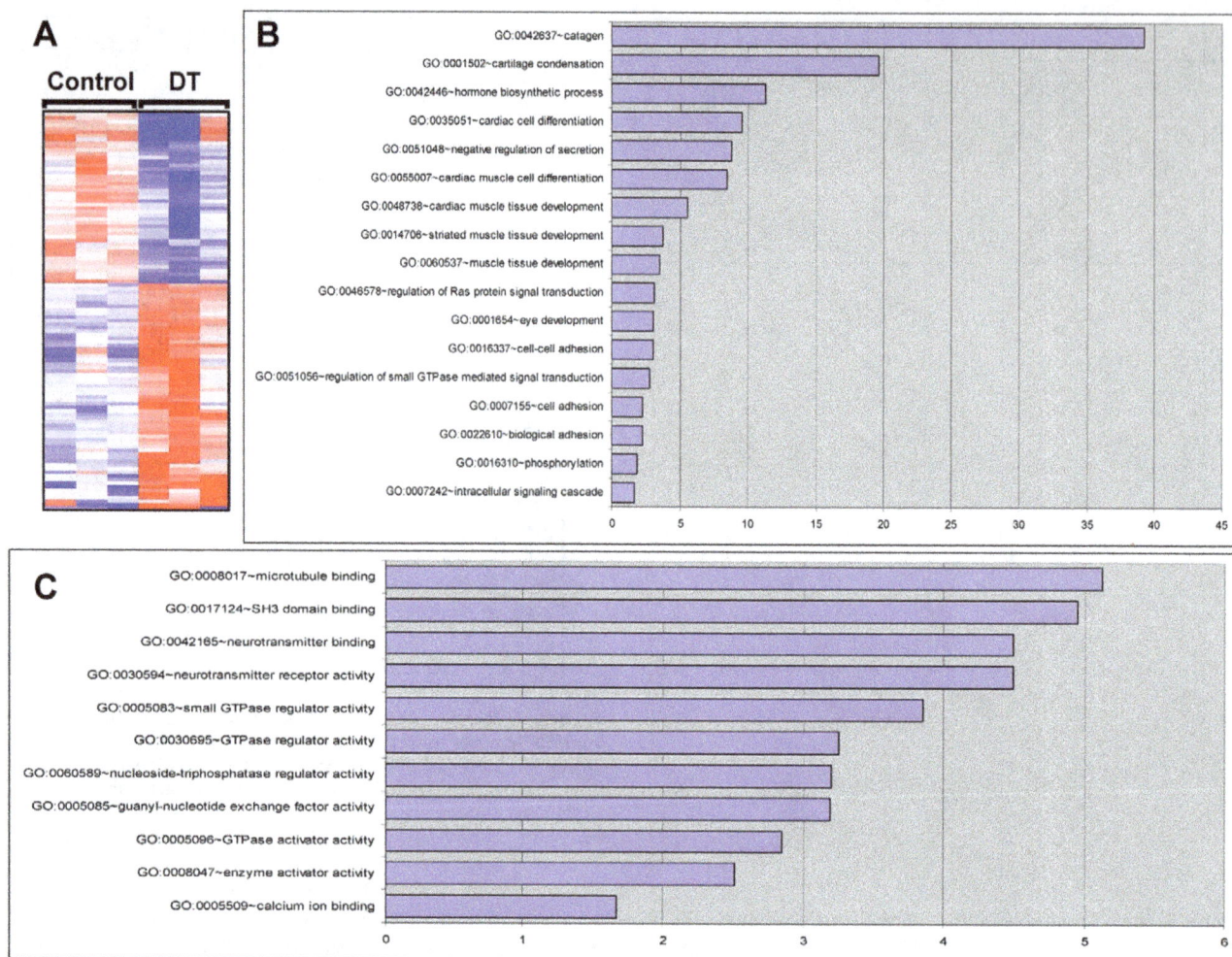

Figure 5. Transcriptome analysis of mycHIF3α expressing lungs. Treescape showing that the transcriptome of the lungs of the mycHIF3α expressing animals are significantly different from that of the control lungs (A). The red color indicates the upregulated genes and the blue color indicates downregulated genes. The expression of the genes presented in the treescape is at least 1.5 fold changed with a false discovery rate (FDR) of 10%. The top 10 biological processes (B) and molecular functions (C) of the differentially expressed genes are shown.

sion of VegfA and VegfC was observed, leading to sub-pleural hemorrhaging [79]. We recently showed that the transgenic expression of an oxygen-insensitive mutant of Hif2α also lead to a late branching defects with enlarged alveoli and altered epithelial differentiation [55]. Contrasting the Hif1α transgenic study, we did not find increased levels of VegfA or endothelial abnormalities, even though the transgenes were expressed in the same manner. This indicates that Hif1α and Hif2α have different effects. In addition, the expression of Hif1α had not changed, whereas Hif3α expression was reduced in our Hif2α transgenic mice [55]. It seems that the effects of Hif1α are more widespread, whereas the number of affected genes by Hif2α is restricted, which is in line with previous reports describing target genes of Hif1α and Hif2α [35,36,80,81,82,83,84,85].

The occurrence of the Hif3α isoforms is well described transcriptionally, but the functional analysis is complicated by the appearance of different splice variants [19,26,42,43,86]. Hif3α isoforms act as negative regulators of the traditional Hif1 (Hif1α/ Hif1ß) and/or Hif2 (Hif2α/Hif1ß) driven hypoxia response by functioning as dominant negative modulators, effectively resulting in the transcriptional competition with Hif1 and Hif2 [16,18,26,42,43]. Gene ablation of Hif3α/NEPAS/IPAS, resulted

in mice that were born alive with enlarged right ventricles and impaired lung remodelling [42]. Furthermore, they showed that expression of endothelin-1 is negatively influenced by Hif3α/ NEPAS, by regulating the binding of Hif1α and Hif2α to the HRE sites if the ET-1 promoter, which may contribute to the observed phenotype. Remarkably, the expression of Vegf, a direct target of Hif1 and Hif2, had not changed, even though the expression of Hif1α and Hif2α was not affected. This hinted at a selective regulation of target genes by NEPAS/Hif3α during pulmonary development. Therefore, we conditionally expressed mycHIF3α in airway epithelial cells during embryonic development in order to further elucidate the role of Hif3α in pulmonary development.

Cellular effects of mycHIF3α transgene expression

Since the NEPAS/Hif3α knockout mice suggested a selective regulation of genes by Hif3α, and our Hif2α transgenic mice showed a selective reduction in Hif3α expression, we conditionally expressed mycHIF3α in airway epithelial cells during embryonic development in order to further elucidate the role of Hif3α in pulmonary development. Analysis of mice expression a transgenic mycHIF3α in lung epithelium revealed a late branching morpho-

Table 1. Significant upregulated genes in the mycHIF3α expressing lungs.

Gene symbol	Gene name	Entrez ID	Fold Change
Dub2a	deubiquitinating enzyme 2a	384701	6,22
Naaladl2	N-acetylated alpha-linked acidic dipeptidase-like 2	635702	2,16
Cldn6	claudin 6	54419	2,14
Hspa1a	heat shock protein 1A	193740	2,14
Fbn2	fibrillin 2	14119	2
ATP6	ATP synthase F0 subunit 6	17705	1,87
Rimklb	ribosomal modification protein rimK-like family member B	108653	1,83
Sema3e	sema domain, immunoglobulin domain (Ig), short basic domain, secreted, (semaphorin) 3	20349	1,8
Tinag	tubulointerstitial nephritis antigen	26944	1,71
Mia1	melanoma inhibitory activity 1	12587	1,68
Plac1	placental specific protein 1	56096	1,68
Cdh16	cadherin 16	12556	1,64
Cnksr2	connector enhancer of kinase suppressor of Ras 2	245684	1,64
Mthfd2l	methylenetetrahydrofolate dehydrogenase (NADP+ dependent) 2-like	665563	1,63
Pcgf1	polycomb group ring finger 1	69837	1,61
Pfn2	profilin 2	18645	1,61
Hspe1	heat shock protein 1 (chaperonin 10)	15528	1,58
Fmod	fibromodulin	14264	1,54
Cdh3	cadherin 3	12560	1,54
Maob	monoamine oxidase B	109731	1,54
Rpl23a	ribosomal protein L23a	268449	1,53
Flrt2	fibronectin leucine rich transmembrane protein 2	399558	1,53
Lgals12	lectin, galactose binding, soluble 12	56072	1,53
Nnat	neuronatin	18111	1,53
Rasef	RAS and EF hand domain containing	242505	1,53
Egfl6	EGF-like-domain, multiple 6	54156	1,53
Ctnnd2	catenin (cadherin associated protein), delta 2	18163	1,52
LOC674930	similar to suppressor of initiator codon mutations, related sequence 1	674930	1,5
Sox2	SRY-box containing gene 2	20674	1,57
Foxp2	forkhead box P2	114142	1,51

genesis defect with a reduced number of alveoli and changes in the differentiation of epithelial cell types.

Surprisingly, no apparent defects are observed early during lung development, even though the transgene is expressed. This may be due to the fact that at these stages of development, putative associating factors of Hif3α, like Hif2α and Hif2α, are not expressed yet. After the pseudoglandular stage of lung development, endogenous Hif2α becomes expressed in the cells positive for mycHIF3α and the effect of the mycHIF3α transgene starts to be noticeable. Histological analysis and gene expression profiling revealed changes in the differentiation of the developing pulmonary epithelium. We found reduced numbers of Clara cells, alveolar type I and type II cells, and in addition, basal cells were observed in atypical spatial positions. The expression pattern of diverse sets of genes was affected, and revealed that mycHIF3α expression mainly affects Hif2-directed transcription, although not all Hif2 target genes are equally affected. We show that expression of mycHIF3α in epithelial cells results in a down regulation of Hif2α, but not of Hif1α. This suggests that Hif3α is not a global regulator of the hypoxic response, but that Hif3α may selectively

function to modulate Hif2α controlled target genes, supporting previous work [42]. The reduction in the expression level of Hif2α late in gestation may be due directly to the presence of mycHIF3α, or due to the impaired differentiation of the type II cells. However, it is clear that mycHIF3α does affect the differentiation of epithelial cells, and this could partly be explained by the aberrant activation of specific genes that are not part of the hypoxic response. Gene expression analysis does not show significant changes in typical hypoxia responsive genes, which indicates that Hif3α may have specific functions beyond the hypoxia response. Therefore, we provide first evidence for novel Hif3α functions beyond the hypoxia response.

The apparent increase in the mesenchymal compartment after the pseudoglandular stage does not seem to be induced by proliferation, as we did not observe an increase in mitotic cells in the mycHIF3α lungs. It may be due to either a delayed development of the double transgenic lungs, or, alternatively, to a specific response in epithelial cells triggered by mycHIF3α. Lysyl oxidase may be activated, which subsequently activates a cascade of proteins, such as Snail, involved in the repression of E-cadherin,

Table 2. Significant downregulated genes in the mycHIF3α expressing lungs.

Gene symbol	Gene name	Entrez ID	Fold Change
Olfr767	olfactory receptor 767	258315	0,45
Ass1	argininosuccinate synthetase 1	11898	0,53
Pgam2	phosphoglycerate mutase 2	56012	0,56
Gipr	gastric inhibitory polypeptide receptor	381853	0,58
Olfr6	olfactory receptor 6	233670	0,6
Igfbp6	insulin-like growth factor binding protein 6	16012	0,61
Nppa	natriuretic peptide precursor type A	230899	0,61
Dio3	deiodinase, iodothyronine type III	107585	0,63
Mphosph6	M phase phosphoprotein 6	68533	0,64
Plscr2	phospholipid scramblase 2	18828	0,64
Ccin	calicin	442829	0,65
Fabp5	fatty acid binding protein 5, epidermal	16592	0,65
Nudcd3	NudC domain containing 3	209586	0,65
Olfr171	olfactory receptor 171	258960	0,65
Rtl1	retrotransposon-like 1	353326	0,66
Rasgrf2	RAS protein-specific guanine nucleotide-releasing factor 2	19418	0,66
Fabp12	fatty acid binding protein 12	75497	0,66
Scnn1a	sodium channel, nonvoltage-gated 1 alpha	20276	0,66
Surfactant related genes			
Scd1	stearoyl-Coenzyme A desaturase 1	20249	0,31
Sftpd	surfactant associated protein D	20390	0,65
Clara cells marker			
Scgb1a1(ccsp)	secretoglobin, family 1A, member 1 (uteroglobin)	22287	0,65
Type I pneumocytes marker			
Aqp5	aquaporin 5	11830	0,65

and ultimately leading to epithelial-mesenchymal transition, as described for metastatic tumors [87,88].

Genes affected by mycHIF3α

The appearance of proximal cells at the expense of distal cells in the mycHIF3α lungs is paralleled by transcriptional changes in several genes, such as Sox2, Rarß and Foxp2. At this point, it remains to be seen whether all effects observed are directly related to mycHIF3α, or that the expression of mycHIF3α affects Hif1α and Hif2α specific complexes, thereby interfering with transcription of specific genes. The increased expression of mycHIF3α could lead to the formation of complexes that normally are not present in the cell, which would shift the balance between Hif2α and Hif3α.

We observed Sox2 positive cells at unusual sites in the lung, which was supported by the aberrant presence of p63 positive basal cells. Previously, we showed that Sox2 directly induces the appearance of basal cells [57]. Since a link was found between Hif2α and Sox2 transcription [68], we analyzed the putative regulation of the Sox2 gene by Hif3α. We show that Hif3α is capable of inducing basal expression of a reporter construct under the control of the Sox2 promoter containing two HRE sites. In addition, we show that HIF3α binds to the conserved HRE sequence in the Sox2 promoter, which suggests that Hif3α may contribute directly to the regulation of Sox2 expression. However, the minimal transcriptional activity of Hif3α, as also shown previously, may explain the appearance of only scattered Sox2

positive cells in the lungs of mycHIF3α mice [16,26,42]. In addition, depletion of individual HIFα genes by siRNA in human ES cells suggested that HIF3α upregulates HIF2α, which subsequently induced the expression of stem cell marker genes, like SOX2 [89]. Although this hypothesis is intriguing, no direct relationship was established, yet. It was also shown that ectopic expression of HIFs in cancer cell lines can induce embryonic stem cell markers, like SOX2 and NANOG [90]. The combination of weak transcriptional activity and the ability to act as a dominant negative modulator of Hif2α may be responsible for the transcriptional regulation of Sox2. These results directly show that through the expression of HIF3α, Sox2+ and p63+ basal cells appear and suggest that the balance between Hif2α and Hif3α may function as a modulator of basal cell emergence [68].

Besides the aberrant induction of Sox2 and p63, the expression domain of *Rarß* was expanded distally in the mycHIF3α transgenic lungs. Rarß knockout mice exhibited premature septation, and formed alveoli twice as fast as wild-type mice [66,67,91]. So, upregulation of *Rarß* in mycHIF3α transgenic mice may in part explain the observed inhibition of pulmonary alveoli formation. We also detected an increase of Foxp2, which is a transcriptional repressor able to inhibit the expression of Ccsp and markers specific for distal epithelial cells, such as Spc and T1α [64,65,92]. Therefore, the reduced numbers of Clara cells (Ccsp+), alveolar type I (Aqp5+) and alveolar type II (Sftpd+) cells could be directly related to the upregulation of Foxp2. Recent findings showed that depletion of cells with CCSP promoter activity was associated with

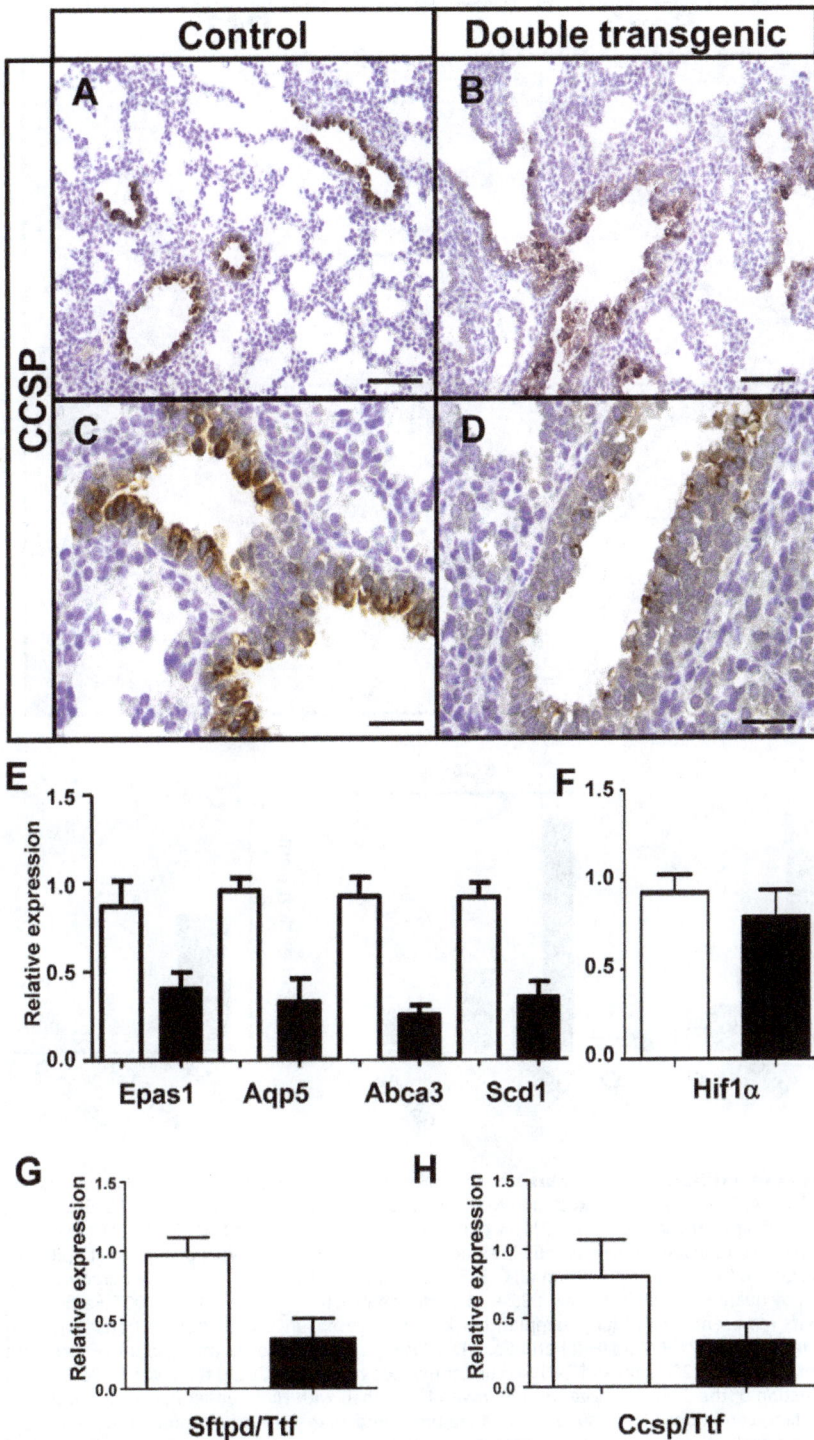

Figure 6. mycHIF3α reduces the number of Clara cells. The expression of the Clara cell marker, Ccsp, was strongly decreased in mycHIF3α transgenic lungs at gestational age E18.5 compared to controls (A and C versus B and D). (E) Alveolar epithelial cell markers are downregulated in Hif3α transgenic lungs at gestational age E18.5 as shown by quantitative PCR. *Epas1* (0,4 $\pm$ 0.1 versus control 0.87 $\pm$ 0.1, n = 3 each, P = 0.012), *Aqp5* (0.33$\pm$ 0.1 versus control 0.96 $\pm$ 0.1, n = 3 each, P = 0.005), *Abca3* (0.25 $\pm$ 0.1 versus control 0.92 $\pm$ 0.1 n = 3 each, P = 0.002), *Scd1*(0.35 $\pm$ 0.1 versus control 0.92 $\pm$ 0.1, n = 3 each, P = 0.001). (F) There is no significant change in the mRNA expression of Hif1α gene (0,8 $\pm$ 0.1 versus control 0.7 $\pm$ 0.1, n = 3 each, P>0.05). Quantification of the number of (G) type II pneumocytes (*Sftpd* over *Ttf1*, 0.36 $\pm$ 0.1 versus control 0.9 $\pm$ 0.1; n = 3, P = 0.01) and (H) Clara cells (*Ccsp* over *Ttf1*, 0.3 $\pm$ 0.1 versus control 0.82 $\pm$ 0.1; n = 5, P = 0.01) showed a significant reduction of in the Hif3α double transgenic animals. White bars represent control lung samples, black bars represent mycHIF3α double transgenic lung samples. Scale bars: 100 μm (A, B) and 50 μm (C,D).

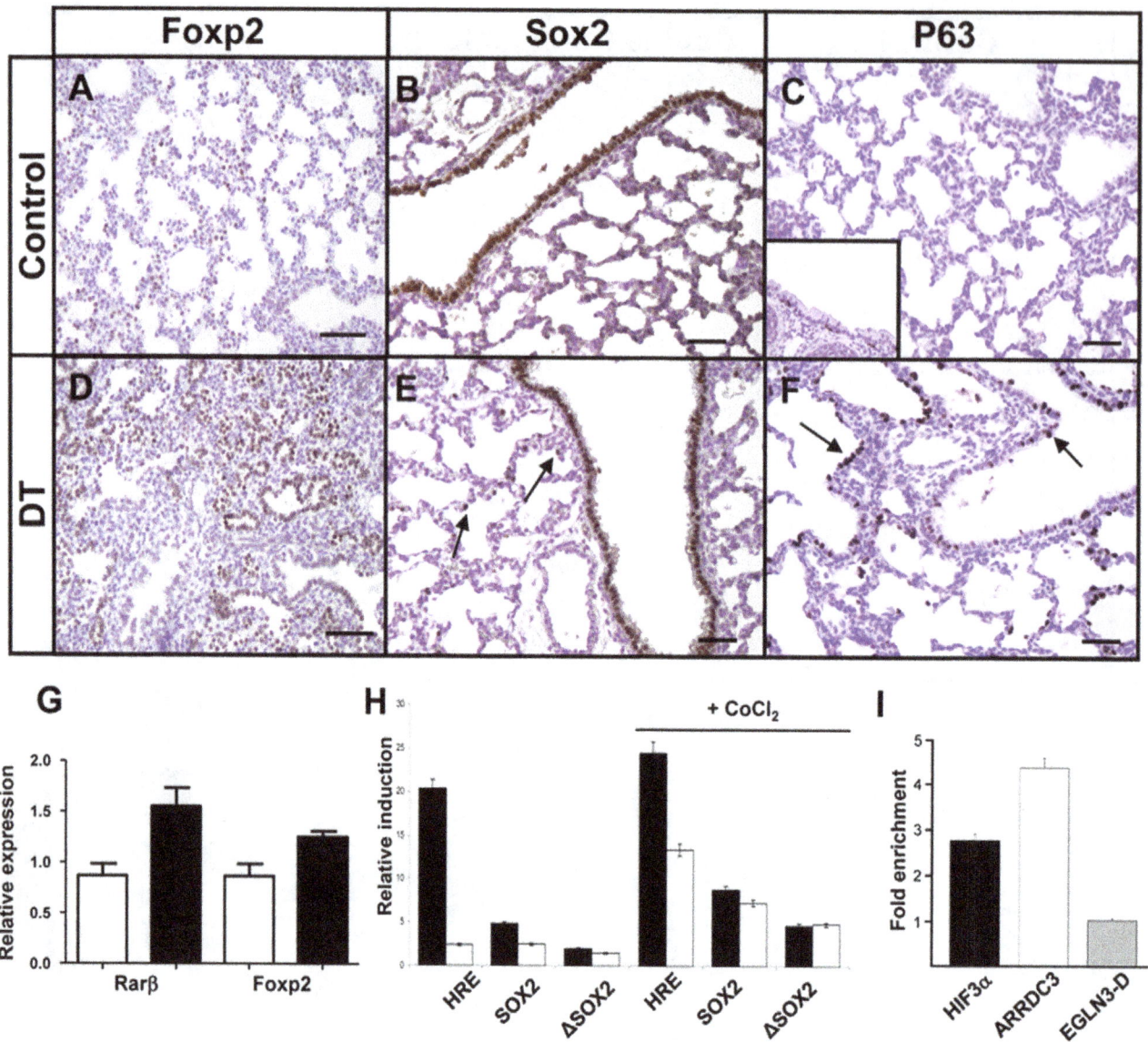

Figure 7. mycHIF3α induces the expression of proximal differentiation markers. mycHIF3α induces an expansion of the Foxp2 positive cells in the double transgenic lungs at gestational age E18.5 (A, D), as well as an expansion towards the distal parts of the lungs of Sox2 (B, E) and p63 (C, F). Sox2 was expressed in both proximal airways and alveolar epithelial cells in mycHIF3α transgenic lungs (*arrows*, E) at PN1. Basal cells are absent in control lungs (C), but are expressed in basal cells of trachea (C, insert). However, p63 is expressed in the proximal airways and alveolar epithelial cells in mycHIF3α transgenic lung (*arrows*, F). *Scale bar*: 200 µm (A and D) and 100 µm (B, C, E, F). (G) *Foxp2* and *Rarβ* are significantly upregulated in Hif3α transgenic lungs at gestational age E18.5 as shown by quantitative PCR. (Foxp2: 1.25 ± 0.1 versus control 0.87 ± 0.1, n = 3, P = 0.007; Rarβ: 1.55 ± 0.1 versus control 0.87 ± 0.1, n = 3, P = 0,009). White bars represent control lung samples, black bars represent mycHIF3α double transgenic lung samples. (H) Hif2α (black bars) and Hif3α (white bars) induce the 9*HRE-Luc (HRE) and Sox2-Luc (Sox2) as measured by the amount of luciferase activity. The fold induction of the HRE promoter is higher with Hif2α (20,3 fold and 24,5 fold under hypoxic conditions-CoCl₂) than with Hif3α (2,4 fold and 13,4 fold under hypoxic conditions-CoCl₂). The induction of the Sox2 promoter is higher with Hif2α than with Hif3α under normoxic conditions (4,8 versus 2,5), but equally strong under hypoxia mimicking conditions (8,8 versus 7,3). Data are presented as the induction (n-fold) relative to cells transfected with the corresponding reporter plasmid and control vector (pcDNA3). The values are the average of two duplicates, and standard deviations are: 0,04 (HRE-Hif2α), 0,02 (Sox2-Hif2α), 0,03 (ΔSox2-Hif2α), 0,08 (HRE-Hif3α), 0,24 (Sox2-Hif3α), 0,06 (ΔSox2-Hif3α), 0,53 (HRE-Hif2α+CoCl2), 0,007 (Sox2-Hif2α+CoCl2), 0,03 (ΔSox2-Hif2α+CoCl2), 0,88 (HRE-Hif3α+CoCl2), 0,02 (Sox2-Hif3α+CoCl2), 0,1 (ΔSox2-Hif3α+CoCl2). (I) Chromatin immunoprecipitation (ChIP) using anti-HIF3α antibody and chromatin isolated from A549 cells. Graph represents the fold enrichment of the HIF3α-specific binding to the conserved HRE of the SOX2 promoter compared to the IgG control ChIP. HIF3α also bound the ARRDC3 HRE region, and the enhancer region D of the EGLN3 gene served as negative control (EGLN3-D).

alveolar hypoplasia and respiratory failure, adding to the idea that Ccsp downregulation as a result of Hif3α-mediated Foxp2 upregulation, directly leads to reduced numbers of Clara cells [93].

The increase d expression of key genes in lung development, which lead to major changes in epithelial differentiation, was confirmed by the loss of expression of other cell type specific markers,, such as *Sftpd*, *Scd1* and *Abca3* for type II cells. At this point it is not clear if the reduced expression of the type II cell markers is the cause, or the result of the loss of type II cells. Previously, we showed a significant downregulation of *Scd1* and

Abca3 in Hif2α expressing transgenic mice, which suffered from respiratory distress and surfactant deficiency [55]. However, the mycHIF3α transgenic mice appeared to produce sufficient levels of Scd1 and Abca3 to support respiration, even though the expression of Hif2α is decreased.

Thus, the increased expression of Sox2, Rarß and Foxp2 in the developing mycHIF3α lungs may directly contribute to the cellular changes observed and explain the phenotypic abnormalities observed in these lungs. The effects may also be cell type specific, as increased HIF3α expression in vascular cells resulted in an antagonistic effect on hypoxia induced HIF1/HIF2 target genes [47].

Concluding remarks

Although we cannot conclude that the dominant negative role of Hif3α as part of the hypoxic response is absent, our previous and current data do suggest that Hif2α and Hif3α have different target genes, during pulmonary development [55]. This is in line with previous findings describing common targets, as well as specific genes induced by Hif1α and Hif2α [80,81,82]. However, these studies used overexpression of Hif1α and Hif2α, which may cause aberrant complexes and loss of target gene specificity, as was reported for certain tumor cells [94]. Using siRNA and chromatin immuno-precipitation approaches, HIF1 and HIF2 target genes were identified [35,36,83,84,85]. Interestingly, it was shown that ETS transcription factors were involved in the regulation of HIF1 and HIF2 driven gene activation in MCF7 cells [83]. Knock down of ELK1 resulted in a reduction of hypoxia induced HIF2 dependent transcription. These data suggested a cooperation between ETS family members and HIF1 and HIF2 in the selection of target genes. An interesting idea is that target selection by HIFs may be cell specifically regulated by additional factors, adding to the complexity of the hypoxic response [8,95]. This is also observed in the analysis of the different transgenic mouse models expressing Hif1α [79], and our studies with HIf2α or Hif3α, showing similarities and differences [55].

Thus, in spite of the limited functional significance of Hif3α/NEPAS in development as a global regulator of the hypoxia response, we demonstrate that Hif3α does contribute by balancing the function of the Hif regulated genes. Furthermore, Hif3α contributes to late branching morphogenesis, alveolar formation and epithelial differentiation. Moreover, the level of *Hif3α*, as well as Hif1α and Hif2α, is tightly regulated to ensure balance between the total number of proximal cells and distal cells.

Supporting Information

Figure S1   Expression of mycHIF3α leads to late branching defect. External appearances of control (A and E) and mycHIF3α transgenic lungs (B and F) at E16.5 and E17.5 showed no apparent differences. Histological analysis of control (C and G) and mycHIF3α transgenic (D and H) lungs showed a gradual decrease in the number of air spaces and aberrant, late branching morphogenesis in mycHIF3α transgenic lungs. Anti-Myc epitope staining confirmed the expression of the mycHIF3α transgene in double transgenic lungs (D and H), which is absent in control lungs (C and G). Scale bars: 2 mm (A, B, E, F) or 200 µm (C, D, G, H).

Acknowledgments

The authors like to thank Professor Jeffrey Whitsett for generously providing the SPC-rtTA mice, Dr Manuel Landazuri for the 9*HRE-Luc plasmid, Dr Victoria Moreno for the pGL3-mpSox2 and pGL3-mpSox2delta plasmids and Dr Carole Pessonnaux for the Hif2α construct.

Author Contributions

Conceived and designed the experiments: YH FG DT RJR. Performed the experiments: YH J-KO MB-vK AB-dM SS WvIJ RJR. Analyzed the data: YH J-KO MB-vK AB-dM SS WvIJ FG RJR. Contributed reagents/materials/analysis tools: YH MB-vK AB-dM SS WvIJ RJR. Wrote the paper: YH FG SS DT RJR.

References

1. Maeda Y, Dave V, Whitsett JA (2007) Transcriptional control of lung morphogenesis. Physiol Rev 87: 219–244.
2. Morrisey EE, Hogan BL (2010) Preparing for the first breath: genetic and cellular mechanisms in lung development. Dev Cell 18: 8–23.
3. Rawlins EL (2010) The building blocks of mammalian lung development. Dev Dyn.
4. Lee YM, Jeong CH, Koo SY, Son MJ, Song HS, et al. (2001) Determination of hypoxic region by hypoxia marker in developing mouse embryos in vivo: a possible signal for vessel development. Dev Dyn 220: 175–186.
5. van Tuyl M, Liu J, Wang J, Kuliszewski M, Tibboel D, et al. (2005) Role of oxygen and vascular development in epithelial branching morphogenesis of the developing mouse lung. Am J Physiol Lung Cell Mol Physiol 288: L167–178.
6. Dunwoodie SL (2009) The role of hypoxia in development of the Mammalian embryo. Dev Cell 17: 755–773.
7. Rajatapiti P, van der Horst IW, de Rooij JD, Tran MG, Maxwell PH, et al. (2008) Expression of hypoxia-inducible factors in normal human lung development. Pediatr Dev Pathol 11: 193–199.
8. Greer SN, Metcalf JL, Wang Y, Ohh M (2012) The updated biology of hypoxia-inducible factor. EMBO J 31: 2448–2460.
9. Semenza GL (2012) Hypoxia-inducible factors in physiology and medicine. Cell 148: 399–408.
10. Jiang BH, Rue E, Wang GL, Roe R, Semenza GL (1996) Dimerization, DNA binding, and transactivation properties of hypoxia-inducible factor 1. J Biol Chem 271: 17771–17778.
11. Wang GL, Jiang BH, Rue EA, Semenza GL (1995) Hypoxia-inducible factor 1 is a basic-helix-loop-helix-PAS heterodimer regulated by cellular O2 tension. Proc Natl Acad Sci U S A 92: 5510–5514.
12. Semenza GL, Jiang BH, Leung SW, Passantino R, Concordet JP, et al. (1996) Hypoxia response elements in the aldolase A, enolase 1, and lactate dehydrogenase A gene promoters contain essential binding sites for hypoxia-inducible factor 1. J Biol Chem 271: 32529–32537.
13. Ema M, Taya S, Yokotani N, Sogawa K, Matsuda Y, et al. (1997) A novel bHLH-PAS factor with close sequence similarity to hypoxia-inducible factor 1alpha regulates the VEGF expression and is potentially involved in lung and vascular development. Proc Natl Acad Sci U S A 94: 4273–4278.
14. Flamme I, Frohlich T, von Reutern M, Kappel A, Damert A, et al. (1997) HRF, a putative basic helix-loop-helix-PAS-domain transcription factor is closely related to hypoxia-inducible factor-1 alpha and developmentally expressed in blood vessels. Mech Dev 63: 51–60.
15. Tian H, McKnight SL, Russell DW (1997) Endothelial PAS domain protein 1 (EPAS1), a transcription factor selectively expressed in endothelial cells. Genes Dev 11: 72–82.
16. Hara S, Hamada J, Kobayashi C, Kondo Y, Imura N (2001) Expression and characterization of hypoxia-inducible factor (HIF)-3alpha in human kidney: suppression of HIF-mediated gene expression by HIF-3alpha. Biochem Biophys Res Commun 287: 808–813.
17. Heidbreder M, Frohlich F, Johren O, Dendorfer A, Qadri F, et al. (2003) Hypoxia rapidly activates HIF-3alpha mRNA expression. FASEB J 17: 1541–1543.
18. Makino Y, Cao R, Svensson K, Bertilsson G, Asman M, et al. (2001) Inhibitory PAS domain protein is a negative regulator of hypoxia-inducible gene expression. Nature 414: 550–554.
19. Makino Y, Kanopka A, Wilson WJ, Tanaka H, Poellinger L (2002) Inhibitory PAS domain protein (IPAS) is a hypoxia-inducible splicing variant of the hypoxia-inducible factor-3alpha locus. J Biol Chem 277: 32405–32408.
20. Maynard MA, Qi H, Chung J, Lee EH, Kondo Y, et al. (2003) Multiple splice variants of the human HIF-3 alpha locus are targets of the von Hippel-Lindau E3 ubiquitin ligase complex. J Biol Chem 278: 11032–11040.
21. Bruick RK, McKnight SL (2001) A conserved family of prolyl-4-hydroxylases that modify HIF. Science 294: 1337–1340.
22. Epstein AC, Gleadle JM, McNeill LA, Hewitson KS, O'Rourke J, et al. (2001) C. elegans EGL-9 and mammalian homologs define a family of dioxygenases that regulate HIF by prolyl hydroxylation. Cell 107: 43–54.
23. Ivan M, Kondo K, Yang H, Kim W, Valiando J, et al. (2001) HIFalpha targeted for VHL-mediated destruction by proline hydroxylation: implications for O2 sensing. Science 292: 464–468.

24. Jaakkola P, Mole DR, Tian YM, Wilson MI, Gielbert J, et al. (2001) Targeting of HIF-alpha to the von Hippel-Lindau ubiquitylation complex by O2-regulated prolyl hydroxylation. Science 292: 468–472.

25. Masson N, Willam C, Maxwell PH, Pugh CW, Ratcliffe PJ (2001) Independent function of two destruction domains in hypoxia-inducible factor-alpha chains activated by prolyl hydroxylation. EMBO J 20: 5197–5206.

26. Gu YZ, Moran SM, Hogenesch JB, Wartman L, Bradfield CA (1998) Molecular characterization and chromosomal localization of a third alpha-class hypoxia inducible factor subunit, HIF3alpha. Gene Expr 7: 205–213.

27. Appelhoff RJ, Tian YM, Raval RR, Turley H, Harris AL, et al. (2004) Differential function of the prolyl hydroxylases PHD1, PHD2, and PHD3 in the regulation of hypoxia-inducible factor. J Biol Chem 279: 38458–38465.

28. Huang LE, Gu J, Schau M, Bunn HF (1998) Regulation of hypoxia-inducible factor 1alpha is mediated by an O2-dependent degradation domain via the ubiquitin-proteasome pathway. Proc Natl Acad Sci U S A 95: 7987–7992.

29. Cockman ME, Masson N, Mole DR, Jaakkola P, Chang GW, et al. (2000) Hypoxia inducible factor-alpha binding and ubiquitylation by the von Hippel-Lindau tumor suppressor protein. J Biol Chem 275: 25733–25741.

30. Ohh M, Park CW, Ivan M, Hoffman MA, Kim TY, et al. (2000) Ubiquitination of hypoxia-inducible factor requires direct binding to the beta-domain of the von Hippel-Lindau protein. Nat Cell Biol 2: 423–427.

31. Tanimoto K, Makino Y, Pereira T, Poellinger L (2000) Mechanism of regulation of the hypoxia-inducible factor-1 alpha by the von Hippel-Lindau tumor suppressor protein. EMBO J 19: 4298–4309.

32. Salceda S, Caro J (1997) Hypoxia-inducible factor 1alpha (HIF-1alpha) protein is rapidly degraded by the ubiquitin-proteasome system under normoxic conditions. Its stabilization by hypoxia depends on redox-induced changes. J Biol Chem 272: 22642–22647.

33. Maxwell PH, Wiesener MS, Chang GW, Clifford SC, Vaux EC, et al. (1999) The tumour suppressor protein VHL targets hypoxia-inducible factors for oxygen-dependent proteolysis. Nature 399: 271–275.

34. Kaelin WG Jr, Ratcliffe PJ (2008) Oxygen sensing by metazoans: the central role of the HIF hydroxylase pathway. Mol Cell 30: 393–402.

35. Mole DR, Blancher C, Copley RR, Pollard PJ, Gleadle JM, et al. (2009) Genome-wide association of hypoxia-inducible factor (HIF)-1alpha and HIF-2alpha DNA binding with expression profiling of hypoxia-inducible transcripts. J Biol Chem 284: 16767–16775.

36. Schodel J, Oikonomopoulos S, Ragoussis J, Pugh CW, Ratcliffe PJ, et al. (2011) High-resolution genome-wide mapping of HIF-binding sites by ChIP-seq. Blood 117: e207–217.

37. Lando D, Peet DJ, Gorman JJ, Whelan DA, Whitelaw ML, et al. (2002) FIH-1 is an asparaginyl hydroxylase enzyme that regulates the transcriptional activity of hypoxia-inducible factor. Genes Dev 16: 1466–1471.

38. Lando D, Peet DJ, Whelan DA, Gorman JJ, Whitelaw ML (2002) Asparagine hydroxylation of the HIF transactivation domain a hypoxic switch. Science 295: 858–861.

39. Mahon PC, Hirota K, Semenza GL (2001) FIH-1: a novel protein that interacts with HIF-1alpha and VHL to mediate repression of HIF-1 transcriptional activity. Genes Dev 15: 2675–2686.

40. Lisy K, Peet DJ (2008) Turn me on: regulating HIF transcriptional activity. Cell Death Differ 15: 642–649.

41. Webb JD, Coleman ML, Pugh CW (2009) Hypoxia, hypoxia-inducible factors (HIF), HIF hydroxylases and oxygen sensing. Cell Mol Life Sci 66: 3539–3554.

42. Yamashita T, Ohneda O, Nagano M, Iemitsu M, Makino Y, et al. (2008) Abnormal heart development and lung remodeling in mice lacking the hypoxia-inducible factor-related basic helix-loop-helix PAS protein NEPAS. Mol Cell Biol 28: 1285–1297.

43. Maynard MA, Evans AJ, Hosomi T, Hara S, Jewett MA, et al. (2005) Human HIF-3alpha4 is a dominant-negative regulator of HIF-1 and is down-regulated in renal cell carcinoma. FASEB J 19: 1396–1406.

44. Makino Y, Uenishi R, Okamoto K, Isoe T, Hosono O, et al. (2007) Transcriptional up-regulation of inhibitory PAS domain protein gene expression by hypoxia-inducible factor 1 (HIF-1): a negative feedback regulatory circuit in HIF-1-mediated signaling in hypoxic cells. J Biol Chem 282: 14073–14082.

45. Li QF, Wang XR, Yang YW, Lin H (2006) Hypoxia upregulates hypoxia inducible factor (HIF)-3alpha expression in lung epithelial cells: characterization and comparison with HIF-1alpha. Cell Res 16: 548–558.

46. Yoshida T, Kuwahara M, Maita K, Harada T (2001) Immunohistochemical study on hypoxia in spontaneous polycystic liver and kidney disease in rats. Exp Toxicol Pathol 53: 123–128.

47. Augstein A, Poitz DM, Braun-Dullaeus RC, Strasser RH, Schmeisser A (2011) Cell-specific and hypoxia-dependent regulation of human HIF-3alpha: inhibition of the expression of HIF target genes in vascular cells. Cell Mol Life Sci 68: 2627–2642.

48. Keith B, Adelman DM, Simon MC (2001) Targeted mutation of the murine arylhydrocarbon receptor nuclear translocator 2 (Arnt2) gene reveals partial redundancy with Arnt. Proc Natl Acad Sci U S A 98: 6692–6697.

49. Maltepe E, Schmidt JV, Baunoch D, Bradfield CA, Simon MC (1997) Abnormal angiogenesis and responses to glucose and oxygen deprivation in mice lacking the protein ARNT. Nature 386: 403–407.

50. Iyer NV, Kotch LE, Agani F, Leung SW, Laughner E, et al. (1998) Cellular and developmental control of O2 homeostasis by hypoxia-inducible factor 1 alpha. Genes Dev 12: 149–162.

51. Compernolle V, Brusselmans K, Acker T, Hoet P, Tjwa M, et al. (2002) Loss of HIF-2alpha and inhibition of VEGF impair fetal lung maturation, whereas treatment with VEGF prevents fatal respiratory distress in premature mice. Nat Med 8: 702–710.

52. Peng J, Zhang L, Drysdale L, Fong GH (2000) The transcription factor EPAS-1/hypoxia-inducible factor 2alpha plays an important role in vascular remodeling. Proc Natl Acad Sci U S A 97: 8386–8391.

53. Scortegagna M, Ding K, Oktay Y, Gaur A, Thurmond F, et al. (2003) Multiple organ pathology, metabolic abnormalities and impaired homeostasis of reactive oxygen species in Epas1-/- mice. Nat Genet 35: 331–340.

54. Tian H, Hammer RE, Matsumoto AM, Russell DW, McKnight SL (1998) The hypoxia-responsive transcription factor EPAS1 is essential for catecholamine homeostasis and protection against heart failure during embryonic development. Genes Dev 12: 3320–3324.

55. Huang Y, Kempen MB, Munck AB, Swagemakers S, Driegen S, et al. (2012) Hypoxia-inducible factor 2alpha plays a critical role in the formation of alveoli and surfactant. Am J Respir Cell Mol Biol 46: 224–232.

56. Tanaka T, Wiesener M, Bernhardt W, Eckardt KU, Warnecke C (2009) The human HIF (hypoxia-inducible factor)-3alpha gene is a HIF-1 target gene and may modulate hypoxic gene induction. Biochem J 424: 143–151.

57. Gontan C, de Munck A, Vermeij M, Grosveld F, Tibboel D, et al. (2008) Sox2 is important for two crucial processes in lung development: branching morphogenesis and epithelial cell differentiation. Dev Biol 317: 296–309.

58. Raghoebir L, Bakker ER, Mills JC, Swagemakers S, Buscop-van Kempen M, et al. (2012) SOX2 redirects the developmental fate of the intestinal epithelium toward a premature gastric phenotype. J Mol Cell Biol.

59. Pescador N, Cuevas Y, Naranjo S, Alcaide M, Villar D, et al. (2005) Identification of a functional hypoxia-responsive element that regulates the expression of the egl nine homologue 3 (egln3/phd3) gene. Biochem J 390: 189–197.

60. Perl AK, Tichelaar JW, Whitsett JA (2002) Conditional gene expression in the respiratory epithelium of the mouse. Transgenic Res 11: 21–29.

61. Zhang F, Pan T, Nielsen LD, Mason RJ (2004) Lipogenesis in fetal rat lung: importance of C/EBPalpha, SREBP-1c, and stearoyl-CoA desaturase. Am J Respir Cell Mol Biol 30: 174–183.

62. Hirche TO, Crouch EC, Espinola M, Brokelman TJ, Mecham RP, et al. (2004) Neutrophil serine proteinases inactivate surfactant protein D by cleaving within a conserved subregion of the carbohydrate recognition domain. J Biol Chem 279: 27688–27698.

63. Stahlman MT, Besnard V, Wert SE, Weaver TE, Dingle S, et al. (2007) Expression of ABCA3 in developing lung and other tissues. J Histochem Cytochem 55: 71–83.

64. Shu W, Lu MM, Zhang Y, Tucker PW, Zhou D, et al. (2007) Foxp2 and Foxp1 cooperatively regulate lung and esophagus development. Development 134: 1991–2000.

65. Shu W, Yang H, Zhang L, Lu MM, Morrisey EE (2001) Characterization of a new subfamily of winged-helix/forkhead (Fox) genes that are expressed in the lung and act as transcriptional repressors. J Biol Chem 276: 27488–27497.

66. Malpel S, Mendelsohn C, Cardoso WV (2000) Regulation of retinoic acid signaling during lung morphogenesis. Development 127: 3057–3067.

67. Wongtrakool C, Malpel S, Gorenstein J, Sedita J, Ramirez MI, et al. (2003) Down-regulation of retinoic acid receptor alpha signaling is required for sacculation and type I cell formation in the developing lung. J Biol Chem 278: 46911–46918.

68. Moreno-Manzano V, Rodriguez-Jimenez FJ, Acena-Bonilla JL, Fustero-Lardies S, Erceg S, et al. (2010) FM19G11, a new hypoxia-inducible factor (HIF) modulator, affects stem cell differentiation status. J Biol Chem 285: 1333–1342.

69. Iyer NV, Leung SW, Semenza GL (1998) The human hypoxia-inducible factor 1alpha gene: HIF1A structure and evolutionary conservation. Genomics 52: 159–165.

70. Kotch LE, Iyer NV, Laughner E, Semenza GL (1999) Defective vascularization of HIF-1alpha-null embryos is not associated with VEGF deficiency but with mesenchymal cell death. Dev Biol 209: 254–267.

71. Compernolle V, Brusselmans K, Franco D, Moorman A, Dewerchin M, et al. (2003) Cardia bifida, defective heart development and abnormal neural crest migration in embryos lacking hypoxia-inducible factor-1alpha. Cardiovasc Res 60: 569–579.

72. Shimoda LA, Semenza GL (2011) HIF and the lung: role of hypoxia-inducible factors in pulmonary development and disease. Am J Respir Crit Care Med 183: 152–156.

73. Beall CM, Cavalleri GL, Deng L, Elston RC, Gao Y, et al. (2010) Natural selection on EPAS1 (HIF2alpha) associated with low hemoglobin concentration in Tibetan highlanders. Proc Natl Acad Sci U S A 107: 11459–11464.

74. Simonson TS, Yang Y, Huff CD, Yun H, Qin G, et al. (2010) Genetic evidence for high-altitude adaptation in Tibet. Science 329: 72–75.

75. Xu S, Li S, Yang Y, Tan J, Lou H, et al. (2011) A genome-wide search for signals of high-altitude adaptation in Tibetans. Mol Biol Evol 28: 1003–1011.

76. Yi X, Liang Y, Huerta-Sanchez E, Jin X, Cuo ZX, et al. (2010) Sequencing of 50 human exomes reveals adaptation to high altitude. Science 329: 75–78.

77. Brusselmans K, Compernolle V, Tjwa M, Wiesener MS, Maxwell PH, et al. (2003) Heterozygous deficiency of hypoxia-inducible factor-2alpha protects mice against pulmonary hypertension and right ventricular dysfunction during prolonged hypoxia. J Clin Invest 111: 1519–1527.

78. Yu AY, Shimoda LA, Iyer NV, Huso DL, Sun X, et al. (1999) Impaired physiological responses to chronic hypoxia in mice partially deficient for hypoxia-inducible factor 1alpha. J Clin Invest 103: 691–696.

79. Bridges JP, Lin S, Ikegami M, Shannon JM (2012) Conditional hypoxia inducible factor-1alpha induction in embryonic pulmonary epithelium impairs maturation and augments lymphangiogenesis. Dev Biol 362: 24–41.

80. Sowter HM, Raval RR, Moore JW, Ratcliffe PJ, Harris AL (2003) Predominant role of hypoxia-inducible transcription factor (Hif)-1alpha versus Hif-2alpha in regulation of the transcriptional response to hypoxia. Cancer Res 63: 6130–6134.

81. Hu CJ, Wang LY, Chodosh LA, Keith B, Simon MC (2003) Differential roles of hypoxia-inducible factor 1alpha (HIF-1alpha) and HIF-2alpha in hypoxic gene regulation. Mol Cell Biol 23: 9361–9374.

82. Wang V, Davis DA, Haque M, Huang LE, Yarchoan R (2005) Differential gene up-regulation by hypoxia-inducible factor-1alpha and hypoxia-inducible factor-2alpha in HEK293T cells. Cancer Res 65: 3299–3306.

83. Aprelikova O, Wood M, Tackett S, Chandramouli GV, Barrett JC (2006) Role of ETS transcription factors in the hypoxia-inducible factor-2 target gene selection. Cancer Res 66: 5641–5647.

84. Elvidge GP, Glenny L, Appelhoff RJ, Ratcliffe PJ, Ragoussis J, et al. (2006) Concordant regulation of gene expression by hypoxia and 2-oxoglutarate-dependent dioxygenase inhibition: the role of HIF-1alpha, HIF-2alpha, and other pathways. J Biol Chem 281: 15215–15226.

85. Warnecke C, Weidemann A, Volke M, Schietke R, Wu X, et al. (2008) The specific contribution of hypoxia-inducible factor-2alpha to hypoxic gene expression in vitro is limited and modulated by cell type-specific and exogenous factors. Exp Cell Res 314: 2016–2027.

86. Torii S, Goto Y, Ishizawa T, Hoshi H, Goryo K, et al. (2011) Pro-apoptotic activity of inhibitory PAS domain protein (IPAS), a negative regulator of HIF-1, through binding to pro-survival Bcl-2 family proteins. Cell Death Differ 18: 1711–1725.

87. Brahimi-Horn MC, Bellot G, Pouyssegur J (2011) Hypoxia and energetic tumour metabolism. Curr Opin Genet Dev 21: 67–72.

88. Pouyssegur J, Dayan F, Mazure NM (2006) Hypoxia signalling in cancer and approaches to enforce tumour regression. Nature 441: 437–443.

89. Forristal CE, Wright KL, Hanley NA, Oreffo RO, Houghton FD (2010) Hypoxia inducible factors regulate pluripotency and proliferation in human embryonic stem cells cultured at reduced oxygen tensions. Reproduction 139: 85–97.

90. Mathieu J, Zhang Z, Zhou W, Wang AJ, Heddleston JM, et al. (2011) HIF induces human embryonic stem cell markers in cancer cells. Cancer Res 71: 4640–4652.

91. Massaro GD, Massaro D, Chan WY, Clerch LB, Ghyselinck N, et al. (2000) Retinoic acid receptor-beta: an endogenous inhibitor of the perinatal formation of pulmonary alveoli. Physiol Genomics 4: 51–57.

92. Zhou B, Zhong Q, Minoo P, Li C, Ann DK, et al. (2008) Foxp2 inhibits Nkx2.1-mediated transcription of SP-C via interactions with the Nkx2.1 homeodomain. Am J Respir Cell Mol Biol 38: 750–758.

93. Londhe VA, Maisonet TM, Lopez B, Jeng JM, Li C, et al. (2011) A subset of epithelial cells with CCSP promoter activity participates in alveolar development. Am J Respir Cell Mol Biol 44: 804–812.

94. Warnecke C, Zaborowska Z, Kurreck J, Erdmann VA, Frei U, et al. (2004) Differentiating the functional role of hypoxia-inducible factor (HIF)-1alpha and HIF-2alpha (EPAS-1) by the use of RNA interference: erythropoietin is a HIF-2alpha target gene in Hep3B and Kelly cells. FASEB J 18: 1462–1464.

95. Loboda A, Jozkowicz A, Dulak J (2012) HIF-1 versus HIF-2--is one more important than the other? Vascul Pharmacol 56: 245–251.

17

Molecular and Genetic Determinants of the NMDA Receptor for Superior Learning and Memory Functions

Stephanie Jacobs[1ⓖ], **Zhenzhong Cui**[1ⓖ], **Ruiben Feng**[1], **Huimin Wang**[2], **Deheng Wang**[3], **Joe Z. Tsien**[1*]

1 Brain and Behavior Discovery Institute and Department of Neurology, Medical College of Georgia at Georgia Regents University, Augusta, Georgia, United States of America, **2** Shanghai Institute of Functional Genomics, East China Normal University, Shanghai, China, **3** Banna Biomedical Research Institute, Xi-Shuang-Ban-Na Prefecture, Yunnan Province, China

Abstract

The opening-duration of the NMDA receptors implements Hebb's synaptic coincidence-detection and is long thought to be the rate-limiting factor underlying superior memory. Here, we investigate the molecular and genetic determinants of the NMDA receptors by testing the "synaptic coincidence-detection time-duration" hypothesis vs. "GluN2B intracellular signaling domain" hypothesis. Accordingly, we generated a series of GluN2A, GluN2B, and GluN2D chimeric subunit transgenic mice in which C-terminal intracellular domains were systematically swapped and overexpressed in the forebrain excitatory neurons. The data presented in the present study supports the second hypothesis, the "GluN2B intracellular signaling domain" hypothesis. Surprisingly, we found that the voltage-gated channel opening-durations through either GluN2A or GluN2B are sufficient and their temporal differences are marginal. In contrast, the C-terminal intracellular domain of the GluN2B subunit is necessary and sufficient for superior performances in long-term novel object recognition and cued fear memories and superior flexibility in fear extinction. Intriguingly, memory enhancement correlates with enhanced long-term potentiation in the 10–100 Hz range while requiring intact long-term depression capacity at the 1–5 Hz range.

Editor: Ya-Ping Tang, Louisiana State University Health Sciences Center, United States of America

Funding: This work was supported by funds from the National Institute of Mental Health (MH060236), National Institute on Aging (AG024022, AG034663 & AG025918), USAMRA00002, and Georgia Research Alliance (all to JZT). The funders had no role in study design, data collection and analysis, decision to publish, or preparation of the manuscript.

* Email: jtsien@gru.edu

ⓖ These authors contributed equally to this work.

Introduction

N-methyl-D-aspartate (NMDA) receptors are known to be the key modulators of synaptic plasticity in the forebrain regions [1–4] and act as the molecular gating switch for learning and memory [5,6]. It is widely accepted that their unique coincidence detection property allows them to impart Hebb's rule on synapses, by requiring the simultaneous pre-synaptic release of glutamate and the depolarization of the postsynaptic membrane to remove the extracellular Mg^{2+} block [7]. NMDA receptors are composed of two GluN1 subunits, as well as two GluN2 subunits [8]. In the adult forebrain regions, GluN2A and GluN2B subunits are the main subunits available in excitatory synapses for receptor complex formation [8,9], and are ideal for coincidence detection due to their strong Mg^{2+} dependency [10–12].

During postnatal brain development, the GluN2B subunits are the predominate subunits expressed specifically in excitatory neurons of the forebrain regions, such as the cortex and hippocampus [9,13]. As the animal develops into adulthood, GluN2B expression decreases while GluN2A expression increases, resulting in an overall decrease in synaptic plasticity. Previously, we have shown that an overexpression of the GluN2B subunit in the forebrain excitatory neurons enhances many forms of learning

and memory in both transgenic mice and rats [14–18]. The prevalent view in the field is that memory enhancement by GluN2B up-regulation is due to its longer opening duration in comparison to that of the GluN2A subunit. This enables the GluN2B-containing NMDA receptors to be better coincidence detectors.

However, the different structural motifs of GluN2 subunits are known to regulate the NMDA receptors' Mg^{2+} dependency, channel opening duration, magnitude of Ca^{2+} influx, as well as, intracellular signaling cascades [10]. For example, the GluN2A and GluN2B subunits have high Mg^{2+} dependency, whereas the GluN2C and GluN2D subunits have much less Mg^{2+} dependency [13]. Thus, the extracellular Mg^{2+} blockade of the GluN2A or GluN2B-containing NMDA receptors suppresses NMDA-mediated Ca^{2+} influx at voltages close to the resting membrane potential allowing the cell to differentiate between correlated synaptic input and uncorrelated activity [7,19,20]. Recent studies have further demonstrated that the C-terminals of GluN2A and GluN2B bind to different downstream signaling molecules. This has led to a greater appreciation for their contribution to synaptic plasticity and behavior [21,22]. To examine the effects of GluN2A on learning and memory, we recently generated CaMKII promoter-driven GluN2A transgenic mice and found profound long-term memory deficits in these mice, while their short-term memories remain unaffected [23]. Therefore, overexpression of GluN2A or GluN2B in the mouse forebrain leads to impaired or enhanced memory function, respectively. These observations have raised several key questions, as to whether enhanced memories in GluN2B transgenic mice or impaired memories in GluN2A transgenic mice were due to their differences in NMDA receptor channel-opening durations or their distinct intracellular signaling processes. Answers to this crucial question can be highly valuable for developing therapeutic strategies for preventing memory loss in patients.

Currently, two hypotheses have been postulated to explain the observed memory enhancement in the GluN2B transgenic animals or memory impairment in GluN2A transgenic mice [4]. One hypothesis, known as the "coincidence-detection" hypothesis, posits that because the GluN2B subunit makes the channel opening duration longer than that of the GluN2A subunit, the GluN2B overexpression allows a greater coincidence detection window, thereby leading to superior memory functions. The shorter coincidence-detection, such as in the GluN2A transgenic mice, underlies impaired long-term memories. The second hypothesis is that the distinct intracellular domain of the GluN2B subunit is responsible for the enhancements observed in the GluN2B transgenic mice [4]. Several recent key observations support this "intracellular domain hypothesis". Biochemical studies have shown that the intracellular C-terminal domains of the GluN2A and GluN2B subunits preferentially interact with different downstream molecules and play distinct roles in synaptic functions [24]. Conversely, studies using genetically truncated GluN2A or GluN2B subunits demonstrate that the C-terminal connections are essential for NMDA receptor function. The truncated subunits often act as functional knockouts of the whole subunit [21]. Although several truncated C-terminal studies have focused on the mechanisms by which the GluN2 subunits mediate the NMDA receptor functions, the structural motifs crucial for learning and memory enhancement remain undefined. It is completely unknown as to whether and what degree the C-terminal domain of the GluN2B would contribute to memory enhancement.

In the present study, we set out to examine the above two hypotheses aimed at determining how and whether the molecular motifs underlying coincidence-detection time duration, or intracellular signaling cascades play a role in enhancing learning and memory. Our strategy is to swap or replace the C-terminal cytoplasmic domain of the GluN2B subunit with the C-terminal domain of the GluN2A subunit, or vice versa. Additionally, we have replaced the C-terminal domain of the GluN2D subunit with the C-terminal domain of the GluN2B subunit effectively reducing the Mg^{2+} dependency of the receptor. We have produced and analyzed five different GluN2 transgenic mouse lines, namely, GluN2A transgenic mice (Tg-GluN2A), Tg-GluN2B$^{2A(CT)}$ transgenic mice, GluN2B transgenic mice (Tg-GluN2B), Tg-GluN2A$^{2B(CT)}$ transgenic mice, and Tg-GluN2D$^{2B(CT)}$ transgenic mice. Our experiments suggest that the C-terminal domain of the GluN2B subunit is necessary and sufficient to produce memory enhancement, as long as it is coupled to Mg^{2+} dependent forms of GluN2 subunits such as GluN2A or GluN2B, while coupling of the GluN2A's C-terminal domain to GluN2B N-terminal and transmembrane domains lead to profound memory impairment. Moreover, coupling of the C-terminal domain of the GluN2B subunit to GluN2D subunit's N-terminal and transmembrane domains lead to memory deficits.

Results

Generation of transgenic mice expressing chimeric GluN2A$^{2B(CTR)}$, GluN2B$^{2A(CTR)}$, or GluN2D$^{2B(CTR)}$ subunits in the forebrain principal neurons

To investigate the potentially distinct roles of the C-terminal domains vs. the N-terminal and membrane domains in GluN2 subunits in mediating memory enhancement, we created constructs encoding chimeric receptors based on GluN2B and GluN2A but with their respective CTDs replaced (denoted as CTR) with each other's (GluN2B$^{2A(CTR)}$ and GluN2A$^{2B(CTR)}$, respectively. We have created three new chimeric GluN2 transgenic mouse lines. We used the same αCaMKII promoter for driving transgene expression in forebrain excitatory neurons as we did for producing the GluN2B [16] and GluN2A transgenic mice [23]. In the first transgenic line, termed Tg-GluN2B$^{2A(CT)}$ chimeric transgenic mice, the C-terminal domain of the GluN2B subunit has been swapped for the counterpart C-terminal domain of the GluN2A and overexpressed in the forebrain excitatory neurons (Figure 1A). This effectively pairs the opening duration of the GluN2B subunit with the signaling domain of the GluN2A subunit. In the second transgenic line, termed Tg-GluN2A$^{2B(CT)}$, the C-terminal domain of the GluN2A subunit has been swapped for the counterpart C-terminal domain of the GluN2B (Figure 1A). This chimeric subunit possesses the GluN2A opening duration but with the signaling domain from the GluN2B subunit. Additionally, to investigate the requirement of the Mg^{2+} dependent synaptic coincidence-detection function for producing GluN2B-mediated intracellular signaling, we created a third transgenic mouse line, namely, Tg-GluN2D$^{2B(CT)}$ mice, in which the C-terminal domain of the GluN2B subunit has been fused to the N-terminal and membrane domain of the GluN2D subunit (which is less Mg^{2+}-dependent) denoted as GluN2D$^{2B(CTR)}$ (Figure 1A).

We confirmed the transgene integration into the genome of their off-spring by Southern Blot analysis using Poly(A) probes (Figure S1A) and Western Blot (Figure S1B). Next, we performed a series of *in situ* hybridization experiments to determine the expression pattern of the transgenes in the mouse brains. As shown, the transgenes are highly enriched in the cortex, striatum, and hippocampus, but not in hindbrain regions such as the cerebellum (Figure 1C). The high expression transgenic mice were crossed with C57BL/6J wild-type mice for at least 8 generations. These chimeric GluN2 transgenic offspring were found to grow and breed normally, having similar adult weights to their wild-type littermates (Figure 1D) (Wt: n = 11, 29.18±0.985 g; Tg-GluN2A$^{2B(CT)}$: n = 7, 30.09±0.990 g; Tg-GluN2B$^{2A(CT)}$: n = 10, 30.34±0.648 g; Tg-GluN2D$^{2B(CT)}$: n = 9, 29.41±0.940 g), and being visually indistinguishable among them. Additionally, we also produced transgenic GluN2A overexpression mice (Tg-GluN2A mice) [23] and transgenic GluN2B overexpression mice (Tg-GluN2B mice) [14–17] for comparisons on learning and memory tests. These two transgenic mouse lines were also maintained on the same genetic background.

The Tg-GluN2A$^{2B(CT)}$, Tg-GluN2B$^{2A(CT)}$, and Tg-GluN2D$^{2B(CT)}$ mice showed no differences in the open field behavioral paradigm, either in time spent in the center verses the periphery (Figure 1E) (center: Wt: n = 7, 117.91±17.006 s; Tg-GluN2A$^{2B(CT)}$: n = 6, 95.95±12.893 s; Tg-GluN2B$^{2A(CT)}$: n = 5, 78.74±15.154 s; Tg-GluN2D$^{2B(CT)}$: n = 6, 79.15±13.483 s; periphery: Wt: 481.86±16.994 s; Tg-GluN2A$^{2B(CT)}$: 503.98±12.901 s Tg-GluN2B$^{2A(CT)}$: 520.96±15.098 s; Tg-GluN2D$^{2B(CT)}$: 520.65±13.527 s), or in locomotor activity (Figure 1F) (center: Wt: 1184.87±218.32 cm; Tg-GluN2A$^{2B(CT)}$:

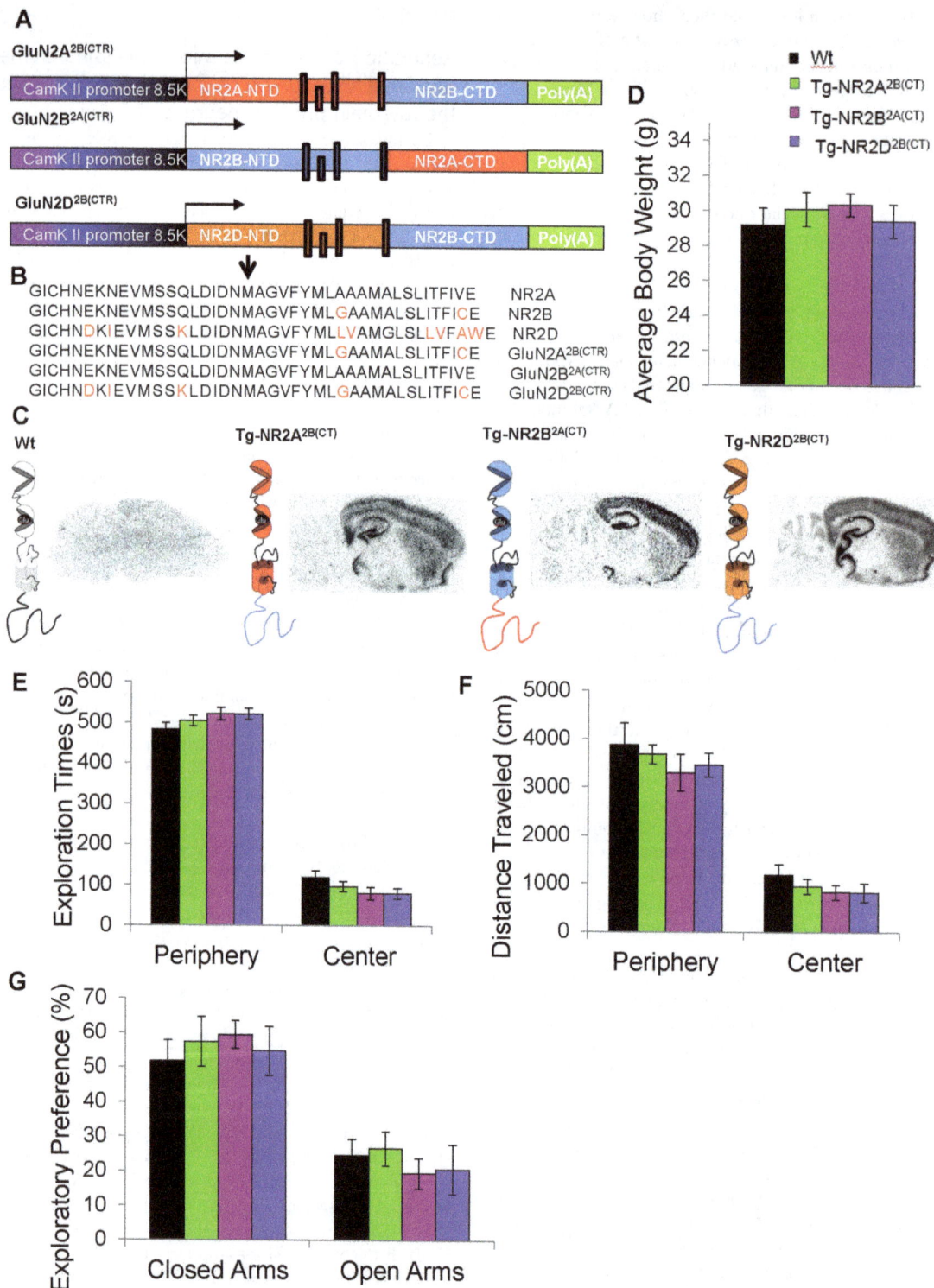

Figure 1. Constructs and basic behavioral assays of the GluN2 chimeric mice. (**A**) Illustration of the constructs used to create the GluN2A[2B(CTR)], GluN2B[2A(CTR)], and GluN2D[2B(CTR)] chimeric subunits. (**B**) A point mutation was made on the cloning vector to induce an Aat II cutting sites to link the N-terminal and membrane domain to the C-terminal domain. After successfully joining the domains, the point mutation was restored to the original sequence. The arrow indicates the fusion position located in trans-membrane domain. (**C**) *In situ* hybridization of the transgene expression in the wild-type mice (Wt), the Tg-GluN2A[2B(CT)] mice, the Tg-GluN2B[2A(CT)] mice and the Tg-GluN2D[2B(CT)] mice using SV-40 probes with a schematic of the receptor subunit expressed in the excitatory neurons. (**D**) No differences were found in the average adult body weight of the wild-type mice, the Tg-GluN2A[2B(CT)], Tg-GluN2B[2A(CT)], and Tg-GluN2D[2B(CT)]mice. (**E**) The chimeric transgenic mice spent similar amounts of time as the wild-type mice in the center verses the periphery of the open field arena. (**F**) The chimeric transgenic mice and the wild-type mice showed similar locomotion in the open field. (**G**) The Tg-GluN2A[2B(CT)], Tg-GluN2B[2A(CT)] and Tg-GluN2D[2B(CT)] mice spent similar amounts of time in the closed arms and the open arms of the elevated plus maze as the wild-type mice.

945.79±158.233 cm; Tg-GluN2B$^{2A(CT)}$: 824.15±147.123 cm; Tg-GluN2D$^{2B(CT)}$: 819.96±195.070 cm; periphery: Wt: 3864.83±460.03 cm; Tg-GluN2A$^{2B(CT)}$: 3675.25±196.950 cm; Tg-GluN2B$^{2A(CT)}$: 3294.38±378.288 cm; Tg-GluN2D$^{2B(CT)}$: 3458.34±249.227 cm). This suggests that these transgenic mice were normal in locomotor activity and anxiety. Additionally, no differences were found in the elevated plus maze paradigm, which also measured for anxiety-like behavior (Figure 1G). Therefore, the chimeric GluN2 mice were indistinguishable from their wild-type littermates in growth, body weights, and these basic behaviors.

Enhancement of long-term object recognition memory in Tg-GluN2A$^{2B(CT)}$ mice but impairments in long-term memory of the Tg-GluN2B$^{2A(CT)}$ and Tg-GluN2D$^{2B(CT)}$ mice

To investigate recognition memory functions in the transgenic mice, we tested the mice in a novel object recognition task for both short-term and long-term memory domains. During training, all transgenic mouse groups showed comparable exploratory behavior and motivation for the task, exploring each object to a similar degree (Figure 2A) (Wt: n = 10, 51.55±3.65%; Tg-GluN2A: n = 10, 50.15±0.932%; Tg-GluN2B: n = 9, 49.17±1.611%; Tg-

GluN2A$^{2B(CT)}$: n = 10, 51.85±3.192%; Tg-GluN2B$^{2A(CT)}$: n = 10, 49.85±0.932%; Tg-GluN2D$^{2B(CT)}$: n = 20, 52.52±2.097%).

At the one hour retention session, the Tg-GluN2A, Tg-GluN2B and Tg-GluN2A$^{2B(CT)}$ mice showed similar interest in the novel object as compared to the wild-type mice (Wt: n = 10, 60.90±4.913%; Tg-GluN2A: n = 11, 59.80±4.270%; Tg-GluN2B: n = 7, 57.38±2.76%; Tg-GluN2A$^{2B(CT)}$: n = 14, 59.15±5.294%) demonstrating no changes in short-term recognition memory. Whereas the Tg-GluN2B$^{2A(CT)}$ and Tg-GluN2D$^{2B(CT)}$ mice show no preference for the novel object (Tg-GluN2B$^{2A(CT)}$: n = 12, 51.15±4.808%; Tg-GluN2D$^{2B(CT)}$: n = 22, 53.85±2.535%), suggesting memory impairment in this test.

At the 24 hour retention session, the Tg-GluN2A mice showed no preference for the novel object and significantly less interest in it than the wild-type mice (Figure 2A) (GluN2A: n = 10, 50.03±3.860%; $F(2, 24) = 7.45$, p = 0.003) as noted previously [23], demonstrating their inability to form a long-term recognition memory. As expected, the Tg-GluN2D$^{2B(CT)}$ mice also showed no preference for the novel object at the 24 hour retention session spending significantly less time exploring the novel object than the wild-type mice (Wt: n = 10, 66.56±3.610%; Tg-GluN2D$^{2B(CT)}$: n = 20, 43.80±2.566%; $F(5, 63) = 6.36$, p = 7.7×10^{-5}). The Tg-GluN2B$^{2A(CT)}$ mice spent only slightly more time with the novel object than the familiar object (n = 12, 55.10±4.821%). However,

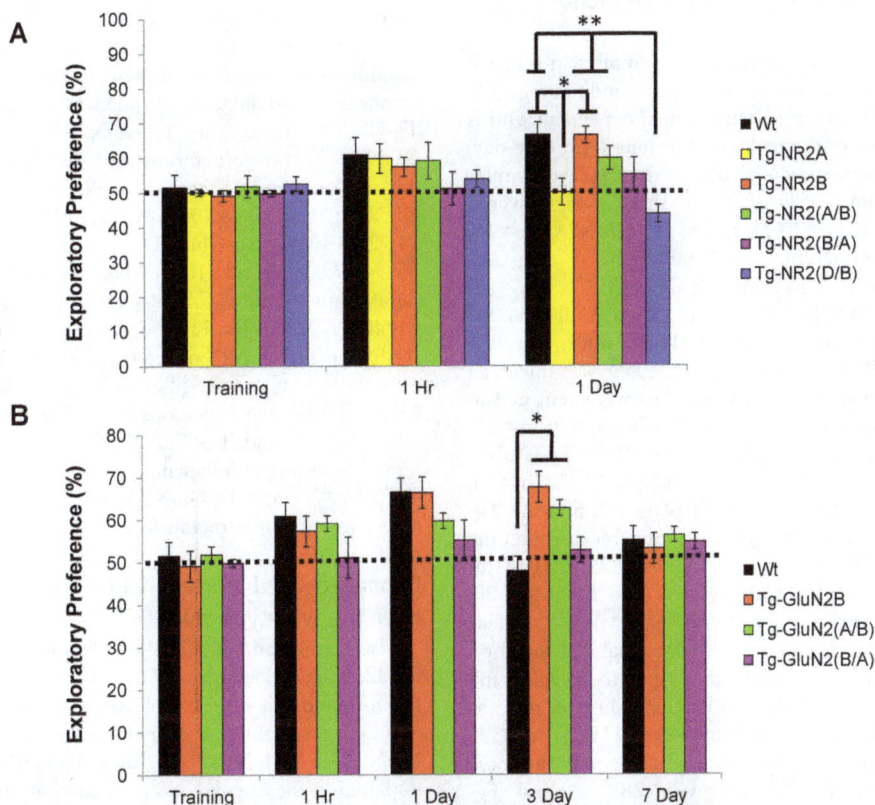

Figure 2. Enhanced long-term recognition memory of the Tg-GluN2A$^{2B(CT)}$ mice and impaired long-term memory on the Tg-GluN2D$^{2B(CT)}$ mice. (A) All groups of mice tested showed similar exploratory behavior in the training session. At the one hour retention session, the Tg-GluN2A, Tg-GluN2B and Tg-GluN2A$^{2B(CT)}$ mice showed similar interest in the novel object as the wild-type mice. Whereas the Tg-GluN2B$^{2A(CT)}$ and Tg-GluN2D$^{2B(CT)}$ mice show almost no preference for the novel object. At the 24 hour retention test, as expected the Tg-GluN2A and Tg-GluN2D$^{2B(CT)}$ mice showed no preference for the novel object. The Tg-GluN2B, The Tg-GluN2A$^{2B(CT)}$ and Tg-GluN2B$^{2A(CT)}$ mice all spent similar amounts of time with the novel object. *p = 0.003, **p = 7.7×10^{-5}. **(B)** In addition to the enhancement seen in the Tg-GluN2A$^{2B(CT)}$ mice at the 24 hour recall session, these mice also showed enhanced recognition NMDA memory even at 3 days post-training over the wild-type mice. *p = 0.003. Whereas the GLUN2A$^{2B(CT)}$ mice show no preference for the novel object at 3 day or 7 days.

the Tg-GluN2B and Tg-GluN2A$^{2B(CT)}$ mice showed similar memory of the novel object to the wild-type mice (Tg-GluN2B: n = 7, 66.44±2.417%; Tg-GluN2A$^{2B(CT)}$: n = 14, 59.77±3.418%). This demonstrates impaired long-term recognition memory in the Tg-GluN2D$^{2B(CT)}$ mice.

To determine the extent of the enhancement in the Tg-GluN2A$^{2B(CT)}$ mice, separate cohorts of mice were further used to test in their ability to retain the memory of the object over three-day and seven-day periods. Remarkably, at the three day retention tests, the Tg-GluN2A$^{2B(CT)}$ mice, like the Tg-GluN2B mice, spent significantly more time investigating the novel object (Figure 2B) (Tg-GluN2B: n = 7, 67.64±4.337%; Tg-GluN2A$^{2B(CT)}$: n = 14, 62.76±2.968%) than the wild-type mice (Wt: n = 10, 47.93±4.045%, $F(2, 28)$ = 7.05, p = 0.003). However, the Tg-GluN2B$^{2A(CT)}$ mice showed no interest in the novel object, spending approximately equal time with both objects (Tg-GluN2B$^{2A(CT)}$: n = 8, 52.74±2.924%). At the seven day retention session, the Wt, GluN2B, Tg-GluN2B$^{2A(CT)}$, and Tg-GluN2A$^{2B(CT)}$ mice spent similar amounts of time investigating the novel object (Wt: n = 10, 55.05±4.096%; Tg-GluN2B: n = 6, 53.12±3.373%; Tg-GluN2A$^{2B(CT)}$: n = 14, 56.36±3.344%; Tg-GluN2B$^{2A(CT)}$: n = 8, 54.75±1.928%; $F(2,35)$ = 0.21, p = 0.93). This demonstrates the significant enhancement in long-term recognition memory in the Tg-GluN2A$^{2B(CT)}$ mice, to a similar degree as the Tg-GluN2B mice did.

Normal contextual fear memory in the chimeric transgenic mice

To investigate the emotional memory in the chimeric transgenic mice, we tested the mice in a contextual fear conditioning task. This type of fear conditioning is hippocampal-dependent and is often used to test short-term (one-hour) and long-term (one-day) time points. In the training session, all of the mice displayed similar freezing responses immediately after the shock was delivered (Figure 3A) (Wt: n = 13, 26.92±6.126%; Tg-GluN2A: n = 15, 25.75±3.146%; Tg-GluN2B: n = 30.00±3.637%; Tg-GluN2A$^{2B(CT)}$: n = 13, 34.61±5.155%; Tg-GluN2B$^{2A(CT)}$: n = 12, 25.00±7.812%; Tg-GluN2D$^{2B(CT)}$: n = 13, 23.18±7.045%). At the one-hour retention session, the wild-type mice, the Tg-GluN2A, Tg-GluN2B$^{2A(CT)}$ and Tg-GluN2D$^{2B(CT)}$ mice all displayed similar freezing responses when they were returned to the shock chamber in the absence of footshock (Wt: n = 13, 26.28±4.444%; Tg-GluN2A: n = 11, 32.22±4.789%; Tg-GluN2B$^{2A(CT)}$: n = 10, 25.28±6.286%; Tg-GluN2D$^{2B(CT)}$: n = 13, 37.51±5.415%). Interestingly, both the Tg-GluN2B and the Tg-GluN2A$^{2B(CT)}$ mice spent significantly more time freezing than the wild-type mice (Tg-GluN2B: n = 7, 45.06±2.823%; Tg-GluN2A$^{2B(CTR)}$: n = 13, 52.56±3.672%; $F(6, 61)$ = 4.98, p = 0.0007). This suggests that Tg-GluN2A$^{2B(CT)}$ mice, similar to Tg-GluN2B, exhibited enhanced 1-hr contextual fear memory.

At the one-day retention session, the Tg-GluN2B mice still displayed significantly more freezing than the wild-type mice as previously reported (Figure 3A) (n = 8, 55.94±4.911%, p = 0.039), suggesting greater long-term contextual fear memory in these mice. However, Tg-GluN2A$^{2B(CT)}$, Tg-GluN2B$^{2A(CT)}$ and Tg-GluN2D$^{2B(CT)}$ mice displayed similar freezing responses as those of the wild-type mice, indicating that all of these chimeric transgenic mice have the normal 1-day hippocampal-dependent contextual fear memories. (Figure 3A) (Tg-GluN2A$^{2B(CT)}$: n = 13, 44.87±4.779%; Tg-GluN2B$^{2A(CT)}$: n = 12, 41.91±7.003%; Tg-GluN2D$^{2B(CT)}$: n = 12, 47.92±7.050%). On the contrary, the Tg-GluN2A mice demonstrated reduced freezing responses during the contextual recall (Wt: n = 14, 43.65±4.382%; Tg-GluN2A: n = 13, 23.54±3.811%, $F(5, 66)$ = 3.65, p = 0.005), suggesting

the deficit in converting short-term contextual fear memory into long-term contextual fear memory due to expression of GluN2A.

Tg-GluN2B$^{2A(CT)}$ and Tg-GluN2D$^{2B(CT)}$ mice showed significant impairments in long-term cued fear memory, whereas Tg-GluN2A$^{2B(CT)}$ showed enhanced memory

To assess whether and how hippocampal-independent forms of memories are affected by the N-terminal and C-terminal domain properties, we used a new cohort of the mice and tested them in cued fear conditioning task which required the mouse to associate an unconditioned stimulus (a shock) with a conditioned stimulus (a tone). In the cued fear conditioning paradigm, all of the mice exhibited little pre-tone freezing responses during the retention tests as they entered a novel chamber (Figure 3B) (Wt: n = 10, 2.22±0.997%; Tg-GluN2A: n = 15, 6.17±1.734%; Tg-GluN2B: n = 7, 6.17±1.251%; Tg-GluN2A$^{2B(CT)}$: n = 11, 2.02±0.758%; Tg-GluN2B$^{2A(CT)}$: n = 11, 1.26±0.576%; Tg-GluN2D$^{2B(CT)}$: n = 12, 3.47±1.373%). Upon the recall tone, all five types of transgenic mice exhibited significant amounts of freezing responses at the one hour retention session, comparable to that of wild-type mice (Wt: n = 10, 71.67±4.843%; Tg-GluN2A: n = 18, 58.08±4.710%; Tg-GluN2B: n = 7, 62.08±6.801%; Tg-GluN2A$^{2B(CT)}$: n = 10, 67.22±4.83%; Tg-GluN2B$^{2A(CT)}$: n = 10, 60.00±4.833%; Tg-GluN2D$^{2B(CT)}$: n = 9, 61.11±4.856%). This shows that the transgenic mouse lines have normal short-term hippocampal-independent emotional memory and all are able to form an association between the tone (CS) and the shock (US).

For one-day cued fear memory retention tests, a second cohort of similarly trained mice was placed into a novel enclosure and an identical tone to the training tone was presented. Interestingly, the Tg-GluN2A mice, the Tg-GluN2B$^{2A(CT)}$ mice and the Tg-GluN2D$^{2B(CT)}$ mice demonstrated significantly less freezing than the wild-type mice (Figure 3B) (Wt: n = 9, 55.02±3.030%; Tg-GluN2A: n = 10,14.75±5.189%; Tg-GluN2B$^{2A(CT)}$: n = 10, 14.72±8.477%; Tg-GluN2D$^{2B(CT)}$: n = 12, 16.67±5.772%; $F(5, 56)$ = 27.03, p < 1.0×10^{-6}). The Tg-GluN2A$^{2B(CT)}$ mice froze significantly more than the wild-type mice (n = 13, 71.58±2.727%). Consistent with the previous studies (Tang et al, 1999), the Tg-GluN2B mice also showed enhanced cued memory over the wild-type mice (Tg-GluN2B: n = 8, 65.75±5.918%, p < 0.05). These data demonstrate that the Tg-GluN2A, Tg-GluN2B$^{2A(CT)}$ and Tg-GluN2D$^{2B(CT)}$ mice have impaired long-term hippocampal-independent fear memory, whereas the Tg-GluN2A$^{2B(CT)}$ and Tg-GluN2B mice exhibited similarly enhanced long-term cued fear memories.

Enhanced cued fear extinction in Tg-GluN2A$^{2B(CT)}$ mice over the wild-type mice

Fear extinction has been widely used as a test for assessing flexible learning behaviors. The extinction of learned fear requires the formation of new flexible relations, instead of forgetting or erasing the established fear memories [25]. Because the Tg-GluN2A, Tg-GluN2B$^{2A(CT)}$, and Tg-GluN2D$^{2B(CT)}$ mice were impaired in the one-day retention session, they were not used for the fear extinction experiment. Instead, we focused our investigation of this form of learning on the Tg-GluN2A$^{2B(CT)}$ mice, the Tg-GluN2B mice, as well as the wild-type mice.

In this fear extinction task, we used a five trial extinction paradigm in which the animals were repeatedly exposed to the training chamber (contextual extinction) or the tone in a novel context (cued extinction) without the delivery of the shock in either context. We first tested the mice in the contextual fear extinction paradigm. We found that the Tg-GluN2B mice initially showed

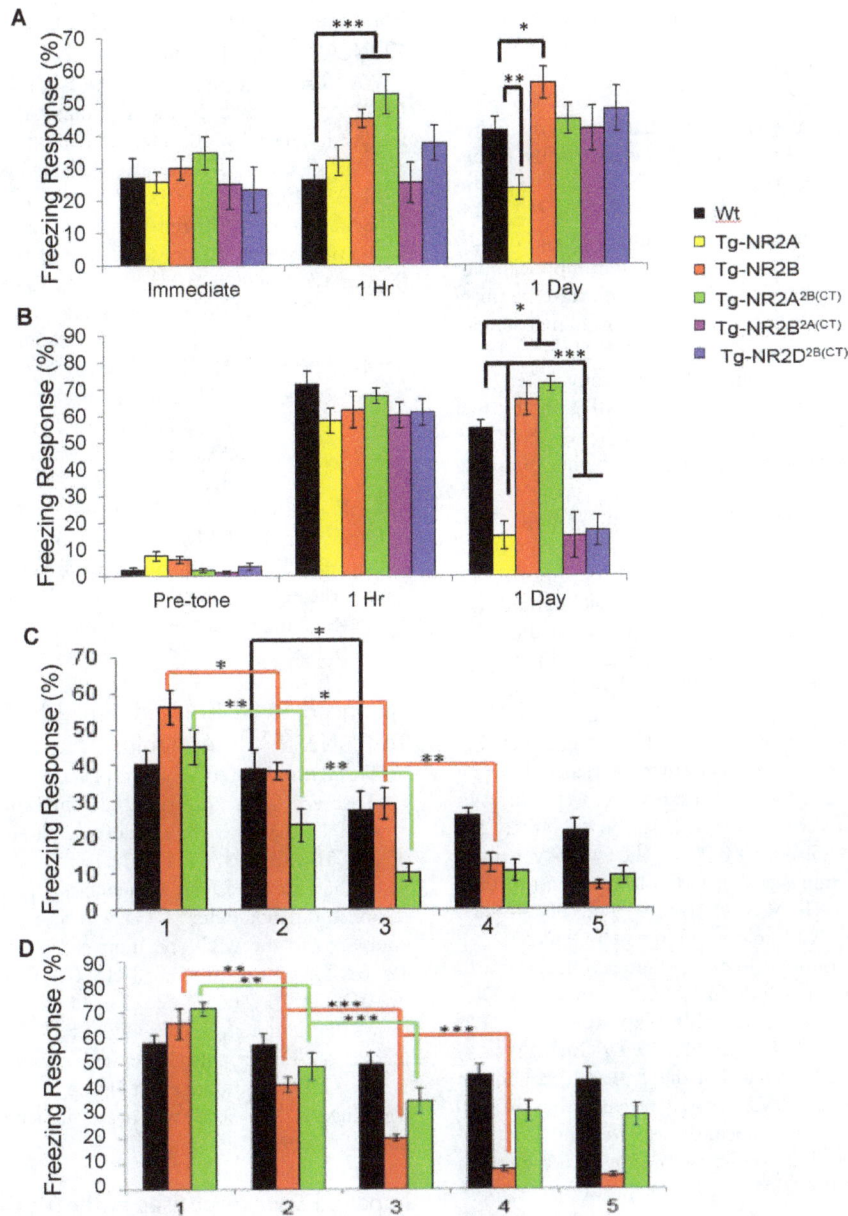

Figure 3. Selectively impaired emotional memory in the Tg-GluN2B$^{2A(CT)}$ and Tg-GluN2D$^{2B(CT)}$ mice. (A) The mice showed similar freezing responses immediately following the US. At the one hour retention session, the wild-type mice, the Tg-GluN2A, Tg-GluN2B$^{2A(CT)}$ and Tg-GluN2D$^{2B(CT)}$ mice all displayed similar freezing responses. Interestingly, both the Tg-GluN2B and the Tg-GluN2A$^{2B(CT)}$ mice spent significantly more time freezing. At the 24 hour recall session only the Tg-GluN2A mice demonstrated a diminished freezing response to the context in which the shock was delivered. *p = 0.005, **p = 0.0007. (B) The mice tested also showed similar pre-tone freezing responses and similar freezing at the one hour contextual recall. At the 24 hour recall session the Tg-GluN2A mice, the Tg-GluN2B$^{2A(CT)}$ mice and the Tg-GluN2D$^{2B(CT)}$ mice demonstrated significantly less freezing than the wild-type mice, whereas the Tg-GluN2A$^{2B(CT)}$ mice and Tg-GluN2B froze significantly more than the wild-type mice. *p < 0.05, **p < 1.0 × 10^{-6}, *** p = 0.0007. (C) The Tg-GluN2A$^{2B(CT)}$ mice showed quicker fear extinction than the wild-type mice in the contextual fear extinction paradigm *p < 0.05, **p < 0.01. (D) The Tg-GluN2A$^{2B(CT)}$ mice showed quicker fear extinction to the CS than the wild-type mice in the contextual fear extinction paradigm *p < 0.05, **p < 0.01, ***p < 0.001.

significantly more freezing than the wild-type mice and the Tg-GluN2A$^{2B(CT)}$ mice 24 hours after the training session (Figure 3C) (1: Wt: n = 15, 40.19 ± 3.852%; Tg-GluN2B: n = 8, 55.94 ± 4.911%; Tg-GluN2A$^{2B(CT)}$: n = 13, 44.87 ± 4.779%). Interestingly, over the extinction trials both the Tg-GluN2B and the Tg-GluN2A$^{2B(CT)}$ mice significantly decreased their freezing responses as early as the second session 2 hours later in comparison to the first trial (2: Tg-GluN2B: 38.29 ± 5.161%; p = 0.014; Tg-GluN2A$^{2B(CT)}$: 23.29 ± 4.409%; p = 0.002), whereas

the wild-type mice did not significantly decrease their freezing response from the first to the second exposure session. It is noted that the Tg-GluN2A$^{2B(CT)}$ mice exhibited significantly less freezing, as determined by ANOVA analysis, than both the Tg-GluN2B mice and their wild-type littermates (Wt: 38.70 ± 5.161%; $F(2, 33) = 3.56$, p = 0.04). All groups of mice spent significantly less time freezing in the third exposure than the second exposure (3: Wt: 27.41 ± 5.034%, p = 0.01; Tg-GluN2B: 29.12 ± 4.294%, p = 0.03; Tg-GluN2A$^{2B(CT)}$: 10.26 ± 2.632%, p = 0.007). The mice continued to decrease their freezing responses in the fourth (4: Wt: 25.74±2.125%; Tg-GluN2B: 12.69±2.576%; Tg-GluN2A$^{2B(CT)}$: 10.47±2.816%, p=0.002) and fifth exposures (5: Wt: 21.48±3.325%; Tg-GluN2B: 6.57±1.300%; Tg-GluN2A$^{2B(CT)}$: 9.19±2.276%). ANOVA analysis indicated that in both the fourth and fifth exposures, the wild-type mice had significantly higher freezing responses than both the Tg-GluN2B and Tg-Glu-N2A$^{2B(CT)}$ mice (4: F(2, 33) =11.91, p=0.0001; 5: F(2, 33) =8.08, p=0.001). These data demonstrate that the Tg-GluN2B mice and the Tg-GluN2A$^{2B(CT)}$ mice had better hippocampal-dependent fear flexibility learning ability than the wild-type mice.

Next, we exposed the mice to the tone in a novel environment for cued fear extinction learning. The Tg-GluN2A$^{2B(CT)}$ mice, similar to the Tg-GluN2B mice, showed significantly faster cued fear extinction than the wild-type mice (Figure 3D). In the first exposure to the tone 24 hours after training, the Tg-GluN2B and Tg-GluN2A$^{2B(CT)}$ mice showed significantly higher freezing responses than the wild-type mice (1: Wt: n=12, 58.10±3.323%; Tg-GluN2B: n=8, 70.87±5.017%; Tg-Glu-N2A$^{2B(CT)}$: n=13, 71.58±2.727%, F(2, 30) =4.97, p=0.01). At the second exposure, two hours after the first exposure, the Tg-GluN2B mice and the Tg-GluN2A$^{2B(CT)}$ mice significantly reduced their freezing responses to the presentation of the tone, whereas the wild-type mice did not (2: Wt: 57.41±4.436%; Tg-GluN2B: 41.46±2.836%, p=0.002; Tg-GluN2A$^{2B(CT)}$: 48.71±5.34%, p=0.003), again suggesting the faster fear extinction in these transgenic mice. Remarkably, both the Tg-GluN2B and Tg-GluN2A$^{2B(CT)}$ mice further decreased freezing from the second to the third exposures as well (3: Wt: 49.54±4.323%; Tg-GluN2B: 20.14±1.278%, p=0.0004; Tg-GluN2A$^{2B(CT)}$: 35.04±4.958%, p=0.0005). The wild-type mice spent significantly more time freezing in the third exposure than the Tg-GluN2B and Tg-GluN2A$^{2B(CT)}$ mice (F(2, 30) =9.87, p=0.0005). The Tg-GluN2B mice and the Tg-GluN2A$^{2B(CT)}$ mice continued to decrease their freezing responses in the fourth (4: Wt: 45.14±4.340%; Tg-GluN2B: 7.75±1.175%; Tg-Glu-N2A$^{2B(CT)}$: 30.77±4.118%) and fifth exposures (5: Wt: 42.82±4.972%; Tg-GluN2B: 5.34±0.939%; Tg-GluN2A$^{2B(CT)}$: 29.27±4.342%). It is worth noting that the Tg-GluN2B had the faster extinction learning. ANOVA analysis revealed that the Tg-GluN2B mice demonstrated significantly less freezing than the Tg-GluN2A$^{2B(CT)}$ mice and the wild-type mice in both the fourth (F(2, 30) =19.32, p=4.0×10^{-6}) and fifth exposures (F(2, 30) =16.22, p=1.7×10^{-5}).

Basic electrophysiological properties in the chimeric GluN2 mice

GluN2A and GluN2B subunits' contribution to synaptic plasticity has been intensely investigated in the CA1 region using both pharmacological and genetic methods [26–29]. We took advantage of the existing knowledge in the literature and investigated and compared how various chimeric transgenic overexpressions would affect the bidirectional control of synaptic plasticity in the CA1 region. To investigate the basic electrophysiological properties in the hippocampus of the Tg-GluN2A$^{2B(CT)}$, Tg-GluN2B$^{2A(CT)}$, and Tg-GluN2D$^{2B(CT)}$ mice, we recorded from the CA1 Schaffer collaterals of the mouse hippocampus. We found the input-output properties (Figure S2A), as well as the paired pulse facilitation (Figure S2B) from each genotype were similar to those of the wild-type controls, thereby demonstrating normal presynaptic function and basal transmissions in these transgenic mice. We then systematically measured the long-term potentiation (LTP) and long-term depression (LTD) in the CA1 slices from each mouse line.

Enhanced 10 Hz induced LTP observed in the Tg-GluN2A$^{2B(CT)}$ CA1 region

We first performed LTP and LTD studies on the Tg-GluN2A$^{2B(CT)}$ mice. In the Tg-GluN2A$^{2B(CT)}$ mice, LTP can be readily induced by 100 Hz stimulation (Figure 4A) (Wt: n=6/3(# of slices/# of animals), 135.2±7.6%; Tg-GluN2A$^{2B(CT)}$: n=7/4, 146.5±8.7%). Interestingly, a significant increase in LTP was observed in the transgenic mice, compared to that of wild-type slices, in response to the 10 Hz frequency stimulation (Figure 4B) (Wt: n=7/4, 103.2±13.0%; Tg-GluN2A$^{2B(CT)}$: n=5/3, 150.1±17.0%). Additionally, a significant difference was further observed at 5 Hz stimulation (Figure 4C) (Wt: n=4/3, 94.4±1.8%; Tg-GluN2A$^{2B(CT)}$: n=6/4, 115.5±3.9%). There is no statistical difference at the 3 Hz stimulation (Figure 4D) (Wt: n=7/3, 70.3±11.3%; Tg-GluN2A$^{2B(CT)}$: n=5/3, 79.2±13.3%) or 1 Hz stimulation (Figure 4E) (Wt: n=7/5, 82.4±1.6%; Tg-GluN2A$^{2B(CT)}$: n=6/3, 95.4±15.6%). Overall, we found that the Tg-GluN2A$^{2B(CT)}$ mice show little difference in the LTD, except at 5 Hz, but show significantly enhanced LTP around at 10 Hz frequency (Figure 4F). These data indicate that the GluN2A$^{2B(CT)}$ overexpression produced synaptic changes that were more similar to that of GluN2B overexpression in the transgenic mice and rats [14,16,18].

Enhanced 10 Hz LTP and diminished 1–3 Hz LTD in the Tg-GluN2B$^{2A(CT)}$ CA1 region

We then measured synaptic plasticity in the Tg-GluN2B$^{2A(CT)}$ CA1 slices. Overexpression of GluN2B$^{2A(CT)}$ significantly increased LTP versus their wild-type littermates at both 100 Hz (Figure 5B) (n=13/6, 176.6±16.2%; Figure 5A) and 10 Hz (n=5/3, 180.7±33.0%) frequencies. Interestingly, while 10 Hz response did not differ, LTD was also significantly impaired as compared to the wild-type hippocampal slices at 5 Hz (n=4/2, 121.6±2.2%; Figure 5C), 3 Hz (n=6/3, 103.6±13.8%; Figure 5D) and 1 Hz (n=21/13, 114.2±6.6%; Figure 5E). This shows that although the Tg-GluN2B$^{2A(CT)}$ mice have significantly increased LTP, they also have significantly blocked 1 Hz and 3 Hz induced LTD (summarized in Figure 5F). This decrease in LTD in Tg-GluN2B$^{2A(CT)}$ slices was more similar to that seen in the Tg-GluN2A mice [23].

Impaired 5 Hz responses in the Tg-GluN2D$^{2B(CT)}$ CA1 region

Finally, we examined the effects of GluN2D$^{2B(CT)}$ overexpression on CA1 plasticity. Since GluN2D has very weak magnesium dependency but much greater opening duration, it would lead to significantly more Ca^{2+} influx into the postsynaptic sites. We performed LTP and LTD measurements on the Tg-GluN2D$^{2B(CT)}$ hippocampal slices. Interestingly, we found no differences at either the 100 Hz (Figure 6A) (n=8/6, 128±8.6%) or the 10 Hz frequency between the transgenic and control littermates (Figure 6B) (n=6/3, 125.5±12.6%). However, at the 5 Hz frequency, a small, but significant LTP was observed in the transgenic slices, in comparison to the LTD induced in the slices from the control littermates (Figure 6C) (n=5/4, 123.5±3.6%). However, there were no significant differences observed in LTD at 1 Hz stimulation (Figure 6E) (n=8/4, 88.6±7.6%), 3 Hz (Figure 6D) (n=7/5, 87.6±12.4%). The summary graphs show the overall similarities between the Tg-GluN2D$^{2B(CT)}$ mice and their wild-type counterparts, except in its 5 Hz frequency response (Figure 6F).

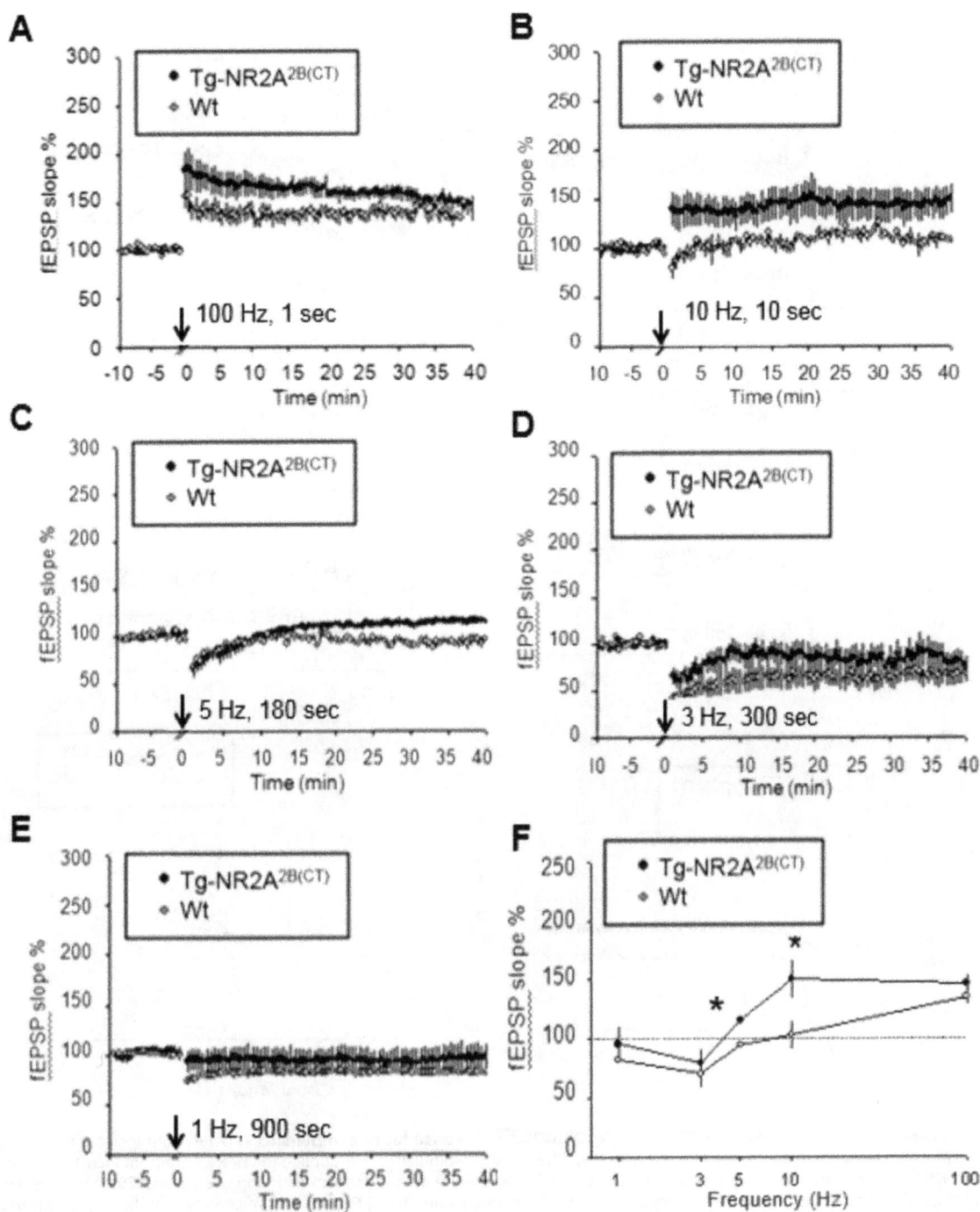

Figure 4. Enhanced LTP in the Tg-GluN2A$^{2B(CT)}$ mouse hippocampal slices. A. Slightly enhanced LTP seen in the Tg-GluN2A$^{2B(CT)}$ mice with a 1 s 100 Hz stimulation. B. Significantly enhanced LTP was seen in the Tg-GluN2A$^{2B(CT)}$ mice when a 10 Hz stimulation was applied from 10 s. C–E. No changes in LTD were seen in the 5 Hz, 3 Hz, or 1 Hz stimulation protocols. F. A summary plot of the % change in fEPSP slope versus the frequencies.

Discussion

The NMDA receptor is widely known as the key coincidence-detector at central synapses to implement Hebb's learning rule. The channel opening-duration and the level of membrane depolarization, determines the amount of Ca^{2+} that influxes into the cell [30,31]. In this study, we have identified the critical molecular motifs of the GluN2 subunits essential for achieving learning and memory enhancement in the adult mouse brain. By systematically analyzing three chimeric GluN2 transgenic mice together with Tg-GluN2A and Tg-GluN2B mice, we have tested two major hypotheses, namely, synaptic coincidence-detection/calcium influx hypothesis vs. GluN2B C-terminal intracellular signaling hypothesis in gating memory enhancement. Our experiments have revealed several novel insights into the relationships between GluN2 subunit motifs, synaptic plasticity, and memory enhancement.

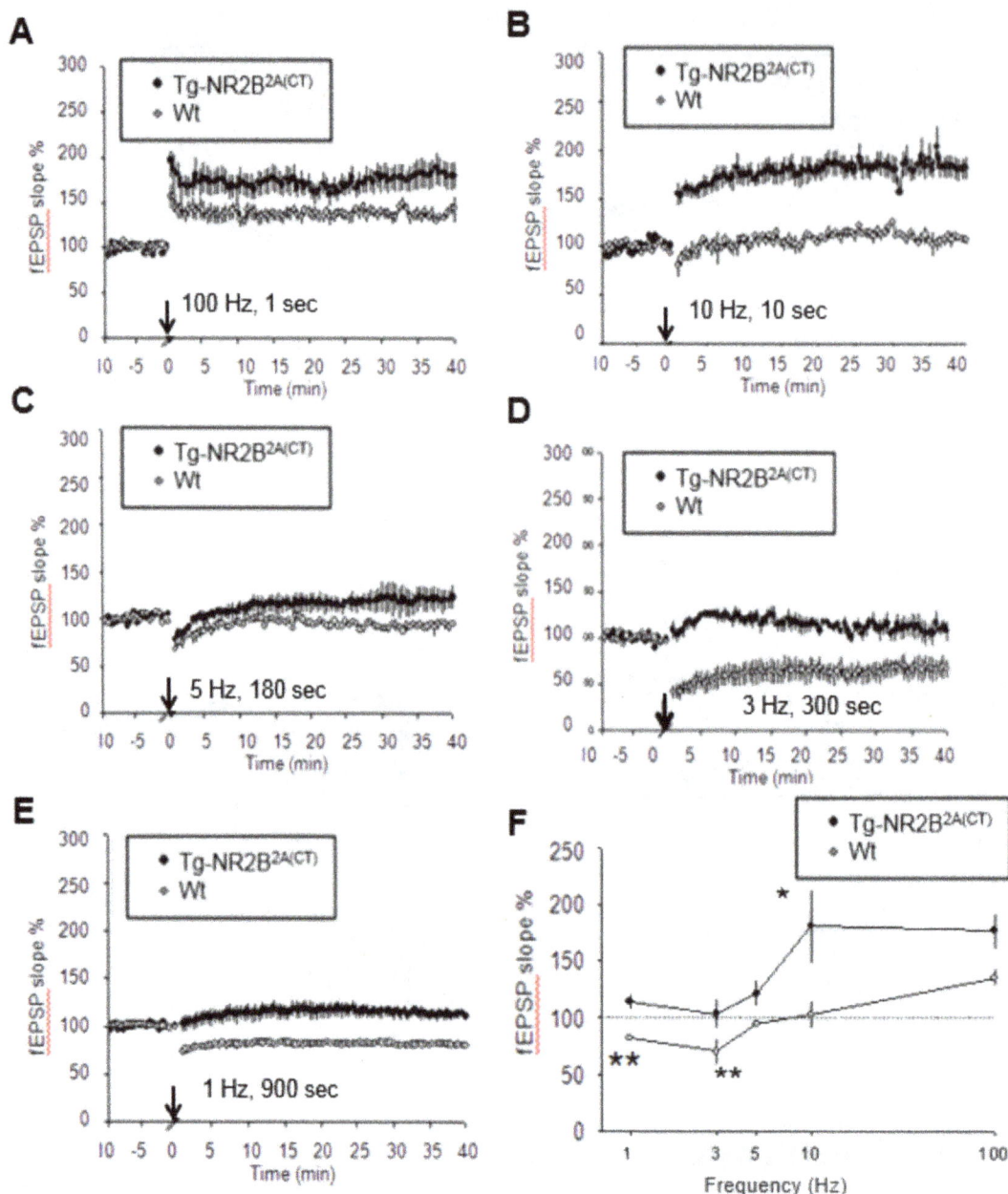

Figure 5. Enhanced LTP and diminished LTD in the Tg-GluN2B$^{2A(CT)}$ mouse hippocampal slices. (**A**) A slight increase in LTP was seen in the 100 Hz stimulation protocol in the Tg-GluN2B$^{2A(CT)}$ mice. (**B**) When a 10 Hz stimulation was applied for 10 s a significant increase in LTP was seen in the Tg-GluN2B$^{2A(CT)}$ mice over their wild-type littermates. (**C**) LTD was diminished at 5 Hz stimulation in the Tg-GluN2B$^{2A(CT)}$ mice. (**D**) LTD was significantly diminished at the 3 Hz stimulation protocol. (**E**) At the 1 Hz stimulation, the Tg-GluN2B$^{2A(CT)}$ mice show significantly diminished LTD. (**F**) A summary plot of the % change in fEPSP slope versus the frequencies.

The "synaptic coincidence-detection" hypothesis reflects the predominant view in the field as the rate-limiting factor in determining learning and memory capability. It posits that because the GluN2B subunit makes the channel opening duration longer than that of the GluN2A subunit, GluN2B overexpression allows a greater coincidence detection window, thereby leading to superior memory functions [4]. Because the N-terminal and transmembrane domains of the GluN2 subunits are known to be crucial for controlling voltage-gating and ion (Ca^{2+}) influx duration, we replaced the GluN2B N-terminal domains with either GluN2A or GluN2D while retaining its wild-type C-terminal intracellular domain. As such, these two chimeric GluN2

subunits possessed the GluN2B intracellular signaling capability but with the other key properties such as the shorter opening duration and the voltage-dependency from the GluN2A and GluN2D, respectively.

As predicted, because GluN2D has greatly reduced Mg^{2+} dependency, which renders synaptic coincidence-detection ineffective, GluN2D$^{2B(CT)}$ transgenic mice indeed exhibited memory deficits in novel object recognition (in both the short-term and long-term form) and long-term cued fear conditioning memory (although the contextual fear memory seemed to be normal). These observations have provided evidence that synaptic coincidence-detection is necessary for producing memory enhancement

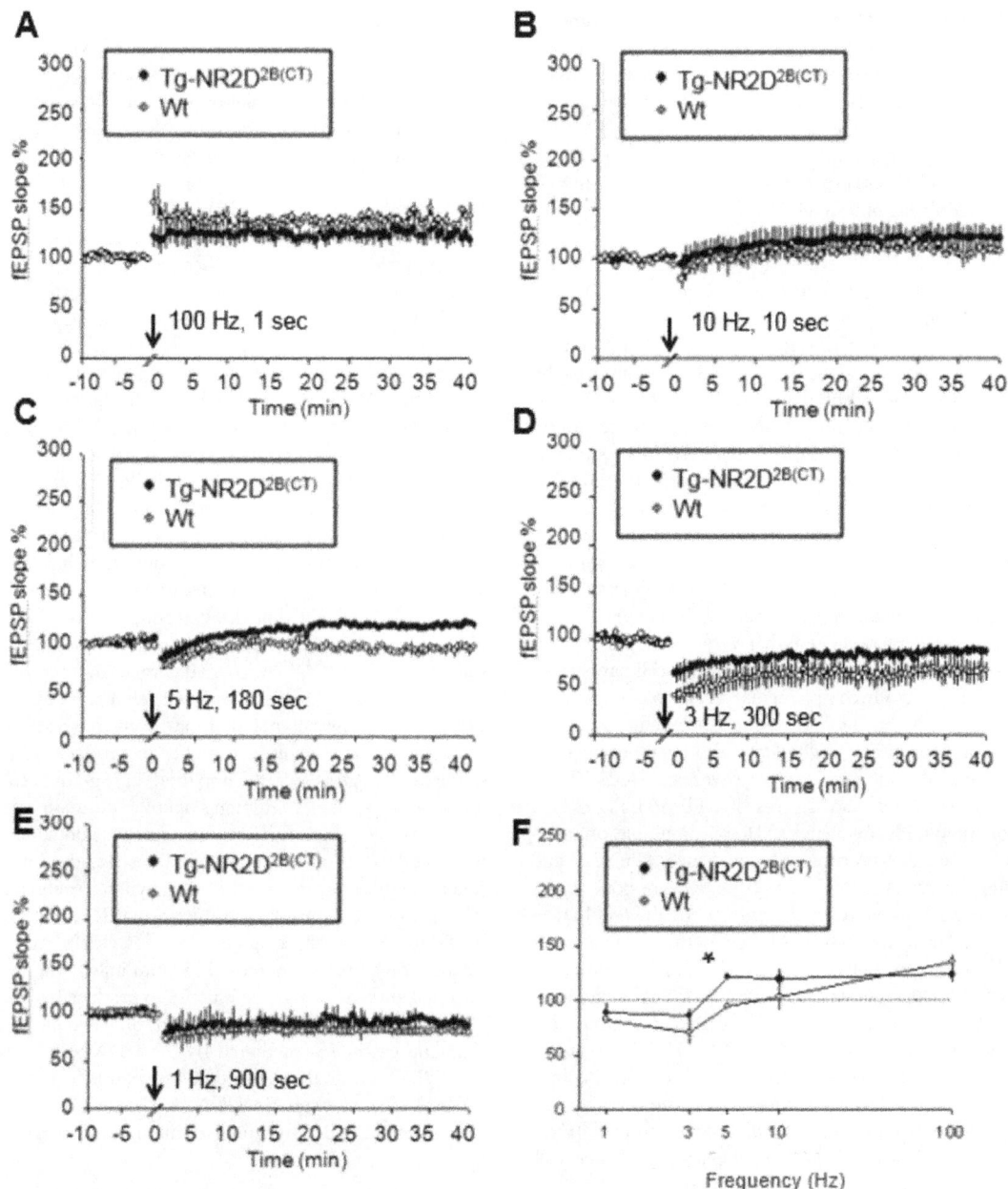

Figure 6. Diminished LTD at the 5 Hz range in the Tg-GluN2D$^{2B(CT)}$ mouse hippocampal slices. (**A**) No changes from the wild-type hippocampal slices were seen in the LTP of the Tg-GluN2D$^{2B(CT)}$ mouse hippocampal slices when a 100 Hz stimulation was applied. (**B**) When a 10 Hz stimulation was applied for 10 s there, again was no significant change in LTP observed in the Tg-GluN2D$^{2B(CT)}$ mice over their wild-type littermates. (**C**) LTD was diminished at 5 Hz stimulation in the Tg-GluN2D$^{2B(CT)}$ mice. (**D**) LTD was not significantly diminished at the 3 Hz stimulation protocol. (**E**) At the 1 Hz stimulation, the Tg-GluN2D$^{2B(CT)}$ mice showed no significant differences in LTD. (**F**) A summary plot of the % change in fEPSP slope versus the frequencies.

via the GluN2B intracellular signaling cascades. Without proper magnesium dependent voltage gating, the presence of the overexpressed GluN2B domain from the chimeric GluN2D$^{2B(CTD)}$ subunit still could not produce optimal synaptic changes for memory enhancement.

On the other hand, we were quite surprised that the Tg-GluN2A$^{2B(CT)}$ transgenic mice exhibited a very similar memory enhancement phenotype to those of the Tg-GluN2B mice. The different channel opening-durations derived from GluN2A and GluN2B subunits' N-terminal and transmembrane domains are not the most critical factor in determining the memory

enhancement, as long as the GluN2B C-terminal domain is transducing the signaling. It is important to note here that Punnakkal et al. found little differences in the whole cell currents of similar chimeric constructs, with only a slight decrease in the peak amplitude of a similar GluN2AB construct from that of the GluN2A wildtype subunit [32]. No changes in the peak amplitude of a similar GluN2BA construct over that of the GluN2B wildtype subunit. Additionally, deactivation times remained unchanged between the wildtype and chimeric receptors. Importantly, they also concluded that the peak opening probability appeared to be determined by the GluN2

N-terminal domain [32]. Therefore, these genetic experiments have shown that Mg^{2+}-dependent coincidence-detection function, but not necessarily the opening-duration difference between GluN2A and GluN2B, is prerequisite for achieving learning and memory enhancement in the adult brain.

Interestingly our present study has provided clear evidence supporting the second hypothesis that is known as the "GluN2B intracellular domain" hypothesis [4,33]. Two separate pieces of evidence came from our behavioral analyses of the Tg-GluN2-B$^{2A(CT)}$ and Tg-GluN2A$^{2B(CT)}$ transgenic mice. First, we found that the Tg-GluN2A$^{2B(CT)}$ mice had enhanced object recognition memory and emotional memory. These phenotypes are very similar to those of the Tg-GluN2B mice (and also Tg-GluN2B rats) (Figure 7). On the contrary, when the C-terminal domain of the GluN2B subunit was replaced by that of the GluN2A subunit, as we did in Tg-GluN2B$^{2A(CT)}$ mice, this swap led to profound memory deficits in novel object recognition test and long-term cued fear memories. These memory deficits mirrored those of Tg-GluN2A mice [23]. These subunits-swap experiments, by extending to learning and memory enhancement, are consistent with other reports that the intracellular domains of the GluN2 subunits play critical roles in mediating different functions, such as synaptic localization, clustering, signal transduction, and behaviors [22,33–39]. Therefore, our studies suggest that both "synaptic coincidence-detection" hypotheses and "GluN2B intracellular signaling" hypothesis are mutually complementary in term of explaining the molecular determinants for memory enhancement.

Two additional conceptual insights have also been obtained on how different molecular motifs of the overexpressed GluN2 subunits regulate the levels and degrees of LTP or LTD over a wide range of stimulation frequencies. Despite multiple pharmacological and knockout approaches to analyzing GluN2A and GluN2B on regulating LTP and LTD [40], few were done under the context of examining its relationship with cognitive enhancement [16,18,41]. Here, we consistently found that GluN2B and GluN2A$^{2B(CT)}$ overexpression enhanced LTP in the range of 10 Hz and/or 100 Hz range without significantly affecting 1 Hz or 3 Hz LTD. It is noteworthy to point out that that Tg-GluN2A$^{2B(CT)}$ seemed to produce larger 100 Hz LTP in the initial 20~30 minutes range than that of wild-type slices, but become indistinguishable by the 40-minutes time points. Interestingly, the Tg-GluN2B overexpression tended to produce much larger 100 Hz induced LTP in comparison to the wild-type mice well

beyond 60 minutes [16]. This indicates the longer time duration of channel opening (thereby more calcium influx) via the GluN2B N-terminal and pore does make larger and more stable LTP in response to 100 Hz stimulation [42,43]. However, at the 10 Hz stimulation range, the amount of calcium influx via the GluN2A N-terminal and pore domains (coupled to GluN2B C-terminal region) can produce the similarly larger LTP in the Tg-GluN2A$^{2B(CT)}$ mice as that of the Tg-GluN2B mice in comparison to that of the wild-type controls. This 10 Hz stimulation frequency can be particularly interesting because we have observed that fear conditioning-induced firing increase in CA1 pyramidal cells is mostly in the range of 5~30 Hz [6,44]. This behaviorally relevant frequency range deserves special investigation for memory enhancement in future experiments both in the hippocampus and other brain regions such as the prefrontal cortex and amygdala. In addition, contrary to LTD produced by 5 Hz stimulation in the wild-type slices, Tg-GluN2D$^{2B(CT)}$ slices exhibited a significant switch to LTP. Taken together, these findings have provided additional support for the notion that the GluN2B C-terminal domain plays a key role in regulating LTP [26,45–48], and more importantly, our study has further defined, for the first time, its essential link to memory enhancement.

While it is evident that the C-terminal of the GluN2B subunit plays a crucial role in producing synaptic potentiation, we found that Tg-GluN2B$^{2A(CT)}$ mice had larger, more robust, LTP not only at 10 or 100 Hz. Intriguingly, such a swap also promoted an overall shift toward potentiation even in response to lower frequencies. As a result, the ability to produce LTD at 1–3 Hz frequency range is greatly impaired in Tg-GluN2B$^{2A(CT)}$ slices. These findings show that longer opening duration achieved by the overexpressed GluN2B N-terminal and pore domains, but coupled to the GluN2A intracellular signaling cascade, brings a greater potentiation but at the cost of losing synaptic depression capacity. This is in stark contrast with the normal LTD in Tg-GluN2B or Tg-GluN2A$^{2B(CT)}$ in response to 1 or 3 Hz stimulation. This strongly suggests that increased calcium influx (via the GluN2B N-terminal and core domains) is useful to produce bigger LTP, but its effect on 1 Hz LTD critically depends on whether the downstream signaling cascade is mediated by the GluN2A C-terminal tail or GluN2B C-terminal tail. In other words, under such circumstance, the presence of chimeric GluN2A C-terminal domain, but not chimeric GluN2B C-terminal domain, can override LTD. This

	LTP	LTD	Novel Object Recognition	Contextual Fear Conditioning	Cued Fear Conditioning
Tg-NR2A$^{2B(CT)}$	Enhanced at 10 Hz	Impaired at 5Hz	Enhanced at 3 days	Enhanced	Normal
Tg-NR2B$^{2A(CT)}$	Enhanced	Significantly Impaired	Impaired at 1 hour Normal at 24 hour	Normal	Impaired at 24 hours
Tg-NR2D$^{2B(CT)}$	Normal	Decreased at 5 Hz	Impaired at 24 hours	Normal	Impaired at 24 hours

Figure 7. Summary of LTP, LTD and behavioral tasks results. The Tg-NRA$^{2B(CT)}$ mice had enhanced LTP, as well as enhanced long-term recognition memory and contextual fear conditioning. The Tg-GluN2B$^{2A(CT)}$ mice have significantly impaired LTD resulting in impaired short-term recognition memory and impaired long-term cued fear conditioning. The Tg-GluN2D$^{2B(CT)}$ mice have decreased LTD at 5 Hz and impaired long-term recognition memory and long-term cued fear memory.

novel insight adds to the notion that GluN2A may have a general ability to drive toward LTP [49–54].

In addition, by taking advantage of the correlational analysis between synaptic changes and memory performances, our present study has uncovered two detailed insights into the memory enhancement strategy: first, bigger LTP would lead to better learning and memory, however, only if the LTD ability remains intact. This is supported by the observation that bigger CA1 LTP is associated with better memory in Tg-GluN2B and Tg-GluN2A$^{2B(CT)}$ mice while their LTD was not altered. Second, if LTP enhancement results in overriding or diminishing LTD capacity, such as those observed in Tg-GluN2B$^{2A(CT)}$, it would also lead to memory deficits. The Tg-GluN2B$^{2A(CT)}$ phenotypes are more similar to the knockout of PSD-95 which also leads to larger LTP, lack of 1 Hz LTD, and memory deficits (i.e. [55,56]). Our recent characterization of Tg-GluN2A mice showed that overexpression of GluN2A results in no change in 100 Hz LTP or 1 Hz, but greatly impaired 3 or 5 Hz LTD. These mice also exhibited long-term memory deficits, while short-term memories remained mostly normal. Our Tg-GluN2D$^{2B(CT)}$ mice, which also showed 5 Hz LTP responses instead of either no change or LTD, as in the wild-type mice, were also profoundly impaired in long-term memory. These observations support the "LTD-memory trace sculpting" hypothesis, that the weakening of uncorrelated synaptic connections would reduce the background "noise" while enabling the stabilization (or crystallization) of the learning-related synaptic patterns [23]. However, it is important to note that our current electrophysiological recordings were limited to the CA1 region. Given the fact that we used the CaMKII promoter to drive the transgenes, electrophysiological analyses should be extended in future experiments into other brain regions such as the amygdala and prefrontal cortex from which cued fear learning and fear extinction are processed. Clearly, simple correlation between CA1 synaptic plasticity and memory are likely not sufficient for counting memory enhancement and thus, any extrapolation should be only taken with great caution. In addition, little is known about how any of the artificial stimulation paradigms for producing LTP or LTD can be translated into real-time memory patterns. Recent successful decoding of real-time fear memory traces in the hippocampal CA1 from the wild-type mice and the forebrain excitatory neuron-specific NMDA receptor inducible knockout mice, have revealed many fundamental insights how the NMDA receptors regulate real-time memory code and memory engrams [6,44]. It would be of great interest to use such brain decoding technologies to investigate the various transgenic mice described here.

In summary, our above experiments have identified the key molecular and genetic determinants that would be necessary and sufficient for achieving superior learning and memory ability in the adult brain. Although transgenetic methods are unlikely to be used for human clinical settings, the C-terminal region of the GluN2B subunit contains many important sites for various molecular interaction including with CaMKII, cdk5, and Kinesin superfamily protein 17 (KIF17) [57–60]. Indeed, manipulations of cdk5 and KIF17 which result in upregulation of GluN2B also resulted in memory enhancement [61,62]. More recently, researchers have taken a novel, dietary approach to up-regulate GluN2B expression in the brain via elevating brain magnesium [63]. They showed that the compound, magnesium threonate, can cross the blood brain barrier efficiently and boost GluN2B expression in the neurons, and subsequent memory improvement in both aging and wild-type mice [64,65]. This compound is currently under clinical trials [66]. Therefore, it is conceivable that knowledge gained from the present study will

be valuable to the current efforts in developing and optimizing memory enhancement strategies.

Methods

Production of Transgenic Mice

We have produced three chimeric GluN2 subunit constructs for the present study. In the first two constructs, the N-terminal and transmembrane domains of GluN2A or GluN2D subunit were fused with the C-terminal domain from the GluN2B subunit, termed GluN2A$^{2B(CTR)}$ and GluN2D$^{2A(CTR)}$, respectively. In the third construct, we also fused the N-terminal and transmembrane domain of GluN2B subunit with the C-terminal domain from the GluN2A subunit, termed GluN2B$^{2A(CTR)}$ (see Figure 1A). The fusion site was located near the end of the fourth transmembrane domain, just before the C-terminal domains begin. For making the constructs, we first introduced a point mutation to create a unique Aat II cutting site for the fusing of the given C-terminal domain (Figure 1B). Upon successful ligation, the point mutation was mutated back to its original sequence. These chimeric transgene constructs were driven by the forebrain-specific αCaMKII promoter for targeting their expression to the excitatory neurons in the forebrain regions such as the cortex and hippocampus.

The chimeric constructs were created by first introducing a point-mutation at the site to create a unique Aat II cutting site for swapping the NT and CT domains. Upon successful ligation the point mutation was swapped back to the original sequence. The modified subunit was targeted for forebrain expression by the CaM-kinase II (CaMKII) promoter as previously described [16,67]. The founding line of transgenic animals was produced by pronuclear injection of a linearized chimeric transgene vector into C57BL/6J zygotes similar to previously described [16,68]. A total of seven independent mouse founder lines (three lines for GluN2A$^{2B(CT)}$ termed "Tg-GluN2A$^{2B(CT)}$" mice, two lines for GluN2B $^{2A(CT)}$ termed "Tg-GluN2B$^{2A(CT)}$" mice, and two lines for the GluN2D$^{2B(CTR)}$ termed "Tg-GluN2D$^{2B(CT)}$" mice). All these lines gave successful germline transmissions. The genotypes of the transgenic mice were determined by PCR analysis of a tail biopsy. The transgene was detected using the SV40 poly(A) sequence, as previously described [16,18,41]. Southern blotting was used to confirm the transgene integration in to the transgenic mouse line. Western blotting of the forebrain regions (cortex and hippocampus) was visualized with either a polyclonal GluN2A C-terminal antibody (Upstate/Millipore) or a polyclonal GluN2B C-terminal antibody (Millipore). For the present electrophysiological and behavioral experiments, we used a high expression line chosen from the Tg-GluN2A$^{2B(CT)}$ mice, the Tg-GluN2B$^{2A(CT)}$ mice, and the Tg-GluN2D$^{2B(CT)}$ mice that have been crossed with C57BL/6J wildtype mice for at least 8 generations. For in situ hybridizations, brains from the transgenic mice and wild-type littermates were isolated and 20 μm sections were prepared using a cryostat. The slices were hybridized to the $[\alpha^{35}S]$ oligonucleotide probe which hybridized to the untranslated artificial intron region in the transgene similarly to previously described [16,23].

Behavioral Experiments

Mice were maintained in a temperature and humidity controlled vivarium with a 12:12 light-dark cycle. All testing was done during the light phase with 3–5 month old animals. Mice were allowed free access to food and water, except during experimental procedures. Mice were extensively handled prior to any testing paradigm. Separate cohorts were used for each study and each recall time point unless otherwise stated. All testing procedures were conducted in sound dampened, dimly lit behavioral rooms. Experimenters were blind to the genotype of the animals. This study was carried out in strict accordance with the recommendations in the Guide for the Care and Use of Laboratory Animals of the National Institutes of Health. The protocols were approved by the Institutional Animal Care and Use Committee of the Georgia Regents University.

Open Field

One cohort of Tg-GluN2A, Tg-GluN2B, Tg-GluN2A$^{2B(CT)}$, Tg-GluN2B$^{2A(CT)}$, and Tg-GluN2D$^{2B(CT)}$ mice and their wild-type littermates were individually placed into a 50 cm L×50 cm W×25 cm H white Plexiglas open field arena. The mouse was allowed to explore for ten minutes. The time that the mouse spent in the center and periphery was determined. The periphery of the open field was considered to be the first four inches along the wall, while the center of the open field was the square inside this area [69]. Additionally, the distance traveled by the mouse was determined using Biobserve Viewer II software.

Elevated Plus Maze

The elevated plus maze consisted of a black Plexiglas "plus" maze approximately 60 cm above the floor, with each arm measuring 30 cm in length and 10 cm wide. Two opposite arms were left open, with the other two arms being enclosed on three sides. The ambient room lighting was 75 lux. The amount of time the mice spent within the enclosed arms was recorded, as well as the amount of time the animal spent in the open arms [69]. The times were used to determine a preference index.

Novel Object Recognition

The behavioral paradigm was the same as previously described [16,70]. The mice were individually habituated to a 50 cm L×50 cm W×25 cm H open field apparatus for 10 minutes a day for three days. On the first testing day, the mice were placed into the open field with two identical objects for 5 minutes. The time they spent exploring each object was recorded. At the described retention time the mice were placed back into the open field arena with one of the familiar objects used in training, and one novel object, and allowed to explore for 5 minutes. The time they spent with each object was recorded and used to determine a preference index. Different groups of mice were used for the each retention session.

Fear Conditioning

An operant chamber (25 cm L×25 cm W×38 cm H) equipped with activity monitors and camera was used. The flooring was a 24 bar shock grid with a speaker, shock generator, and photo-beam scanner (MedAssociates). The chamber was located in a sound damping isolation box. The apparatus was thoroughly cleaned with 70% ethanol between mice to avoid any olfactory cues. Freezing was monitored by the software and confirmed by the experimenter.

Testing procedures were similar to those previously described [16,71,72]. Animals were habituated to the testing environment for 5 minutes one day before testing. On the day of training the mice were placed into the chamber and allowed to explore for 5 minutes. Then the mice were exposed to a conditioned stimulus (CS, 85 dB tone at 2800 Hz), with the unconditioned stimulus (US, a scrambled foot shock at 0.75 mA) occurring the last 2 seconds of the CS. The mice were allowed to stay in the chamber for 30 seconds after the CS/US pairing to monitor immediate freezing.

To test the contextual freezing exhibited by the animal, at the described time (1 hour or 24 hours) the trained mice were placed back into the shock chamber for 5 minutes while their freezing response was monitored. The mice were then placed into a novel chamber and monitored for their freezing response (pre-tone) for 3 minutes before the onset of the CS tone for 3 minutes. During the tone the animal's freezing response was monitored to test the cued fear retention.

Freezing was judged as the complete immobility of the animal, except for movement necessary for respiration. The mice were then returned to their home cage for either 1 hour or 24 hours. At the described time the mice were returned to the chamber for measurement of the contextual freezing. The mice were then placed in a novel chamber and the tone was delivered for 3 min, during which their cued freezing response was monitored. To test the fear extinction of the animals the same recall testing paradigm was repeated at 2 hour intervals for four additional trials.

Statistical analysis of behavioral data

All behavioral data are presented as mean ± SEM. Significance was determined by ANOVA analysis with Tukey-Kramer, or a Student's t-test. P values of <0.05 were considered significant.

Hippocampal Slice Recordings

Transverse slices of the hippocampus were rapidly prepared from wild-type and Tg-GluN2A$^{2B(CT)}$, Tg-GluN2B$^{2A(CT)}$, and Tg-GluN2D$^{2B(CT)}$ mouse lines (3~6 months old) and maintained in an interface chamber at 28°C and were subfused with artificial cerebral spinal fluid (ACSF, 124 mM NaCl, 4.4 mM KCl, 2.0 mM CaCl$_2$, 1.0 mM MgSO$_4$, 25 mM NaCHO$_3$, 1.0 mM Na$_2$HPO$_4$ and 10 mM glucose) and bubbled with 95% O$_2$ and 5% CO$_2$. Slices were kept in the recording chamber for at least two hours. A bipolar tungsten stimulating electrode was placed in the stratum radiatum in the CA1 region. A glass microelectrode (3–12 MΩ) filled with ACSF was used to measure the extracellular field potentials in the stratum radiatum. Test response elicited at 0.02 Hz. Current intensity (0.5–1.2 mA) which produced 30% of maximal response was used for studies of PPF and synaptic plasticity at different frequencies. Various interpulse intervals (20–400 msec) were used for measuring PPF. Low-frequency stimulation of (5 Hz for 3 min, 3 Hz for 300 s, or 1 Hz for 900s) was then used to produce depotentiation [73]. Long term potentiation was induced by tetanic stimulation (100 Hz for 1 s and 10 Hz stimulation for 10 s). Data are expressed as mean ± SEM. One-way ANOVA (with Duncan's multiple range test for *post hoc* comparison) and Student's t-test were used for statistical analysis. The detailed procedures were the same as described (Tang et al. 1999; Shimizu, et al. 2000; Wang et al. 2003; Wang, et al. 2008).

Supporting Information

Figure S1 Conformation of the integration of the transgene. (**A**) Southern Blot analysis of Tg-GluN2(A/B), Tg-GluN2(B/A), and Tg-GluN2(D/B) mice. The numbers indicate the positive control and the copy number. (**B**) Western Blot analysis of the chimeric animals showing enhanced expression of the GluN2A C-terminal domain in the Tg-GluN2A, Tg-GluN2(B/A) mice and no enhancement of the GluN2A C-terminus in the Tg-GluN2(A/B) and Tg-GluN2(D/B) mice. (**C**) Western Blot analysis of the chimeric animals showing enhanced expression of the GluN2B C-terminal tail in the Tg-GluN2(A/B), and Tg-GluN2(D/B) mice of the expression in the wild-type mice.

Figure S2 Electrophysiology of hippocampal slices. (A) There were no significant differences in the basal synaptic transmission as seen in the CA3-CA1 input-output curve between the wildtype mice and the transgenic mice. **(B)**. The paired-pulse facilitation was unchanged between the wildtype and the chimeric transgenic mice indicating that the presynaptic function is unchanged.

Acknowledgments

The authors would like to thank Philip Wang and Shuqin Zhang for his valuable assistances with behavioral experiments, as well as Fengying Huang for animal colony maintenance.

Author Contributions

Conceived and designed the experiments: ZC JZT. Performed the experiments: SJ ZC RF HW DW. Analyzed the data: SJ ZC RF HW DW. Contributed reagents/materials/analysis tools: JZT. Wrote the paper: SJ JZT.

References

1. Bliss TV, Collingridge GL (1993) A synaptic model of memory: long-term potentiation in the hippocampus. Nature 361: 31–39.
2. Stevens CF, Sullivan J (1998) Synaptic plasticity. Curr Biol 8: R151–153.
3. Bear MF, Malenka RC (1994) Synaptic plasticity: LTP and LTD. Curr Opin Neurobiol 4: 389–399.
4. Tsien JZ (2000) Linking Hebb's coincidence-detection to memory formation. Curr Opin Neurobiol 10: 266–273.
5. Wang H, Hu Y, Tsien JZ (2006) Molecular and systems mechanisms of memory consolidation and storage. Prog Neurobiol 79: 123–135.
6. Zhang H, Chen G, Kuang H, Tsien JZ (2013) Mapping and Deciphering Neural Codes of NMDA Receptor-Dependent Fear Memory Engrams in the Hippocampus. PLoS One 8: e79454.
7. Mayer ML, Westbrook GL, Guthrie PB (1984) Voltage-dependent block by Mg2+ of NMDA responses in spinal cord neurones. Nature 309: 261–263.
8. Monyer H, Sprengel R, Schoepfer R, Herb A, Higuchi M, et al. (1992) Heteromeric NMDA receptors: molecular and functional distinction of subtypes. Science 256: 1217–1221.
9. Sheng M, Cummings J, Roldan LA, Jan YN, Jan LY (1994) Changing subunit composition of heteromeric NMDA receptors during development of rat cortex. Nature 368: 144–147.
10. Cull-Candy SG, Leszkiewicz DN (2004) Role of distinct NMDA receptor subtypes at central synapses. Sci STKE 2004: re16.
11. Erreger K, Dravid SM, Banke TG, Wyllie DJ, Traynelis SF (2005) Subunit-specific gating controls rat NR1/NR2A and NR1/NR2B NMDA channel kinetics and synaptic signalling profiles. J Physiol 563: 345–358.
12. Dingledine R, Borges K, Bowie D, Traynelis SF (1999) The glutamate receptor ion channels. Pharmacol Rev 51: 7–61.
13. Monyer H, Burnashev N, Laurie DJ, Sakmann B, Seeburg PH (1994) Developmental and regional expression in the rat brain and functional properties of four NMDA receptors. Neuron 12: 529–540.
14. Cui Y, Jin J, Zhang X, Xu H, Yang L, et al. (2011) Forebrain NR2B overexpression facilitating the prefrontal cortex long-term potentiation and enhancing working memory function in mice. PLoS One 6: e20312.
15. Jacobs SA, Tsien JZ (2012) Genetic overexpression of NR2B subunit enhances social recognition memory for different strains and species. PLoS One 7: e36387.
16. Tang YP, Shimizu E, Dube GR, Rampon C, Kerchner GA, et al. (1999) Genetic enhancement of learning and memory in mice. Nature 401: 63–69.
17. Tang YP, Wang H, Feng R, Kyin M, Tsien JZ (2001) Differential effects of enrichment on learning and memory function in NR2B transgenic mice. Neuropharmacology 41: 779–790.
18. Wang D, Cui Z, Zeng Q, Kuang H, Wang LP, et al. (2009) Genetic enhancement of memory and long-term potentiation but not CA1 long-term depression in NR2B transgenic rats. PLoS One 4: e7486.
19. Nowak L, Bregestovski P, Ascher P, Herbet A, Prochiantz A (1984) Magnesium gates glutamate-activated channels in mouse central neurones. Nature 307: 462–465.
20. Gielen M, Siegler Retchless B, Mony L, Johnson JW, Paoletti P (2009) Mechanism of differential control of NMDA receptor activity by NR2 subunits. Nature 459: 703–707.
21. Sprengel R, Suchanek B, Amico C, Brusa R, Burnashev N, et al. (1998) Importance of the intracellular domain of NR2 subunits for NMDA receptor function in vivo. Cell 92: 279–289.
22. Ryan TJ, Kopanitsa MV, Indersmitten T, Nithianantharajah J, Afinowi NO, et al. (2013) Evolution of GluN2A/B cytoplasmic domains diversified vertebrate synaptic plasticity and behavior. Nat Neurosci 16: 25–32.
23. Cui Z, Feng R, Jacobs S, Duan Y, Wang H, et al. (2013) Increased NR2A: NR2B ratio compresses long-term depression range and constrains long-term memory. Sci Rep 3: 1036.
24. Kennedy MB, Beale HC, Carlisle HJ, Washburn LR (2005) Integration of biochemical signalling in spines. Nat Rev Neurosci 6: 423–434.
25. Falls WA, Miserendino MJ, Davis M (1992) Extinction of fear-potentiated startle: blockade by infusion of an NMDA antagonist into the amygdala. J Neurosci 12: 854–863.
26. Bartlett TE, Bannister NJ, Collett VJ, Dargan SL, Massey PV, et al. (2007) Differential roles of NR2A and NR2B-containing NMDA receptors in LTP and LTD in the CA1 region of two-week old rat hippocampus. Neuropharmacology 52: 60–70.
27. Hrabetova S, Serrano P, Blace N, Tse HW, Skifter DA, et al. (2000) Distinct NMDA receptor subpopulations contribute to long-term potentiation and long-term depression induction. J Neurosci 20: RC81.
28. Xu Z, Chen RQ, Gu QH, Yan JZ, Wang SH, et al. (2009) Metaplastic regulation of long-term potentiation/long-term depression threshold by activity-dependent changes of NR2A/NR2B ratio. J Neurosci 29: 8764–8773.
29. Peng Y, Zhao J, Gu QH, Chen RQ, Xu Z, et al. (2010) Distinct trafficking and expression mechanisms underlie LTP and LTD of NMDA receptor-mediated synaptic responses. Hippocampus 20: 646–658.
30. Wigstrom H, Gustafsson B (1986) Postsynaptic control of hippocampal long-term potentiation. J Physiol (Paris) 81: 228–236.
31. Lynch MA, Littleton JM (1983) Possible association of alcohol tolerance with increased synaptic Ca2+ sensitivity. Nature 303: 175–176.
32. Punnakkal P, Jendritza P, Kohr G (2012) Influence of the intracellular GluN2 C-terminal domain on NMDA receptor function. Neuropharmacology 62: 1985–1992.
33. Halt AR, Dallapiazza RF, Zhou Y, Stein IS, Qian H, et al. (2012) CaMKII binding to GluN2B is critical during memory consolidation. EMBO J 31: 1203–1216.
34. Sheng M (1996) PDZs and receptor/channel clustering: rounding up the latest suspects. Neuron 17: 575–578.
35. Kennedy MB (1997) The postsynaptic density at glutamatergic synapses. Trends Neurosci 20: 264–268.
36. Kennedy PR, Bakay RA (1997) Activity of single action potentials in monkey motor cortex during long-term task learning. Brain Res 760: 251–254.
37. Kornau HC, Seeburg PH, Kennedy MB (1997) Interaction of ion channels and receptors with PDZ domain proteins. Curr Opin Neurobiol 7: 368–373.
38. Steigerwald F, Schulz TW, Schenker LT, Kennedy MB, Seeburg PH, et al. (2000) C-Terminal truncation of NR2A subunits impairs synaptic but not extrasynaptic localization of NMDA receptors. J Neurosci 20: 4573–4581.
39. Martel MA, Ryan TJ, Bell KF, Fowler JH, McMahon A, et al. (2012) The subtype of GluN2 C-terminal domain determines the response to excitotoxic insults. Neuron 74: 543–556.
40. Shipton OA, Paulsen O (2014) GluN2A and GluN2B subunit-containing NMDA receptors in hippocampal plasticity. Philos Trans R Soc Lond B Biol Sci 369: 20130163.
41. Cao X, Cui Z, Feng R, Tang YP, Qin Z, et al. (2007) Maintenance of superior learning and memory function in NR2B transgenic mice during ageing. Eur J Neurosci 25: 1815–1822.
42. Zhang XL, Sullivan JA, Moskal JR, Stanton PK (2008) A NMDA receptor glycine site partial agonist, GLYX-13, simultaneously enhances LTP and reduces LTD at Schaffer collateral-CA1 synapses in hippocampus. Neuropharmacology 55: 1238–1250.
43. Berberich S, Jensen V, Hvalby O, Seeburg PH, Kohr G (2007) The role of NMDAR subtypes and charge transfer during hippocampal LTP induction. Neuropharmacology 52: 77–86.
44. Chen G, Wang LP, Tsien JZ. (2009). Neural population-level memory traces in the mouse hippocampus. PLoS One. 4(12): e8256.
45. Clayton DA, Mesches MH, Alvarez E, Bickford PC, Browning MD (2002) A hippocampal NR2B deficit can mimic age-related changes in long-term potentiation and spatial learning in the Fischer 344 rat. J Neurosci 22: 3628–3637.
46. Kohr G, Jensen V, Koester HJ, Mihaljevic AL, Utvik JK, et al. (2003) Intracellular domains of NMDA receptor subtypes are determinants for long-term potentiation induction. J Neurosci 23: 10791–10799.
47. Foster KA, McLaughlin N, Edbauer D, Phillips M, Bolton A, et al. (2010) Distinct roles of NR2A and NR2B cytoplasmic tails in long-term potentiation. J Neurosci 30: 2676–2685.
48. Gardoni F, Mauceri D, Malinverno M, Polli F, Costa C, et al. (2009) Decreased NR2B subunit synaptic levels cause impaired long-term potentiation but not long-term depression. J Neurosci 29: 669–677.
49. Liu HN, Kurotani T, Ren M, Yamada K, Yoshimura Y, et al. (2004) Presynaptic activity and Ca2+ entry are required for the maintenance of NMDA receptor-independent LTP at visual cortical excitatory synapses. J Neurophysiol 92: 1077–1087.
50. Erreger K, Chen PE, Wyllie DJ, Traynelis SF (2004) Glutamate receptor gating. Crit Rev Neurobiol 16: 187–224.

51. Berberich S, Punnakkal P, Jensen V, Pawlak V, Seeburg PH, et al. (2005) Lack of NMDA receptor subtype selectivity for hippocampal long-term potentiation. J Neurosci 25: 6907–6910.
52. Li YH, Han TZ (2008) Glycine modulates synaptic NR2A- and NR2B-containing NMDA receptor-mediated responses in the rat visual cortex. Brain Res 1190: 49–55.
53. Santucci DM, Raghavachari S (2008) The effects of NR2 subunit-dependent NMDA receptor kinetics on synaptic transmission and CaMKII activation. PLoS Comput Biol 4: e1000208.
54. Gerkin RC, Lau PM, Nauen DW, Wang YT, Bi GQ (2007) Modular competition driven by NMDA receptor subtypes in spike-timing-dependent plasticity. J Neurophysiol 97: 2851–2862.
55. Migaud M, Charlesworth P, Dempster M, Webster LC, Watabe AM, et al. (1998) Enhanced long-term potentiation and impaired learning in mice with mutant postsynaptic density-95 protein. Nature 396: 433–439.
56. Carlisle HJ, Fink AE, Grant SG, O'Dell TJ (2008) Opposing effects of PSD-93 and PSD-95 on long-term potentiation and spike timing-dependent plasticity. J Physiol 586: 5885–5900.
57. Gho M, McDonald K, Ganetzky B, Saxton WM (1992) Effects of kinesin mutations on neuronal functions. Science 258: 313–316.
58. Setou M, Nakagawa T, Seog DH, Hirokawa N (2000) Kinesin superfamily motor protein KIF17 and mLin-10 in NMDA receptor-containing vesicle transport. Science 288: 1796–1802.
59. Hawasli AH, Bibb JA (2007) Alternative roles for Cdk5 in learning and synaptic plasticity. Biotechnol J 2: 941–948.
60. Yin X, Takei Y, Kido MA, Hirokawa N (2011) Molecular motor KIF17 is fundamental for memory and learning via differential support of synaptic NR2A/2B levels. Neuron 70: 310–325.
61. Hawasli AH, Benavides DR, Nguyen C, Kansy JW, Hayashi K, et al. (2007) Cyclin-dependent kinase 5 governs learning and synaptic plasticity via control of NMDAR degradation. Nat Neurosci 10: 880–886.
62. Wong RW, Setou M, Teng J, Takei Y, Hirokawa N (2002) Overexpression of motor protein KIF17 enhances spatial and working memory in transgenic mice. Proc Natl Acad Sci U S A 99: 14500–14505.
63. Slutsky I, Abumaria N, Wu LJ, Huang C, Zhang L, et al. (2010) Enhancement of learning and memory by elevating brain magnesium. Neuron 65: 165–177.
64. Abumaria N, Yin B, Zhang L, Li XY, Chen T, et al. (2011) Effects of elevation of brain magnesium on fear conditioning, fear extinction, and synaptic plasticity in the infralimbic prefrontal cortex and lateral amygdala. J Neurosci 31: 14871–14881.
65. Abumaria N, Luo L, Ahn M, Liu G (2013) Magnesium supplement enhances spatial-context pattern separation and prevents fear overgeneralization. Behav Pharmacol 24: 255–263.
66. Cyranoski D (2012) Testing magnesium's brain-boosting effects. Nature News: Nature Publishing Group.
67. Tsien JZ, Chen DF, Gerber D, Tom C, Mercer EH, et al. (1996) Subregion- and cell type-restricted gene knockout in mouse brain. Cell 87: 1317–1326.
68. Tsien JZ, Huerta PT, Tonegawa S (1996) The essential role of hippocampal CA1 NMDA receptor-dependent synaptic plasticity in spatial memory. Cell 87: 1327–1338.
69. Wang LP, Li F, Wang D, Xie K, Shen X, et al. (2011) NMDA receptors in dopaminergic neurons are crucial for habit learning. Neuron 72: 1055–1066.
70. Wang H, Feng R, Phillip Wang L, Li F, Cao X, et al. (2008) CaMKII activation state underlies synaptic labile phase of LTP and short-term memory formation. Curr Biol 18: 1546–1554.
71. Cui Z, Wang H, Tan Y, Zaia KA, Zhang S, et al. (2004) Inducible and reversible NR1 knockout reveals crucial role of the NMDA receptor in preserving remote memories in the brain. Neuron 41: 781–793.
72. Rampon C, Jiang CH, Dong H, Tang YP, Lockhart DJ, et al. (2000) Effects of environmental enrichment on gene expression in the brain. Proc Natl Acad Sci U S A 97: 12880–12884.
73. Brandon EP, Zhuo M, Huang YY, Qi M, Gerhold KA, et al. (1995) Hippocampal long-term depression and depotentiation are defective in mice carrying a targeted disruption of the gene encoding the RI beta subunit of cAMP-dependent protein kinase. Proc Natl Acad Sci U S A 92: 8851–8855.

18

Comparative Proteomics of Milk Fat Globule Membrane Proteins from Transgenic Cloned Cattle

Shunchao Sui[1☉], Jie Zhao[1☉], Jianwu Wang[2,3], Ran Zhang[1], Chengdong Guo[1], Tian Yu[3], Ning Li[1]*

1 State Key Laboratory of Agrobiotechnology, China Agricultural University, Beijing, PR China, 2 Beijing GenProtein Biotech Company Ltd., Beijing, PR China, 3 Wuxi Kingenew Biotechnology Co., Ltd., Wuxi, Jiangsu Province, China

Abstract

The use of transgenic livestock is providing new methods for obtaining pharmaceutically useful proteins. However, the protein expression profiles of the transgenic animals, including expression of milk fat globule membrane (MFGM) proteins, have not been well characterized. In this study, we compared the MFGM protein expression profile of the colostrum and mature milk from three lines of transgenic cloned (TC) cattle, i.e., expressing recombinant human α-lactalbumin (TC-LA), lactoferrin (TC-LF) or lysozyme (TC-LZ) in the mammary gland, with those from cloned non-transgenic (C) and conventionally bred normal animals (N). We identified 1, 225 proteins in milk MFGM, 166 of which were specifically expressed only in the TC-LA group, 265 only in the TC-LF group, and 184 only in the TC-LZ group. There were 43 proteins expressed only in the transgenic cloned animals, but the concentrations of these proteins were below the detection limit of silver staining. Functional analysis also showed that the 43 proteins had no obvious influence on the bovine mammary gland. Quantitative comparison revealed that MFGM proteins were up- or down-regulated more than twofold in the TC and C groups compared to N group: 126 in colostrum and 77 in mature milk of the TC-LA group; 157 in colostrum and 222 in mature milk of the TC-LF group; 49 in colostrum and 98 in mature milk of the TC-LZ group; 98 in colostrum and 132 in mature milk in the C group. These up- and down-regulated proteins in the transgenic animals were not associated with a particular biological function or pathway, which appears that expression of certain exogenous proteins has no general deleterious effects on the cattle mammary gland.

Editor: Silvia Mazzuca, Università della Calabria, Italy

Funding: This project was supported by the "863" High-Tech Research Development Project (project grant numbers 2010AA10A105 and 2011AA100601). The funders had no role in study design, data collection and analysis, decision to publish, or preparation of the manuscript.

Competing Interests: Jianwu Wang of the authors is employed by the commercial company Beijing GenProtein Biotech Co., Ltd., who has a role in performing the experiment, analyzing the data and preparing the manuscript. Tian Yu of the authors is employed by the commercial company Wuxi Kingenew Biotechnology Co., Ltd., who has a role in performing the experiment, analyzing the data. There are no further patents, products in development or marketed products to declare.

* Email: ninglicaulab@gmail.com

☉ These authors contributed equally to this work.

Introduction

The technology of using genetically modified bovine lines to produce recombinant proteins in milk has flourished since the 1990s, based on the historic breakthrough of somatic cell cloning technology [1,2]. Milk has been proven to be an excellent vehicle for producing and delivering recombinant human proteins [3]. The cow mammary gland can be modified to synthesize foreign proteins in transgenic cloned bovine milk, which may be useful for human consumption and assimilation because of its similarities of composition to human milk [4]. The advantages of milk-based system also include the easiness of animal housing and maintenance and lower cost of harvesting proteins than the cell bioreactor [5]. However, we have little information about how transgenic modification and cloning process influence bovine endogenous gene expression, milk characteristics and health. A further consideration is how these changes of products from these animals might affect the health of consumers. To address these issues, we examined both the protein expression profiles and composition of milk from transgenic cloned cows compared with conventionally bred animals. Many works have been done to analyze the compositions of meat and milk from transgenic cloned cattle, but the majority of these studies have focused on a few major constituents rather than a comprehensive analysis of proteins expression, especially of low-abundance proteins [6–8].

The concerns about consumption of food from transgenic cloned animals were the hot spot in recent research. The FDA has raised the instructions for the food from cloned animal and the studies have showed that the milk and meats from cloned cattle were as safe as the conventionally bred cattle. However, there were no instructions for food products from transgenic cloned animals, we have to use the instructions of cloned animals to examine the food products from transgenic cloned cattle. Milk proteins can be divided into three classes, namely casein, whey proteins, and MFGM proteins. Compared to other classes, MFGM proteins are the least abundant, making up 1–4% of total milk proteins, but they include thousands of different proteins [9], which were important composition of the milk proteins.

The MFGM is a protein-rich lipid bilayer that surrounds lipids in milk. Although the MFGM has been studied for 50–60 years, previous studies have focused primarily on membrane globule formation, intracellular transition, and secretion [10]. It is believed that the MFGM originates from the apical plasma membrane of mammary epithelial cells [11,12]. Precursor microlipid droplets from the endoplasmic reticulum fuse to each other and travel to the apical cytoplasm where they are surrounded by the apical plasma membrane and then are secreted into the alveolar lumen [10,13–15]. Therefore, we can obtain information about mammary epithelial cell health by analyzing the protein composition of the MFGM.

In addition, MFGM proteins have many beneficial bioactivities, e.g., as antibiotics and anticancer agents, MUC1 is one of the most abundant proteins in the MFGM and is thought to protect exposed surfaces from physical damage and invasive pathogenic microorganisms [16,17]. For example, bovine lactophorin C-terminal fragment and PAS6/7, both MFGM constituents, inhibit replication of human rotavirus and prevent gastroenteritis [18]. MFGM proteins also have commercial value in cheese and butter production.

Technological advances in proteomics, especially quantitative proteomics, have enabled an increased understanding and characterization of milk proteins [19]. So far, proteomics has been successfully applied to studies of human milk whey proteins, bovine milk whey proteins, human MFGM proteins, bovine MFGM proteins, and goat MFGM proteins [20–23].

We genetically engineered three transgenic bovine lines that specifically express recombinant human α-lactalbumin (TC-LA group), lactoferrin (TC-LF group), and lysozyme (TC-LZ group) [24–26], from which the compositions of the milk include the human proteins with higher expression level compared to that from conventionally bred animals. Prior research for milk whey proteins and the nutritional components from transgenic cloned cattle have indicated that the expression of exogenous proteins did not significantly change the milk whey protein profile, and the mean values for the majority of the measured parameters were within the normal range [27].

In this work, we compared the MFGM proteins of colostrum and mature milk from these three healthy transgenic cloned lines to those from cloned and conventionally bred animals. The MFGM proteins of the five groups were characterized using 2D-PAGE and 2D LC-MS/MS coupled with the iTRAQ proteomic strategy. Our objectives were to: (1) characterize the proteins of the MFGM in transgenic cloned cattle; (2) compare the relative expression of MFGM proteins among transgenic cloned cattle milk; and (3) investigate the health of mammary epithelial cells in the transgenic cloned cattle. The study will supplement the former study by completing the evaluation for the milk products from transgenic cloned cattle and revealing the situation of the mammary epithelial cells from transgenic cloned cattle. This study provides the complement evidence of the milk composition analysis for the transgenic cloned bovine, and the results may indirectly reflect the health of the mammary epithelial cells of the transgenic cloned bovine.

Materials and Methods

Milk Sample Collection

The protocol was approved by the Institutional Animal Care and Use Committee of China Agricultural University (ID: SKLAB-2010-05-01). The three transgenic cattle lines TC-LZ (n = 10), TC-LA (n = 4), and TC-LF (n = 3) have been described [24–26] and were compared to cloned cattle (C, n = 3) which were also cloned by somatic cell nuclear transfer (SCNT), and conventionally bred cattle (N, n = 9) of a similar genetic background. The cattle were similar in age and lactation period and were housed under the same conditions. The milk collection were carried out as described previously [27]. The colostrum was obtained during the initial three days of lactation and mature milk was obtained on the 30th, 60th and 90th day after lactation and the milk was collected from the cows twice daily and was pooled to form one daily sample. Milk samples were collected by milking machine, and samples were immediately stored at −20°C until further analysis.

Protein Extraction

Milk sample (50 ml) from each cow was centrifuged at 4000×g to separate milk fat globule cream from whole milk. The separated cream was rinsed in 25 ml phosphate buffered saline twice and 25 ml deionized water once. Then the milk fat globule proteins were obtained by methanol/trichloromethane precipitation as follows. The cream (2 ml) was then stored at −80°C, and upon thawing 10 ml methanol and 10 ml trichloromethane were added. The samples were centrifuged at 50,000×g for 30 min at 4°C (Avanti J-26XP, Beckman Coulter, Indianapolis, USA), and the supernatant was discarded. The precipitated proteins were solubilized in a solution containing 7 M urea, 2 M thiourea, 0.025 M DTT, 2% (w/v), and 1% CHAPS [20]. The relative concentration of protein in each extract was determined by measuring the amount of whey protein with the 2D-Quant quantitate kit (GE Healthcare). MFGM protein samples from each cow were pooled together to generate one individual sample and then individual samples in the same groups (TC-LZ, TC-LA, TC-LF, C and N) were pooled again to generate one group sample according to the equal protein mass.

In-solution digestion

Soluble proteins (200 µg) were digested in-solution using the filter aided sample preparation (FASP) method [28]. Each protein extract (30 µl) was mixed with 200 µl of buffer containing 8 M urea in 0.1 M Tris-HCl (pH 8.5) in the filter unit and centrifuged at 14,000×g for 40 min and repeated one time. The flow-through was discarded, and iodoacetamide solution was added to the filter and mixed at 600 rpm for 1 min then incubated without mixing for 5 min. The filtered units were centrifuged at 14,000×g for 30 min, then 100 µl of buffer containing 8 M urea in 0.1 M Tris-HCl (pH 8.0) was added to the filtered unit, then the solution was centrifuged again at 14,000×g for 40 min and repeated twice. We then added 40 µl of the above-mentioned buffer containing endoproteinase Lys-C (enzyme to protein ratio 1:50, w/w) and mixed at 600 rpm in thermo-mixer for 1 min. The units were incubated overnight then transferred to new collection tubes, and then 120 µl of 0.05 M NH₄HCO₃ in water with trypsin (enzyme to protein ratio 1:100 w/w) was added and mixed at 600 rpm for 1 min. The units were incubated at room temperature for 4 h and then centrifuged at 14,000×g for 40 min. Finally, 50 µl of 0.5 M NaCl was added and the units were centrifuged at 14,000×g for 20 min.

The extracted peptides were desalted using a 1.3-ml C18 solid-phase extraction column (Sep-Pak® cartridge; Waters Corporation, Milford, USA). The peptides were dried using a vacuum centrifuge and then resuspended in 5 mM ammonium formate containing 5% acetonitrile, pH 3.0.

LC-MS/MS Analysis

LC-MS/MS was performed on a nano Acquity UPLC system (Waters Corporation) connected to an LTQ Orbitrap XL mass spectrometer (Thermo Electron Corp., Bremen, Germany) equipped with an online nano-electrospray ion source (Michrom Bioresources, Auburn, USA). The MS was operated in data-dependent mode to switch automatically between MS and MS/MS acquisition. Survey full-scan MS spectra (m/z 300–1800) were acquired in the Obitrap with one microscan and with a mass resolution of 60,000 at m/z 400, followed by MS/MS of the eight most-intense peptide ions in the LTQ analyzer. LC-MS/MS spectra were acquired using SEQUEST (v.28 revision 12, Thermo Electron Corp) against the International Protein Index bovine database version 3.54. A decoy database containing the reverse sequences was appended to the database. The search parameters were set as follows: partial trypsin cleavage with two missed cleavage sites was considered; the peptide mass tolerance was 1.4 Da; and the fragment ion tolerance was 1 Da. Trans Proteomic Pipeline software (revision 4.20; Institute of Systems Biology, Seattle, WA) was then used to identify proteins based on corresponding peptide sequences with ≥95% confidence. The peptide results were filtered by PeptideProphet with a p value> 0.90 and a ProteinProphet probability = 0.95 [29,30].

Quantitative Analysis of MFGM Proteins by iTRAQ LC-MS/MS

The digested peptides of the five groups were transferred to vials containing individual iTRAQ reagents (Applied Biosystems) following the iTRAQ standard protocol for the 8-plex kit. The N group was labeled with iTRAQ113, the TC-LF with iTRAQ114, TC-LZ with iTRAQ115, TC-LA with iTRAQ116, and the C group with iTRAQ117.

The iTRAQ-labeled samples were pooled and the SCX HPLC experiment was performed on a Shimadzu 20AD 5-μm SCX polysulfoethyl column (2.1 mm×100 mm, The Nest Group, Inc., MA) as the first dimension. Each collected components of the processed SCX fractions ran via RP-LC ESI-MS/MS on an Applied Biosystems Q-Star Elite XL mass spectrometer in which the RPLC column was a ZORBAX 300SB-C18 (5 μm, 0.1 mm×150 mm, Microm, USA). The Q-Star Elite XL mass spectrometer was operated in the smart information-dependent acquisition activated mode with automatic collision energy and automatic MS/MS accumulation. Survey full-scan MS spectra (m/z 400–1800) were acquired with one microscan and a mass resolution of 60,000 at m/z 400, followed by MS/MS of the four most-intense peptide ions in the analyzer. The relative abundance of the MFGM proteins in the different samples was derived from the ionic peak areas of the iTRAQ reporter.

iTRAQ identification and quantification analysis of the MFGM proteins were obtained using Protein Pilot 3.0 (Applied Biosystems, USA) with the following user-defined parameters: Sample Type, iTRAQ 8-plex (Peptide Labeled); Cysteine alkylation, MMTS; Digestion, Trypsin; Specify Processing, Quantitate; Database, IPI v3.62 bovine; Search Effort, thorough ID; Results Quality, Detected Protein Threshold [Unused ProtScore (Conf)] >1.3 (95%); Run False Discovery Rate Analysis.

Results

Overview of MFGM Proteome in the TC Groups

MFGM proteins were characterized in three TC bovine lines and in C group and N group using 2D nano-LC-MS/MS and a quantitative proteomics method (iTRAQ). LC-MS/MS identified 1225 proteins among the five groups of MFGM samples, 939 proteins in colostrum, and 910 proteins in mature milk (Table 1). There were 637 proteins that were present in TC-LA group, 721 in the TC-LF group, 720 in the TC-LZ group, 527 in the C group,

and 668 in the N group. iTRAQ identified 851 MFGM proteins in colostrum and 775 MFGM proteins in mature milk of all the bovine lines. The identified proteins from the three TC groups were compared to those from the C and N groups. The TC-LA group contained 166 proteins that were not found in the control lines, the TC-LF group contained 265 proteins, and the TC-LZ group contained 184 proteins (Figure 1). We found that 43 of these proteins were shared by all the three TC groups (Table 2). The molecular mass of this subset was between 10 kDa and 90 kDa, and the pI values were between 4 and 10. DAVID Bioinformatics Resources was used to analyze the function of the 43 common proteins in GO terms [31]. We found that these proteins were enriched in the biological processes of protein transport, cellular component of the Golgi apparatus, and the molecular function of GTP binding. However, the functions of these specific expressed proteins did not cluster in any particular pathway (Figure 2).

Quantitative Comparison of MFGM Proteins

We obtained relative MFGM protein quantity information from the five groups using iTRAQ. Only proteins with consistent quantitative results in two independent iTRAQ runs were used for further analysis of expression and function. A total of 459 proteins from colostrum MFGM and 426 proteins from mature MFGM were identified. Statistical analysis of expression levels in the three TC groups and C group compared with the N group revealed more than twofold differences (p≤0.05) in the expression of many of these proteins (Figure 3). In this regard, 126 proteins in the colostrum and 77 proteins in the mature milk of the TC-LA group had a greater than twofold difference in expression, likewise, in the TC-LF group there were 157 proteins in colostrum and 222 proteins in mature milk, in the TC-LZ group there were 49 proteins in colostrum and 98 proteins in mature milk, and in the cloned control group there were 98 proteins in colostrum and 132 proteins in mature milk.

Comparison between the MFGM proteins among the five groups also revealed many differences. The unsupervised hierarchical cluster analysis (HCA) and the principal component analysis (PCA) were used to assess the variability among the 5 groups. We found that the TC-LA, TC-LZ, and C groups had similar protein compositions in colostrum and mature milk. The TC-LF group was distinguishable from the other two TC groups (Figure 4). The colostrum of the TC-LF and TC-LA groups clustered together as that of the TC-LZ and C groups did. The results indicated that in the colostrum the TC-LZ was closer to the C group and in the mature milk the TC-LZ and TC-LA were closer to the C group. Both in colostrum and mature milk, the TC-LF groups were showed a greater difference from the other TC groups. In colostrum, 11 proteins were differentially expressed in the TC-LA group compared to the control groups, 25 in the TC-LF group, and 6 in the TC-LZ group. In mature milk, there were 8 proteins that were differentially expressed compared to the controls in the TC-LA group, 25 in the TC-LF group, and 5 in the TC-LZ group. In TC groups, the proteins with expression changed simultaneously were not appeared. Several of these differentially expressed proteins, such as fibrinogen α chain, α-S2-casein, κ-casein, β-lactoglobulin, and lactoferrin have been explored in other studies of whey and MFGM [20,32–34].

Our identified MFGM proteins were subjected to cluster analysis to evaluate the relationship between TC animals and conventionally bred controls. In the TC-LA group, six proteins were notably down-regulated, including α-2-antiplasmin (−4.9 fold), β-lactoglobulin (−4.8 fold), serum albumin (−4.9 fold), platelet glycoprotein 4 (−4.1 fold), lactoferrin (−4.2 fold), serine hydroxymethyltransferase, mitochondrial (−4.1 fold), and a

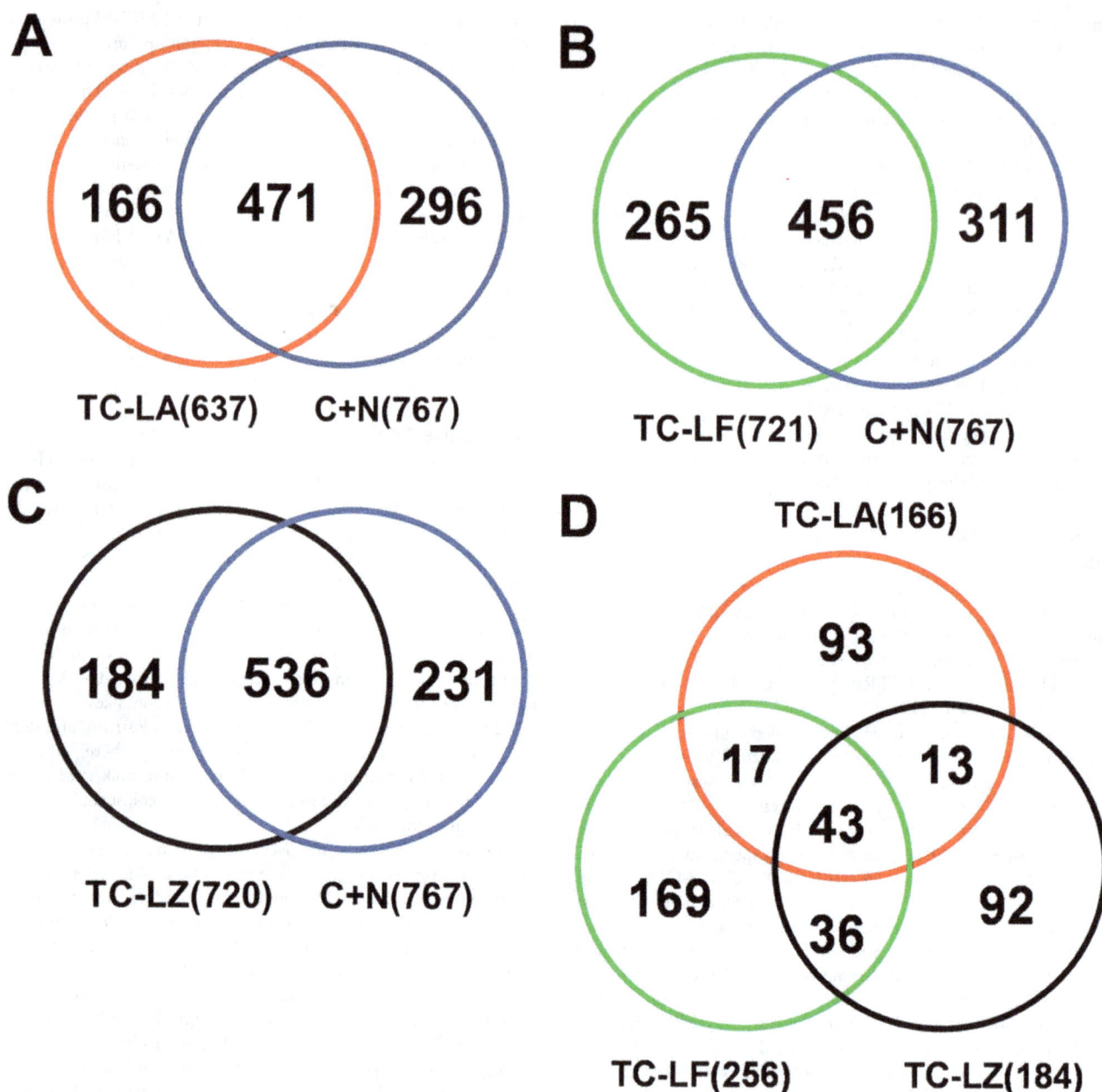

Figure 1. Comparison of MFGM proteins from the different TC and control lines. Cloned and conventionally bred control animals [(C+N)] were compared with (A) MFGM proteins from the TC-LA line, (B) MFGM proteins from the TC-LF line, and (C) MFGM proteins from the TC-LZ line. (D) The intersection of differentially expressed MFGM proteins from the three TC groups yielded 43 proteins in common.

Table 1. Proteins identified in the MFGM of colostrum and mature milk of the TC-LZ, TC-LA, and TC-LF transgenic cloned lines and of the cloned and conventionally bred normal controls.

	TC-LZ	TC-LA	TC-LF	Cloned	Normal	Total
Colostrum	620	516	411	220	562	939
Mature milk	444	393	567	464	353	910
In total	720	637	721	527	668	1225

Table 2. Proteins specifically expressed in transgenic cloned cattle MFGM.

No	Accession	Description	MW	PeptideLength	Isoelectric Point
1	IPI00686092.1	Peroxiredoxin-1	22194.98	199	8.6084
2	IPI00687416.2	Rras2 Protein	23384.41	204	5.8076
3	IPI00687550.1	3'(2'),5'-Bisphosphate Nucleotidase 1	33306.63	308	5.1936
4	IPI00689149.2	Oxysterol-Binding Protein	84236.69	746	6.2374
5	IPI00692733.2	Transmembrane Protein 30A	40656.91	361	8.6073
6	IPI00693697.3	Similar To Adiponutrin	50423.45	455	7.5812
7	IPI00695308.5	Mucin-16 (Fragment)	31335.98	282	9.0396
8	IPI00695670.1	Upf0585 Protein C16Orf13 Homolog	22699.17	204	7.9092
9	IPI00695741.2	62 Kda Protein	62274.23	569	5.3891
10	IPI00696647.1	Similar To Pincher Isoform 1	60985.31	540	7.6711
11	IPI00697565.3	Ras-Related Protein Rab-21	24130.66	222	8.0481
12	IPI00698430.1	Eukaryotic Translation Initiation Factor 5 Isoform 7	48939.49	429	5.2609
13	IPI00700391.1	Similar To V-Crk Sarcoma Virus Ct10 Oncogene Homolog Isoform 1	33809.31	304	5.2451
14	IPI00703351.2	Similar To Aldehyde Dehydrogenase 3B1 Isoform 1	53661.38	486	8.4692
15	IPI00703423.1	B-Cell Receptor-Associated Protein 31	27883.64	245	9.765
16	IPI00704666.1	Trna Methyltransferase 112 Homolog	14309.9	125	5.8553
17	IPI00704752.1	Ras-Related Protein Rab-7A	23528.72	207	6.5881
18	IPI00706451.3	Adp-Ribosylation Factor 4	20514.36	180	5.9819
19	IPI00706632.1	Elongation Factor 1-Gamma	50345.45	440	6.5236
20	IPI00706968.3	Nuclear Factor Of Kappa Light Polypeptide Gene Enhancer In B-Cells Inhibitor, Beta	37543.32	357	4.2979
21	IPI00707103.4	Synaptosomal-Associated Protein 29	28471.85	258	5.1001
22	IPI00707616.3	Cidea Protein	24524.03	219	9.7273
23	IPI00708591.1	Znrf2 Protein	23870.75	238	7.1119
24	IPI00710727.1	Transitional Endoplasmic Reticulum Atpase	89272.24	806	4.8908
25	IPI00711304.4	Similar To Inositol Polyphosphate-5-Phosphatase A	70475.79	637	8.6043
26	IPI00713743.1	Inosine-5'-Monophosphate Dehydrogenase 2	55726.54	514	7.339
27	IPI00714515.1	Fgr Protein	59292.95	527	5.3694
28	IPI00714621.3	Uncharacterized Protein C1Orf93 Homolog	21482.83	201	6.2609
29	IPI00714818.1	Eukaryotic Translation Initiation Factor 3 Subunit H	39879.94	352	6.653
30	IPI00714992.4	Cgmp-Dependant Type Ii Protein Kinase	87033.26	762	8.6701
31	IPI00715153.2	Fas-Associated Factor 2	52629.57	445	5.4387
32	IPI00716638.7	Wd Repeat-Containing Protein 1	66214.06	606	6.7147
33	IPI00717376.3	Similar To Cytoskeleton-Associated Protein 4	64489.56	589	5.9276
34	IPI00718291.4	Rab14 Protein	23881.49	215	6.1266
35	IPI00718671.5	Synaptosomal-Associated Protein	23212.11	211	4.4574
36	IPI00729755.2	Acyl-Coa Synthetase Long-Chain Family Member 1	78228.82	699	7.2276
37	IPI00730045.3	Similar To Rififylin Isoform 4	39655.22	356	6.1033
38	IPI00733615.1	Dnaj Homolog Subfamily A Member 2	45731.44	412	6.4421
39	IPI00867128.1	Phosphoserine Aminotransferase	40502.07	370	7.7533

Table 2. Cont.

No	Accession	Description	MW	PeptideLength	Isoelectric Point
40	IPI00867311.1	Scamp2 Protein	36678.97	328	6.1107
41	IPI00906554.1	Guanine Nucleotide-Binding Protein Subunit Alpha-11	42042.75	359	6.029
42	IPI00924133.1	Similar To Ciliary Neurotrophic Factor Receptor Alpha Precursor (Cntfr Alpha) Isoform 3	40661.3	372	6.9436
43	IPI00944416.1	Cyb5B Protein	16267.5	146	4.8233

putative uncharacterized protein (−4.4 fold). In the TC-LF group, 10 proteins were notably different from the control group, including apolipoprotein A-II (−4.4 fold), ATP-binding cassette sub-family G (WHITE) member 2 (+4.3 fold), α-1-acid glycoprotein (−4.9 fold) similar to solute carrier family 39 (zinc transporter) member 8 (+4.1 fold), α-S2-casein (+4.2 fold), MUC 1 (+4.9 fold), acyl-CoA synthetase long-chain family member 1 (+4.8 fold), and vitamin D–binding protein (−5.2 fold). In the TC-LZ, four proteins were notably different from the control group, namely fibrinogen α chain (+2.8 fold), β-lactoglobulin (−2.7 fold), cidea protein (−3.1 fold), and similar to ATPase type 13A4 (−2.9 fold). In the C group, six proteins were notably different from the conventionally bred group, namely fibrinogen α chain (+3.1 fold), α-S2-casein (+5.3 fold), α-S1-casein (+4 fold), Rit1 protein (−3.6 fold), α-lactalbumin (−4.8 fold), and nucleobindin-1 (+3.3 fold).

Function and Pathway Analysis of Differentially Expressed Proteins

The Ingenuity Pathway Analysis (IPA, Ingenuity Systems, Inc. Redwood City) software was used to analyze the functions and networks of the differentially expressed proteins. To further examine the specifically expressed proteins, they were categorized according to their biological process, cellular component or molecular function as annotated in the Gene Ontology (GO) database. Differentially expressed MFGM proteins from TC-LA colostrum were mainly enriched in the biological processes of response to wounding and the acute inflammatory response, in the cellular components of the extracellular region, and in the molecular function of endopeptidase inhibitor activity. Differentially expressed MFGM proteins from TC-LA mature milk were

mainly enriched in cytoskeleton organization, membrane-bounded vesicles, and unfolded protein binding. Differentially expressed MFGM proteins from TC-LF colostrum were mainly enriched in response to organic substance, the extracellular region, and endopeptidase inhibitor activity. Differentially expressed MFGM proteins from TC-LF mature milk were mainly enriched in small GTPase–mediated signal transduction, pigment granules, and GTPase activity. Differentially expressed MFGM proteins from TC-LZ colostrum were mainly enriched in response to wounding, the extracellular region, and cell-surface binding. Finally, differentially expressed MFGM proteins from TC-LZ mature milk were mainly enriched in cell motility, the actin cytoskeleton, and actin binding (Figure 5).

In the TC-LA group, there were 18 proteins that were changed above 3 folds in colostrum and mature milk. The pathway analysis showed that the changed MFGM proteins in the TC-LA colostrum most involved in the network which have the functions of cellular movement, immune cell trafficking, cell-to-cell signaling and interaction. And in the mature milk are cell death, cell-to-cell signaling and interaction, hematological system development and function (Figure 6A, B).

In TC-LF group, there were 46 proteins that were changed in colostrum and mature milk significantly. The IPA software showed that the changed MFGM proteins in mature milk involved in the network of which top functions are cellular assembly and organization, cellular function and maintenance, dermatological diseases and conditions. And the MFGM proteins from colostrum involved in the network of which top functions are lipid metabolism, molecular transport, and small molecule biochemistry (Figure 6C, D).

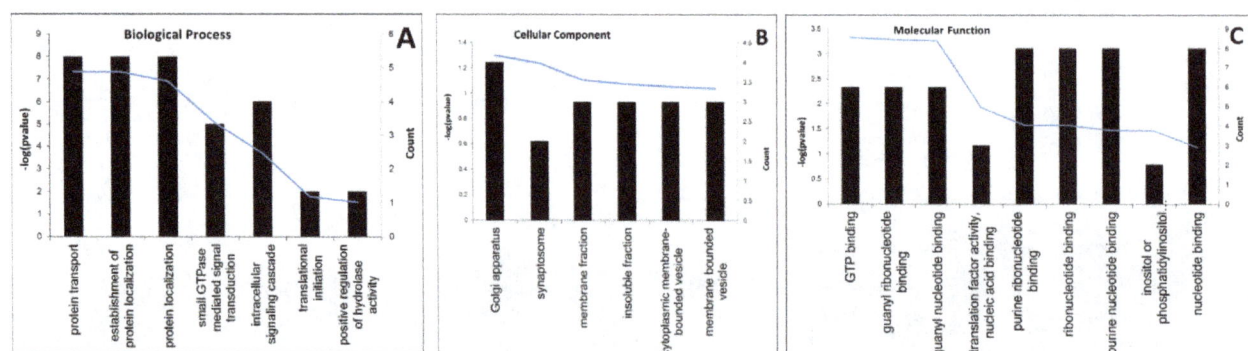

Figure 2. Functional analysis of MFGM proteins specifically expressed in the transgenic cloned animals. Analysis of proteins enriched in the transgenic cloned groups according to: (A) biological process, (B) cellular component, and (C) molecular function. Blue curves indicate the degree of enrichment of gene function.

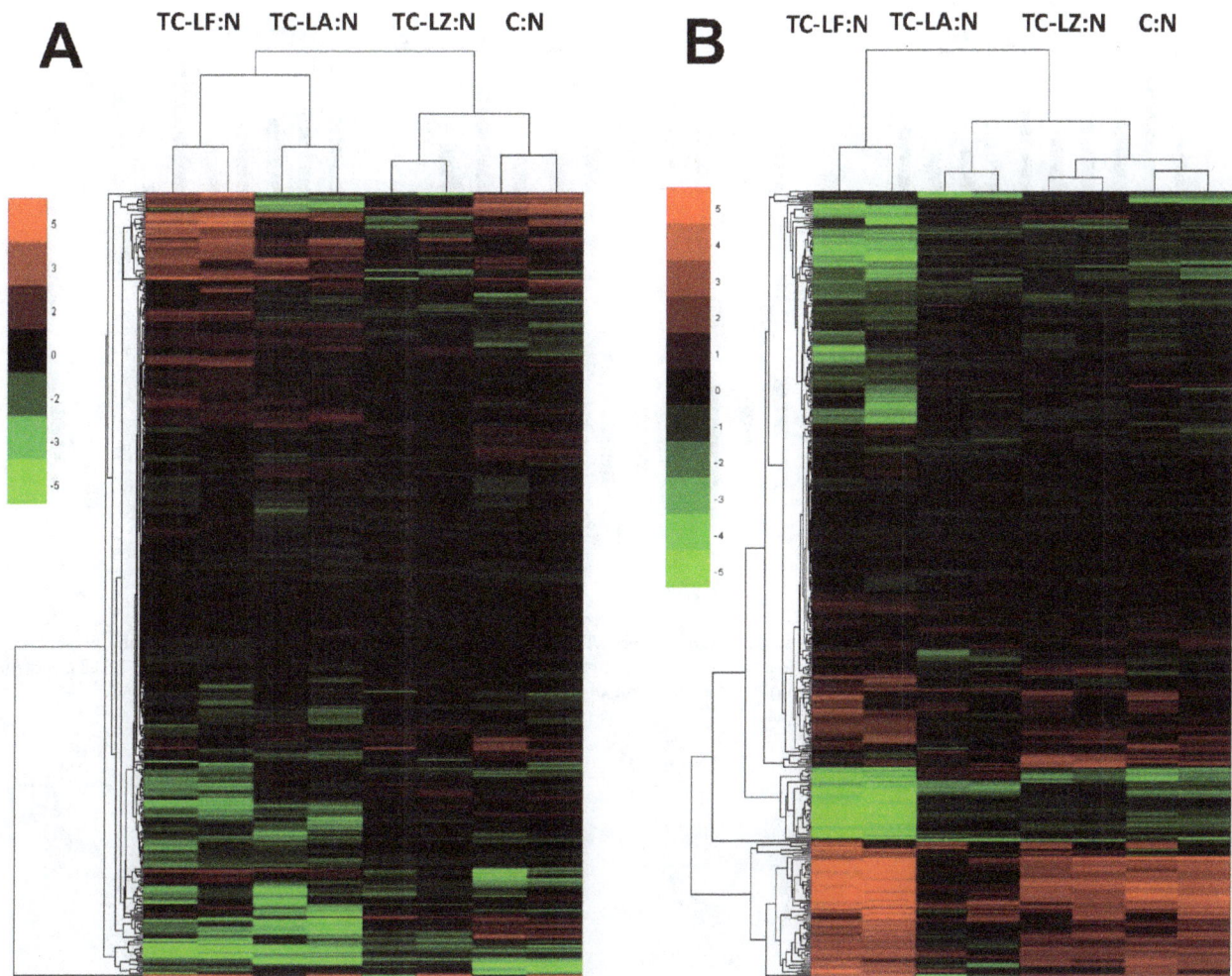

Figure 3. Cluster analysis of MFGM proteins. (A) Analysis of MFGM proteins in colostrum of the three TC groups compared to cloned [(C)] and conventionally bred [(N)] controls. (B) Analysis of MFGM proteins in mature milk of the three transgenic groups compared to controls.

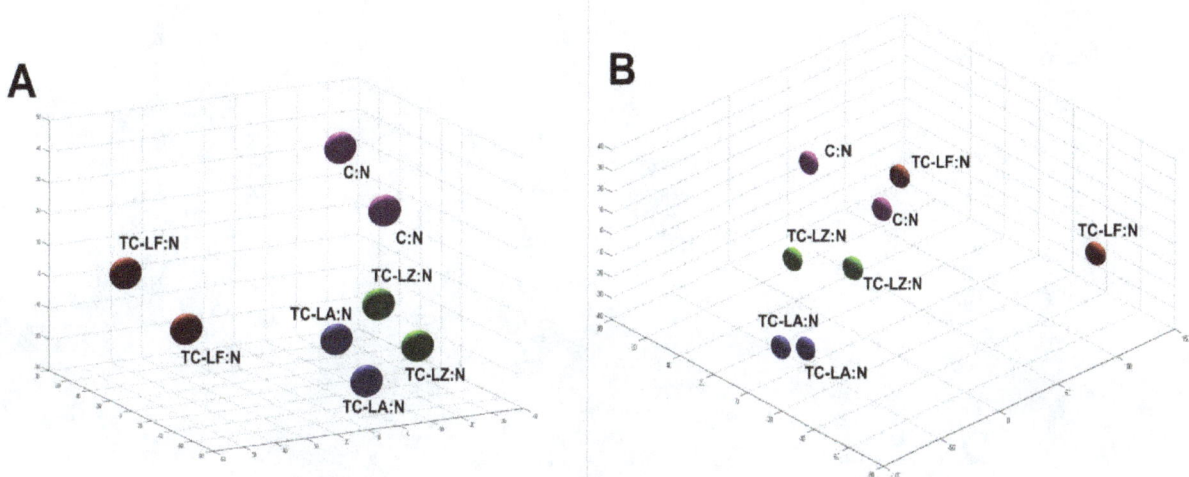

Figure 4. Principal component analysis of MFGM proteins from the three transgenic groups, the cloned group, and the conventionally bred control group. (A) Analysis of MFGM proteins in colostrum. (B) Analysis of MFGM proteins in mature milk.

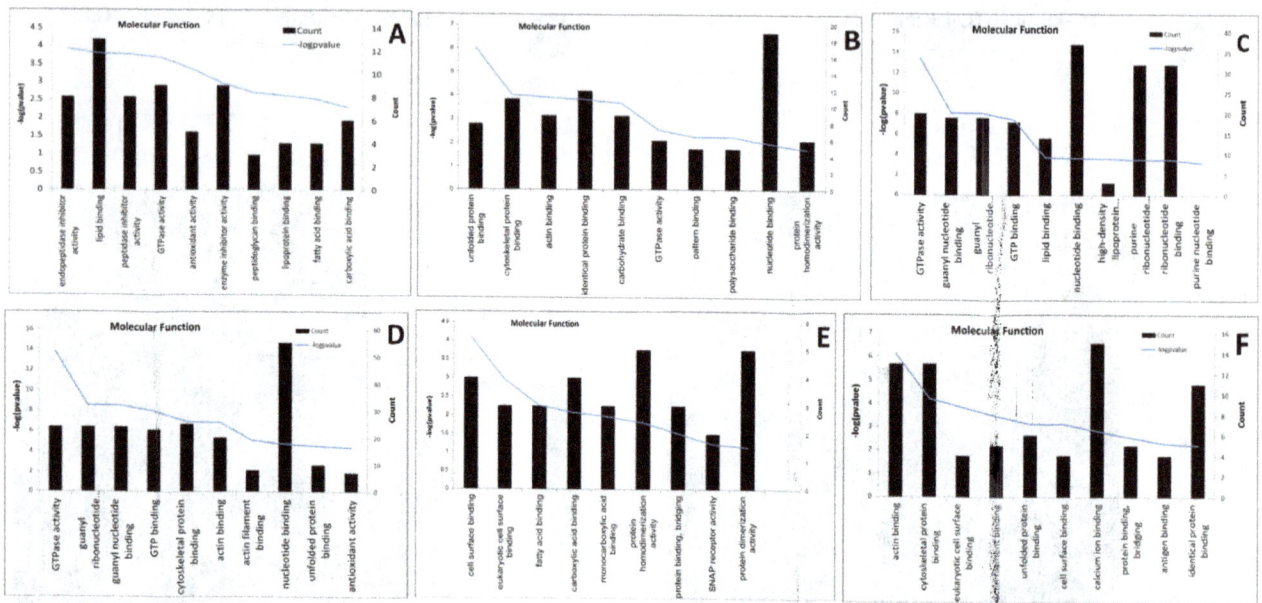

Figure 5. Molecular function analysis of differentially expressed MFGM proteins in the transgenic groups. Enrichment analysis of differentially expressed MFGM proteins in (A) TC-LA colostrum, (B) TC-LA mature milk, (C) TC-LF colostrum, (D) TC-LF mature milk, (E) TC-LZ colostrum, and (F) TC-LZ mature milk. Blue curves indicate the degree of enrichment of gene function.

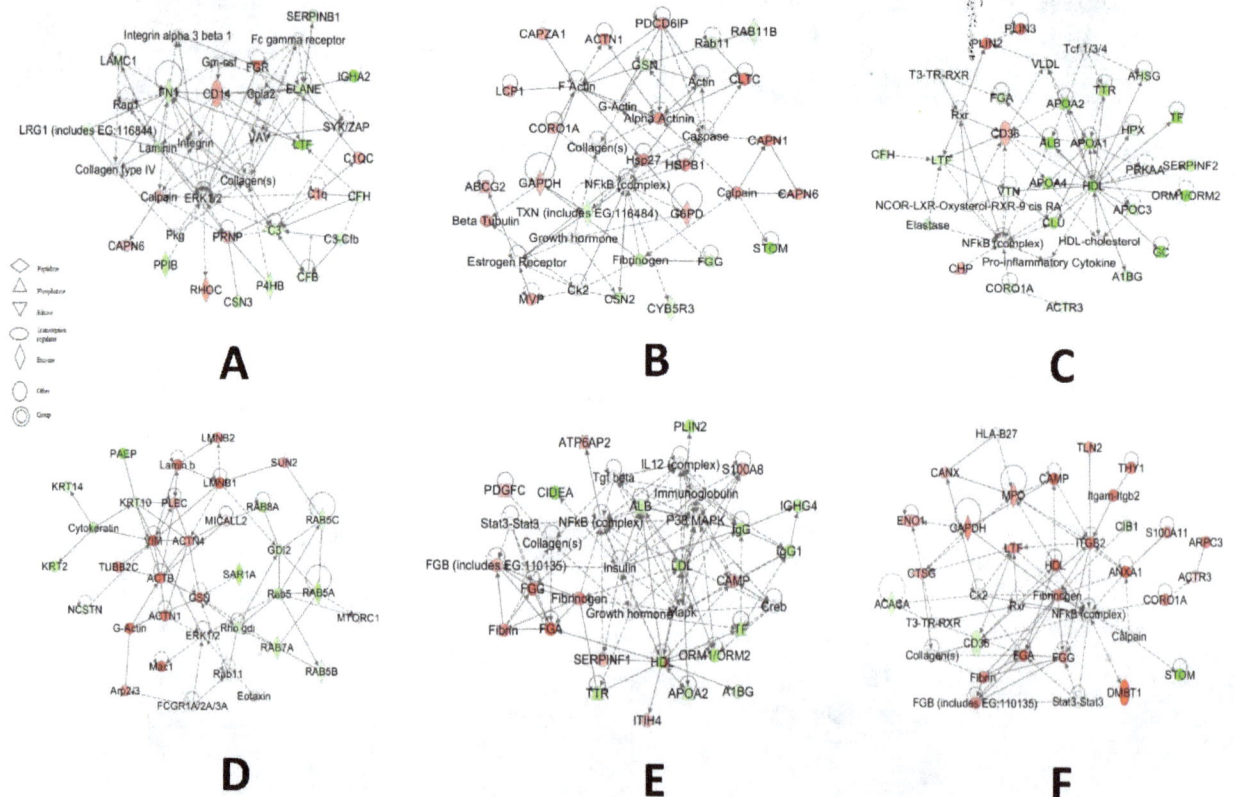

Figure 6. Pathway analysis of differentially expressed proteins from the transgenic groups. IPA defined pathways of the differentially expressed MFGM proteins from: (A) TC-LA colostrum, (B) TC-LA mature milk, (C) TC-LF colostrum, (D) TC-LF mature milk, (E) TC-LZ colostrum, and (F) TC-LZ mature milk.

In group of TC-LZ, there were 10 proteins that were altered obviously in colostrum and mature milk. The IPA software showed that the changed MFGM proteins in TC-LZ mature milk involved in the network of which the top functions are cell-to-cell signaling and interaction, tissue development, cell death. And in colostrum MFGM proteins, the top functions are inflammatory response, cell-to-cell signaling and interaction, cell signaling (Figure 6E, F).

The MFGM proteins of C group were also compared with the N group, the results showed that the changed proteins in the colostrum involved in the functions of cell-to-cell signaling and interaction, tissue development, hematological system development and function. In mature MFGM, the changed proteins were in the network of functions: cellular movement, cellular assembly and organization, cellular function and maintenance. Differentially expressed MFGM proteins in colostrum from the C group were enriched in the biological processes of response to wounding and the acute inflammatory response, in the cellular components of the extracellular region, and in the molecular function of GTPase activity. Differentially expressed MFGM proteins in mature milk of the cloned control animals were enriched in the biological process of glucose catabolism, in the cellular component of the actin cytoskeleton, and in the molecular function of cytoskeletal protein binding (Figure 7).

Discussion

Proteomics is one of the most popular methods to analyze complex samples. Recent advances in proteomics make the identification of low-abundance proteins and analysis of complex proteins such as high-molecular-weight transmembrane proteins possible, such analyses are not possible with standard 2D-PAGE methods alone. The combination of LC-MS/MS and iTRAQ or methods with the iTRAQ technologies are suitable for analysis of hydrophobic macro-molecules like plasma membrane proteins and are highly sensitive, fast, and reliable. Comparing with traditional 2D-PAGE method, the methods based on mass spectrometry are more suitable for MFGM proteins analysis. Data from several previous studies of MFGM proteins in humans [23], cattle [35] and goats [36] demonstrated that the amount of the identified proteins by this method was larger than other methods and the iTRAQ method was suitable to analyze the MFGM proteins with the same sensitivity compared to LC-MS/MS.

Applying these methods, we examined the effects of exogenous gene expression and cloning techniques on bovine MFGM proteins, an neglected area of research. Comparing three TC bovine lines expressing different exogenous proteins with C and N group, we identified 43 proteins that were specifically expressed in the transgenic animals below the detection limit for 2D-PAGE and silver staining. Considering both the concentration and functional

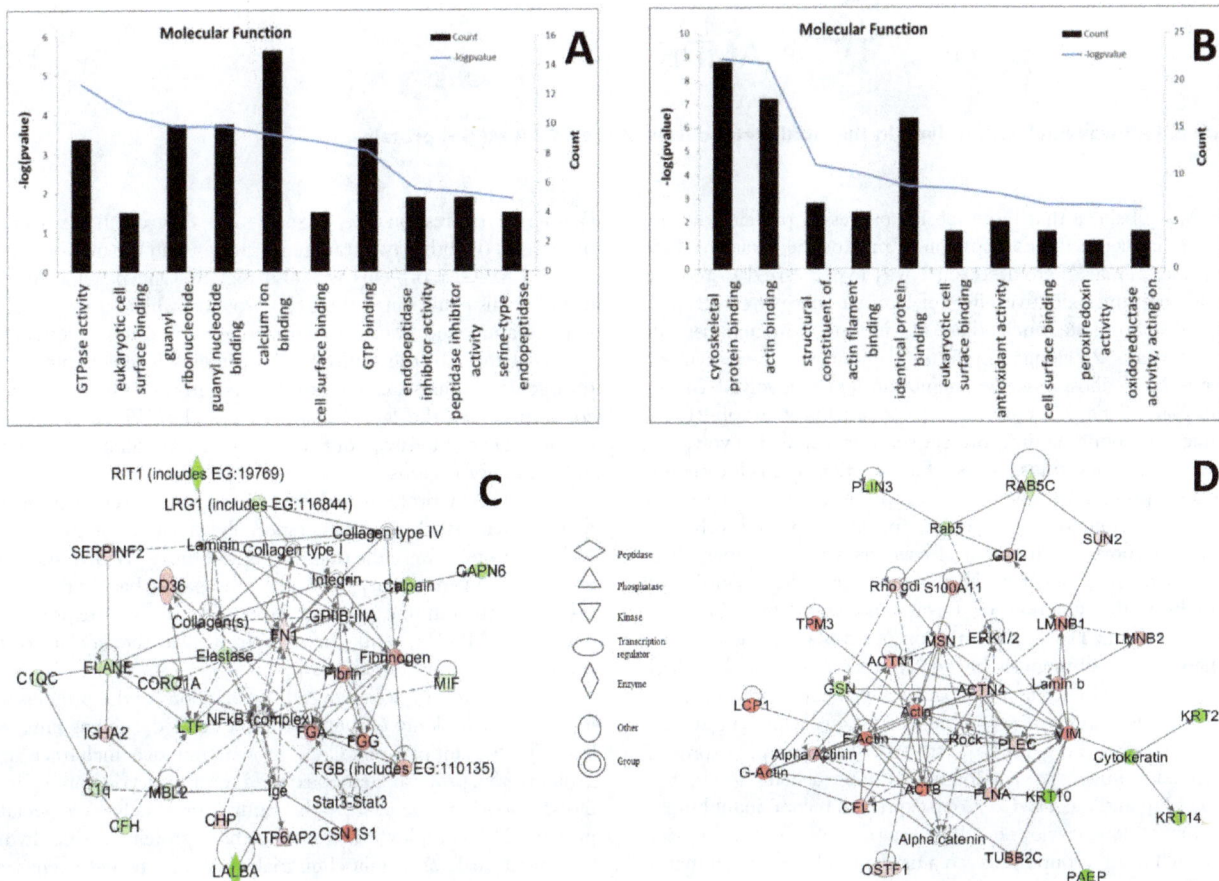

Figure 7. Molecular function and pathway analysis of differentially expressed MFGM proteins from the cloned control group. (A, B): Molecular function of MFGM proteins from colostrum (A) and mature milk (B). (C, D): IPA defined pathways of MFGM proteins from colostrum (C) and mature milk (D).

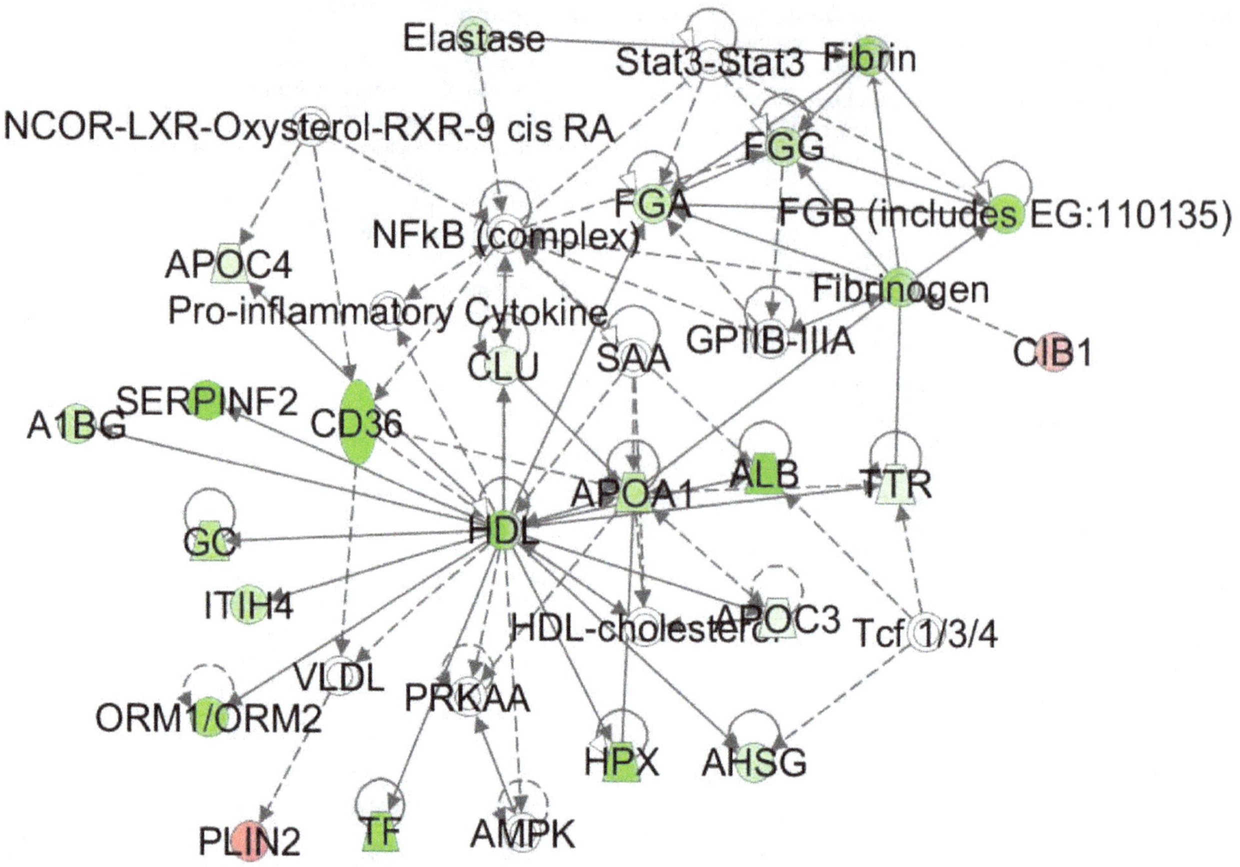

Figure 8. Pathway analysis involved in the Lipid Metabolisms of the TC-LA MFGM proteins.

analysis, we believe that these specific expressed proteins did not adversely influence the composition of milk or the health of these transgenic animals. Peroxiredoxin-1 was hypothesized to responsible for the anti-oxidative effect of laminar share stress [37]. 3′ (2′), 5′-bisphosphate nucleotidase 1, inosine-5′-monophosphate dehydrogenase 2, elongation factor 1-gamma, sdp-ribosylation factor 4 and phosphoserine aminotransferase involved in the metabolism of the cell, the protein ras-related protein rab-21 and guanine nucleotide-binding protein subunit alpha-11 involved in the mediated signal transduction. The function of b-cell receptor-associated protein 31 involves in apoptotic cleavage of cellular proteins, the function of ras-related protein rab-7A involves in protein transport, the function of acyl-coa synthetase long-chain family member 1 involves in the fatty acid degradation and metabolism, the function of Eukaryotic Translation Initiation Factor 3 Subunit H was in regulation of translational initiation.

There were distinguishable differences between the three transgenic lines in terms of exogenous protein expression. The TC-LF group expressed recombinant human lactoferrin at high levels, i.e., 2.5–3.8 g/l [25], the TC-LA group, expressed recombinant human α-lactalbumin at intermediate levels, i.e., 1.5 g/l [26], and the TC-LZ group expressed recombinant human lysozyme at low levels, i.e., 13–28 mg/l [24]. The expression profile of TC-LF group displayed a marked difference with that of other TC groups but the range of variation was comparable to the C group. Comprehensive analysis of the relative quantities of MFGM proteins in the transgenic and control groups led to the hypothesis that MFGM protein expression was influenced by the quantity of the exogenous proteins. In the TC-LF group,

lactoferrin expression was higher than that of the exogenous proteins in the other two transgenic groups, and the number of the affected MFGM proteins was also higher. One possible mechanism for this could entail the S100 proteins, which are involved in inflammatory and antimicrobial responses. We found low abundance of the S100A8 and S100A9 proteins in colostrum of the transgenic animals relative to the N group. This may be a consequence of the fact that LF, LA, and LZ (i.e., which were expressed exogenously in our study animals) are all involved in the inflammatory response.

The MFGM proteins are important in cattle cultivation and quality of livestock products, especially in cheese industry. The MFGM proteins were the major composition in MFGM, they play a core role in mammary epithelial cell against bacteria, so these proteins and immune globulin proteins are highly expressed in colostrum MFGM, and the colostrum was beneficial to the acquired immunity in new born.

In TC-LA group, we analyzed the functions of the proteins and found out that plenty of them were not defined, such as similar to lipocalin, α-2-antiplasmin, Cd5L protein, putative uncharacterized protein, canx protein. Hemopexinand serotransferrin coming from blood majorly influenced by the cattle individually. Actin-related protein 2/3 complex subunit 3, rab35 protein, serine hydroxymethyltransferase, mitochondrial, l-lactate dehydrogenase B Chain are involved in ubiquitin ligase [38]. Apolipoprotein E plays a key role in regulating plasma levels of lipoproteins and its level was not associated with high-density lipoprotein fractions [39]. The former studies of the composition of milk from transgenic cloned cattle indicated that the fatty acids of TC-LA colostrum were less than that of the N group especially in monounsaturated fatty acids [27]. The pathway analysis found the down-regulated lipid metabolism (Figue 8), which presumably was involved in monounsaturated fatty acids synthesis.

In TC-LF group, some proteins have high abundance in milk such as κ-Casein, α-S2-Casein, serotransferrin, α-lactalbumin, serum albumin and lactoperoxidase, so a small amount of residual proteins will cause detectable change of concentrations. Plenty proteins with the function of lipid metabolism were differentially expressed. Bovine MUC1 prevents binding of bacteria to human intestinal cells and has a role in preventing the binding of common enteropathogenic bacteria to human intestinal epithelial surfaces [40]. The expression of fatty acids of mature TC-LF milk was less than that of other four groups, especially in saturated fatty acids [27]. Plenty proteins which involve in lipid metabolism were expressed lowly in TC-LF mature MFGM, such as apolipoprotein A-II, which is one of the major proteins in HDL and its main function is to modulate the lipid binding and lecithin-cholesterol acyltransferase activities of HDL by promoting the dissociation of apo A-I from HDL [41]. ATP-binding cassette, sub-family G (White), member 2 have the function of maintaining lipid homeostasis in the mammary gland [42]. Fatty acid-binding protein (FABP) regulates the channeling of fatty acids toward copious milk fat synthesis in bovine mammary gland [43]. The expression level of CIDE-A protein was regulated by insulin and/ or fatty acids in mammary epithelial cells, and thereby played an important role in lipid and energy metabolism [44]. Acyl-coa synthetase long-chain family member 1 (ACSL1) which involves in fatty acid analysis [43] was differentially expressed in lactation,. ACSL1, FABP and CD 36 were suggested to have the function of channel long chain fatty acids toward esterification into milk triacylglycerol [43,45].

The studies showed that both the bovine and human α-lactalbumin had the functions of feeds back on the mammary gland to regulate involution [46]. In our study, the bovine α-Lactalbumin and the exogenous human α-Lactalbumin were also expressed in mammary gland and the concentration was higher than the normal bovine. But we did not notice that the α-lactalbumin could induce the apoptotic activity, and our former study have confirmed the biological activity of recombination human α-lactalbumin [26]. In this study, over expression of the α-lactalbumin did not influence the expression of MFGM proteins, which indicated that the mammary gland epithelial cell would not be affected from the apoptotic activity.

In summary, the study uses proteomics methods to analyze transgenic cloned cattle MFGM proteins for the first time. The expression of exogenous proteins did not significantly change the MFGM protein profiles, and the relative quantity expression of MFGM proteins were all within the normal ranges. The differences among the TC groups were not greater than those between the N group and C group. The data from this study improves the understanding of the bovine milk proteome and provides data for the assessment of the food safety of transgenic cloned animals.

Acknowledgments

We thank Jin He for the valuable comments and suggestions during the paper preparation.

Author Contributions

Conceived and designed the experiments: NL. Performed the experiments: SS JZ JW TY CG. Analyzed the data: SS JZ JW TY. Contributed reagents/materials/analysis tools: NL RZ. Contributed to the writing of the manuscript: SS JZ.

References

1. Simons JP, McClenaghan M, Clark AJ (1987) Alteration of the quality of milk by expression of sheep beta-lactoglobulin in transgenic mice. Nature 328: 530–532.
2. Gordon K, Lee E, Vitale JA, Smith AE, Westphal H, et al. (1992) Production of human tissue plasminogen activator in transgenic mouse milk. 1987. Biotechnology 24: 425–428.
3. Rudolph NS (1999) Biopharmaceutical production in transgenic livestock. Trends Biotechnol 17: 367–374.
4. Houdebine LM (2000) Transgenic animal bioreactors. Transgenic Res 9: 305–320.
5. Wheeler MB, Walters EM, Clark SG (2003) Transgenic animals in biomedicine and agriculture: outlook for the future. Anim Reprod Sci 79: 265–289.
6. Laible G, Brophy B, Knighton D, Wells DN (2007) Compositional analysis of dairy products derived from clones and cloned transgenic cattle. Theriogenology 67: 166–177.
7. Baldassarre H, Hockley DK, Olaniyan B, Brochu E, Zhao X, et al. (2008) Milk composition studies in transgenic goats expressing recombinant human butyrylcholinesterase in the mammary gland. Transgenic Res 17: 863–872.
8. Baldassarre H, Schirm M, Deslauriers J, Turcotte C, Bordignon V (2009) Protein profile and alpha-lactalbumin concentration in the milk of standard and transgenic goats expressing recombinant human butyrylcholinesterase. Transgenic Res 18: 621–632.
9. Quaranta S, Giuffrida MG, Cavaletto M, Giunta C, Godovac-Zimmermann J, et al. (2001) Human proteome enhancement: high-recovery method and improved two-dimensional map of colostral fat globule membrane proteins. Electrophoresis 22: 1810–1818.
10. Keenan TW (2001) Milk lipid globules and their surrounding membrane: a brief history and perspectives for future research. J Mammary Gland Biol Neoplasia 6: 365–371.
11. Keenan TW (2001) Assembly and secretion of the lipid globules of milk. Adv Exp Med Biol 501: 125–136.
12. Auty MA, Twomey M, Guinee TP, Mulvihill DM (2001) Development and application of confocal scanning laser microscopy methods for studying the distribution of fat and protein in selected dairy products. J Dairy Res 68: 417–427.
13. Mather IH, Keenan TW (1998) Origin and secretion of milk lipids. J Mammary Gland Biol Neoplasia 3: 259–273.
14. Heid HW, Keenan TW (2005) Intracellular origin and secretion of milk fat globules. Eur J Cell Biol 84: 245–258.
15. Wu CC, Howell KE, Neville MC, Yates JR 3rd, McManaman JL (2000) Proteomics reveal a link between the endoplasmic reticulum and lipid secretory mechanisms in mammary epithelial cells. Electrophoresis 21: 3470–3482.
16. Patton S, Gendler SJ, Spicer AP (1995) The epithelial mucin, MUC1, of milk, mammary gland and other tissues. Biochim Biophys Acta 1241: 407–423.
17. Guri A, Griffiths M, Khursigara CM, Corredig M (2012) The effect of milk fat globules on adherence and internalization of Salmonella Enteritidis to HT-29 cells. J Dairy Sci 95: 6937–6945.
18. Inagaki M, Nagai S, Yabe T, Nagaoka S, Minamoto N, et al. (2010) The bovine lactophorin C-terminal fragment and PAS6/7 were both potent in the inhibition of human rotavirus replication in cultured epithelial cells and the prevention of experimental gastroenteritis. Biosci Biotechnol Biochem 74: 1386–1390.
19. Moore JB, Weeks ME (2011) Proteomics and systems biology: current and future applications in the nutritional sciences. Adv Nutr 2: 355–364.
20. Bianchi L, Puglia M, Landi C, Matteoni S, Perini D, et al. (2009) Solubilization methods and reference 2-DE map of cow milk fat globules. J Proteomics 72: 853–864.
21. Picariello G, Ferranti P, Mamone G, Klouckova I, Mechref Y, et al. (2012) Gel-free shotgun proteomic analysis of human milk. J Chromatogr A 1227: 219–233.
22. Spertino S, Cipriani V, De Angelis C, Giuffrida MG, Marsano F, et al. (2012) Proteome profile and biological activity of caprine, bovine and human milk fat globules. Mol Biosyst 8: 967–974.
23. Liao Y, Alvarado R, Phinney B, Lonnerdal B (2011) Proteomic characterization of human milk fat globule membrane proteins during a 12 month lactation period. J Proteome Res 10: 3530–3541.
24. Yang B, Wang J, Tang B, Liu Y, Guo C, et al. (2011) Characterization of bioactive recombinant human lysozyme expressed in milk of cloned transgenic cattle. PLoS One 6: e17593.
25. Yang P, Wang J, Gong G, Sun X, Zhang R, et al. (2008) Cattle mammary bioreactor generated by a novel procedure of transgenic cloning for large-scale production of functional human lactoferrin. PLoS One 3: e3453.
26. Wang J, Yang P, Tang B, Sun X, Zhang R, et al. (2008) Expression and characterization of bioactive recombinant human alpha-lactalbumin in the milk of transgenic cloned cows. J Dairy Sci 91: 4466–4476.
27. Zhang R, Guo C, Sui S, Yu T, Wang J, et al. (2012) Comprehensive assessment of milk composition in transgenic cloned cattle. PLoS One 7: e49697.

28. Wisniewski JR, Zougman A, Nagaraj N, Mann M (2009) Universal sample preparation method for proteome analysis. Nat Methods 6: 359–362.

29. Nesvizhskii AI, Keller A, Kolker E, Aebersold R (2003) A statistical model for identifying proteins by tandem mass spectrometry. Anal Chem 75: 4646–4658.

30. Keller A, Nesvizhskii AI, Kolker E, Aebersold R (2002) Empirical statistical model to estimate the accuracy of peptide identifications made by MS/MS and database search. Anal Chem 74: 5383–5392.

31. Huang da W, Sherman BT, Lempicki RA (2009) Bioinformatics enrichment tools: paths toward the comprehensive functional analysis of large gene lists. Nucleic Acids Res 37: 1–13.

32. D'Amato A, Bachi A, Fasoli E, Boschetti E, Peltre G, et al. (2009) In-depth exploration of cow's whey proteome via combinatorial peptide ligand libraries. J Proteome Res 8: 3925–3936.

33. Aziz A, Zhang W, Li J, Loukas A, McManus DP, et al. (2011) Proteomic characterisation of Echinococcus granulosus hydatid cyst fluid from sheep, cattle and humans. J Proteomics 74: 1560–1572.

34. Affolter M, Grass L, Vanrobaeys F, Casado B, Kussmann M (2010) Qualitative and quantitative profiling of the bovine milk fat globule membrane proteome. J Proteomics 73: 1079–1088.

35. Lu J, Boeren S, de Vries SC, van Valenberg HJ, Vervoort J, et al. (2011) Filter-aided sample preparation with dimethyl labeling to identify and quantify milk fat globule membrane proteins. J Proteomics 75: 34–43.

36. Addis MF, Pisanu S, Ghisaura S, Pagnozzi D, Marogna G, et al. (2011) Proteomics and pathway analyses of the milk fat globule in sheep naturally infected by Mycoplasma agalactiae provide indications of the in vivo response of the mammary epithelium to bacterial infection. Infect Immun 79: 3833–3845.

37. Mowbray AL, Kang DH, Rhee SG, Kang SW, Jo H (2008) Laminar shear stress up-regulates peroxiredoxins (PRX) in endothelial cells: PRX 1 as a mechan-osensitive antioxidant. J Biol Chem 283: 1622–1627.

38. Lee KA, Hammerle LP, Andrews PS, Stokes MP, Mustelin T, et al. (2011) Ubiquitin ligase substrate identification through quantitative proteomics at both the protein and peptide levels. J Biol Chem 286: 41530–41538.

39. Takahashi Y, Itoh F, Oohashi T, Miyamoto T (2003) Distribution of apolipoprotein E among lipoprotein fractions in the lactating cow. Comp Biochem Physiol B Biochem Mol Biol 136: 905–912.

40. Parker P, Sando L, Pearson R, Kongsuwan K, Tellam RL, et al. (2010) Bovine Muc1 inhibits binding of enteric bacteria to Caco-2 cells. Glycoconj J 27: 89–97.

41. Mahley RW, Innerarity TL, Rall SC Jr, Weisgraber KH (1984) Plasma lipoproteins: apolipoprotein structure and function. J Lipid Res 25: 1277–1294.

42. Viturro E, Farke C, Meyer HH, Albrecht C (2006) Identification, sequence analysis and mRNA tissue distribution of the bovine sterol transporters ABCG5 and ABCG8. J Dairy Sci 89: 553–561.

43. Bionaz M, Loor JJ (2008) ACSL1, AGPAT6, FABP3, LPIN1, and SLC27A6 are the most abundant isoforms in bovine mammary tissue and their expression is affected by stage of lactation. J Nutr 138: 1019–1024.

44. Yonezawa T, Haga S, Kobayashi Y, Katoh K, Obara Y (2009) Saturated fatty acids stimulate and insulin suppresses CIDE-A expression in bovine mammary epithelial cells. Biochem Biophys Res Commun 384: 535–539.

45. Spitsberg VL, Matitashvili E, Gorewit RC (1995) Association and coexpression of fatty-acid-binding protein and glycoprotein CD36 in the bovine mammary gland. Eur J Biochem 230: 872–878.

46. Sharp JA, Lefevre C, Nicholas KR (2008) Lack of functional alpha-lactalbumin prevents involution in Cape fur seals and identifies the protein as an apoptotic milk factor in mammary gland involution. BMC Biol 6: 48.

PERMISSIONS

LIST OF CONTRIBUTORS

Julian Zimmermann and Marius Krauthausen
Department of Neurology, Universitätsklinikum Bonn, Bonn, Germany

Markus J. Hofer
Department of Neuropathology, University Clinic of Marburg and Giessen, Marburg, Germany

Michael T. Heneka
Department of Neurology, Universitätsklinikum Bonn, Bonn, Germany
Clinical Neuroscience Unit, University of Bonn, Bonn, Germany

Iain L. Campbell
School of Molecular Bioscience, University of Sydney, Sydney, Australia

Marcus Müller
Department of Neurology, Universitätsklinikum Bonn, Bonn, Germany
School of Molecular Bioscience, University of Sydney, Sydney, Australia

Takako Hattori and Mayumi Yao
Department of Biochemistry and Molecular Dentistry, Okayama University Dental School, Okayama, Japan

Takashi Yamashiro
Department of Orthodontics, Okayama University Graduate School of Medicine, Dentistry, and Pharmaceutical Sciences, Okayama University Dental School, Okayama, Japan

Nao Tomita and Shinsuke Itoh
Department of Biochemistry and Molecular Dentistry, Okayama University Dental School, Okayama, Japan
Department of Orthodontics, Okayama University Graduate School of Medicine, Dentistry, and Pharmaceutical Sciences, Okayama University Dental School, Okayama, Japan

Eriko Aoyama
Biodental Research Center, Okayama University Dental School, Okayama, Japan

Masaharu Takigawa
Department of Biochemistry and Molecular Dentistry, Okayama University Dental School, Okayama, Japan

Biodental Research Center, Okayama University Dental School, Okayama, Japan

Wan-chi Lin, Jeffrey W. Schmidt, Bradley A. Creamer and Aleata A. Triplett
Eppley Institute for Research in Cancer and Allied Diseases, University of Nebraska Medical Center, Omaha, Nebraska, United States of America

Kay-Uwe Wagner
Eppley Institute for Research in Cancer and Allied Diseases, University of Nebraska Medical Center, Omaha, Nebraska, United States of America
Department of Pathology and Microbiology, University of Nebraska Medical Center, Omaha, Nebraska, United States of America

Xitiz Chamling
Department of Pediatrics, University of Iowa Interdisciplinary Program of Genetics, Iowa City, Iowa, United States of America

Kevin Bugge and Charles Searby
Department of Pediatrics, University of Iowa Interdisciplinary Program of Genetics, Iowa City, Iowa, United States of America
Howard Hughes Medical Institute, Chevy Chase, Maryland, United States of America

Seongjin Seo and Arlene V. Drack
Department of Ophthalmology and Visual Sciences, University of Iowa Carver College of Medicine, Iowa City, Iowa, United States of America

Val C. Sheffield
Department of Pediatrics, University of Iowa Interdisciplinary Program of Genetics, Iowa City, Iowa, United States of America
Department of Ophthalmology and Visual Sciences, University of Iowa Carver College of Medicine, Iowa City, Iowa, United States of America
Howard Hughes Medical Institute, Chevy Chase, Maryland, United States of America

Kamal Rahmouni and Deng F. Guo
Department of Internal Medicine, University of Iowa Carver College of Medicine, Iowa City, Iowa, United States of America
Department of Pharmacology, University of Iowa Carver College of Medicine, Iowa City, Iowa, United States of America

Novruz B. Ahmedli
Jules Stein Eye Institute, University of California Los Angeles, Los Angeles, California, United States of America

Alejandra Young
Jules Stein Eye Institute, University of California Los Angeles, Los Angeles, California, United States of America
Molecular Biology Institute, University of California Los Angeles, Los Angeles, California, United States of America

Debora B. Farber
Jules Stein Eye Institute, University of California Los Angeles, Los Angeles, California, United States of America
Molecular Biology Institute, University of California Los Angeles, Los Angeles, California, United States of America
Brain Research Institute, University of California Los Angeles, Los Angeles, California, United States of America

Meisheng Jiang and Ying Wang
Department of Molecular and Medical Pharmacology, University of California Los Angeles, Los Angeles, California, United States of America

Linda Vuong, Daniel E. Brobst and Ivana Ivanovic
Department of Cell Biology, University of Oklahoma Health Sciences Center, Oklahoma City, Oklahoma, United States of America

Muayyad R. Al-Ubaidi
Department of Cell Biology, University of Oklahoma Health Sciences Center, Oklahoma City, Oklahoma, United States of America
Oklahoma Center for Neurosciences, University of Oklahoma Health Sciences Center, Oklahoma City, Oklahoma, United States of America

David M. Sherry
Department of Cell Biology, University of Oklahoma Health Sciences Center, Oklahoma City, Oklahoma, United States of America
Oklahoma Center for Neurosciences, University of Oklahoma Health Sciences Center, Oklahoma City, Oklahoma, United States of America
Department of Pharmaceutical Sciences, University of Oklahoma Health Sciences Center, Oklahoma City, Oklahoma, United States of America

Jun Liu., Yan Luo., Hengtao Ge., Chengquan Han, Hui Zhang, Yongsheng Wang, Jianmin Su,
Fusheng Quan, Mingqing Gao and Yong Zhang

College of Veterinary Medicine, Northwest A&F University, Key Laboratory of Animal Biotechnology of the Ministry of Agriculture, Yangling, Shaanxi, China

Manuela Cervelli, Gabriella Bellavia, Sandra Moreno and Paolo Mariottini
Dipartimento di Biologia, Universitá"Roma Tre," Rome, Italy

Marcello D'Amelio, Virve Cavallucci and Francesco Cecconi
Laboratory of Molecular Neuroembryology, Istituto di Ricovero e Cura a Carattere Scientifico (IRCCS) Fondazione Santa Lucia, Rome, Italy

Joachim Berger
Faculty of Medicine, Nursing and Health Sciences, Monash University, Clayton, Australia

Mauro Piacentini and Roberta Nardacci
Istituto Nazioale per le Malattie Infettive, IRCCS "L. Spallanzani," Rome, Italy

Manuela Marcoli and Guido Maura
Dipartimento di Farmacia, Sez. Farmacologia e Tossicologia, Centro di Eccellenza per la Ricerca Biomedica CEBR, Universitá di Genova, Genoa, Italy

Roberto Amendola
Agenzia nazionale per le nuove tecnologie, l'energia e lo sviluppo economico sostenibile (ENEA), Il Centro Ricerche Casaccia, Sezione Tossicologia e Scienze Biomediche (BAS-BIOTECMED), Rome, Italy

Mohammad Haeri, Peter D. Calvert, Eduardo Solessio and Barry E. Knox
Departments of Neuroscience and Physiology, Biochemistry and Molecular Biology, and Ophthalmology, SUNY Upstate Medical University, Syracuse, New York, United States of America

Edward N. Pugh, Jr.
Center for Neuroscience, University of California, Davis, California, United States of America

Roopali Chaudhary, Christina C. Pierre, Daria Wojtal and Juliet M. Daniel
Department of Biology, McMaster University, Hamilton, Ontario, Canada

Kyster Nanan
Department of Pathology & Molecular Medicine, Queen's University, Kingston, Ontario, Canada

Simona Morone
Department of Medical Sciences, University of Torino, Torino, Italy

Christopher Pinelli and Geoffrey A. Wood
Department of Pathobiology, University of Guelph, Guelph, Ontario, Canada

Sylvie Robine
Departmentof Morphogenesis and Intracellular Signalling, Institut Curie-CNRS, Paris, France

Géraldine H. Petit
Neuronal Survival Unit, Wallenberg Neuroscience Center, Department of Experimental Medical Science, BMC B11, Lund University, Lund, Sweden

Elijahu Berkovich and Cheryl Fitzer- Attas
Teva Pharmaceutical Industries Ltd., Global Innovative Products, Petach-Tikva, Israel

Mark Hickery
H. Lundbeck A/S, Neurology, Copenhagen, Denmark

Pekka Kallunki and Karina Fog
H. Lundbeck A/S, Neurodegeneration-1, Copenhagen, Denmark

Patrik Brundin
Neuronal Survival Unit, Wallenberg Neuroscience Center, Department of Experimental Medical Science, BMC B11, Lund University, Lund, Sweden

Van Andel Research Institute, Center for Neurodegenerative Science, Grand Rapids, Michigan, United States of America

Ryuji Iida, Paul W. Kincade and José Alberola-lla
Immunobiology and Cancer Program, Oklahoma Medical Research Foundation, Oklahoma City, Oklahoma, United States of America

Robert S. Welner
Beth Israel Deaconess Medical Center, Boston, Massachusetts, United States of America

Wanke Zhao and Zhizhuang Joe Zhao
Department of Pathology, University of Oklahoma Health Sciences Center, Oklahoma City, Oklahoma, United States of America

Kay L. Medina
Department of Immunology, Mayo Clinic, Rochester, Minnesota, United States of America

Dennis J. Wu, Stephanie M. Stanford and Novella Rapini
Division of Cellular Biology, La Jolla Institute for Allergy and Immunology, La Jolla, California, United States of America

Wenbo Zhou, Kristy Sawatzke and Erik Peterson
Center for Immunology, Department of Medicine, University of Minnesota, Minneapolis, Minnesota, United States of America

Sarah Enouz and Dietmar Zehn
3 Swiss Vaccine Research Institute, Epalinges, and Division of Immunology and Allergy, Department of Medicine, Lausanne University Hospital, Lausanne, Switzerland

Nunzio Bottini
Division of Cellular Biology, La Jolla Institute for Allergy and Immunology, La Jolla, California, United States of America
Institute for Genetic Medicine, University of Southern California, Los Angeles, California, United States of America

Valeria Orrú and Edoardo Fiorillo
Institute for Genetic Medicine, University of Southern California, Los Angeles, California, United States of America
Istituto di Ricerca Genetica e Biomedica (IRGB), CNR, Monserrato, Italy

Christian J. Maine and Linda A. Sherman
Department of Immunology, The Scripps Research Institute, La Jolla, California, United States of America

Isaac Engel and Mitch Kronenberg
Division of Developmental Immunology, La Jolla Institute for Allergy and Immunology, La Jolla, California, United States of America

Marcelle Altshuler
Department of Neuroscience and Center for Translational Research in Neurodegenerative Disease, University of Florida, Gainesville, Florida, United States of America

Ashley Cannon, Dennis W. Dickson and Leonard Petrucelli
Department of Neuroscience, Mayo Clinic, Jacksonville, Florida, United States of America

Jada Lewis and Simon D'Alton
Department of Neuroscience and Center for Translational Research in Neurodegenerative Disease, University of Florida, Gainesville, Florida, United States of America
Department of Neuroscience, Mayo Clinic, Jacksonville, Florida, United States of America

Weiling Zheng, Zhen Li, Anh Tuan Nguyen, Caixia Li and Zhiyuan Gong
Department of Biological Sciences, National University of Singapore, Singapore, Singapore

Alexander Emelyanov
Institute of Cell and Molecular Biology, Singapore, Singapore

Yadi Huang, Joshua Kapere Ochieng, Marjon Buscop-van Kempen, Anne Boerema-de Munck and Dick Tibboel
Department of Pediatric Surgery, Erasmus MC-Sophia Children's Hospital, Rotterdam, The Netherlands

Sigrid Swagemakers
Department of Bioinformatics, Erasmus MC, Rotterdam, The Netherlands
Department of Genetics, Erasmus MC, Rotterdam, The Netherlands

Robbert J. Rottier
Department of Pediatric Surgery, Erasmus MC-Sophia Children's Hospital, Rotterdam, The Netherlands
Department of Cell Biology, Erasmus MC, Rotterdam, The Netherland

Wilfred van IJcken
Department of Biomics, Erasmus MC, Rotterdam, The Netherlands

Frank Grosveld
Department of Cell Biology, Erasmus MC, Rotterdam, The Netherland

Stephanie Jacobs, Zhenzhong Cui, Ruiben Feng and Joe Z. Tsien
Brain and Behavior Discovery Institute and Department of Neurology, Medical College of Georgia at Georgia Regents University, Augusta, Georgia, United States of America

Huimin Wang
Shanghai Institute of Functional Genomics, East China Normal University, Shanghai, China

Deheng Wang
Banna Biomedical Research Institute, Xi-Shuang-Ban-Na Prefecture, Yunnan Province, China

Shunchao Sui, Jie Zhao, Ran Zhang, Chengdong Guo and Ning Li
State Key Laboratory of Agrobiotechnology, China Agricultural University, Beijing, PR China

Jianwu Wang
Beijing GenProtein Biotech Company Ltd., Beijing, PR China

Wuxi Kingenew Biotechnology Co., Ltd., Wuxi, Jiangsu Province, China

Tian Yu
Wuxi Kingenew Biotechnology Co., Ltd., Wuxi, Jiangsu Province, China

Index

www.ingramcontent.com/pod-product-compliance
Lightning Source LLC
Chambersburg PA
CBHW080518200326
41458CB00012B/4257